Progress in Botany

Volume 69

Series Editors
Ulrich Lüttge, TU Darmstadt, Institut für Botanik,
FB Biologie (10), Schnittspahnstraße 3–5,
64287 Darmstadt, Germany
Wolfram Beyschlag, Fakultät für Biologie, Lehrstuhl für
Experimentelle Ökologie und Ökosystembiologie,
Universität Bielefeld, Universitätsstraße 25, 33615 Bielefeld,
Germany
Jin Murata, Botanical Gardens, Graduate School of Science,
University of Tokyo, 3-7-1 Hakusan, Bunkyo-ku,
Tokyo 112-0001, Japan

Ulrich Lüttge • Wolfram Beyschlag • Jin Murata
Editors

Progress in Botany 69

Springer

Professor Dr. U. Lüttge
TU Darmstadt, Institut für Botanik
FB Biologie (10)
Schnittspahnstraße 3–5
64287 Darmstadt
Germany
luettge@bio.tu-darmstadt.de

Professor Dr. J. Murata
Botanical Gardens
Graduate School of Science
University of Tokyo
3-7-1 Hakusan
Bunkyo-ku, Tokyo 112-0001
Japan
murata@ns.bg.s.u-tokyo.ac.jp

Professor Dr. W. Beyschlag
Fakultät für Biologie
Lehrstuhl für Experimentelle Ökologie
und Ökosystembiologie
Universität Bielefeld
Universitätsstraße 25
33615 Bielefeld
Germany
w.beyschlag@uni-bielefeld.de

ISBN 978-3-642-09204-6 e-ISBN 978-3-540-72954-9

Progress in Botany ISSN 0340–4773

Cover design: WMXDesign GmbH, Heidelberg

Printed on acid-free paper

9 8 7 6 5 4 3 2 1

springer.com

Contents

Contributors

Markus Albert
ZMBP–AG Felix,
University of Tübingen, Auf der Morgenstelle 5, D-72076 Tübingen, Germany,
markus.albert@zmbp.uni-tuebingen.de

Margarete Baier
Department of Biochemistry and Physiology of Plants, W5/117
Faculty of Biology, University of Bielefeld, Universitaetstrasse 25,
D-33615 Bielefeld, Germany

Jarrett J. Barber
Department of Statistics, Department of Botany, Box 3165,
University of Wyoming 1000 E. University Ave. Laramie, WY 82071, USA

Xana Belastegui-Macadam
ZMBP–AG Felix
University of Tübingen, Auf der Morgenstelle 5, D-72076 Tübingen, Germany

Marc Bleischwitz
ZMBP–AG Felix,
University of Tübingen, Auf der Morgenstelle 5, D-72076 Tübingen, Germany

Stephan Blossfeld
Department of Ecological Plant Physiology & Geobotany,
University of Düsseldorf, Universitätsstrasse 1, D-40225 Düsseldorf, Germany

Ralph Bock
Max-Planck-Institut für Molekulare Pflanzenphysiologie, Am Mühlenberg 1,
D-14476 Potsdam-Golm, Germany, rbock@mpimp-golm.mpg.de

Anne M. Borland
School of Biology and Psychology, Institute for Research on Environment
and Sustainability, Devonshire Building, Newcastle University, Newcastle Upon
Tyne, NE1 7RU, UK, a.m.borland@ncl.ac.uk

Karl-Josef Dietz
Department of Biochemistry and Physiology of Plants, W5/117
Faculty of Biology, University of Bielefeld, Universitaetstrasse 25,
D-33615 Bielefeld, Germany

Elisabeth Esch
Institut für Pflanzengenetik, Universität Hannover, Herrenhäuser Strasse 2,
D-30419 Hannover, Germany

Burkhard Frenzel
Institut für Botanik und Botanischer Garten 210, Universität Hohenheim,
D-70593 Stuttgart, Germany, bfrenzel@uni-hohenheim.de

Dirk Gansert
Department of Ecology & Ecosystem Research, University of Göttingen,
Untere Karspüle 2, D-37073 Göttingen, Germany, dganser@gwdg.de

Dortje Golldack
Department of Biochemistry and Physiology of Plants,
W5/117 Faculty of Biology, University of Bielefeld, Universitaetstrasse 25,
D-33615 Bielefeld, Germany, dortje.golldack@uni-bielefeld.de

Martin Heil
Departamento de Ingeniería Genética, CINVESTAV–Irapuato,
Km. 9.6 Libramiento Norte, CP 36821, Irapuato, Guanajuato, Mexico,
mheil@ira.cinvestav.mx

Vera Hemleben
ZMBP, General Genetics, University of Tübingen, Auf der Morgenstelle 28,
D-72076 Tübingen, Germany

Dirk Hölscher
Max-Planck-Institute for Chemical Ecology, Beutenberg Campus Hans-Knöll-Str.
8, D-07745 Jena, Germany

Renate Horn
Institut für Biowissenschaften, Abtl. Pflanzengenetik, Universität Rostock, Albert-
Einstein-Strasse 3, D-18051 Rostock, Germany, renate.horn
@uni-rostock.de

Christina Roenn Ingvardsen
Institute of Genetics and Biotechnology, Research Centre Flakkebjerg,
University of Aarhus, Forsoegsvej 1, DK-4200 Slagelse, Denmark. Christina.
Ingvardsen@agrsci.dk

Carolyn Y. Ivans
Campbell Scientific, Inc., Logan, UT 84321, USA

Ralf Kaldenhoff
ZMBP–AG Felix
University of Tübingen, Auf der Morgenstelle 5, D-72076 Tübingen, Germany

Andrea Kandlbinder
Department of Biochemistry and Physiology of Plants, W5/117 Faculty of Biology, University of Bielefeld, Universitaetstrasse 25, D-33615 Bielefeld, Germany

Paul Kardol
Center for Terrestrial Ecology, Netherlands Institute of Ecology (NIOO-KNAW), PO Box 40, 6666 ZG Heteren, The Netherlands

Manfred Kluge
Institute of Botany, Darmstadt University of Technology, Schnittspahnstrasse 10, D-64287 Darmstadt, Germany, klugemanfred@t-online.de

Andrew Kulmatiski
Department of Wildland Resources and the Ecology Center Utah State University Logan, UT 84322, USA, andrewkulmatiski@hotmail.com

Inara R. Leal
Departamento de Botânica, Universidade Federal de Pernambuco, Av. Prof. Moraes Rego, 50670-901 Recife, Pernambuco, Brasil

A. Joshua Leffler
Department of Wildland Resources and the Ecology Center, Utah State University, Logan, UT 84322-5230, USA

Thomas Lübberstedt
Institute of Genetics and Biotechnology, Research Centre Flakkebjerg, University of Aarhus, Forsoegsvej 1, DK-4200 Slagelse, Denmark.

Sabine Lüthje
Biozentrum Klein Flottbek und Botanischer Garten, Universität Hamburg, Ohnhorststrasse 18, D-22609 Hamburg, Germany, s.luthje@botanik.uni-hamburg.de

Sebastian T. Meyer
Plant Ecology & Systematics, University of Kaiserslautern, PO Box 3049, D-67663 Kaiserslautern, Germany

Ewa Niewiadomska
Institute of Plant Physiology Polish Academy of Sciences, Niezapominajek 21, 30-239 Krakow, Poland

Kiona Ogle
Department of Botany and Department of Statistics, Department of Botany, Box 3165, University of Wyoming 1000 E. University Ave. Laramie, WY 82071, USA, kogle@uwyo.edu

Cai-Hong Pang
College of Life Sciences, Shandong Normal University, Jinan 250014, Shandong Province, PR China

Michael S. Peek
Department of Biology, William Paterson University, Wayne, NJ 07470, USA

Ronald J. Ryel
Department of Wildland Resources and the Ecology Center, Utah State University, Logan, UT, 84322-5230, USA, ron.ryel@usu.edu

Britt Schejbel
Institute of Genetics and Biotechnology, Research Centre Flakkebjerg, University of Aarhus, Forsoegsvej 1, DK-4200 Slagelse, Denmark.

Bernd Schneider
Max-Planck-Institute for Chemical Ecology, Beutenberg Campus Hans-Knöll-Str. 8, D-07745 Jena, Germany, schneider@ice.mpg.de

Mark Aurel Schöttler
Max-Planck-Institut für Molekulare Pflanzenphysiologie, Am Mühlenberg 1, D-14476 Potsdam-Golm, Germany

Marcelo Tabarelli
Departamento de Botânica, Universidade Federal de Pernambuco, Av. Prof. Moraes Rego, 50670-901 Recife, Pernambuco, Brasil

Bao-Shan Wang
College of Life Sciences, Shandong Normal University, Jinan 250014, Shandong Province, PR China, bswang@sdnu.edu.cn

Rainer Wirth
Plant Ecology & Systematics, University of Kaiserslautern, PO Box 3049, D-67663 Kaiserslautern, Germany, wirth@rhrk.uni-kl.de

Ulrike Zentgraf
ZMBP, General Genetics, University of Tübingen, Auf der Morgenstelle 28, D-72076 Tübingen, Germany, ulrike.zentgraf@zmbp.uni-tuebingen.de

Review

Curriculum Vitae

1936	Manfred Kluge was born in Zschopau (Saxonia; Germany)
1957	Graduation from the Gymnasium
1957–1962	Undergraduate studies at the University of Frankfurt and the Darmstadt University of Technology in Biology, Chemistry and Geography. Examination for Teachers Profession (Gymnasium) in 1962
1962–1964	Graduate student in Botany with Professor Hubert Ziegler at the Darmstadt University of Technology
1964	Doctor's degree (Faculty for Chemistry, Biology and Geosciences of the Darmstadt University of Technology), Scientific Assistant at the Institute of Botany in Darmstadt
1966	Married to Helga Müller, also graduate student of Hubert Ziegler
1969	Habilitation (Botany) at the Darmstadt University of Technology, Lecturer at the Institute of Botany; birth of son Christoph
1970	Birth of son Andreas, Postdoctoral Research Fellow at the Australian National University, Canberra
1971–1973	Associate Professor at the Institute of Botany of the Technical University Munich
1974	Full Professor of Botany, Faculty Biology of Darmstadt University of Technology
2002	Retirement

1985 Guest Professor at the Botany Department of the National University Singapore. 1989, 1991 and 1993 Guest Docent of the German Academic Exchange Service (DAAD) at the Botany Department of the University Antananarivo (Madagascar). Repeatedly elected Director of the Institute of Botany and the Botanical Gardens of the Darmstadt University of Technology and Dean of the Faculty of Biology of this University. Chairman of the DFG Schwerpunktsprogramm "Biochemical fundaments of ecological adaptation in plants"(1974–1979), and the DFG Graduate School 340 "Communication in biological systems: from the molecule to the organism in its environment"; Vice Chairman of the DFG Center of Excellence 199 "Molecular ecophysiology of plants: acquisition of resources, membrane transport, regulation of resource consumption". Tudor of the Studienstiftung des Deutschen Volkes at the Darmstadt University of Technology (1976–2003).

Ecophysiology: Migrations Between Different Levels of Scaling

Manfred Kluge

Contents

Abstract Causal explanation of the mechanisms behind ecological adaptation in plants requires research on different levels of scaling, ranging from molecules to plants in their natural environment. In the following article this is illustrated by a description of the author's scientific life-work mainly with respect to the crassulacean acid metabolism (CAM) and research on *Geosiphon*, an endosymbiotic consortium of a fungus and a cyanobacterium.

Introduction

The title chapters of *Progress in Botany* provide a forum where in the past celebrities of the plant sciences explored important fields of their scientific life-work. Thus, the invitation by the Editors of *Progress in Botany* to write the title chapter of

Manfred Kluge
Institute of Botany, Darmstadt University of Technology, Schnittspahnstrasse 10, D-64287 Darmstadt, Germany
e-mail: klugemanfred@t-online.de

volume 69 honors me beyond measure. Simultaneously it confronts me with a dilemma because, with the title chapter of the preceding volume, Barry Osmond provided an excellent survey of the world of crassulacean acid metabolism (CAM), a topic which concerns also the major part of my scientific work. Therefore I cannot avoid that CAM plays an important role also in my treatise, but I want to consider this topic from a different point of view and to include also aspects other than CAM.

In his paper Barry explored CAM in all its diversity of recent and future aspects and the implications of CAM research for other disciplines of the plant sciences, and he expressed his "indebtedness to very many companions in CAM research". The latter holds very much true also for me and even concerns largely the same circle of personalities to whom I am indebted and whom I have to thank. Thus, in order to avoid repetition and hoping that it will not be misunderstood as immodesty, I will try to illustrate my personal adventure of nearly four decades of research on CAM. In particular I would like to show the logic which pushed me during my involvement in CAM to exciting migrations through the various levels of scaling, ranging from the plants in their environment down to the world of molecules.

From my beginning I considered CAM primarily as a matter of ecophysiology, irrespective of its implication also for other disciplines of plant sciences (compare Osmond 2007). Ulrich Lüttge (1997) clearly characterized the hierarchy of approaches required aiming to extend ecophysiology beyond the descriptive level. One has:

- "To start by depicting habitats and plants physiognomically,
- To deduce problems from such observations in the field as are suited for physiological, biochemical and biophysical and perhaps even molecular experimentation in the laboratory, and
- To return from the laboratory to the field with increasing sophisticated technologies for measurements and analyses applicable to field conditions."

My migrations included all these steps and thus are largely symptomatic for the path one has to follow when aiming to really understand even single facets of the breathtaking diversity of mechanisms evolved in plants to cope with the demands of their environments.

1 First Steps Towards Biology

My interest in biology emerged from two important experiences of my youth. When I was 12 years old, as a Christmas gift I got a little microscope. It opened to me a new exciting world of sometimes bizarre beauty; and to put anything under the microscope which looked somehow interesting became besides playing the piano a stern hobby.

The next, and this time a more serious impetus in the direction of biology, came from my school days. I graduated from a gymnasium practicing so-called "reformed senior classes" where I enjoyed excellent instruction in biology which not only comprised the prescribed topics of the gymnasial biology curriculum for the final two years, i.e. genetics and phylogeny, but also introduced us, based on the work of

L. v. Bertalanffy, to theoretical biology. Thus, I began to understand organisms as a steady state – maintaining dynamic open systems. Important for the direction of my later work was the introduction to classic ecology. It provided me first insights into the structure of ecosystems, food webs and matter cycling, and I began to suspect that, because of its complexity, ecology is a particularly demanding discipline within the biosciences. For me this seemed to be more attractive than deterrent.

For the development of the capability of scientific thinking and working it was very important that, in the "reformed senior class" system for the matriculation examination, a thesis based on own independent work had to be written. Within the frame of this thesis I conducted a limnological study on a pond where I investigated during the cycle of a whole year the changes in the biocenosis parallel to the alterations in water temperature, oxygen content, pH values and other environmental parameters. This study was a veritable introduction into research work. It provided me with a considerable knowledge of species, strengthened skills in handling microscopic and analytical techniques, developed the ability for independent scientific working and gave me the first feeling of satisfaction to find something that nobody had known before. Altogether, in particular the reformed senior classes of the gymnasium laid a sound fundament for my later studies, thus I have all reason to look gratefully back to my school days.

After the matriculation examination in spring 1957 it was clear to me to become teacher at a gymnasium, less because this was already tradition in my family, rather because I myself felt, strengthened by the example of my father, that teaching can be a passion and source of deep satisfaction. Briefly I was not sure whether I should choose music or biology as the main topic of my studies for the teacher's profession, but finally I decided myself in favor of science and kept music as a hobby. Thus, for the summer semester 1957 I enrolled in the Faculty of Natural Sciences of the University at Frankfurt.

2 Academic Studies and Entrance into Professional Activity

As topics for the academic studies for teacher's profession beside biology I chose chemistry and geography. In Frankfurt my studies were focused on zoology, but in 1959 I changed to the Darmstadt University of Technology where for the following three semesters I was engaged mainly in botany.

In Darmstadt the biology departments were quite small, and in particular the Institute of Botany was excelled by a close contact between the at this time sole professor, the scientific staff and the students, creating a nearly family-type and stimulating atmosphere. When I came to Darmstadt, Otto Stocker, one of the great pioneers and celebrities of experimental plant ecology, was just retired and Hubert Ziegler had become his successor. Although being still very young, H. Ziegler had already a great reputation as researcher on phloem transport, but his wide spectrum of interest included also physiological aspects of plant ecology, so that in Darmstadt the long tradition in this field was not only continued rather but, due to the introduction of modern biochemical methodology, extended to the quest for the biochemical fundaments of ecological adaptation in plants.

H. Ziegler was an excellent, electrifying teacher, and due to him my main interest soon shifted to plant physiology and biochemistry, although on purpose I tried to avoid too early and too narrow a specialization and to retain curiosity towards the diversity of the biological world. Also in this respect H. Ziegler was a great mentor and exemplary for me.

In 1961 I won a scholarship of the Studienstiftung des Deutschen Volkes, and I began to work in H. Ziegler's group on the chemical composition of phloem exudates of trees. After I passed the teacher's examination in 1962, H. Ziegler accepted me as a candidate for a doctorate. He was a great mentor and really a friend to his pupils. Within the frame of my doctoral thesis I continued my investigation on phloem transport. In summer 1964 I finished my studies by taking the Doctor's degree. Moreover, that year became particularly important for me, because a new PhD student, Helga Müller, joined the group of Hubert Ziegler to conduct comparative studies on the carbohydrate composition of sieve tube saps. She not only appreciated my help in the time-consuming gathering of phloem exudates from the trees of the Botanical Garden but also found a partner who was happy to accompany her excellent playing of the flute on the harpsichord. It became a partnership for the whole life. Thank you, Helga.

3 Research on CAM

3.1 Initial Physiological Studies at the Level of the Entire Plant

After I took the Doctor's degree H. Ziegler encouraged me to enter an academic career and offered me the position of Scientific Assistant at his institute which I happily accepted. Looking for my own field of research, accidentally I read a paper by Nishida (1963) which trapped me so much that it determined the topic of my main future scientific work: the crassulacean acid metabolism (CAM). In 1965, when I began my investigations on CAM, at the level of science the phenomenon had already been known for 150 years, but it was still considered more or less as a physiological curiosity exhibited by certain specialists among plants. In the mentioned publication Nishida showed by viscous flow porometry that CAM plants perform a pattern of stomatal movement inverse to that of non-CAM plants, i.e. CAM plants open the stomata at night and close them during the day. This pattern coincides with typical CAM CO_2 exchange, namely net CO_2 uptake during the night (Phase I, denotation after Osmond 1978), a short period of enhanced net CO_2 uptake after the onset of the light period (Phase II), depression of net CO_2 uptake during several hours of the day (Phase III), and recommencement of net CO_2 uptake towards the end of the day (Phase IV). However, the question remained open whether the unusual stomatal behavior is the primary event determining the unique course of CAM CO_2 exchange, or whether this CO_2 exchange is otherwise controlled.

The attempt to answer this question was the beginning of my love story with CAM lasting now more than 40 years.

My start into CAM research was facilitated by several lucky circumstances. First, in 1965, because of its reputation as a physiological curiosity (Osmond 1978), CAM kept world-wide only few researchers busy. Thus, speaking in terms of ecology, for a young researcher CAM provided still a kind of niche without too much competitive pressure. Triggered by the discovery of C4-photosynthesis in 1966, this situation later changed considerably. Second, I had access to an efficient system of modern infrared gas analyzers already established by O. L. Lange when he worked from 1961 until 1963 in Darmstadt. This system allowed one to measure continuously net CO_2 exchange and transpiration by plants under controlled conditions. Finally, in the mid-1960s young scientists still at the beginning of their work like me had a good chance to get research funds from the Deutsche Forschungsgemeinschaft just by presenting a convincing research program, without the pressure that prior results, best published in the journals with highest impact factors, had to be shown.

Furnished with the mentioned equipment and a DFG grant of 5000 DM for three years, I started a study aiming to provide a causal explanation of the CAM gas exchange phenomena.

I began my investigations with a simple, now I would say brutal, experiment in one of the classic CAM plants, *Kalanchoë daigremontiana*, where fortunately it is possible to strip off the abaxial epidermis of the planar leaves remaining attached to entire plant, without substantial damage to the leaf mesophyll. Net CO_2 exchange and transpiration (the latter as an index for the stomatal behavior) were measured under constant water vapor pressure deficit of the ambient air before and after the leaf epidermis had been stripped off. The result was very clear (Fig. 1). As to be expected from the results by Nishida (1963), in the intact leaves both during day and night there was a perfect coincidence between net CO_2 exchange and the stomatal movements indicated by transpiration, whereas in the leaves without epidermis the typical CAM pattern of CO_2 exchange was preserved whilst expectedly transpiration stayed constant. My interpretation was "that not the stomatal behavior is the primary event determining the CO_2 exchange in CAM plants rather than that the CO_2 exchange of the mesophyll cells, which itself is the direct consequence of CAM, determines whether the stomata close or open at a given time" (Kluge and Ting 1978). However, the strict synchronization of CO_2 exchange and stomatal movement in the intact leaves implied that there must be a coupling between the mesophyll-inherent CO_2 exchange and the movements of the stomatal guard cells. I found a plausible explanation on the basis of the pioneering publication by K. Raschke (1966) showing in *Zea mays* that the stomata are links in a regulatory circuit which stabilizes the CO_2 concentration in the subepidermal intercellular spaces ($[CO_2]_{int}$). A consequence of such a circuit is that an increase in the internal CO_2 concentration leads to a closure of the stomata, and vice versa. On the basis of Raschke's regulatory circuit model I proposed that also in CAM plants $[CO_2]_{int}$ is the signal factor. For me it was just logic that CAM plants open their stomata at night because, due to dark CO_2 fixation, $[CO_2]_{int}$ in the mesophyll should be low, and they close them during the day because malic acid decarboxylation would lead

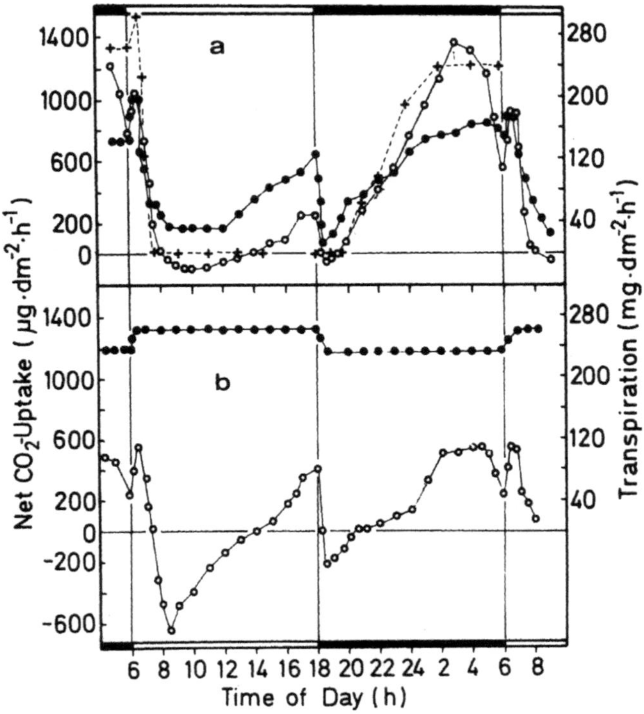

Fig. 1 Net CO_2 exchange and transpiration in *Kalanchoe daigremontiana*. **a** Coincidence of CO_2 exchange and transpiration in an intact plant. ○—○ net CO_2 exchange, ●—● transpiration, +—+ porometric values (relative units) obtained in the same species under the same conditions by Nishida (1963). **b** CO_2 exchange and transpiration in the same plant as **a**, however with the lower epidermis of all leaves having been stripped off (from Kluge and Fischer 1967)

to high $[CO_2]_{int}$. In the early years of my struggling with CAM I made a great effort to verify this hypothesis by following two strategies, namely (a) trying to measure $[CO_2]_{int}$ directly and to correlate it with the actual status of the stomatal aperture, and (b) trying to influence the duration of stomatal closure during the day (Phase III) by experimentally manipulating the source for a high internal CO_2 concentration, i.e. the malic acid pool built up during the previous night.

Already in our first CAM experiment (Fig. 1b) we observed a massive CO_2 outburst occurring upon removal of the epidermal diffusion barrier during Phase III. This outburst was drastically lower or even absent with intact leaves. Thus, for us this outburst was indicative of massive CO_2 enrichment built up behind the closed stomata. Actually, this observation was the first hint in favor of the now widely accepted view that CAM is also a CO_2 enrichment mechanism (see e.g. Osmond 2007). Later it could be demonstrated by direct measurements in air samples (Cockburn et al. 1979) taken along the day/night cycle from the succulent *Kalanchoë* leaves that, in the night, when the stomata were shown to be open,

$[CO_2]_{int}$ dropped close to zero, while during Phase III, when the stomata were closed, the leaf internal CO_2 was enriched by a factor of ten with respect to the ambient air (Kluge et al. 1981a).

The next step was to corroborate systematically the hypothesis that, optimal water status provided, the duration of stomatal closure in CAM plants during Phase III is mainly determined by the production of CO_2 from the previously synthesized and stored malic acid. I could directly prove this assumption by ^{14}C-pulse-chase experiments (Kluge 1968a).Subsequently a large series of gas exchange measurements followed which were paralleled by estimation of biochemical parameters in the plants under variation of the environmental conditions. These experiments not only uncovered the extraordinary phenotypic flexibility of CAM performance, but also showed that all experimental treatments which prevent filling of the malic acid storage pool during the night and all treatments which accelerate the consumption of the previously stored malic acid during the day shorten the duration of Phase III. After this we were quite sure that Phase III of CAM is a function of malate consumption (Kluge 1968a).

It was an interesting side-issue of the above-mentioned studies to find that application of CO_2-free air during the night inhibits not only the nocturnal synthesis of malate, but also the depletion of starch (Kluge 1969a). This showed that the breakdown of starch to generate phosphoenol pyruvate (PEP) as a CO_2 acceptor of the PEP carboxylase (PEPCase) reaction must be coupled rather directly to the availability of the substrate CO_2. Looking back I regret that neither we nor other researchers followed this finding further immediately, because presumably it could have guided us quite early into the investigation of carbohydrate metabolism in CAM which currently collects more attention (Osmond 2007).

We also observed that, at the beginning of the light period, a lag-phase in the onset of malic acid consumption occurred which coincided with Phase II. It was for me the first hint that, under the experimental condition applied, at the onset of light malate stored in the vacuole is not immediately disposable for decarboxylation. The duration of the lag phase turned out to depend on the actual vacuolar malate concentration and temperature. Thus it was reasonable to conclude that these two factors might be important for the determination of the malate release from the vacuole. Some years later I came back to these findings (Kluge et al. 1991d), and the investigation of temperature effects on the tonoplast became an important aspect in my research on CAM.

From the beginning I guessed that the pattern of the CAM gas exchange must be meaningful for the carbon and water balance of the concerned plants growing in the field (Kluge and Fischer 1967) so that CAM can reasonably be interpreted in terms of ecophysiological adaptation to environments where water is in short supply. I had the dream to get once the opportunity to verify this interpretation by studying the behavior of CAM plants in their natural habitats. A couple of years later this dream became reality. However, first we wanted to prove the ecophysiological relevance of CAM at least in the laboratory, and thus we conducted the (to my knowledge) first stress experiment on a CAM plant (*Kalanchoë daigremontiana*; Kluge and Fischer 1967). It turned out that just a few days cessation of watering were sufficient to completely suppress opening of the stomata during the day (i.e. Phases II, IV), whereas nocturnal opening of the stomata with net CO_2 uptake continued, although on a somewhat lower level. All stress effects on CAM gas exchange disappeared within one hour after re-watering the plant. These findings suggested that CAM allows the plants to retain a positive carbon balance even if carbon acquisition during the day is prevented by water deficiency stress.

Because of compassion with our experimental plants we were not cruel enough to extent withholding of water for longer than five days. This was a pity because otherwise we would have discovered that, upon long-lasting water deficiency stress, CAM plants finally close their stomata also during the night but continue CAM behind the closed stomata by nocturnal recycling of respiratory CO_2 ("CAM idling"). Szarek and Ting (1975), who discovered CAM idling, observed that in the field it is ecophysiologically highly relevant for CAM plants which have to overcome long-lasting drought seasons.

Our stress experiments showed us also that the $[CO_2]_{int}$ regulatory circuit controlling the aperture of stomata operates undisturbed only if the plants do not suffer water deficiency stress. The conclusion was that drought factors which were still unknown to us at that time override the CO_2 regulatory mechanism. Today it is textbook knowledge that a drop in the plant water potential, with abscisic acid as molecular messenger, leads to hydroactive closure of the stomata dominating the other regulation circuits.

3.2 Investigations at the Metabolic Level: the Problem of CAM Regulation

Orderly functioning of CAM requires that during the day the malic acid derived free CO_2 is directed quantitatively to the Calvin cycle for final assimilation. From the beginning it was a striking question for me how it is prevented that during Phase III the malate derived intermediary CO_2 is trapped by PEPCase and thus futilely recycled into the malate pool instead of entering correctly the photosynthetic pathway. By means of $^{14}CO_2$ tracer experiments I could show that in the light during Phase III the PEPCase-mediated primary CO_2 fixation is switched off. In contrast, during the night even under illumination CO_2 is nearly exclusively fixed by PEPCase thus practically not available to the Calvin cycle (Kluge 1969b; Fig. 2). These in vivo changes in the activity of the primary CO_2 fixation turned out to be correlated with the diurnal malic acid rhythm of CAM. In other words: the PEPCase pathway is in an active state when the malate level in the CAM performing cells is low, and vice versa (Kluge 1971). This correlation was strictly retained when the course and the amplitude of the malic acid fluctuations were experimentally manipulated.

Looking for an explanation of the diel changes in the activity of the primary CO_2 fixation, the lucky circumstance came to my assistance that, at the same time when we discovered the in vivo down-regulation of PEPCase during Phase III, Queiroz (1966) found that in vitro this enzyme can be inhibited by L-malate. This led me to

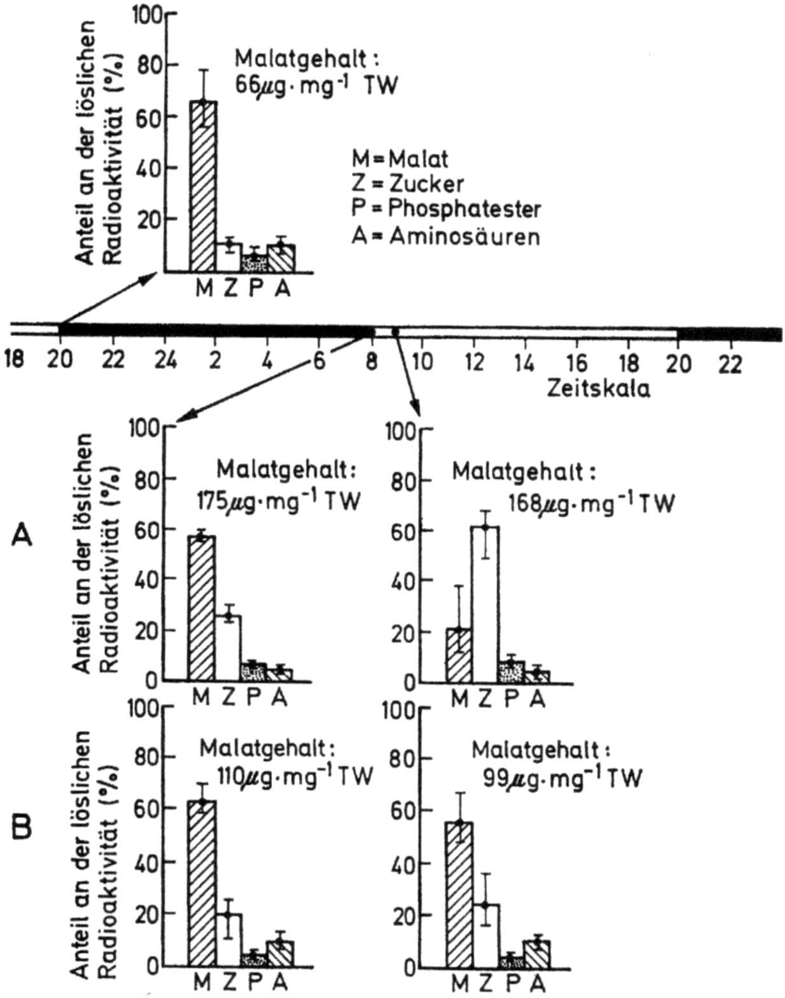

Fig. 2 *Kalanchoe tubiflora*: evidence for the inhibition in vivo of primary CO_2 fixation by PEPCase during Phase III of CAM. Indicative of this inhibition were changes in the labeling patterns in the soluble fraction after 15min $^{14}CO_2$ fixation in the light by phyllodia detached from the plant and labeled at the time points indicated by the *arrows*. The values indicate arithmetic means of three independent extractions. *Graph at top of diagram*: labelling pattern obtained at the beginning of the night. **A** Phyllodia with uninhibited nocturnal malic acid accumulation. In contrast to the labeling pattern obtained during the night (*closed horizontal bar*), one hour after the beginning of the light period (*open horizontal bar*) most of the tracer was trapped by sugars instead of malate, indicating that PEPCase was switched off. **B** Phyllodia with inhibited nocturnal malic acid accumulation. The labeling pattern in the light remained unchanged and indicative of PEPcase being still active. *TW* Dry weight, *M* malate, *Z* sugars (mainly sucrose), *P* phosphate esters, *A* amino acids (from Kluge 1971)

assume that feed-back inhibition of PEPCase by the nocturnally synthesized and stored malate might be important for suppressing CO_2 fixation by PEPCase during Phase III. However, it was also evident to me that an enzymatic feed-back inhibition alone cannot provide the complete explanation. Logically, there must be a factor which during the night keeps the malate level low in the cytoplasm, otherwise the PEPCase located there would be inhibited. In contrast, during Phase III the cytoplasmatic malate level has to be high to bring about feed-back inhibition of the enzyme. I proposed that diel changes in the source-sink behavior of the storage compartment for malate, notably the vacuole, could be the factor which determines the fluctuation of the cytoplasmatic malate level important for the feed-back control of PEPCase (Kluge 1969b). Without having already clear ideas about the mechanism behind this, I imagined that in CAM the vacuole acts with respect to malate in analogy to a siphon which is a highly attractive sink, thus keeping the cytoplasm free of malate, as long it is not filled, but which overflows thus enriching the cytoplasm with malate when it reaches a critical state of filling. In this model the sink situation corresponds to the CAM processes during Phase I and the overflow to the processes during Phase III (Kluge 1969a).

Although the model of co-operation between the vacuole and feed-back inhibition of PEPcase in CAM regulation appeared to be altogether conclusive, I was aware of the many questions remaining open and that the effort to answer them required to learn more: (a) on the structural, kinetic and regulatory properties of PEPCase, and (b) on the mechanisms controlling the reversible transport of malic acid across the tonoplast. Consequently mainly of these two complexes of questions determined my research on CAM during the following years.

3.3 Biographic Insert: Visiting Research Fellow at the ANU in Canberra, Associate Professor in Munich, Return to Darmstadt

The outlines in Sections 1.3.1 and 1.3.2 describe my struggle with CAM until summer 1969 when I finished my qualification for an academic career by Habilitation at the Darmstadt University of Technology. Thereafter I became a Lecturer in the Institute of Botany of this university, and I held this position until spring 1971. During this time I won a stipend from the Deutsche Forschungsgemeinschaft which allowed me to spend six months as Visiting Research Fellow in the group of Ralf Slatyer and Barry Osmond (Research School of Biological Sciences, ANU, Canberra). Barry introduced me to enzymology, and our work was mainly dedicated to the PEPCase of CAM plants. We estimated kinetic and regulatory properties of the semi-purified enzyme and found first indications that PEPCase is allosterically regulated, mainly by malate as negative effector (Kluge and Osmond 1972), We also demonstrated that pyruvate-P_i-dikinase, previously supposed to be a marker enzyme of C4-plants, is active also in CAM plants and permits in the light their incorporation of malate-derived pyruvate into carbohydrates (Kluge and Osmond 1971).

One of the highlights of my stay in Australia was the international meeting "Photosynthesis and Photorespiration" held in November 1970 in Canberra. Since this meeting the view has emerged and is now generally accepted that CAM represents an ecophysiologically relevant mode of photosynthetic carbon assimilation equivalent to C3- and C4-photosynthesis. Thus, by no means could CAM be any longer considered as a physiological curiosity featured by some specialists among higher plants, and in the following decades this new understanding initiated a boom in international CAM research. In this context it is worth mentioning that later we could demonstrate that even classic C3 plants such as spinach can be driven experimentally to a photosynthetic pathway analogous to CAM, with CO_2 fixed in darkness into malate entering the Calvin cycle during light (Böcher and Kluge 1978).

I owe to my stay in Australia also my first impressions of tropical rain forests, coral reefs and deserts, really unforgettable key experiences for me as an ecologically interested biologist.

After my stay in Canberra I went to the Technical University Munich and became Associate Professor at the Institute of Botany. The Director of this institute was Hubert Ziegler who, briefly before I departed to Australia, had moved from Darmstadt to Munich. There I stayed until 1974, and my work during this time strengthened my conviction that transport across the tonoplast is a key factor in the metabolic regulation of CAM. To get deeper insights into this matter I looked to contact experts in cellular transport. In this situation I was offered by the Darmstadt University of Technology the position of a Full Professor of Botany, freshly established parallel to the traditional chair held by Ulrich Lüttge as a follower of H. Ziegler. Ulrich, a leading transport physiologist, was already successfully active in CAM research. So, the call to Darmstadt opened to me the ideal possibility for co-operation and to learn more on vacuolar transport. I was the more pleased since I already knew Ulrich from our common time in Darmstadt as pupils of H. Ziegler and I was sure that we were scientifically and personally on the same wavelength.

Thus, my family and I decided to follow the call to Darmstadt, a decision which we never repented. Compared with other universities, in Darmstadt the Faculty of Biology was quite small, but it provided me such a beneficial environment for my work that in 1978 I refused to follow a call to the University of Essen. In particular the friendship and effective co-operation with Ulrich in research, teaching and administration belong to the most valuable experiences of my scientific career. This co-operation was the fundament that over the years could be founded first a DFG Researcher Group, later the DFG Center of Excellence 199 and finally the DFG Graduate School 340 at the Faculty of Biology in Darmstadt.

Towards the end of my stay in Munich the international boom in CAM research had increased the knowledge on CAM to such an extent that a comprehensive consideration of the many new results was highly desired. This was for me the impetus to publish the first monograph on CAM (Kluge and Ting 1978) which was then followed by the book by Winter and Smith (1996).

3.4 Investigations at the Level of Enzymes: the PEP Case

The central importance of PEPCase not only for CAM and C4-photosynthesis but also for many other metabolic processes in plants has triggered internationally a vast number of investigations on this enzyme, covering the whole range from its function and regulation to its molecular structure and genomic expression. Here I want to stress briefly our own involvement in the PEPCase story.

As already outlined, aiming to understand the role of PEPCase in the regulation of the complex CAM carbon flow, it is necessary to study the kinetic and regulatory properties of the enzyme. I began such studies during my sabbatical in Canberra. There we worked with plant extracts freed from low molecular substances by Sepadex filtration aiming to measure pH optimum curves of enzyme activity, K_m values and K_i values. Usually professional enzymologists refuse to take such kinetic data into account if they are not derived from measurements on the highly purified enzyme protein. Of course, such data are available also from PEPCase of CAM plants. However, it turned out that sometimes it can be useful not to obey the experts too much. Namely, it was discovered independently and nearly simultaneously by three laboratories (including ours), that PEPcase extracted from the plants at different times during the diurnal CAM cycle showed different K_i values for malate, provided the enzyme extracts were quickly desalted by passing through molecular sieves and measuring strictly within a few minutes after the plant material had been collected (v. Willert et al. 1979; Kluge et al. 1980; Winter 1980). Upon purification of the protein the differences vanished. In close co-operation with Jeanne Brulfert and Orlando Queiroz (CNRS, Gif-sur-Yvette, France) we then found that not only the sensitivity of PEPCase towards malate varies along the diurnal CAM cycle but also its sensitivity towards the activator glucose-6-phospahte (G-6-P) and the K_m for the substrate PEP. Thus, in the early 1980s it became clear that, during the diurnal CAM cycle, PEPCase switches reversibly from a physiologically active night form to a physiologically inactive day form. In vivo measurements of gas exchange, $^{14}CO_2$ labeling patterns and malic acid fluctuations parallel to the enzyme extraction supported the view that the changing properties of PEPCase are relevant for up- and down-regulation of the enzyme in vivo and with it for the regulation of the carbon flow in CAM (Kluge et al. 1981b).

It is worth mentioning that our work on the PEPCase of CAM plants founded a long-lasting co-operation and friendship with French colleagues around Orlando Queiroz in Gif-sur-Yvette, in particular with Jeanne Brulfert to whom I owe important stimulation and input for my work. Jeanne and I organized in 1983 a first international workshop on PEPCase which assembled world-wide most of the researches involved in this enzyme. During this workshop the idea was first discussed that the diurnal changes in the regulatory properties of CAM PEPCase might be due to post-translational modification of the enzyme protein. Indeed, Nimmo et al. (1986) then discovered that in the PEPCase of *Bryophyllum fedtschenkoi* the night-form is phosphorylated, not however the day-form. Soon after we found the same in other *Kalanchoë* species (Brulfert et al. 1986) and could demonstrate that the protein is predominantly labeled with $^{32}PO_4$ in serine residues. The discovery of reversible PEPCase phosphorylation was also per se interesting because in the mid-1980s only few plant enzymes were known to be regulated by this way, and it was the only known example of a plant enzyme where protein phosphorylation leads to up- instead of down-regulation. Now it is generally accepted that not only in CAM is the reversible phosphorylation of PEPCase part of a complex regulatory network which includes also the involved PEPCase specific protein kinases and phosphatases (Nimmo 2000).

We have also conducted studies on PEPCase on the molecular level. Already by means of serological techniques we could show that, in the leaves of the facultaltive CAM-plant *Kalanchoë blossfeldiana* upon photoperiodic induction of CAM, a new isoform of PEPCase appeared additionally to that already present in the C3-state of the plant (Brulfert et al. 1982). This finding was then specified on the level of PEPCase gene sequences (Gehrig et al. 1995). We also compared in the CAM performing orchid *Vanilla planifolia* the amino acid sequences of the PEPCase isoforms (deduced from the c-DNA) present in the different organs of the plant (Gehrig et al. 1998a, b). We found that in all organs a constitutive PEPCase isoform (a "house-keeping" form) is expressed functioning most likely in context with anaplerotic reactions, whereas in the CAM-performing organs (leaves and stem) besides the housekeeping isoform an additional one occurs specifically responsible for the start reaction of CAM. Finally, a comparison of PEPCase nucleotide and aminoacid sequences in 50 plant species comprising the whole range from Bryophyta to Spermatophyta (Gehrig et al. 1998a, b, 2001a, b) convinced us that PEPCase sequences are valuable markers for the reconstitution of phylogenetic relations between organisms and metabolic pathways. In this context it was interesting to learn that in the spermatophytes the PEPCase isoforms supposed to be functionally related to CAM are widely dispersed over the different levels of taxa which support the view (Cushman and Bohnert 1999) that CAM is of polyphyletic origin.

3.5 Investigations at the Level of Cellular Compartmentation: the Role of the Vacuole and of Tonoplast Properties in CAM

As outlined in Section 1.3.2 already the results of my early studies on CAM forced me to the conclusion that diel oscillation in the PEPcase activity is the consequence of the rhythmic changes of the source/sink behavior of the vacuole for malate. This assumption pushed me into a long-lasting friendly controversy with Orlando Queiroz. Orlando was convinced that the diurnal CAM carbon flow is principally regulated by an endogenous circadian rhythm of the PEPC activity (Queiroz 1979), following a master oscillator intrinsic to the enzyme system. Finally he believed to have evidence "that an enzyme molecule behaves by itself as an oscillator shifting between different conformations" (Queiroz and Queiroz-Claret 1992).

This controversy forced me to look for more theoretical and experimental arguments in favor of my conviction that in CAM the vacuole is the basic oscillator.

Stimulated by discussions with the great cyberneticist W. Oppelt and his colleague H. Tolle, both professors of the Department of Regulation Techniques of the Technical University Darmstadt, and in co-operation with them we developed a dynamic computer model of the metabolic and regulatory processes in CAM (Nungesser et al. 1984). The model was based on a minimalistic metabolic scheme of CAM and allowed a quite realistic simulation of its typical phenomena such as diurnal acid fluctuation and gas exchange patterns. Later Freder Beck (Department of Physics, Darmstadt University of Technology) in co-operation with Ulrich Lüttge and his team developed a more sophisticated and thus more efficient theoretical model, which also identified the vacuole as master switch governing the circadian rhythmicity of CAM (Lüttge 2000).

Already our first computer model, and even more so the version developed by Freder Beck, clearly showed that proper simulation of the diurnal CAM behavior can be achieved by only inserting the changing flux behavior of the vacuole as an input variable into the model. Other approaches, where rhythmic changes of enzyme activities instead of the vacuole were supposed to represent the oscillator, failed nearly completely.

At the experimental level we could directly demonstrate by ^{14}C-tracer studies the changing source-sink behavior of the vacuole for malate (Fig. 3). Then the question arose for determining the factors controlling this changing behavior. The mechanisms behind the loading of the vacuole were studied by Ulrich Lüttge and his group (see e.g. Lüttge 2000, 2004; Osmond 2007). I was more interested in the mechanisms controlling the efflux of malate. It had been shown that this efflux is thermodynamically passive and it was assumed that at the end of the night, with vacuolar pH being low, malic acid efflux proceeds presumably by diffusion of the undissociated molecule through the lipid phase of the membrane (Lüttge and Smith 1984). Looking for the factors which might influence this diffusion I paid attention mainly to cell water relations and temperature.

1. *Water relations*. Still during my time in Munich we could show that the efflux of malate from the vacuole might be under the control of osmotic gradients (Kluge and Heininger 1973). In co-operation with Ulrich later in Darmstadt these investigations were continued and confirmed the view that osmotic gradients between the vacuole and its environment, thus finally the turgor, are important factors determining the acidification and deacidification of the vacuole (Lüttge et al. 1975).
2. *Temperature*. For a long time it was known that the gas exchange and carbon flow of CAM are phenotypically affected by temperature. As the literature shows, this phenomenon can be explained partially at the level of the enzymes involved in CAM. In this context we studied the temperature response of PEPCase in relation to its allosteric effectors (Buchanan-Bollig and Kluge 1984). However, there were also observations suggesting that the vacuole is the target of temperature effects (Kluge and Ting 1978). For instance, one of my PhD students, Detlev Ritz, found by ^{13}C mass-spectrometric analysis that a high temperature during the night prevents the storage of malate, while its synthesis is still possible (Ritz et al. 1987). Then we could show that temperature, via the

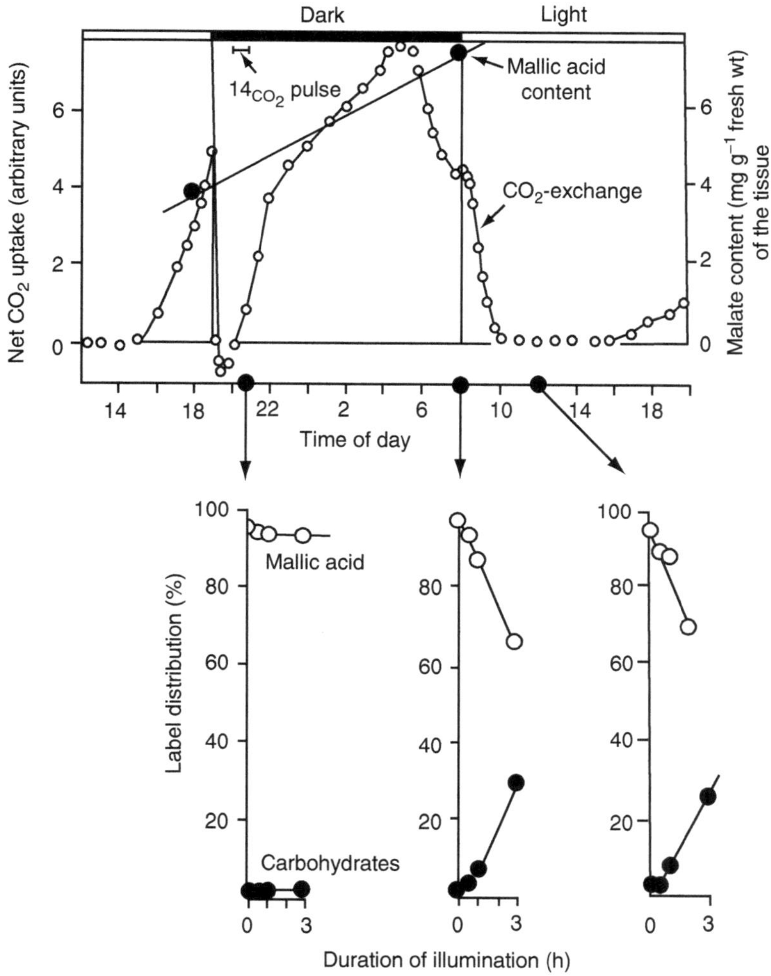

Fig. 3 ^{14}C pulse-chase experiments to show changes in the storage behavior of the vacuole during the night and the first hours of the following day. About 50 phyllodia of *Kalanchoe tubiflora* were detached from a single plant and enclosed in a glass cuvette to follow the CAM gas exchange and the nocturnal malic acid accumulation (*upper panel*). At the beginning of the night 30 phyllodia were exposed for 1 h in darkness to $^{14}CO_2$ as indicated in the upper panel ("$^{14}CO_2$ pulse"). After the pulse the labeled phyllodia were transferred back to normal air and remained in darkness. At the time points indicated by the *arrows* subsamples were taken from the labeled phyllodia and exposed to light for 3 h. During this time the phyllodia were continuously analyzed with respect to changes in the label distribution. In the *lower panel* it can be seen that with empty vacuoles at the beginning of the night even under illumination the labeled malate is retained in the vacuole, thus its label is not transferred to other metabolites. In contrast, at the end of the night, when the vacuole is filled up, upon illumination malate-derived label appears quickly in carbohydrates as photosynthetic end products. The upper panel shows the net CO_2 exchange during the CAM cycle and the nocturnal malic acid accumulation by the detached phyllodia. *Closed bar at top of the graph*: duration of the night (from Kluge 1977)

malic acid efflux from the vacuole, affects the carboxlation pathways of CAM (Friemert et al. 1988), and clear-cut evidence came also from studies on temperature effects on the onset of malic acid release from the vacuole at the beginning of the light period (Kluge et al. 1991d). This release showed at 10 °C a lag-phase of more than two hours, while at temperatures higher than 30 °C no lag-phase occurred, suggesting that at high temperatures malic acid easily leaks out from the vacuole and thus becomes instantaneously available for decarboxylation. The Arrhenius plot of the temperature effects on the duration of the lag-phase showed a break in the slope at about 15 °C. We interpreted this physiological effect in terms of phase separation of the tonoplast lipids beginning in the range of 15 °C. At temperatures higher than this critical value the tonoplast becomes increasingly fluid, thus leaky for malic acid. At temperatures lower than the phase separation range the opposite holds true (Kluge et al. 1991d).

The next step was to verify this hypothesis by measuring the fluidity of the tonoplast as a function of temperature directly. Thus, we had to shift our investigation to the level of the isolated vacuolar membrane. For this approach again co-operation with the group of Ulrich Lüttge was important because there we learned how to isolate, to purify and to characterize the tonoplasts from CAM plant cells. We learned in the laboratory of H.-J. Galla (Department of Organic Chemistry, Darmstadt University of Technology) the biophysical techniques required to measure the thermotropic phase behavior of membrane lipids (electron power magnetic resonance spectroscopy, measurement of fluorescence depolarization). The results of our studies were fully consistent with the predictions derived from physiological experiments (Kluge et al. 1991d). Namely, we found that in *K. daigremontiana*, grown at "standard conditions" (25 °C day, 17 °C night), below 10 °C the tonoplast is in a rigid state. With increasing temperature up to 52 °C tested so far, the membrane becomes steadily more fluid, with a break in the curve in the range of 14–18 °C, suggesting spontaneous phase separation (Fig. 4, upper panel; Kluge et al. 1991b). Moreover, we discovered that the fluidity of the tonoplast can adapt to the growth temperature (Kliemchen et al. 1989), a phenomenon called "homeoviscous adaptation" (HVA) and at that time not known to occur in plant biomembranes, except the thylacoids of *Nerium oleander*. Concerning the tonoplast of *K. daigremontiana*, the HVA to high temperatures means that in the acclimated plants much higher temperature than in the controls is required to reach a given degree of membrane fluidity, thus to elicit the phase transition in the membrane lipids. It was a great satisfaction for us to see that, indeed, in the plants acclimated to the higher temperature, thus having a more rigid tonoplast, the efflux from the vacuole at a given temperature was significantly slower with respect to the not-acclimated controls (Fig. 4). HVA of the tonoplast occurs also upon acclimation to low temperatures, but as to be expected in the other direction (Kliemchen et al. 1989). Finally, we could make clear that HVA of the tonoplast is due to changes in the membrane protein and phospholipid composition (Schomburg and Kluge 1994; Behzadipour et al. 1998).

Our data on the temperature response of tonoplast fluidity provided the experimental input needed by Ulrich Lüttge, Freder Beck and their groups for the development of

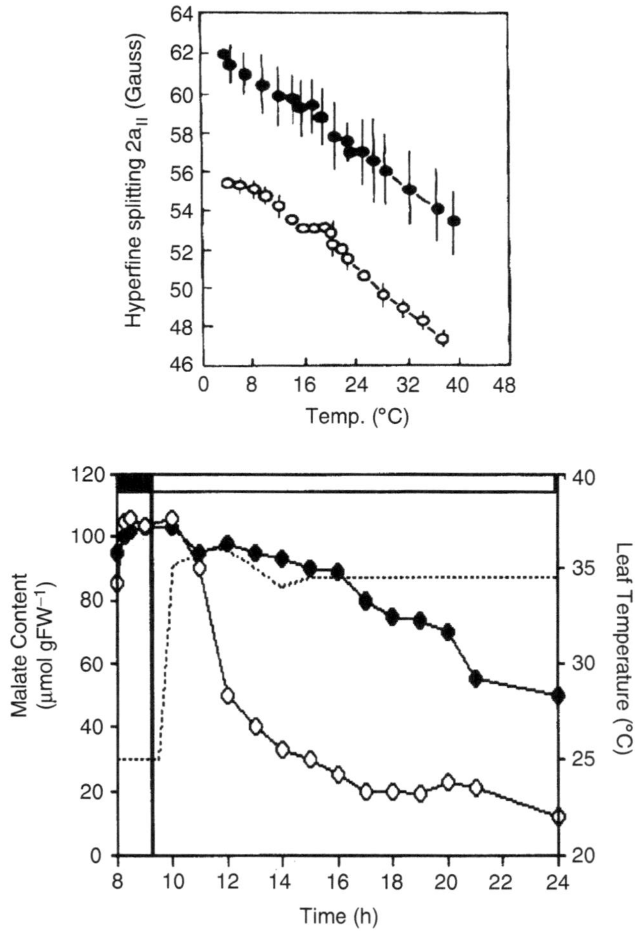

Fig. 4 Thermotropic changes of membrane fluidity and homeoviscous adaptation in the tonoplast in *Kalanchoe daigremontiana* upon acclimatization of the plants to higher growth temperature. *Upper panel* (from Kliemchen et al. 1993) shows the tonoplast fluidity, indicated by the electron power magnetic resonance (EPR) parameter "hyperfine splitting" ($2a_{II}$) in spin-labeled tonoplast preparations, in relation to the incubation temperature during EPR spectroscopy. The lower the $2a_{II}$ value, the more fluid is the tonoplast. *Open symbols*: tonoplasts from plants grown under standard conditions (25 °C day, 17 °C night), mean values of two independent repetitions of the experiment. *Closed symbols*: tonoplasts from plants grown for four weeks at "high temperature, HT" conditions (34 °C day, 25 °C night). Values represent arithmetic means with standard deviation of seven independent experiments. *Lower panel* (redrawn from Kliemchen et al. 1993): Course of malic acid depletion during the light period of CAM in *K. daigremontiana* under the HT conditions mentioned above. *Closed symbols*: plants grown already for four weeks under HT. *Open symbols*: plants grown under standard conditions and exposed without acclimation to HT. The figure shows that, in the plants having a more rigid tonoplast due to HT acclimation, depletion of the vacuolar malate in light is retarded compared with the plants grown at standard temperature thus having more fluid tonoplasts. *Dotted line*: leaf temperature. *Black bar on top of the graph*: Last hour of the dark period; *open bar*: light period

their theoretical model of the master switch determining the circadian rhythmicity CAM. For instance, we could show that in *Kalanchoë daigremontiana* HVA of the tonoplast prevents the endogenous CAM rhythms from being abolished by high temperatures, which is the case in not-acclimated plants (Grams et al. 1995). This supports the prediction from the theoretical model that temperature affects the biological clock controlling CAM by changing the state of order in the tonoplast.

Thus, particularly in the context with the investigation on the vacuolar malate transport, the great benefit of our close co-operation with Ulrich's group was evident and became a supporting pillar for the DFG Center of Excellence 199.

3.6 Investigations at the Biotope Level: Performance of CAM in the Field

From the beginning of my work on CAM I was certain that even the most plausible hypothesis on the ecophysiological relevance of CAM delineated from studies in the laboratory or climate chambers essentially requires verification by field studies, in the case of CAM necessarily sometimes at quite exotic places. Some dyed-in-the-grain biochemists and molecular biologists considered our plans for field studies as "ecotourism", which made the problem of fund raising no easier.

Field studies on CAM are still a challenging task, although today considerably facilitated by advanced experimental technology. Modestly we began our field work in the surroundings of Darmstadt where at dry sites the mini-succulents *Sedum acre* and *S. mite* are abundant. By means of continuous in situ gas exchange measurements on plants enclosed in a fully climate-conditioned cuvette paralleled by biochemical analyses, in situ $^{14}CO_2$ tracer experiments and estimation of the carbon isotope composition ($\delta^{13}C$ values) we could show that the above-mentioned species are C3-CAM intermediates which are permanently capable of CAM performance, but make use of this pathway only if water is in short supply (Schuber and Kluge 1981). Stimulated by P.N. Avadhani (National University of Singapore) we investigated CAM behavior in epiphytic ferns and orchids growing in tropical forests near Singapore (Ong et al. 1986; Goh and Kluge 1989; Kluge et al. 1989).

In 1989 I had the opportunity to join the team of Ulrich Lüttge for field work on St. John (US Virgin Islands) dealing with the tropical CAM-tree *Clusia rosea* (Ball et al. 1991).

For me a highlight of our field work was our studies on CAM in plants of the vegetation of Madagascar. In this context once again I have to mention gratefully Jeanne Brulfert. To her I owe not only my contacts at the University of Antanarivo where in 1989, 1991 and 1993 I had the opportunity to give lectures and courses in plant ecophysiology for graduate students. Without her scientific input, enthusiasm and practical support our research in Madagascar would have been less successful. Also the late Didier Ravelomanana (Botany Department at Antananarivo) and other Malagasy scientists and graduate students contributed substantially to our work.

Although we conducted in Madagascar also the usual in situ measurements of CAM performance by means of portable CO_2/H_2O porometers, a suitable screening method for large-scale analysis was required. Therefore we used mainly the estimation of the stable carbon isotope composition ($\delta^{13}C$ values) to get information on the abundance, mode of behaviour and ecophysiological relevance of CAM in the plants we thought were interesting. Moreover, $\delta^{13}C$ estimation addressing such ecological questions can be successfully applied to plant material obtained from herbaria. For instance, supplementary to the material collected by ourselves in the field, we analyzed among many other species nearly all Malagasy and Continental-African *Kalanchoë* samples deposed at the Musée National d'Histoire Naturelle de Paris. I still remember the absolute silence and the specific smell which surrounded us during our many days of work in the huge herbarium of this museum, an initially unexpected facet of field studies. Hubert Ziegler in Munich and Eliane Deleens in Orsay were essential partners in our work by carrying out the $\delta^{13}C$ analysis in the samples we sent to them. The predictions for CAM behavior of given species derived from their $\delta^{13}C$ values were in many cases verified by experiments on living plants in the phytotron.

Our $\delta^{13}C$ screening on plants of the Malagasy vegetation comprised all species of Didiereaceae, many Euphorbiaceae, Orchidaceae and other families (Kluge et al. 1995). For two reasons, however, our main interest was focused on the comparison of CAM modes in the genus *Kalanchoë* (Kluge et al. 1991a). First, historically CAM was discovered in a species of this genus (*K. pinnata*) and many of the fundamental insights into the biochemistry and physiology of CAM have derived from studies on *Kalanchoës*, for instance *K. daigremontiana*. Second, most of the species are endemic to Madagascar, they show an extraordinary diversity of growth forms, and they are abundant in all of the very different climatic zones and in climatically very different microhabitats of the island. Thus, we were interested to find out whether the mode of CAM performance of the various *Kalanchoë* species could be related to the specific habitat preference and to the morphological characteristics of the plants.

First we realized that most likely all species of the genus *Kalanchoë* are capable of CAM, but that there are considerable differences among the species in the way CAM is conducted. Although potentially CAM performers, some species behave as C3 plants with CO_2 acquisition mainly during the day, thus low water-saving effectiveness. Many other species show stereotypically the water-conserving mode of "strong" CAM with opening of the stomata and CO_2 fixation taking place only during the night, irrespective of water availability. Finally, there are species performing highly flexible CAM with CO_2 fixation both during night and day, with the relative contribution of the two modes depending on the actual water availability.

We were excited by the finding that the differences in CAM behavior are mirrored at the biotope level (Fig. 5). In arid zones and dry microhabitats, the *Kalanchoë* species performing water-saving strong CAM mode are abundant. In contrast, species of the C3 type prefer the humid climate zones or humid, shady habitats in otherwise dry regions, and finally, the climatic zones and habitats with frequently changing water availability are the sites preferred by the species performing highly flexible CAM (Kluge et al. 1991a, 1993; Kluge and Brulfert 1996).

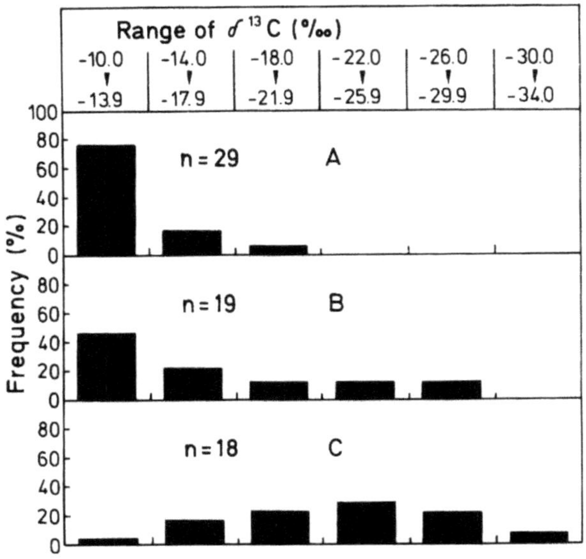

Fig. 5 Quantitative evaluation of $\delta^{13}C$ distribution in *Kalanchoë* species over the vegetation zones of Madagascar. The frequency is expressed as a percentage of the total number of species (*n*) analyzed for the given zone. **A** Xerophilous bush, **B** evergreen dry forest and savannah, **C** evergreen rain forest and montane forest. Where CAM is taking place, $\delta^{13}C$ values in the range of –10‰ to –16‰ are indicative of CO_2 uptake exclusively during the night, and those more negative than –25‰ indicate CO_2 uptake entirely during the day, i.e. a C3 pattern of CO_2 acquisition. $\delta^{13}C$ values between these extreme ranges indicate that the plants in question acquired external carbon in a changing proportion during both night and day (Kluge et al. 1991)

It was also fascinating for us to realize that the various modes of CAM behavior in the *Kalanchoë* species are clearly related to the intragenic taxonomic position of the species and their growth forms. This taxonomic approach was triggered by the lucky circumstance that Dr. Lucile-Allorge Boiteau (CNRS, Gif-sur-Yvette, France) generously made accessible to us a posthumous draft of a monograph on the Malagasy *Kalanchoës* by her father, the late Pierre Boiteau, formerly Director of the Botanical Gardens Zimbazaza (Antananarivo) and one of the leading authorities on the Malagasy succulents. P. Boiteau divided the genus *Kalanchoë* into Section I (Kitchingia), Section II (Bryophyllum) and Section III (Eukalanchoë). It turned out that the C3 performers among the *Kalanchoës* belong altogether to the ancestral Section I and the first two groups of Section II. They represent herbaceous, quite thin-leafed plants. The strong CAM performers belong mainly to the phylogenetically more derived groups of Section II and Section III. The species of these taxa exhibit pronounced leaf succulence. We also included the *Kalanchoë* species of continental Africa in our screening. They belong to the Section III, and $\delta^{13}C$ analysis predictably showed that CAM with predominantly nocturnal CO_2 uptake is the rule.

It is worth mentioning that, encouraged by our ecophysiological studies on the genus *Kalanchoë*, Lucile Allorge-Boiteau finished and finally published P. Boiteau's posthumous monograph (Boiteau and Allorge-Boiteau 1995).

With the results of our field work in mind we turned back to the laboratory to find out whether the grouping of the *Kalanchoë* species based on the ecophysiological and taxonomic criteria is mirrored also on the molecular level. Hans Gehrig, a post-doc in my group, investigated in about half of the known *Kalanchoë* species DNA polymorphism derived from random-amplified polymorphic DNA (RAPD)-PCR, and he compared the nucleotide sequences of internal transcribed spacer (ITS)-1 and ITS-2 regions (Gehrig et al. 1997, 2001a, b). He found out that within the Crassulaceae the genus *Kalanchoë* forms a monophyletic clade. Within this clade the species form three main clusters which coincide with the three intragenic sections of the genus *Kalanchoë* distinguishable by classic taxonomical and ecophysiological criteria, including the modes of CAM performance. ITS analysis showed that Section I is phylogenetically the most ancestral, Section III however the most derived one, thus confirming the propositions by P. Boiteau derived from classic taxonomic criteria. Another interesting finding of the molecular study was that the Section III species are divided into two sister groups, with one comprising only Malagasy species and the other all African species of the genus. The African sister group appears more derived with respect to the Malagasy Section III sister group. Thus, our molecular data provided strong support in favor of the view that the center of phylogenetic radiation of the genus *Kalanchoë* is located in Madagascar, from where the species spread into continental Africa (Boiteau and Allorge-Boiteau 1995).

Our field studies in Madagascar on the genus *Kalanchoë* and other Malagasy CAM plants verified very clearly the selective advantage of CAM. They permitted the conclusion that CAM, because of its high phenotypic plasticity and genotypic variability, allows those species furnished with this mode of photosynthesis to cope both with short-term fluctuations and long-lasting constellations of environmental demands. Thus, CAM enables these plants to conquer successfully highly diverse ecological niches along gradients in the physical environment and resource availability (Kluge et al. 2001; Kluge 2005). In this context a comparison with the CAM diversity in all 11 species of Didiereacea and in all eight Malagasy species of leafless orchids was very instructive to us and supported our interpretation of the results obtained with *Kalanchoë*s (Kluge et al. 2001). Both the Didiereacean species and the leafless orchids are specialists restricted to sites where water is long-lasting of short supply, namely in the case of the Didiereacea the thorn bush in the semi-arid South and South-West of Madagascar, and in the case of leafless orchids epiphytic habitats. As predictable from the uniform environmental conditions of these habitats, the $\delta^{13}C$ of these plants were in all cases indicative of inflexible CAM with CO_2 fixation taking place exclusively during the night.

4 Fascinated by a Curiosity: *Geosiphon pyriformis*

During my later years at the university besides the research on CAM I was trapped by *Geosiphon pyriformis*, an organism far away from the world of succulent higher plants. It was due to my teaching that I became involved in the organism.

For a better understanding of the following some introductory explanations might be helpful. *Geosiphon* is until now the only known example of a coenocytic soil fungus that lives in endosymbiotic (!) association with a cyanobacterium, i.e. *Nostoc punctiforme*. The fungal hyphae grow in the upper layers of soil together with free-living *Nostoc* cells. If the tip of such a hypha meets *Nostoc*, some cells are incorporated into the hypha by a kind of phagocytosis, and upon this act the fungus forms an unicellular pear-shaped "bladder" up to 2 mm long where the *Nostoc* cells are accommodated, multiply and become physiologically active (Fig. 6). *Geosiphon* is apparently very rare and, because of the unpretending appearance of the bladders, difficult to find in nature. That is the reason why relatively little was known on this organism when we began our work on it in 1990. It has to be mentioned that in the early 1960s Eberhard Schnepf had investigated already the ultrastructure of *Geosiphon* bladders (Schnepf 1964). This pioneering study led to a general theory of cell compartmentation and provided strong arguments in favor of the endosymbiosis theory of the eukaryotic cell. After E. Schnepf's studies the organism was forgotten for more than 20 years, until Dieter Mollenhauer rediscovered it on an arable site near Bibergemünd (Spessart Mountains; Germany; Mollenhauer 1988). This is at present the only known site world-wide where *Geosiphon* is naturally abundant.

Many years before I became directly involved in research on *Geosiphon*, within the frame of my teaching I had already covered it in lectures on "Symbiosis and Parasitism in Plants" as an especially fascinating example of endocytosis. Then, in 1986 I needed a photograph of *Geosiphon* for the textbook on botany I was going to write with Ulrich Lüttge (see Section 5). With Dieter Mollenhauer I made an excursion to the field where he had rediscovered the organism. We sampled soil carrying some dozens of bladders and went back to the laboratory to take the needed photographs. Dieter's wife Resi, gifted with admirable patience and subtle intuition for the in vitro cultivation of algae, refused to discard the precious sample and succeeded in establishing a culture of *Geosiphon* (Mollenhauer and Mollenhauer 1988), although previous attempts in this direction by other researchers always had failed. The possibility of growing *Geosiphon* in the laboratory was the breakthrough opening the way for experimental work on this organism. We used this chance and started a close co-operation between my group and Dieter Mollenhauer. Sharing the tasks, we began to investigate at first physiological and metabolic aspects, while Dieter and his wife studied the processes during the initiation and further development of the symbiotic consortium.

Looking for financial support for research on *Geosiphon*, still at the beginning of our work, we were immediately confronted with the question by referees: "And where is your molecular approach?" To me this type of question seems to be symptomatic for certain trends in modern plant science trying to put the cart before the horse. It was always my opinion, and I was reconfirmed by the further development of the *Geosiphon* project, that first careful work at higher levels of the scale is required to uncover clear starting points for meaningful molecular approaches. Nevertheless, finally our standpoint was accepted and the project could be anchored in the DFG Center of Excellence (SFB) 199 and the DFG

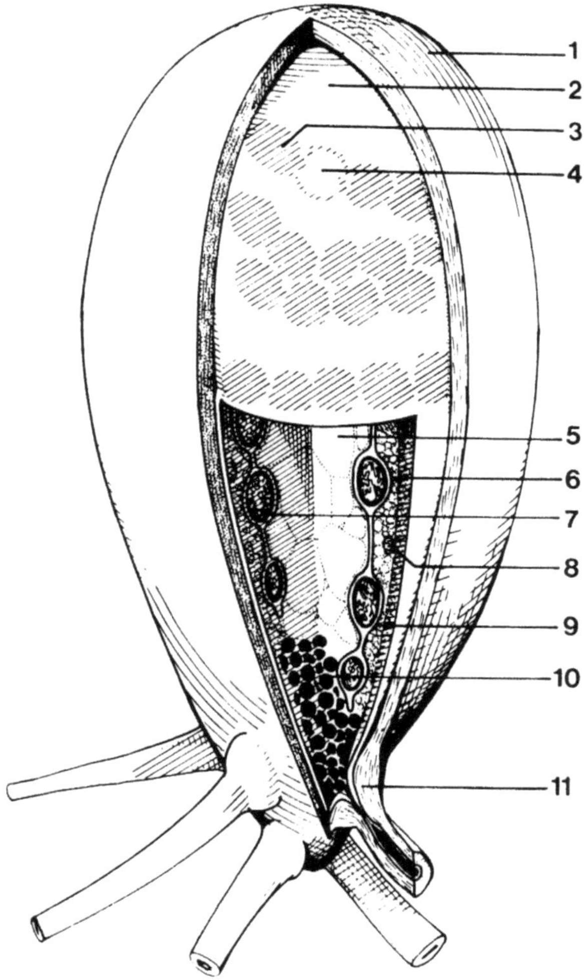

Fig. 6 Organization of a *Geosiphon* bladder (semi-schematic): *1* cell wall of the bladder, *2* plasmalemma of the fungus (outside view), *3* endocytotic *Nostoc* seen through the plasmalemma and outer layer of the fungal cytoplasm, *4* heterocyte in a *Nostoc* filament, *5* central fungal cytoplasm with large vacuoles, *6* peripheral fungal cytoplasm with small vacuoles, *7* section of a vegetative *Nostoc* cell, *8* mitochondrion, *9* pericyanobacterial membrane, i.e. the membrane of the symbiosome, *10* glycogen granules, lipid droplets, microbodies and other inclusions, *11* hypha of the mature bladder with narrow lumen. The function of the constriction is unknown (from Kluge et al. 1994)

Graduate School 340 "Communication in biological systems: from the molecule to the organism in its environment". A graduate student, Arthur Schüssler, joined us to work under the supervision of E. Schnepf (University of Heidelberg) and myself on a doctoral thesis. Arthur became particularly important for the development of the *Geosiphon* research project and now, after the retirement of the initiators, he continues it.

An extended discussion of all results obtained over the years on *Geosiphon* is beyond the scope of this chapter (for reviews see Kluge et al. 1997, 2002; Schüßler and Kluge 2001). In the following just some highlights are mentioned.

We found that inside the *Geosiphon* bladders the *Nostoc* cells are photosynthetically active (Kluge et al. 1991c), with higher efficiency as compared with free-living *Nostoc* cells of the same strain (Bilger et al, 1994). Due to its endosymbionts *Geosiphon* is most likely capable of nitrogen fixation (Kluge et. al. 1992), and it enriches phosphate in the fungal cytoplasm, not however in *Nostoc* cells (Maetz et al. 1999).

A. Schüßler studied the wall of the symbiosome bordering endosymbiotic *Nostoc* cells inside bladders against the fungal cytoplasma. He showed that this interface is produced by the fungus and that its structure is strikingly similar to that bordering the fungus against the plant root cell in arbuscular mycorrhiza (AM). Thus, Arthur postulated that the internal structure of the *Geosiphon* bladder is comparable with the within-root situation of AM. This and other arguments make *Geosiphon* a promising model system for studies on AM (Schüßler 2002) and shifted the project unexpectedly in the direction of AM research.

D. and R. Mollenhauer uncovered the complete development of the *Geosiphon* consortium from the first contacts between the future symbiotic partners via incorporation of the *Nostoc* cells by the fungus in the formation of the bladders (Mollenhauer et al. 1996). In co-operation with the "Institut für den wissenschaftlichen Film" (Göttingen) they documented this process by an excellent video movie (reference: VHS C 1955) which is now commercially available for teaching and research. Tragically Resi Mollenhauer was not allowed during her life to see the publication of the video movie and the whole success of the *Geosiphon* project which owed so much to her, because in 1996 after a long-lasting illness she died.

It was also found out that only immobile *Nostoc* primordia, but not motile hormogonia are susceptible to recognition and incorporation by the fungus (Mollenhauer et al. 1996). Then A. Schüßler showed that the extracellular slime of the primordia contains mannose. This sugar is absent in the other stages of the *Nostoc* life cycle (Schüßler et al. 1997) so that it is tempting to speculate but remains to be shown that mannose has something to do with the specific partner recognition required to establish the *Geosiphon* consortium.

Finally, after a couple of years of patient basic research on *Geosiphon*, the time was ripe for approaches at the molecular level. A. Schüßler and his co-workers started gene expression studies. In this context Holger David succeeded in establishing a cDNA library derived from fungal mRNA of the *Geosiphon* bladders. This led to the first molecular characterization of a carbohydrate transporter from a symbiotic glomeromycotan fungus and resulted in a paper published in *Nature* (Schüßler et al. 2006). The gate is now open for expression studies on further *Geosiphon* genes.

The question about the taxonomic position of *Geosiphon* initiated a study of the small subunit (SSU) rRNA genes of *Geosiphon* (Gehrig et al. 1996), which then led to a comprehensive analysis of the molecular taxonomy of AM fungi (Schüßler et al.

2001). This triggered a revision of fungal taxonomy by the description of a fifth phylum, the Glomeromycota, which comprises all AM forming fungi. *Geosiphon* turned out to represent an ancestral lineage within the Glomeromycota. Finally, the molecular analysis of the taxonomic position of *Geosiphon* provided important arguments for the reconstruction of the early evolution of Fungi using a six-gene phylogeny. This study by an international consortium including A. Schüßler was also published in *Nature* (James et al. 2006).

The discovery leads back to ecology in that *Geosiphon* belongs taxonomically to the AM fungi, which are obligatory symbiotic. Because of its close relationship to the classic AM fungi, most likely (however remaining to be shown) the mycelium of *Geosiphon* interacts in situ not only with *Nostoc* to form the endosymbiotic consortium rather but also with the roots of higher plants. Moreover, in nature *Geosiphon* grows always together with the hornwort *Anthoceros punctatus* and the liverwort *Blasia* spp. It is known that *Anthoceros* at *Geosiphon* stands forms with a glomeromycotan fungus an AM-like symbiosis. In contrast, D. Mollenhauer showed that, at the natural stand, *Nostoc punctiforme* not only interacts with *Geosiphon* but is also a symbiotic partner of *Anthoceros* and *Blasia*. Thus, most likely within its habitat *Geosiphon* could be a link in an ecological network (Kluge et al. 2002) which integrates *Nostoc*, *Anthoceros*, *Blasia* and higher plants. This network opens interesting questions for future ecological research.

5 Teaching: Mainly Great Pleasure, Sometimes Frustration

After graduation I decided myself deliberately in favor of a university career because it promised the possibility to combine two passions: research and teaching. Looking back I think it was the right decision, although the politically motivated over-regulation of the German university system which began in the late 1960s became increasingly conter-productive for the genuine tasks of a university professor. How often I wished to have the time wasted in endless and ineffective debates in the institutions of the academic self-administration to be available for more intense work with the students. Also in this context I mention gratefully the close co-operation with Ulrich Lüttge which was very important to overcome the frictions born by the university system of today, thus facilitating the effort to maintain efficiency in the work of the Institute of Botany in Darmstadt at the highest level.

Teaching, including the basic lectures in General Biology and General Botany, always meant pleasure for me and were a permanent source of inspiration, never an inconvenient obligation. In particular teaching to undergraduates I experienced as a challenging and important task, because today the freshmen coming from the gymnasium begin their study of biology with very different functional prerequisites, sometimes even without clear motivation for the topic. Of course, occasionally there was also frustration when all efforts to raise real interest failed. With the lecture "General Botany" Ulrich Lüttge and I took turns from one year to the other and collected by this way a lot of common experience on the difficulties and demands of teaching this topic under the allowance of the current curriculum. On the basis of this experience we wrote our textbook "Botanik" (Lüttge et al. 2005), dedicated in particular to undergraduate students. Writing such a book requires a permanent dialogue and reciprocal tuning between the authors, and the success of the opus tells us that here we were on the right path.

Retrospectively sometimes I ask myself if it would not have been better to focus the research onto a more narrow topic instead of migrating between the various levels of scaling; for instance, to stay at the level of biochemistry and to investigate the exciting problem of PEPCase and its regulation cascade in all its ramifications instead of having always the whole CAM syndrome in mind. Given my propensity, the holistic approach was more satisfying for me and corresponded more to my understanding of the essence of ecophysiology as outlined in the introduction. Probably I would try to follow anew the same line, but I am aware that under the challenges of modern biology and current research politics this would be very difficult, if not impossible. With regard to teaching at a botanical institute with a relatively small teaching staff like Darmstadt an utmost broad spectrum of scientific interest and experience is essential to cope with the requirements of the curriculum. Also in this context my holistic orientation was very helpful.

Acknowledgements In addition to the editors, who honored me so much by the invitation to write this article, I have to thank all the personalities who helped me to the privilege that I was allowed to unify passion with profession. Above all I thank my family who kept me free for my work and patiently tolerated that often I was not enough available, thus making life easy for me. I am deeply grateful to my doctoral tutor Hubert Ziegler who strongly influenced my way of thinking and approaching scientific problems. My friends P.N. ("Danny") Avadhani, Jeanne Brulfert, Ulrich Lüttge, Dieter Mollenhauer, Barry Osmond, Orlando Queiroz, Irvin P. Ting and many others I thank for the important inspiration and effort I received from them; and I gratefully mention all the dedicated members of my administrative and technical staff, and the many guest scientists and students who joined my group over the years. Finally I want to apologize to all those colleagues, postdoctoral coworkers and graduate students whose merits I could not mention in the present treatise because of the limitated space. My deep thanks are due also to them.

References

Ball E, Hann J, Kluge M, Lee H S J, Lüttge U, Orthen B, Popp M, Schmitt A and Ting IP (1991) Ecophysiological comportment of the tropical CAM-tree *Clusia* in the field. II. Modes of photosynthesis in trees and seedlings. New Phytol 117:483–491

Behzadipour M, Ratajczak R, Faist K, Pawlitschek P, Trémolières A, Kluge M (1998) Phenotypic adaptation of tonoplast fluidity to growth temperature in the CAM Plant *Kalanchoë daigremontiana* Ham. et Per. is accompanied by changes in the membrane phospholipid and protein composition. J Membr Biol 166:61–70

Bilger W, Büdel B, Mollenhauer R, Mollenhauer D (1994) Photosynthetic activity of two developmental stages of a *Nostoc* strain (Cyanobacteria) isolated from *Geosiphon pyriforme* (Mycota). J Phycol 30:225–230

Boiteau P, Allorge-Boiteau L (1995) *Kalanchoe* (Crassulaceés) de Madagascar. Karthala, Paris

Böcher M, Kluge M (1978) Der C_4-Weg der CO_2-Fixierung bei *Spinacea oleracea*. II. Pulse-chase Experimente mit suspendierten Blattstreifen. Z Pflanzenphysiol 86:405–421

Brulfert J, Müller D, Kluge M, Queiroz O (1982) Photoperiodism and crassulacean acid metabolism I: Immunological and kinetic evidences for different patterns of phosphoenolpyruvate carboxylase isoforms in photoperiodically inducible and non-inducible crassulacean acid metabolism plants. Planta 154:326–331

Brulfert J, Vidal J, Le Marechal P, Gadal P, Queiroz O, Kluge M, Krüger I (1986) Phosphorylation–dephosphorylation process as a probable mechanism for the diurnal regulatory changes of phosphoenolpyruvate carboxylase in CAM plants. Biochem Biophys Res Commun 136:151–159

Buchanan-Bollig IC, Kluge M, Müller D (1984) Kinetic changes with temperature of phosphoenolpyruvate carboxylase from a CAM plant. Plant Cell Environment 7:63–70

Cockburn W, Ting IP, Sternberg CO (1979) Relationship between stomatal behavior and internal carbon dioxide concentration in crassulacean acid metabolism plants. Plant Physiol 63:1029–1032

Cushman JG, Bohnert B (1999) Molecular genetics of crassulacean acid metabolism. Annu Rev Plant Physiol Plant Mol Biol 50:305–332

Friemert V, Heininger D, Kluge M, Ziegler H (1988) Temperature effects on malic-acid efflux from the vacuoles and on the carboxylation pathways in CAM Plants. Planta 174:453–461

Gehrig H, Taybi T, Kluge M, Brulfert J (1995) Identification of multiple PEPC iso-genes in leaves of the facultative crassulacean acid metabolism (CAM) plant *Kalanchoë blossfeldiana* Poelln. cv Tom Thumb. FEBS Lett 377:399–402

Gehrig H, Schüßler A, Kluge M (1996) *Geosiphon pyriforme*, a fungus forming endocytobiosis with *Nostoc* (Cyanobacteria), is an ancestral member of the Glomales: evidence by SSUrRNA analysis. J Mol Evol 43:71–78

Gehrig H, Rösicke H, Kluge M (1997) Detection of DNA polymorphism in the genus *Kalanchoë* by RAPD-PCR fingerprint and its relationships to infrageneric taxonomic position and ecophysiological photosynthetic behaviour of the species. Plant Sci 125:41–51

Gehrig H, Faist K, Kluge M (1998a) Identification of phosphoenolpyruvate carboxylase isoforms in leaf, stem and roots of the obligate CAM plant *Vanilla planifolia* Salib. (Orchidaceae): a physiological and molecular approach. Plant Mol Biol 38:1215–1223

Gehrig H, Heute V, Kluge M (1998b) Towards a better knowledge of the molecular evolution of phosphoenolpyruvate carboxylase by comparison of partial c-DNA sequences. J Mol Evol 46:107–114

Gehrig H, Gaußmann O, Marx H, Schwarzott D, Kluge M (2001a) Molecular phylogeny of the genus *Kalanchoe* (Crassulaceae) inferred from nucleotide sequences of the ITS 1 and ITS 2 region. Plant Sci 160:827–835

Gehrig H, Heute V, Kluge M (2001b) New partial sequences of phosphoenolpyruvatecarboxylase as molecular phylogenetic markers. Mol Phyl Evol 20:262–274

Goh JC, Kluge M (1989) Gas exchange and water relations in epiphytic orchids. Ecol Stud 76:139–166

Grams TEE, Kluge M, Lüttge U (1995) High temperature adapted plants *Kalanchoe daigremontiana* show changes in the temperature dependence of the endogenous CAM rhythm. J Exp Bot 46:1927–1929

James TY, et al (2006) Reconstructing the early evolution of fungi using a six-gene phylogeny. Nature 443:818–822

Kliemchen A, Schomburg M, Galla H-J, Lüttge U, Kluge M (1989) Phenotypic changes in the fluidity of the tonoplast membrane of crassulacean acid metabolism plants in response to temperature and salinity stress. Planta 189:403–409

Kluge M (1968a) Untersuchungen über den Gaswechsel von *Bryophyllum* während der Lichtperiode. I. Zum Problem der CO_2-Abgabe. Planta 80:255–263

Kluge M (1968b) Untersuchungen über den Gaswechsel von *Bryophyllum* während der Lichtperiode. II. Beziehungen zwischen dem Malatgehalt des Gewebes und der CO_2-Aufnahme. Planta 80:359–377

Kluge M (1969a) Zur Analyse des CO_2-Austausches von Bryophyllum. II. Hemmung des nächtlichen Stärkeabbaus in CO_2-verarmter Atmosphäre. Planta 86:142–150

Kluge M (1969b) Veränderliche Markierungsmuster bei $^{14}CO_2$-Fütterung von *Bryophyllum tubiflorum* zu verschiedenen Zeitpunkten der Hell-Dunkel-Periode. I. Die Fütterung unter Belichtung. Planta 88:113–129

Kluge M (1971) Veränderliche Markierungsmuster bei $^{14}CO_2$-Fütterung von *Bryophyllum tubiflorum* zu verschiedenen Zeitpunkten der Hell-Dunkel-Periode. II. Beziehungen zwischen dem Malatgehalt des Gewebes und dem Markiermuster nach $^{14}CO_2$-Lichfixierung. Planta 98:20–30

Kluge M (1977) Regulation of carbon dioxide fixation in plants. Proc Symp Soc Exp Biol 31:155–175

Kluge M (2005) Options of photosynthesis in the genus *Kalanchoe*: an approach integrating the levels of biochemistry, ecophysiology, molecular taxonomy and phytogeographic distribution of the species. Nova Acta Leopold 92:195–205

Kluge M, Brulfert J (1996) Crassulacean acid metabolism in the genus *Kalanchoë*: ecological, physiological and biological approaches. In: Winter K, Smith AP, Smith JAC (eds) Crassulacean acid metabolism. biochemistry, ecophysiology and evolution. Springer, Berlin Heidelberg New York, pp 324–335

Kluge M, Fischer K (1967) Über Zusammenhänge zwischen dem CO_2-Austausch und der Abgabe von Wasserdampf durch *Bryophyllum daigremontianum*. Planta 77:212–223

Kluge M, Heininger B (1973) Untersuchungen an *Bryophyllum* über den Efflux von Malat aus den Vakuolen der assimilierenden Zellen und mögliche Einflüsse dieses Vorgangs auf den CAM. Planta 113:333–343

Kluge M, Osmond CB (1971) Pyruvat-P_i-dikinase in crassulacean acid metabolism. Naturwissenschaften 58:414–415

Kluge M, Osmond CB (1972) Studies on PEP-carboxylase and other enzymes involved in crassulacean acid metabolism of *Bryophyllum tubiflorum* and *Sedum praealtum*. Z Pflanzenphysiol 66:97–105

Kluge M, Ting IP (1978) Crassulacean acid metabolism. Ecol Stud 30:3

Kluge M, Böcher M, Jungnickel G (1980) Metabolic control of crassulacean acid metabolism: evidence for diurnal changing sensitivity against inhibition by malate of PEP-carboxylase in *Kalanchoë tubiflora* Hamet. Z Pflanzenphysiol 97:197–204

Kluge M, Böhlke C, Queiroz O (1981a) Crassulacean acid metabolism in *Kalanchoë*: variations of the intercellular CO_2 concentration during a normal CAM cycle, and during cycles with either light or darkness only. Planta 152:87–92

Kluge M, Brulfert J, Queiroz O (1981b) Further evidence for diurnal changes in regulatory properties of PEP-carboxylase in crassulacean acid metabolism (CAM). Plant Cell Environment 4:251–256

Kluge M, Friemert V, Ong BL, Brulfert J, Goh JC (1989) In situ studies of crassulacean acid metabolism in *Drymoglossum piloselloides*, an epiphytic fern of the humid tropics. J Exp Bot 40:441–452

Kluge M, Brulfert J, Ravelomanana D, Lipp J, Ziegler H (1991a) Crassulacean acid metabolism in *Kalanchoë* species collected at various climatic zones of Madagascar: a survey by $\delta^{13}C$ analysis. Oecologia 88:407–414

Kluge M, Kliemchen A, Galla H-J (1991b) Temperature effects on crassulacean acid metabolism: EPR spectroscopic studies on the thermotropic phase behaviour of the tonoplast membranes *Kalanchoë daigremontiana*. Bot Acta 104:355–360

Kluge M, Mollenhauer D, Mollenhauer R (1991c) Photosynthetic carbon assimilation in *Geosiphon pyriforme* (Kützing) F. v. Wettstein, an endosymbiotic association of fungus with a cyanobacterium. Planta 185:311–315

Kluge M, Wolf H, Fischer A (1991d) Crassulacean acid metabolism: temperature effects on the lag-phase in the photosynthetic oxygen evolution occurring at the outset of the light period. Plant Physiol Biochem 29:83–90

Kluge M, Mollenhauer D, Mollenhauer R, Kape R (1992) *Geosiphon pyriforme,* an endosymbiotic consortium of a fungus and a cyanobacterium (*Nostoc*), fixes nitrogen. Bot Acta 105:343–344

Kluge M, Brulfert J, Lipp J, Ravelomanana D, Ziegler H (1993) A comparative study by $\delta^{13}C$-analysis of crassulacean acid metabolism (CAM) in *Kalanchoë daigremontiana* (Crassulaceae) species of Africa and Madagascar. Bot Acta 106:320–324

Kluge M, Mollenhauer D, Mollenhauer R (1994) *Geosiphon pyriforme* (Kützing) von Wettstein, a promising system for studying endocyanoses. Prog Bot 55:5

Kluge M, Brulfert J, Rauh W, Ravalomanana D, Ziegler H (1995) Ecophysiological studies on the vegetation of Madagascar: $\delta^{13}C$ and δD survey for incidence of crassulacean acid metabolism (CAM) among orchids from montane forests and succulents from the xerophytic thorn-bush. Isotopes Environ Health Stud 31:191–210

Kluge M, Gehrig H, Mollenhauer D, Mollenhauer R, Schnepf E, Schüßler A (1997) News on *Geosiphon pyriforme*, an endocytobiotic consortium of a fungus with a cyanobacterium. In: Schenk HEA, Herrmann R, Jeon KW, Müller NE, Schwemmler W (eds) Eukaryotism and symbiosis: intertaxonic combination versus symbiotic adaptation. Springer, Berlin Heidelberg New York, pp 469–476

Kluge M, Razanoelisoa B, Brulfert J (2001) Implications of genotypic diversity and phenotypic plasticity in the ecophysiological success of CAM plants, examined by studies on the vegetation of Madagascar. Plant Biol 3:214–222

Kluge M, Mollenhauer D, Wolf E, Schüßler A (2002) The *Geosiphon/Nostoc* endocytobiosis. In: Rai AN, Bergmann B, Rasmussen M (eds) Cyanobacterian symbiosis. Kluwer, Dordrecht, pp 19–30

Lüttge U (1997) Physiological ecology of tropical plants. Springer, Berlin Heidelberg New York

Lüttge U (2000) The tonoplast functioning as the master switch for circadian regulation of crassulacean acid metabolism. Planta 211:761–769

Lüttge U (2004) Ecophysiology of crassulacean acid metabolism (CAM). Ann Bot 93:629–652

Lüttge U, Smith JAC (1984) Mechanism of passive malic-acid efflux from vacuoles of the CAM plant *Kalanchoe daigremontiana*. J Membr Biol 81:149–158

Lüttge U, Kluge M, Ball E (1975) Effects of osmotic gradients on vacuolar malic storage. A basic principle in oscillatory behaviour of crassulacean acid metabolism. Plant Physiol 56:613–616

Lüttge U, Kluge M, Bauer G (2005) Botanik, 5th edn. Wiley–VCH, Weinheim

Maetz M, Schüßler A, Wallianos A, Traxel A (1999) Subcellular trace element distribution in *Geosiphon pyriforme*. Nucl Instrum Methods B 150:200–207

Mollenhauer D (1988) Weitere Untersuchungen an *Geosiphon pryriforme*- einer Lebensgemeins chaft von Pilz und Blaualge. Nat Mus 118:289–309

Mollenhauer D, Mollenhauer R (1988) *Geosiphon* cultures ahead. Endocyt Res 5:69–73

Mollenhauer D, Mollenhauer R, Kluge M (1996) Studies on initiation and development of the partner association in Geosiphon pyriforme (Kütz) v. Wettstein, a unique endocytobiotic system of a fungus (Glomales) and the cyanobacterium *Nostoc punctiforme*. (Kütz)Harriot. Protoplasma 193:3–9

Nimmo GA, Nimmo HG, Hamilton ID, Fewson CA, Wilkins MB (1986) Purification of the phosphorylated night form and dephosphorylated day form of phosphoenolpyruvate carboxylase from *Bryophyllum fedtschenkoi*. Biochem J 239:213–220

Nimmo HG (2000) The regulation of phosphoenolpyruvate carboxylase in CAM plants. Trends Plant Sci 5:75–80

Nishida K (1963) Studies on stomatal movement of crassulacean plants in relation to the acid metabolism: Physiol Plant 16:281–298

Nungesser D, Kluge M, Tolle H, Oppelt W (1984) A dynamic computer model of the metabolic and regulatory processes in CAM. Planta 162:204–214

Ong BL, Kluge M, Friemert V (1986) Crassulacean acid metabolism in the epiphytic ferns *Drymoglossum piloselloides* and *Pyrrosia longifolia*: studies on the response to environmental signals. Plant Cell Environ 9:547–557

Osmond CB (1978) Crassulaceen acid metabolism: a curiosity in context. Annu Rev Plant Physiol 29:379–414

Osmond CB (2007) Crassulaceae acid metabolism: now and then. Prog Bot 68:3–32

Queiroz O (1966) Recherche d'un modéle enzymatique pour le déterminisme de la désacidification diurne chez les Crassulacees. C R Acad Sci 265:1928–1931
Queiroz O (1979) CAM: rhythms of enzyme capacity and activity as adaptive mechanisms. In: Gibbs M, Latzko E (eds) Encyclopedia of plant physiology, photosynthesis II, vol 6. Springer, Berlin Heidelberg New York, pp 263–270
Queiroz O, Queiroz-Claret C (1992) Seasonal and daily control of enzyme synthesis and activity by circadian clocks. In: Tuitou Y, Hans E (eds) Biological in clinical and laboratory medicine. Springer, Berlin heidelberg New York, pp 78–89
Raschke K (1966) Die Reaktionen des CO_2-Regelsystems in den Schließzellen von *Zea mays* auf weißes Licht. Planta 68:111–140
Ritz D, Kluge M, Veith HJ (1987) Effect of temperature and CO_2-concentration on malate accumulation during CAM: evidence by ^{13}C-mass spectrometry. Plant Physiol Biochem 25:391–399
Schnepf E (1964) Zur Feinstruktur von *Geosiphon pyriforme*. Arch Mikrobiol 49:112–131
Schomburg M, Kluge M (1994) Phenotypic adaptation to elevated temperatures of tonoplast fluidity in the CAM plant *Kalanchoë daigremontiana* is caused by membrane proteine. Bot Acta 107:328–332
Schuber M, Kluge M (1981) In situ studies on crassulacean acid metabolism in *Sedum acre* L. and *Sedum mite* Gil. Oecologia 50:82–87
Schüßler A (2002) Molecular pylogeny, taxonomy and evolution of *Geosiphon pyriformis* and arbuscular myccorrhizal fungi. Plant Soil 244:75–83
Schüßler A, Kluge M (2001) *Geosiphon pyriforme*, an endocytosymbiosis between fungus and cyanobacteria, and its meaning as a model system for AM research. In: Hock B (ed) The Mycota IX. Springer, Berlin Heidelberg New York, pp 151–161
Schüßler A, Mollenhauer D, Schnepf E, Kluge M (1994) *Geosiphon pyriforme*, an endosymbiotic association of fungus and cyanobacteria: the spore structure resembles that of mycorrhizal (VAM) fungi. Bot Acta 107:36–45
Schüßler A, Meyer T, Gehrig H, Kluge M (1997) Variations of lectin binding sites in extracellular glycoconjugates during the life cycle of *Nostoc punctiforme*, a potentially endosymbiotic cyanobacterium. Eur J Phycol 32:233–239
Schüßler A, Schwarzott D, Walker C (2001) A new fungal phylum, the *Glomeromycota*: phylogeny and evolution. Mycol Res 105:1413–1421
Schüßler A, Martin H, Cohen D, Fitz M, Wipf D (2006) Characterization of a carbohydrate transporter from symbiotic glomeromycotan fungi. Nature 444:933–936
Szarek SR, Ting IP (1975) Physiological responses to rainfall in *Opuntia basilaris*. Am J Bot 62:602–609
Willert DJ von, Willert K von (1979) Light modulation of the activity of the PEP-carboxylase in CAM plants of the Mesembryanthemaceae. Z Pflanzenphysiol 95:42–49
Winter K (1980) Day/night changes in the sensitivity of phosphoenolpyruvate carboxylase to malate during crassulacean acid metabolism. Plant Physiol 65:792–796
Winter K, Smith JAC (1996) Crassulacean acid metabolism; biochemistry, ecophysiology and evolution. (Ecological Studies, vol 114). Springer, Berlin Heidelberg New York

Genetics

Variability of Recombination Rates in Higher Plants

Elisabeth Esch and Renate Horn(✉)

Contents

Abstract Recombination and chiasma frequency, like other features of meiosis, are subject to various genetic control mechanisms. Here, we give an overview of the genetic and environmental factors as well as the genomic structures that play a role for the variability of recombination rates in plant genomes. Suppressed or greatly reduced recombination is observed in chromosomal regions that contain repetitive sequences such as e.g. centromeres, pericentromeric and telomeric regions as well as between sex chromosomes. DNA recombination plays a major role in genome evolution and crop improvement. To estimate the variability of recombination rates, both genetic and physical distances are required (measured in centiMorgans and usually kilobases, respectively) and so far, these data are only available for a limited number of model species. Progress in whole genome sequencing project in the coming years will allow further analyses of more species and a comparison of recombination rates and their variability between different species. Additional molecular studies of recombination in plants are necessary to

Renate Horn
Institut für Biowissenschaften, Abtl. Pflanzengenetik, Universität Rostock, Albert-Einstein-Strasbe 3, D-18051 Rostock, Germany
e-mail: renate.horn@uni-rostock.de

U. Lüttge et al. (eds.), *Progress in Botany 69*.

improve our understanding of the genetic mechanisms behind the variability of recombination rates in plant genomes that is far from being comprehensive.

1 Introduction

Recombination and chiasma frequency, like other features of meiosis, are subject to stringent genetic control (reviewed e.g. by Anderson and Stack 2002; Jones et al. 2003; Hamant et al. 2006). In numerous studies in different plant species, variability of recombination frequencies between different genetic backgrounds was shown; and several genes that are essential for normal meiotic levels of recombination have been identified and characterised by the use of morphological and molecular markers as well as by cytogenetic methods, and recently by molecular approaches using forward and reverse genetics.

Recombination varies among genomic regions in plants, but so far genome-wide patterns of recombination are available for only a few plant species, e.g. *Arabidopsis* and rice. Genetic improvement in breeding programmes depends on the generation and selection of new recombinant progenies from intervarietal or interspecific crosses (Wang et al. 1995). The degree of genetic variation obtained in the segregating generations of a particular cross depends on the random reassortment of chromosomes and the amount of recombination between chromosomes. The latter can vary in species and can be influenced by both environment and genotype.

We like to give an overview about the genetic and environmental factors as well as the genomic structures that play a role for the variability of recombination rates in plant genomes. There is initial evidence that recombination rates might be both greater and more variable in plants than some animals, thereby contributing to the increased variability of genomic characteristics of plants (Gaut et al. 2007).

2 Genes and Genetic Background Modifying Recombination

2.1 General Control

Genetic control of recombination is supposed to be of two general types: coarse or fine (Simchen and Stamberg 1969). The numerous genes comprising the coarse system were defined to rigidly control the occurrence and the progression of meiosis. The allelic differences in these genes are seen as a complete lack of recombination. They uniformly affect the entire genome and are supposed to occur extremely rare. The fine control of recombination is much more complex and requires the basic mechanism of recombination is functionally in order. The allelic differences are seen as wide variations in recombination frequencies. The effects of individual elements of the fine control system are localised to small regions of the genome. One gene can control several segments spread over the genome and, vice versa, recombination frequency in a particular region can be regulated by several genes explaining continuous distributions of recombination frequencies.

Another way to classify the genetic control of recombination is focusing on the position of the gene(s) compared to the region(s) affected. The variability in recombination frequencies within a species among genetic backgrounds can be attributed to two general classes of genetic factors: *cis* and *trans* (reviewed by Yandeau-Nelson et al. 2006; Timmermans et al. 1997). *Cis*-acting elements are genetic factors that affect recombination in the region in which the factor resides. In contrast, *trans*-acting factors are genetic modifiers that are not closely linked to the interval in which they affect recombination. *Trans*-acting modifiers of meiotic recombination have been further divided into two categories: those that affect recombination across the entire genome (i.e. global modifiers) and those that act only on specific genetic intervals (i.e. region-specific modifiers).

2.2 *Polygenic Quantitative Control*

There are a number of reports indicating a polygenic quantitative control of recombination frequency in plant populations.

One of the earliest reports on the variability of recombination frequencies and its genetic control in plants was in 1927 by Rasmusson. In pea, he found significant differences in recombination values between pairs of morphological markers and supposed as cause different genotypes. A review of reports on variation in recombination values revealed that variation had been discovered in all organisms thoroughly investigated (Rasmusson 1927). It was suggested to treat recombination rates as any other quantitative trait.

In rye, genotypic control of chiasma frequency was proposed from the variation found in inbred lines, where the differences in chiasma frequencies between lines were highly significant compared with the differences within lines (Rees 1955). From the continuous distribution of mean chiasma frequencies a polygenic control was concluded. Further studies indicated that predominately additive genetic variance accounted for the genetic variability in chiasma frequency present in rye (Rees and Thompson 1956).

Differences in chiasma frequencies between five *Hordeum* subspecies were shown to be controlled mainly by genes with additive effects and, to a lesser extent, by genes with non-additive effects including dominance (Gale and Rees 1970). No epistatic effects were detected. A large part of the variation in chiasma frequency was attributed to a major gene, or to a cluster of polygenes, located on chromosome 2.

Genotypically controlled significant differences in chiasma frequencies between populations of different origin were reported within *Lolium perenne*, *L. multiflorum* and *Festuca pratensis* (Rees and Dale 1974).

Significant variability for recombination frequency showing a continuous variation has also been found in *Raphanus sativus* L. (Dayal 1977) and soybean (*Glycine max* (L.) Merr.) (Pfeiffer and Vogt 1990), and has been interpreted as indicating polygenic control.

In maize (*Zea mays* L.), heterogeneity in recombination frequencies has been well documented using phenotypic and molecular markers. In a study using F_2 and S_1 families, wide variation in recombination frequencies was documented (Tulsieram et al. 1992). The families were randomly derived from three maize populations: a corn belt synthetic and two composite populations of corn belt × exotic germplasm. Recombination values differed significantly among populations and among families in each population for all three chromosomal regions studied. Two regions showed continuous variation for recombination frequencies, presumably indicating a genetic control by several to many loci, each having small effects. The range in variability in recombination among genotypes was also established in the Iowa Stiff Stalk Synthetic (BSSS) population (Fatmi et al. 1993). Variation in recombination frequency in five regions on three chromosomes was measured. Differences between the maximum and minimum recombination frequencies ranged from two- to 11-fold. Several intervals on different chromosomes were shown to have either negatively or positively correlated recombination frequencies. At each chromosome region, individual genotypes with recombination frequencies at least two standard deviations above or below the population mean were isolated. Recombination rates displayed continuous variation and were normally distributed, implying the segregation of multiple loci controlling recombination and, thus, a quantitative genetic control of this trait (Fatmi et al. 1993). In a subsequent study (Hadad et al. 1996) to verify the differences among individuals in the Iowa Stiff Stalk Synthetic maize population, the recombination frequencies were repeatably divergent among those individuals which were selected based on high or low recombination frequencies on specific chromosomes. The broad-sense heritability estimates derived from the regression of six S_1 lines on six S_0 individuals ranged from 0.69 to 0.20. Investigation of the variability of recombination distances and linkage heterogeneity has been extended to a wide range of maize inbreds, wide crosses and maize × teosinte hybrids (Williams et al. 1995). A large variability, often two-fold, was found in the recombination rates, especially among the tropical maize open pollinated races. In all regions except one, there was continuous variation for recombination frequency.

(Yandeau-Nelson et al. 2006) analysed the influence of different genetic backgrounds on the variability of recombination in a relatively small genetic and physical interval. The effects of modifiers on recombination across the about 140 kb and about 0.1 cM maize *a1-sh2* multigenic interval of chromosome 3L were studied in three genetically diverse backgrounds into which a sequence identical *a1-sh2* interval had been introgressed. Each of the genetic backgrounds exhibited a statistically different recombination rate between *a1* and *sh2* due to the action of *trans*-acting modifier(s). The recombination rates varied almost two-fold. The *trans*-acting factors also affected the distribution of recombination breakpoints within the *a1-sh2* interval among genetic backgrounds. Recombination rates across two genetic intervals on chromosome 1 did not exhibit the same relationships among backgrounds as was observed in *a1-sh2*. This suggested that at least some detected *trans*-acting factors did not equally affect recombination across the genome and were instead region specific (Yandeau-Nelson et al. 2006).

In summary, the recombination modification system for maize is characterised by large amounts of genetic variability in recombination rates among genetic backgrounds. The genetic control is suggested to be polygenic, although the genetic factors involved remain to be identified.

In *Arabidopsis thaliana*, using a selective antibiotic marker-based system allowing to detect very small differences (Barth et al. 2001) and by analysing chiasma frequency, variation in meiotic recombination frequency depending on the genetic background has been revealed (Sanchez-Moran et al. 2002). In a comparison of seven marker pairs in four *Arabidopsis* ecotypes and their F_1 crosses, recombination frequencies for most marker pairs were almost identical between ecotypes (Barth et al. 2001). However, substantial changes in recombination frequencies were detected for two marker pairs among crosses of ecotypes as well as F_1 crosses. In general, the recombination frequencies were smaller for all crosses involving the Mt-0 background. In response to an altered genetic background, some genomic regions showed pronounced changes in local recombination frequencies, while in other regions only small changes were detected. Mean chiasma frequency showed significant variation between eight geographically and ecologically diverse accessions (Sanchez-Moran et al. 2002), but there was no significant variation between plants within the accessions. Cvi and Ler were the two accessions having the lowest overall mean chiasma frequencies. The analysis also revealed that the pattern of chiasma distribution between arms and among chromosomes was not consistent over accessions. Chromosome 4 was the least variable while chromosome 2 was the most variable.

Selection for increased or decreased recombination frequency has also been used to document polygenic quantitative genetic control of recombination. In lima bean (*Phaseolus lunatus*), Allard (1963) studied the effect of selection over four generations for high and low recombination values at three pairs of linked marker loci. While there was little or no response to downward selection, a sharp response to selection in the high direction was observed. The steadiness and rapidity of the responses indicated that some of the polygenes involved may have had fairly large effects.

Smiryaev et al. (2000) investigated the possibility of selection for higher frequency of meiotic recombination in tomato. The selection resulted in an increase of the recombination rate by 40% for the marker interval analysed.

2.3 Single Gene Control

Single genes controlling meiotic recombination are also known in plants. Disrupting these genes usually results in extreme phenotypes that have very depressed levels of recombination or show a complete breakdown of the recombination process (reviewed by Anderson and Stack 2002). Here we focus on genes that cause variation in recombination frequency, but do not prevent recombination. Specific alleles that regulate the rate and distribution of crossover have been identified in plant species.

One of the best studied is a major nuclear gene called *Rm1* (recombination modulator 1) in *Petunia hybrida* able to noticeably modify meiotic recombination rates (Cornu et al. 1989). *Rm1* is located on chromosome II. The regulation exerted by *Rm1* is not directed only to a particular chromosome. It acts in *trans* on all chromosomes of petunia, the effects varying according to the chromosome segment studied. The presence of *Rm1* in the heterozygous state enhanced the recombination rates along the seven chromosomes in 13 of 15 tested pairs of markers. As for example, the presence of *Rm1* increases recombination almost 30-fold on the *An2-Rt* segment along chromosome VI (Cornu et al. 1989; Robert et al. 1991). But also a decrease in recombination rate was observed for a single pair of moderately linked genes on chromosome III (Cornu et al. 1989). *Rm1* affects recombination in both male and female gametogenesis, although the effect may differ in degree of expression. The effects of *Rm1* seem to be higher within chromosome regions that normally recombine at very low rates. The regions more sensitive to the action of *Rm1* contain areas near the centromere, known to be predominantly heterochromatic (Cornu et al. 1989). *Rm1* might affect the stage of heterochromatin, within which recombinations are normally rare (Robert et al. 1991). The gene *Rm1* shows incomplete dominance (Robert et al. 1991). Additionally, minor modifier genes are involved in the regulation of recombination. Some of these genes may act independently of or epistatic to *Rm1*. Like *Rm1*, they have effects that differ according to the chromosome segment analysed. The effect of these genes may be specific to some chromosome segments, and some of the modifier genes are probably recessive. To determine whether *Rm1* is directly involved in recombination, or whether it plays a role in chromosome synapsis, synaptonemal complex (SC) formation was investigated in pollen mother cells of *Rm1 Rm1* and *rm1 rm1* plants (Abirached-Darmency et al. 1992). No morphological difference in their SC structure could be found. In the absence of *Rm1*, the chiasma frequency was reduced, corresponding to incomplete synapsis in pachytene. SC formation seemed to be more regular and more efficient in the presence of *Rm1*. It was proposed that *Rm1* probably corresponds to a regulatory system with multiple effects, including the course of synapsis and the extent and quality of SC formation (Abirached-Darmency et al. 1992).

In *Arabidopsis thaliana* also a single gene mutation affecting an increase in recombination has been described (Masson and Paszkowski 1997). The recessive mutant *xrs4* demonstrates enhanced meiotic recombination. The meiotic recombination efficiency for the analysed markers was significantly increased up to approximately 2.5-fold. The enhanced meiotic recombination was accompanied by a deficiency in somatic recombination. A corresponding phenotype has not been reported in other systems and was interpreted as an indication for a novel, plant-specific regulatory circuit linking mitotic and meiotic recombination.

Monogenic control of recombination frequencies has also been described in maize. Tulsieram et al. (1992) found for one of their analysed chromosomal segments (*Pgm1-Adh1*) a discontinuous distribution of recombination frequencies and suggested a single gene with major effects to be involved. Timmermans et al. (1997) revealed a dominant inherited *cis*-acting factor causing an approximately two-fold increase of the rate of recombination in the *Sh1-Bz1* interval. This factor did not influence recombination in adjacent regions. Another *cis*-modifier was identified in the *a1-sh2* interval, causing a three-fold variation in recombination frequency (Yao and Schnable 2005). This modifier not only influenced the rate of recombination but also the distributions of recombination breakpoints across the *a1-sh2* interval. Maize meiotic mutant *desynaptic* (*dy*), a single recessive gene, was shown to be a recombination modifier gene based on cytogenetic and segregation analyses (Ji et al. 1999). Genetic map length and chiasma frequencies throughout the genome were reduced in *dy* homozygous plants. The notable exception was increased recombination along the NOR-bearing region of chromosome 6S, a region that typically does not recombine. In genetic backgrounds with high chiasma frequency, *dy* could reduce recombination without causing non-disjunction, univalents, and pollen abortion. *Desynaptic* was proposed to be a point defect in crossover control.

A rather unique mutation was described in *Hypochoeris radicata* (Parker 1975). The desynapsis of this mutant specifically affects only one of the four chromosomes and is controlled by a single recessive gene.

In recent years, the *Arabidopsis* sequencing effort together with improved methods for cytological and molecular cytogenetic analysis had a major impact on the identification of recombination related genes in plants (Anderson and Stack 2002; Jones et al. 2003). Some of the meiotic genes recently identified using the molecular tools of reverse genetics show a quantitative effect on recombination or chiasmata. In *Arabidopsis Atmsh4* mutants chiasma frequency is greatly reduced to around 15% of wild-type frequency (Higgins et al. 2004). In this detailed characterisation, AtMSH4 was proposed to have a role in an early step of crossover formation. *Arabidopsis* mutants missing the ZYP1 protein, a protein with structural similarity to the synaptonemal complex (SC) transverse filament proteins known in other species, showed a slight reduction of recombination such that 70–80% of wild-type chiasma frequency was maintained (Higgins et al. 2005). The distribution of chiasmata was unaffected in *zyp1* null mutants. In the characterisation of a series of allelic mutants of the *A. thaliana MER3* gene (also named *RCK*; Chen et al. 2005) a large decrease of about 75% in meiotic crossovers was detected by cytological and genetic analysis (Mercier et al. 2005). The analysis of mutants of the *Arabidopsis AtMLH3* revealed that a loss of gene function resulted in reduction in chiasma frequency of around 60% and was accompanied by a substantial delay of +25 h in prophase I progression (Higgins et al. 2004). Despite having a relatively low mean chiasma frequency per cell, the *Atmlh3* mutant showed an unusually wide range of cell chiasma frequencies compared with wild-type meiosis. This observation suggested that chiasma formation is much less controlled in the mutant. A role of AtMLH3 in the resolution of double Holliday junctions in favour to form crossovers during meiosis was proposed.

A wealth of evidence testifies the widespread genotypic control of recombination in plants, and various systems of polygenic and monogenic control with segment- and chromosome-specific effects as well as those affecting the entire genome have been established. However, the molecular basis of the observed variability in recombination frequencies is still little understood and is amenable to further studies.

3 Influence of Breeding and Reproduction

Rees and Dale (1974) first proposed a higher recombination rate in domesticated plants relative to their wild progenitors. This hypothesis was tested years later by Ross-Ibarra (2004) who analysed data on chiasma frequencies available from almost a century of literature on plant cytogenetics (>600 plant species). The results from both an across-species analysis and a paired comparison of domesticates to their progenitor taxa strongly suggested that the domestication process generally increases the recombination rate of a species. Domesticated taxa showed a higher overall recombination rate than non-domesticated taxa. None of the other characteristics investigated for their correlation with recombination rate, e.g. mating system, life form, weediness, interacted significantly with domestication.

Rees and Dale (1974) found in *Festuca* and *Lolium* that short-lived populations had higher chiasma frequencies than longer-lived more perennial ones, and they concluded that the variation in chiasma frequency of populations is of an adaptive nature, allied in some way to longevity.

It seems to be fairly well established that selfing species tend to have higher chiasma frequencies than their outcrossing relatives (e.g. reviewed by Gibbs et al. 1975; Hansson et al. 2006). In the analysis of a large plant data set, plants that regularly self-fertilise had higher chiasma frequencies than outcrossing plants (Ross-Ibarra 2004). More recently, using genetic markers to measure recombination frequencies, a comparison of *Arabidopsis thaliana* and *A. lyrata* suggested that recombination is more frequent across comparable physical distances in the inbreeder, *A. thaliana*, as predicted (Hansson et al. 2006).

Several studies report that inbreeding leads to a decrease in recombination, e.g. in rye (Rees 1955; Rees and Thompson 1956), radish (Dayal 1977) and *Lolium perenne* (Karp and Jones 1982, 1983). In the latter species, over generations of inbreeding there was a decrease in mean chiasma frequency and an increase in cell and bivalent variances (Karp and Jones 1982). Also the pattern of chiasma localisation changed (Karp and Jones 1983). The restriction of chiasmata to the ends of the chromosome arms was progressively lost with inbreeding and chiasmata occurred in more interstitial and proximal regions.

In addition, heterotic effects for recombination have been observed in heterozygous F_1 individuals showing enhanced chiasma/recombination frequencies compared with the homozygous parents, e.g. in rye (Rees and Thompson 1956), radish (Dayal 1977) and *A. thaliana* (Barth et al. 2001).

4 Inducible Changes in Recombination

Substantial variability in recombination frequency has been attributed to several non-genetic factors. Many studies have been carried out especially on the effects of temperature on chiasma frequency. Some claimed a negative correlation between temperature and chiasma frequency, others a positive correlation, a bimodal response, or no effect at all (reviewed by Wilson 1959a; Lin 1982). Wilson (1959a) attributed these different results to two factors. The first was that the times for which cells were exposed to various temperatures were arbitrarily determined. The second was the heterogeneity among individuals of a species with respect to their chiasma frequencies. From the results obtained by different authors in various organisms Lin (1982) followed that the most consistent effects of temperature on chiasma formation is that, as temperature increases, chiasma frequency decreases. Berkemeier and Linnert (1987) inferred that chiasma formation and meiotic recombination is known to be maximal at optimal living conditions, but is lowered by all kinds of alterations, especially those of altered temperature. For example, in *Endymion nonscriptus* high temperature (20 °C) leads to reduced chiasma frequency, while in *Hyacinthus orientalis* decreasing temperature below 20 °C results in a progressive reduction in chiasma frequency (Elliot 1955). In both Liliaceae the effect was shown to be due to failure of chromosome pairing at pachytene. Wilson (1959b) then found a definite correlation between increasing temperatures and decreasing mean chiasma frequencies in *E. nonscriptus*. In *Rhoeo spathacea* var. *variegata* a prolonged treatment (36–60 h) at high temperature (37 °C) significantly reduced the chiasma frequency (Lin 1982). A heat treatment of 34 °C for 24 h was found to lower chiasma frequency in *Vicia faba* (Berkemeier and Linnert 1987). Also in *Allium ursinum* there was a considerable reduction in chiasma frequency concomitant with increasing duration of treatment (35 °C for 30 h or longer; Loidl 1989). A high proportion of cells arrested at leptotene and showed no synapsis at all. In *Arabidopsis thaliana,* different temperature regimes affected recombination frequencies in different ways depending on the genomic region investigated (Barth et al. 2000). However, temperature effects identified in this study were most likely related to stress symptoms observed for the plants grown at 32 °C.

Different results on the effect of environment on recombination have been reported. For example, in *Hordeum* (Gale and Rees 1970) significant variation in chiasma frequency due to season was found. While in soybean (Pfeiffer and Vogt 1990) and maize (Tulsieram et al. 1992) recombination frequencies were not significantly different in response to environments.

Sequence of meiotic development also has an effect on chiasma frequency. During the development of tulip anthers meiosis starts at the base and progresses to the tip, and the first cells to pass through meiosis have the lowest chiasma frequency (Couzin and Fox 1974). These findings parallel those of Rees and Naylor (1960) within rye anthers, in that meiosis starts in a wave spreading from the point of entry of the vascular bundle. However, in rye those cells which entered meiosis first had a higher chiasma frequency. A possible causative factor is the nutritional status of the cells in the different parts of the anther.

Haque and Godward (1985) provided a comprehensive summary on the effect of radiation on chiasma frequency: increase, decrease, and no change in chiasma frequency following irradiation have been reported by many authors in a number of plant species. More recently, UV-A irradiation was shown to have no effect on recombination in *Arabidopsis thaliana* (Barth et al. 2000).

The possibility to influence recombination with nutrient salts was also investigated. The control of calcium and phosphorus levels in nutrient cultures in soybeans did not apparently modify linkage in the considered marker pair (Hanson 1961). Bennett and Rees (1970) found a slight increase in chiasma frequencies in rye after a ten-fold increase in phosphate concentrations in the culture of growing seedlings. In *Arabidopsis*, recombination was enhanced after phosphate treatment in one genomic region but suppressed in another genomic region (Barth et al. 2000).

Although many non-genetic variables affect recombination, attempts to modify recombination in plants through environmental manipulations have to a large extent produced only non-predictable changes.

5 Chromosome Regions of Suppressed Recombination

5.1 Centromeric and Telomeric Regions

In addition to the origins of replication of DNA molecules, telomeres and centromers are the most important functional elements in eukaryotic chromosomes (Murata 2002). The former are essential for protecting chromosomes end-to-end fusion and exonucleolytic degradation, and the latter for precise chromatid segregation at mitosis and meiosis. Although centromere-specific DNA sequences have been isolated in a wide range of plant species (reviewed by Murata 2002; Hall et al. 2004), almost no conservation was found in their DNA sequences. Exceptions are cereal centromeres, which contain common *Ty3/gypsy*-type retrotransposon-like sequences. Compared with budding yeast (*Saccharomyces cervisiae*) centromeres, that consist of only about 125 bp of unique sequence (Clark 1998), higher plant centromeres are much more complex (Ma and Bennetzen 2006). These large heterochromatic regions consist of large arrays of satellite repeats that are usually arranged in a tandem head-to-tail fashion, intermixed with additional repeats, including transposable elements. In all plant centromeres investigated, the sizes of the satellite repeat units (also called monomers) are relatively consistent, ranging from about 120 bp to 180 bp, with the exception of wheat with 537 bp (reviewed by Hall et al. 2004). Recombination is eliminated or at least greatly reduced in chromosomal regions that contain repetitive sequences such as centromeres, pericentromeric heterochromatin, nucleolar organizer regions (NOR), and telomeres (Sherman and Stack 1995; Copenhaver et al. 1998; Künzel et al. 2000).

In rice, Cheng et al. (2001b) used FISH to localize 18 RFLP-anchored bacterial artificial chromosome (BAC) clones together with the centromeric sequence along the length of chromosome 10. They found a reduced crossing over in the heterochromatic short arm and around the centromere compared with the long arm. However, suppression of crossing over near the centromere was less than in tomato, wheat, and barley (Cheng et al. 2001b). A contig of 1.16 Mb from rice centromere 4 (CEN4) revealed arrays of the CentO satellite (155 bp or 164 bp) interspersed with the *Ty3/gypsy* retroelement CRR (for centromere-specific retrotransposon of rice) as well as flanking regions that are rich in transposons, retroelements, and pseudogenes (Cheng et al. 2002). Using FISH technology, a dramatic variation in the amount of CentO satellite DNA between different rice chromosomes and between the corresponding chromosomes of different varieties of rice was revealed (Cheng et al. 2002). Wu et al. (2003) constructed physical maps of rice chromosomes 1, 2, and 6–9 with P1-derived artificial chromosome (PAC) and BAC clones. Six centromere-specific repeats were analysed to facilitate characterisation of the chromosomal recombination frequency and the genomic composition and structure of the centromeric regions. The relative ratio of the recombination rate within the genomic regions along each chromosome varied markedly, with a range of 0–5.05 cM per 100 kb. Chromosomal recombination at each centromere and its flanking regions, known as pericentromeric regions, was completely suppressed. The physical sizes of these genetically mapped centromeric regions (without any recombination) were estimated to be 3.93, 3.3, 5.59, 3.84, 2.6, and 3.38 Mb for chromosomes 1, 2, 6, 7, 8, and 9, respectively, corresponding to 11.4% of the entire size of the six chromosomes (Wu et al. 2003). Recently, Ma and Bennetzen (2006) presented a detailed analysis of the retrotransposons present in a 1.97-Mb sequence that included centromere 8 (CEN8) of *japonica* rice. Thirty-three long-terminal repeat (LTR) retrotransposon families (including 11 previously unknown) were identified in the CEN8 region, totalling 245 elements and fragments that account for 67% of the region. Assuming that 18–22% of the rice genome is composed of LTR retrotransposons (Ma et al. 2004b), LTR retrotransposons are at least three- or four-fold enriched in centromeric region compared with most non-centromeric regions (Ma and Bennetzen 2006). Given that homologous recombination during meiosis is highly repressed or completely inhibited in all rice centromeres (Wu et al. 2003), it is not surprising that a low relative abundance of solo LTR was observed in the CEN8 region (Ma and Bennetzen 2006). A suppression of homologous recombination would be expected to also inhibit unequal recombination events. The observation that LTR retrotransposons in the CEN8 region are older on average than non-centromeric regions suggests that the other events that remove LTR retrotransposon sequences from the genome, primarily small deletions caused by illegitimate (i.e. non-homologous) recombination (Ma et al. 2004b), may also be suppressed, allowing a longer time for intact elements to persist and for solo LTR to accumulate (Ma and Bennetzen 2006).

Arabidopsis centromeres contain 2.8–4.0 Mb tracts of tandemly repeated 178 bp satellites, occasionally interrupted by insertion of *Athila*, a *Ty3/gypsy* retroelement (Copenhaver et al. 1999; Kumekawa et al. 2000, 2001; Hosouchi et al. 2002). The middle repetitive regions that flank the satellites also contain *Athila* and other retroelements, along with 5S ribosomal DNA (rDNA) tracts, transposable elements, and pseudogenes. *Arabidopsis* satellites show no homology to CentO or CentC

(Cheng et al. 2002). Drouaud et al. (2006) demonstrated that genetic recombination rates varied along the chromosome 4 of *Arabidopsis* from 0 cM/Mb near the centromere to 20 cM/Mb on the short arm next to the NOR region, with a chromosome average of 4.6 cM/Mb. There was almost no genetic recombination in the centromeric region and no clustering of cold intervals was observed outside the centromeric region.

Other structural features that influence crossing over include chromosomal knobs and telomeres. Knobs are primarily composed of repetitive sequences e.g. in *Arabidopsis* (*Arabidopsis* Sequencing Consortium 2000), maize (Ananiev et al. 1998) and rice (Cheng et al. 2001b) and would be expected to suppress recombination at that location in a manner similar to that for heterochromatin. Using FISH technology, Fransz et al. (2000) could demonstrate that the heterochromatic knob on the short arm of *Arabidopsis* chromosome 4 was highly condensed and corresponded to a cold spot of recombination on genetic maps. In maize, Hiatt et al. (2002) found that heterochromatic knobs, which can serve as alternatives to the conventional centromeres, contain two satellites (measuring 180 bp and 350 bp) that differ from the centromere-specific CentC (156 bp) of maize. Apart from consisting largely of the CentC satellite sequences, the maize centromere region contains the centromere-specific retrotransoposon of maize (CRM), a retroelement with homology to CRR from rice (Nagaki et al. 2003). In maize cells carrying the abnormal chromosome (Ab10), knobs function as meiotic centromeres (Rhoades 1942, Yu et al 1997). Since this phenomen is suppressed by the *smd1* (suppressor of meiotic drive) mutation, it was assumed that the Smd1 product is related to the activation of ectopic centromeres (Dawe and Cande 1996).

In contrast to the centromere sequences, the telomere DNA sequences which have been investigated so far were conserved, except for *Allium* and related species (Murata 2002). In plants, telomere DNA sequences were first isolated from *Arabidopsis thaliana* (Richards and Ausubel 1988). The telomere DNA is primarily composed of tandem repeats of the sequence 5´-TTTAGGG-3´ and cross-hybridised to telomeres of human as well as maize. Isolation of telomere DNA from other plant species such as *Cicer arietinum* (Gortner et al. 1998), *Lycopersicum esculentum* (Ganal et al. 1991), *Nicotiana tabacum* (Suzuki et al. 1994), *Triticum aestivum* (Bucholc and Buchowicz 1995), and *Zea mays* (Gardiner et al. 1996) revealed that the sequences were almost identical to *A. thaliana*. However, *Arabidopsis*-like telomeres are not present in all plants. In the monocot plant order Aspargales, two evolutionary switch-points in telomere sequence are known (Fajkus et al. 2005). The first leads to a telomere motif based on a TTAGGG-type repeat and the second, in the ancestor to *Allium*, leads to a telomere that has yet to be fully described. Probably many other groups with "unusual" telomeres will be found in the future (Fajkus et al. 2005).

Repeats present in telomeric and subtelomeric regions generally suppress meiotic crossing over at the telomeres, as demonstrated for tomato by recombination nodule (RN) mapping (Sherman and Stack 1995) and FISH using diakinensis chromosomes (Zhong et al. 1998).

Plant artificial chromosomes will be useful, e.g. for testing centromere function and identifying critical centromere elements (Hall et al. 2004). Functional assays and comparison of centromere DNA sequences have far-reaching implications, including an improvement in our understanding of centromere evolution and its influence on speciation allowing analysis of *cis*- and *trans*-acting factors that are important for proper inheritance of chromosomes, and permitting enhanced plant transformation using artificial chromosomes (Hall et al. 2004).

5.2 Sex Chromosomes

Although higher plants can regulate sex by either autosomal genes or sex chromosomes (Charlesworth 2002), few flowering plants have evolved heteromorphic sex chromosomes analogous to those of animals (Vyskot and Hobza 2004). The structural evolution of sex chromosomes poses fundamental questions for evolutionary and reproduction biology. Sex chromosomes are supposed to have evolved from autosomes. The Y chromosomes of plants as well as animals often consist of huge blocks of largely non-recombining DNA. This lack of recombination is thought to lead to the degeneration of the Y chromosome. The degeneration is characterised by a loss of functional genes, except those with male-specific function, and the accumulation of repetitive DNA sequences, both transposable elements and tandem arrays of satellite DNA sequences (Charlesworth et al. 1994, 2005).

Papaya (*Carica papaya* L., $2n=18$), is a polygamous angiosperm with male, female, and hermaphroditic forms. Papaya contains a primitive Y chromosome, with a male-specific region that accounts for only about 10% of the chromosome but has undergone severe recombination suppression and DNA sequence degeneration (Liu et al. 2004). The papaya sex locus has been genetically mapped to linkage group 1 (LG1). A high-density genetic map contained 1501 amplified fragment length polymorphism (AFLP) and morphological markers from 54 F_2 plants (Ma et al. 2004a). The map revealed severe suppression of recombination around the sex determination locus with a total of 225 markers (i.e. 15% of all markers mapped on the genome and 66% of the 342 markers on LG1) co-segregating with the sex locus. A mosaic arrangement of conserved (X-like) and diverged (Y-like) sequences showed that degradation of the Y chromosome is distributed across the non-recombining region (Liu et al. 2004). The male-specific region (MSY) showed 37.7% lower gene density, 27.6% higher retroelement density, and 188.9% higher repeat density. The severe suppression of recombination and extensive divergence between homologues in the region containing the papaya sex-determining genes indicates that is is an incipient sex chromosome. The discovery of an incipient Y chromosome in papaya, of which 10% is a non-recombining, rapidly evolving, sex-determining region flanked by normal autosome-like regions that comprise the remaining 90% of the chromosome, provides direct evidence for the theory that the sex chromosomes evolved from autosomes (Liu et al. 2004).

The most well known species with sex chromosomes are white campion (*Silene latifolia*) and common sorrel (*Rumex acetosa*). White campion is characterised by clearly different, large heteromorphic sex chromosomes, X and Y in males and two X in females. However, a question that remains unexplained is why the Y chromosome in *S. latifolia* is about 1.4-fold larger than the X chromosome. Comparison of nucleotide sequences from X and Y chromosomes has shown that the Y chromosome, although mostly euchromatic, might be genetically degenerate, at least to some extent (Guttman and Charlesworth 1998). There are three regions on the (male) Y chromosome that control sexual development. These include genes encoding gynoecium suppression, stamen promotion, and male fertility (Westergaard 1958; Negrutiu et al. 2001). A novel tandem repeat unit called TRAYC (tandem repeat accumulated on the Y chromosome) has accumulated on the Y chromosome of *S. latifolia* (Hobza et al. 2006). Its presence demonstrates that processes of satellite accumulation are at work even in this early stage of sex chromosome evolution. The presence of TRAYC in other species of the *Elisanthe* section suggests that this repeat spread after the sex chromosomes evolved but before speciation within this section (Hobza et al. 2006). TRAYC accumulation is most prominent near the centromere of the Y chromosome. Hobza et al. (2006) propose a role of the centromere as the starting point for the cessation of recombination between the X and Y chromosome. The prominent role of the centromeric region as a starting point for X-Y recombination arrest is also supported by the finding of an accumulation of RAYSI satellite sequences in the centromeric region of the Y chromosome in *Rumex acetosa* (Shibata et al. 1999, 2000; Navajas-Perez et al. 2006). *R. acetosa* is a dioecious species with a multiple chromosome system in which the females are XX and the males are XY_1Y_2. The two different Y chromosomes, both constitutively heterochromatic, are necessary for male fertility but not for the development of male flowers. During male meiosis, a sex trivalent occurs and both Y chromosomes pair only with one end of the X chromosome (Ruiz Rejon 2003). The RAYSI satellite DNA family is specific for the multiplex sex chromosome system, whereas it is absent from the genomes of other dioecious *Rumex* species having an XX/XY sex chromosome system (Navajas-Perez et al. 2006).

The characterisation and comparison of these plant models will enable us to study the stages leading to the evolution of non-recombining sex chromosomes in plants (Vyskot and Hobza 2004).

5.3 Introgressions from Wild Species

Wide interspecific crosses have played a major role in the evolution of crop species, because crops like e.g. rapeseed (*Brassica napus*), wheat (*Triticum aestivum*), oat (*Avena sativa*), and cotton (*Gossypium barbadense*, *G. hirsutum*) represent allopolyploids (Jencezweski and Alix 2004). Alloploid species contain more than two sets of chromosomes related by ancestral homology (termed homoeologous) that makes correct chromosome segregation at meiosis more complicated. Chromosome pairing between homologous (as opposed to homoeologous) chromosomes must be achieved to insure proper segregation at meiosis. Newly formed allopolyploids usually display homoeologous pairing and multivalent configurations that significantly reduce fertility. However, the fertility of neopolyploids increases rapidly, owing largely to selection against meiotic configurations that generate unbalanced gametes (Ramsey and Schemske (2002). Jencesweski and Alix (2004) compiled evidence for the presence of homoeologous pairing suppressors in several allopolyploid species that contribute to meiotic stability of these species.

Correct segregation of chromosomes during meiosis requires a restriction of pairing between homoeologous chromosomes. Otherwise, homoeologous associations generate complex meiotic configurations that lead to the production of unbalanced and aneuploid gametes, aneuploid progenies, chromosome rearrangements (Benavente et al. 2001; Sanchez-Moran et al. 2001), and hence to reduced fertility (Gottschalk 1978; Gillies 1989; Ramsey and Schemske 2002). Precise control of chromosome pairing is therefore a prerequisite for the meiotic and reproductive stability of polyploids (Jenczewski and Alix 2004). The process by which homologous chromosomes associate to segregate correctly at anaphase I involves a series of successive events, to which the term chromosome pairing has been applied: chromosome recognition, alignment, synapsis, and crossing over (Martinez-Perez et al. 1999; Schwarzacher 1997).

Bread wheat, *Triticum aestivum*, is an allohexaploid species (AABBDD, $2n = 6x = 42$). Although wheat homoeologous chromosomes retain residual homology (Jauhar et al. 1991), they almost never pair at metaphase I of meiosis in the hexaploid and corresponding haploid forms of *T. aestivum* (Jenczewski and Alix 2004). The first evidence for the presence of a strong pairing suppressor in hexaploid wheat was adduced simultaneously by Sears and Okamoto (1958) and Riley and Chapman (1958). In hybrids (AA^mBD) between *T. aestivum* and *T. monococcum* (A^mA^m, $2n = 2x = 14$), Sears and Okamoto (1958) found enhanced pairing in metaphase I of meiosis in the absence of chromosome 5B. In nulli-5B haploids of *T. aestivum*, Riley and Chapman (1958) observed numerous bivalents and trivalents. This change in chromosome pairing was attributed to the removal of the genetic system located on $5B^L$ (Riley 1960; Riley et al. 1960) and named *Ph1* (*Pairing homoeologous 1*; Wall et al. 1971). Since the discovery of the role of the *Ph1* gene in wheat, there has been much interest to overcome its function in order to transfer desirable genes from wild species to homoeologous wheat chromosomes. A *Ph1*-defective mutant of *T. aestivum* (*ph1b*), which was obtained by X-ray irradiation, displayed extensive homoeologous pairing at the metaphase I of meiosis (Sears 1977). Corresponding to bread wheat, another mutation of *Ph1* (*ph1c*) was induced by X-radiation of *T. turgidum* ssp. *durum* (Giorgi 1978). The transfer of the dominant gene Ph^I (*Ph* inhibitor) from *Aegilops speltoides* to the hexaploid wheat cultivar Chinese Spring also opened new possibilities (Chen et al. 1994). This gene suppresses the effect of the *Ph1* locus and permits homoeologous recombination between wheat and alien chromosomes in F_1 hybrids (Chen et al. 1994; Aghaee-Sarbarzeh et al. 2000). Recently, Aghaee-Sarbarzeh et al. (2002) reported the successful transfer and characterisation of leaf and stripe rust resistance genes from *A. triuncialis* and *A. geniculata* both to bread wheat via *Ph*^{I}-induced homoeologous recombination. Jenczewski and Alix (2004) showed that very good evidence for the presence and activity of pairing control genes have been adduced in *Avena sativa*, *Festuca arundinacea*, *Brassica napus*, *Gossipium hirsutum*, and *G. bardadense*, and in amphidiploids related to the diploid species *Lolium perenne*, *L. multiflorum*, and *L. rigidum*. More circumstantial evidence has been obtained for polyploids in the genera *Aegilops*, *Hordeum*, *Nicotiana*, and *Coffea*. Another aspect involving interest of plant breeding in interspecific crosses is the introgression of new resistances against abiotic or biotic stress from wild species into cultivated species. However, the introduction of alien chromosome segments via recombination often also results in undesirable genes from the wild species being included in the transfer, causing a depression of yield and quality (Friebe et al. 1996). In addition, during the past decades, the prospect of commercial release of genetically engineered plants has put new emphasis on gene flow from crop species to weeds (Chevre et al. 2007). It depends on the genomic structure of the species concerned, as well as on the degree of their genome homology for intergenomic recombination to occur.

In plants, sequence divergence was assumed to lead to a decrease in recombination. Liharska et al. (1996), investigating the impact of foreign chromatin on recombination in a genetically defined interval on chromosome 6 of tomato (*Lycopersicum esculentum*), could elegantly demonstrate this. When the entire chromosome arm of tomato was replaced with that of the related species *Lycopersicum peruvianum*, the distance between the flanking markers decreased more than six-fold. When foreign chromatin was introduced from another closer-related species (*Lycopersicum pimpinellifolium*), the decrease was much more moderate (20%). These results indicate that increasing sequence polymorphisms both within and between species could significantly reduce recombination. Recently, Li et al. (2004, 2006) developed a GUS-based reporter system for monitoring homoeologous recombination in *Arabidopsis*. The constructs contain two overlapping copies of a large intron that act as recombination substrate. Each copy is 626 bp long and has 589 bp of overlap with the other copy. After recombination between these two copies, a functional GUS gene with a single copy of the intron is generated and leads to the production of GUS following the splicing of the intron (Li et al. 2006). Sequence divergence between the introns markedly reduced the frequency of somatic recombination. As little as 0.5% divergence lead to a 4.1-fold reduction. Increasing divergence to 2%, 4%, and 9% led to 9.6-, 11.7-, and 20.3-fold reductions in the frequency of recombination, respectively (Li et al. 2006). The authors assumed that such a range is probably representative of the degrees of sequence divergence encountered in interspecific crosses in plants.

The use of the Ogu_INRA cytoplasmic male sterility (cms) in rapeseed (*Brassica napus*) for hybrid breeding requires efficient restorer lines. Random amplified polymorphic DNA (RAPD) and restriction fragment length polymorphism (RFLP) markers tightly linked to the restorer gene *Rfo* have been identified. (Delourme et al. 1994, 1995). However, considerably reduced recombination at the introgressed radish genome segment in the *Brassica* restorer line has made improvement of the line very difficult. Primard-Brisset et al. (2005) successfully used gamma ray irradiation to induce chromosome breakage in order to force non-spontaneous recombination between a *Rfo*-carrying introgression from radish and the rapeseed homologous chromosome from a low glucosinolate content *B. napus* line. The new line R2000 recovered some of the corresponding *B. oleracea* rapeseed chromosome that were originally replaced by the radish introgression line in all homozygous restored rapeseed lines.

A high level of homology remains between the genomes of *B. oleracea* and *B. rapa* and those of *B. napus*, allowing regular pairing and recombination during meiosis between them (Parkin et al. 1995). The AAC hybrids, resulting from crosses between *B. napus* (AACC) and its progenitor *B. rapa* (AA), are ideal structures to simultaneously assess homologous pairing and recombination occurring between the A genomes in *B. rapa* and *B. napus*, and homoeologous pairing and recombination between the A and C genomes or within C chromosomes (Leflon et al. 2006). In these AAC hybrids, Leflon et al. (2006) studied the pairing and recombination pattern of A and C chromosomes during meiosis. In 71% of the pollen mother cells, homologous A chromosomes paired regularly, and usually one chromosome of each pair was transmitted to the progeny. C chromosomes remained mainly univalent, but were involved in homoeologous pairing in 21.5% of the cells, and 13.0% of the transmitted C chromosomes were either recombined or broken. The rate of transmission of C chromosomes depended on the identity of the particular chromomsome and on the way the hybrid was crossed, as the male or the female parent, to *B. napus* or to *B. rapa*. Gene transfers in triploid hybrids are favoured between the A genomes of *B. rapa* and *B. napus*, but also occur between the A and C genomes, though at lower rates.

Oriental or brown mustard (*Brassica juncea*) is an amphidiploid derived from spontaneous hybridisation between *B. rapa* (genome AA, $2n = 20$) and *B. nigra* (genome BB, $2n = 16$; U 1935). *B. juncea* (genome AABB, $2n = 36$) is heat- and drought-tolerant, resistant to blackleg disease, and more resistant to seed shattering than *B. napus* and *B. rapa* (Woods et al. 1991). However, the high erucic acid and glucosinolate contents of. *B. juncea* preclude its use as an edible oilseed crop in Canada (Cheng et al. 2001a). Addressing this problem, a line 1058 was derived from a BC_1F_3 plant of the cross (*B. juncea* × *B. rapa*) × *B. juncea*, which shows a low 3-butenyl glucosinolate content and a reduction in allyl glucosinolate (Cheng et al. 2001a). Cytological studies revealed that line 1058 is missing one pair of B-genome chromosomes. The reduction in 3-butenyl-glucosinolate is likely due to the introgression of genes for low glucosinolate from *B. rapa* canola into *B. juncea*, whereas the low allyl glucosinolate content could be the result of the missing pair of B-genome chromosomes carrying the genes for allyl glucosinolate biosynthesis (Cheng et al. 2001a). The nullisomic genomic constitution of line 1058 opens up possibilities for chromosome engineering (Cheng et al. 2001a). Substitution of the homoeologous C-genome chromosomes for the missing B chromosome could improve the fertility and alter agronomic and quality aspects of line 1058.

Studies on interspecific hybrids demonstrate the need and importance of cytologically monitoring the genomic constitutions of breeding lines produced by interspecific crosses. Although a large number of transfers with useful alien genes have been produced, few have been exploited commercially. Reduced recombination rates in introgressed regions often makes it very difficult to further minimise the introgressed region in order to eliminate undesirable traits closely linked to the favoured trait. In addition, the breeding material has often difficulties to compensate well for the loss of genetic material due to deletions caused by the intergenomic recombination.

6 Conclusions

Recent studies have shown that DNA recombination plays a major role in genome evolution. Moreover, the technique of cross-breeding, one of the most basic and practical methods used to improve crop varieties, is based on homologous chromosome recombination. However, our knowledge of recombination rates and patterns in plants is far from comprehensive (Gaut et al. 2007). To estimate recombination rates, both genetic and physical distances are required (measured in cM and usually kb, respectively) and so far, these data are only available for a limited number of model species at the moment. Progress in whole genome sequencing projects in the coming years will allow further analyses of more species and a comparison of recombination rates and their variability between different species. FISH localisation of mapped DNA sequences to pachytene chromosomes and synaptonemal complexes in plants will be an important tool for integrating recombination maps with physical maps in the context of chromosome structure (Anderson and Stack 2002). However, one difficulty in integrating genetic and physical maps in plants is to determine how well the two different approaches reflect the biological reality of crossing over.

Further studies of recombination in plants are necessary to improve our understanding for the mechanism behind the variability of recombination rates in plant genomes.

References

Abirached-Darmency M, Tarenghi E, Dejong JH (1992) The effect on meiotic synapsis of a recombination modulator in *Petunia hybrida*. Genome 35:443–453

Aghae-Sarbarzeh M, Singh H, Dhaliwal HS (2000) *Ph*^{I} gene derived from *Aegilops speltoides* induces homoeologous chromosome pairing in wide crosses of *Triticum aestivum*. J Hered 91:418–421

Aghae-Sarbarzeh M, Ferrahi M, Singh S, Singh H, Friebe B, Gill BS, Dhaliwal HS (2002) *Ph*^{I}-induced transfer of leaf and stripe-rust resistance genes from *Aegilops triuncialis* and *Ae. geniculata* to bread wheat. Euphytica 127:377–382

Allard RW (1963) Evidence for genetic restriction of recombination in lima bean. Genetics 48:1389

Anderson LK, Stack SM (2002) Meiotic recombination in plants. Curr Genomics 3:507–525

Ananiev EV, Phillips RL, Rines HW (1998) Complex structure of knob DNA on maize chromosome 9: retrotransposon invasion in heterochromatin. Genetics 149:2025–2037

Arabidopsis Sequencing Consortium (2000) The complete sequence of a heterochromatic island from higher eukaryote. Cell 100:377–386

Barth S, Melchinger AE, Devezi-Savula B, Lübberstedt T (2000) A high-throughput system for genome-wide measurement of genetic recombination in *Arabidopsis thaliana* based on transgenic markers. Funct Integr Genomics 1:200–206

Barth S, Melchinger AE, Devezi-Savula B, Lübberstedt T (2001) Influence of genetic background and heterozygosity on meiotic recombination in *Arabidopsis thaliana*. Genome 44:971–978

Benavente E, Alix K, Dusautoir JC, Orellana J, David JL (2001) Early evolution of the chromosomal structure of *Triticum turgidum–Aegilops ovata* amphidiploids carrying and lacking the *Ph1* gene. Theor Appl Genet 103:1123–1128

Bennett MD, Rees H (1970) Induced variation in chiasma frequency in rye in response to phosphate treatments. Genet Res 16:325–331

Berkemeier J, Linnert G (1987) Effects of high and low temperature on meiotic recombination in normal and oligochiasmatic lines of *Vicia faba*. Biol Zentralbl 106:219–230

Bucholc M, Buchowicz J (1995) An extrachromosomal fragment of telomeric DNA in wheat. Plant Mol Biol 27:435–439

Charlesworth B (2002) The evolution of chromosomal sex determination. In: Short R (ed) The genetics and biology of sex determination. Wiley, Chichester, pp 207–219

Charlesworth B, Sniegowski P, Stephan W (1994) The evolutionary dynamics of repetitive DNA in eukaryotes. Nature 371:215–220

Charlesworth D, Charlesworth B, Marais G (2005) Steps in the evolution of heteromorphic sex chromosomes. Heredity 95:118–128

Chen CB, Zhang W, Timofejeva L, Gerardin Y, Ma H (2005) The *Arabidopsis ROCK-N-ROLLERS* gene encodes a homolog of the yeast ATP-dependent DNA helicase MER3 and is required for normal meiotic crossover formation. Plant J 43:321–334

Chen PD, Tsujimoto H, Gill BS (1994) Transfer of *Ph^I* gene promoting homoeologous pairing from *Triticum speltoides* into common wheat and their utilization in alien genetic introgression. Theor Appl Genet 88:97–101

Cheng BF, Seguin-Swartz G, Somers DJ, Rakow G (2001a) Low glucosinolate *Brassica juncea* breeding line revealed to be nullisomic. Genome 44:738–741

Cheng Z, Presting GG, Buell CR, Wing RA, Jiang J (2001b) High-resolution pachytene chromosome mapping of bacterial artificial chromosomes anchored by genetic markers reveals the centromere location and the distribution of genetic recombination along chromosome 10 of rice. Genetics 157:1749–1757

Cheng Z, Dong F, Langdon T, Quyang S, Buell CR, Gu M, Blattner FR, Jiang J (2002) Functional rice centromeres are marked by a satellite repeat and centromere-specific retrotransposon. Plant Cell 14:1691–1704

Chevre AM, Adamczyk K, Eber F, Huteau V, Coriton O, Letanneur, Laredo C, Jenczewski E, Monod H (2007) Modelling gene flow between oilseed rape and wild radish. I. Evolution of chromosome structure. Theor Appl Genet 114:209–221

Clarke L (1998) Centromeres: proteins, protein complexes, and repeated domains at centromeres of simple eukaryotes. Curr Opin Genet Dev 8:212–218

Cornu A, Farcy E, Mousset C (1989) A genetic basis for variations in meiotic recombination in *Petunia hybrida*. Genome 32:46–53

Copenhaver GP, Browne WE, Preuss D (1998) Assaying genome-wide recombination and centromere functions with *Arabidopsis* tetrads. Proc Acad Natl Sci USA 95:247–252

Copenhaver GP, Nickel K, Kuromori T, Benito MI, Kaul S, Lin X, Bevan M, Murphy G, Harris B, Parnell LD, McCombie WR, Martienssen RA, Marra M, Preuss D (1999) Genetic definition and sequence analysis of *Arabidopsis* centromeres. Science 286:2468–2474

Couzin DA, Fox DP (1974) Variation in chiasma frequency during tulip anther development. Chromosoma 46:173–179

Dawe RK, Cande WZ (1996) Induction of centromeric activity in maize by suppressor of meiotic drive 1. Proc Natl Acad Sci USA 93:8512–8517
Dayal N (1977) Cytogenetical studies in inbred lines of radish (*Raphanus sativus* L. var *radicola* Pers.) and their hybrids. II. Genetic regulation of chiasma frequency. Cytologia 42:273–278
Delourme R, Bouchereau A, Hubert N, Renard M, Landry BS (1994) Identification of RAPD markers linked to a fertility restorer gene for the Ogura radish cytoplasmic male sterility of rapeseed (*Brassica napus* L.). Theor Appl Genet 88:741–748
Delourme R, Eber F, Renard M (1995) Breeding double low restorer lines in radish cytoplasmic male sterility of rapeseed (*Brassica napus* L.). Proc Int Rapeseed Congr 9:6–8
Drouaud J, Camilleri C, Bourguignon PY, Canaguier A, Berard A, Vezon D, Giancola S, Brunel D, Colot V, Prum B, Quesneville H, Mezard C (2006) Variation in crossing-over rates across chromosome 4 of *Arabidopsis thaliana* reveals the presence of meiotic recombination "hot spots". Genome Res 16:106–114
Elliot CG (1955) The effect of temperature on chiasma frequency. Heredity 9:385–398
Fajkus J, Sykorova E, Leitch AR (2005) Telomeres in evolution and evolution of telomeres. Chromosome Res 13:469–479
Fatmi A, Poneleit CG, Pfeiffer TW (1993) Variability of recombination frequencies in the Iowa Stiff Stalk Synthetic (*Zea mays* L.). Theor Appl Genet 86:859–866
Fransz PF, Amstrong S, Jong JH de, Parnell LD, van Drunen C, Dean C, Zabel P, Bisseling T, Jones GH (2000) Integrated cytogenetic map of chromosome arm 4S of *A. thaliana*: structural organization of heterochromatic knob and centromere region. Cell 100:367–376
Friebe B, Jiang J, Raupp WJ, McIntosh RA, Gill BS (1996) Characterization of wheat-alien translocations conferring resistance to disease and pests: current status. Euphytica 91:59–87
Gale MD, Rees H (1970) Genes controlling chiasma frequency in *Hordeum*. Heredity 25:393–410
Ganal MW, Lapitan NL, Tanksley SD (1991) Macrostructure of the tomato telomeres. Plant Cell 3:87–94
Gardiner JM, Coe EH, Chao S (1996) Cloning maize telomeres by complementation in *Saccharomyces cervisiae*. Genome 39:736–748
Gaut BS, Wright SI, Rizzon C, Dvorak J, Anderson LK (2007) Recombination: an underappreciated factor in the evolution of plant genomes. Nat Genet 8:77–84
Gibbs PE, Milne C, Carrillo MV (1975) Correlation between breeding system and recombination index in five species of *Senecio*. New Phytol 75:619–626
Gillies CB (1989) Chromosome pairing and fertility in polyploids. In: Gillies CB (ed) Fertility and chromosome pairing: recent studies in plants and animals. CRC, Boca Raton, pp 137–176
Giorgi B (1978) A homoeologous pairing mutant isolated in *Triticum durum* cv. Capelli. Mut Breed Newsl 11:4–5
Gortner G, Nenno M, Weising K, Zink D, Nagl W, Kahl G (1998) Chromosomal localization and distribution of simple sequence repeats and the *Arabidopsis*-type telomere sequence in the genome of *Cicer arietinum* L. Chromosome Res 6:97–104
Gottschalk W (1978) Open problems in polyploid research. Nucleus 21:99–112
Guttman DS, Charlesworth D (1998) An X-linked gene with a degenerate Y-linked homologue in a dioecious plant. Nature 393:263–266
Hadad RG, Pfeiffer TW, Poneleit CG (1996) Repeatability and heritability of divergent recombination frequencies in the Iowa Stiff Stalk Synthetic (*Zea mays* L.). Theor Appl Genet 93:990–996
Hall AE, Keith KC, Hall SE, Copenhaver GP, Preuss D (2004) The rapidly evolving field of plant centromeres. Curr Opin Plant Biol 7:108–114
Hamant O, Ma H, Cande WZ (2006) Genetics of meiotic prophase I in plants. Annu Rev Plant Biol 57:267–302
Hanson WD (1961) The effect of calcium and phosphorus nutrition on genetic recombination in the soybean. Crop Sci 1:384
Hansson B, Kawabe A, Preuss S, Kuittinen H, Charlesworth D (2006) Comparative gene mapping in *Arabidopsis lyrata* chromosomes 1 and 2 and the corresponding *A. thaliana* chromosome 1: recombination rates, rearrangements and centromere location. Genet Res 87:75–85

Haque MZ, Godward MBE (1985) Comparison between 2 genera, species and cultivars in Lactuceae. 2. Effects of seed-irradiation on chiasma frequency. Cytologia 50:841–853
Hiatt EN, Kentner EK, Dawe RK (2002) Independently regulated neocentromere activity of two classes of tandem repeat arrays. Plant Cell 14:407–420
Higgins JD, Armstrong SJ, Franklin FCH, Jones GH (2004) The *Arabidopsis MutS* homolog *AtMSH4* functions at an early step in recombination: evidence for two classes of recombination in *Arabidopsis*. Gene Dev 18:2557–2570
Higgins JD, Sanchez-Moran E, Armstrong SJ, Jones GH, Franklin FCH (2005) The *Arabidopsis* synaptonemal complex protein ZYP1 is required for chromosome synapsis and normal fidelity of crossing over. Genes Dev 19:2488–2500
Hobza R, Lengerova M, Svoboda J, Kubekova H, Kejnovsky E, Vyskot B (2006) An accumulation of tandem DNA repeats on the Y chromosome in *Silene latifolia* during early stages of sex chromosome evolution. Chromosoma 115:376–382
Hosouchi T, Kumekawa N, Tsuruoka H, Kotani H (2002) Physical map-based sizes of the centromeric regions of the *Arabidopsis thaliana* chromosomes 1, 2 and 3. DNA Res 9:117–121
Jauhar PP, Riera-Lizarazu O, Dewey WG, Gill BS, Crane CF, Bennett JH (1991) Chromosome pairing relationships among the A, B and D genomes of wheat. Theor Appl Genet 82:441–449
Jenczweski E, Alix K (2004) From diploids to alloploids: the emergence of efficient pairing control gene in plants. Crit Rev Plant Sci 23:21–45
Ji YF, Stelly DM, De Donato M, Goodman MM, Williams CG (1999) A candidate recombination modifier gene for *Zea mays* L. Genetics 151:821–830
Jones GH, Armstrong SJ, Caryl AP, Franklin FCH (2003) Meiotic chromosome synapsis and recombination in *Arabidopsis thaliana*; an integration of cytological and molecular approaches. Chromosome Res 11:205–215
Karp A, Jones RN (1982) Cytogenetics of *Lolium perenne*. 1. Chiasma frequency variation in inbred lines. Theor Appl Genet 62:177–183
Karp A, Jones RN (1983) Cytogenetics of *Lolium perenne*. 2. Chiasma distribution in inbred lines. Theor Appl Genet 64:137–145
Künzel G, Korzun L, Meister A (2000) Cytologically integrated physical restriction fragment length polymorphism maps for the barley genome based on translocation breakpoints. Genetics 154:397–412
Kumekawa N, Hosouchi T, Tsuruoka H, Kotani H (2000) The size and sequence organization of the centromeric region of *Arabidopsis thaliana* chromosome 5. DNA Res 7:315–321
Kumekawa N, Hosouchi T, Tsunruoka H, Kotani H (2001) The size and sequence organization of centromeric region of *Arabidopsis thaliana* chromosome 4. DNA Res 8:285–290
Leflon M, Eber F, Letanneur JC, Chelysheva L, Coriton O, Huteau V, Ryder CD, Barker G, Jenczewski E, Chevre AM (2006) Pairing and recombination at meiosis of *Brassica rapa* (AA) x *Brassica napus* (AACC) hybrids. Theor Appl Genet 113:1467–1480
Li L, Santerre-Ayotte S, Boivin EB, Jean M, Belzile F (2004) A novel reporter for intrachromosomal homoeologous recombination in *Arabidopsis thaliana*. Plant J 40:1007–1015
Li L, Jean M, Belzile F (2006) The impact of sequence divergence and DNA mismatch repair on homoeologous recombination in *Arabidopsis*. Plant J 45:908–916
Liharska T, Koornneef M, Wordragen MV, Kammen AV, Zabel P (1996) Tomato chromosome 6: effect of alien chromosomal segments on recombination frequencies. Genome 39:485–491
Lin YJ (1982) Temperature and chiasma formation in *Rhoeo spathacea* var. *variegata*. Genetica 60:25–30
Liu Z, Moore PH, Ma H, Ackerman CM, Ragiba M, Yu Q, Pearl HM, Kim MS, Charlton JW, Stiles JI, Zee FT, Paterson AH, Ming R (2004) A primitive Y chromosome in papaya marks incipient sex chromosome evolution. Nature 427:348–352
Loidl J (1989) Effects of elevated temperature on meiotic chromosome synapsis in *Allium ursinum*. Chromosoma 97:449–458

Ma H, Moore PH, Liu Z, Kim MS, Yu Q, Fitch MM, Sekioka T, Paterson AH, Ming R (2004a) High-density linkage mapping revealed suppression of recombination at the sex determination locus in papaya. Genetics 166:419–436
Ma J, Bennetzen JL (2006) Recombination, rearrangement, reshuffling, and divergence in a centromeric region of rice. Proc Natl Acad Sci 103:383–388
Ma J, Devos KM, Bennetzen JL (2004b) Analyses of LTR-retrotransposon structures reveal recent and rapid genomic DNA loss in rice. Genome Res 14:860–869
Martinez-Perez E, Shaw P, Readres S, Aragon-Alcaide L, Miller T, Moore G (1999) Homologous chromosome pairing in wheat. J Cell Sci 112:1761–1769
Masson JE, Paszkowski J (1997) *Arabidopsis thaliana* mutants altered in homologous recombination. Proc Natl Acad Sci USA 94:11731–11735
Mercier R, Jolivet S, Vezon D, Huppe E, Chelysheva L, Giovanni M, Nogué F, Doutriaux MP, Horlow C, Grelon M, Mézard C (2005) Two meiotic crossover classes cohabit in *Arabidopsis*: one is dependent on *MER3*, whereas the other one is not. Curr Biol 15:692–701
Murata (2002) Telomeres and centromeres in plants. Curr Genomics 3:527–538
Nagaki K, Song J, Stupar RM, Parokonny AS, Yuan Q, Quyang S, Liu J, Hsiao J, Jones KM, Dawe RK, Buell CR, Jiang J (2003) Molecular and cytological analyses of large tracks of centromeric DNA reveal structure and evolutionary dynamics of maize. Genetics 163:759–770
Navajas-Peres R, Schwarzacher T, Herran R de la, Rejon CR, Rejon MR, Garrido-Ramos MA (2006) The origin and evolution of the variability in a Y-specific satellite-DNA of *Rumex acetosa* and its relatives. Gene 368:61–71
Negrutiu I, Vyskot B, Barbacar N, Georgiev S, Moneger F (2001) Dioecious plants. A key to the early events of sex chromosome evolution. Plant Physiol 127:1418–1424
Parker JS (1975) Chromosome-specific control of chiasma formation. Chromosoma 49:391–406
Parkin IAP, Sharpe AG, Keith DJ, Lydiate DJ (1995) Identification of the A and C genomes of amphidiploid *Brassica napus* (oilseed rape). Genome 38:1122–1131
Pfeiffer TW, Vogt SD (1990) Variability for recombination frequencies in the AP12 soybean population. Crop Sci 30:545–549
Primard-Brisset C, Poupard JP, Horvais R, Eber F, Pelletier G, Renard M, Delourme R (2005) A new recombined double low restorer line for the *Ogu*-INRA cms in rapeseed (*Brassica napus* L.) Theor Appl Genet 111:736–746
Ramsey J, Schemke DW (2002) Neopolyploidy in flowering plants. Annu Rev Ecol Syst 33:589–639
Rasmusson J (1927) Genetically changed linkage values in *Pisum*. Hereditas 10:1–152
Rees H (1955) Genotypic control of chromosome behaviour in rye. I. Inbred lines. Heredity 9:93–116
Rees H, Dale PJ (1974) Chiasmata and variability in *Lolium* and *Festuca* populations. Chromosoma 47:335–351
Rees H, Naylor B (1960) Developmental variation in chromosome behaviour. Heredity 15:17–27
Rees H, Thompson JB (1956) Genotypic control of chromosome behaviour in rye. III. Chiasma frequency in homozygotes and heterozygotes. Heredity 10:409–424
Rhoades MM (1942) Preferential segregation in maize. Genetics 27:395–407
Richards EJ, Ausubel FM (1988) Isolation of a higher eukaryotic telomere from *Arabidopsis thaliana*. Cell 53:127–136
Riley R (1960) The diploidization of polyploid wheat. Heredity 15:407–429
Riley R, Chapman V (1958) Genetic control of the cytologically diploid behaviour of hexaploid wheat. Nature 13:713–715
Riley R, Chapman V, Kimber G (1960) Position of the gene determining the diploid-like behaviour of polyploid wheat. Nat Lond 186:259–260
Robert N, Farcy E, Cornu A (1991) Genetic control of meiotic recombination in *Petunia hybrida*: dosage effect of gene *Rm1* on segments *Hf1-Lg1* and *An2-Rt*; role of modifiers. Genome 34:515–523

Ross-Ibarra J (2004) The evolution of recombination under domestication: a test of two hypotheses. Am Nat 163:105–112

Ruiz Rejon M (2003) Sex chromosomes in plants. (Encyclopedia of plant and crop sciences. Dekker agropedia, vol 4) Dekker, New York, pp 1148–1151

Sanchez-Moran E, Benavente E, Orellana J (2001) Analysis of karyotypic stability of homoeologous-pairing (*ph*) mutants in allopolyploid wheat. Chromosoma 110:371–377

Sanchez-Moran E, Armstrong SJ, Santos JL, Franklin FCH, Jones GH (2002) Variation in chiasma frequency among eight accessions of *Arabidopsis thaliana*. Genetics 162:1415–1422

Schwarzacher T (1997) Three stages of meiotic homologous chromsome pairing in wheat: cognition, alignment and synapsis. Sex Plant Prod 10:324–331

Sears ER (1977) An induced mutant with homoeologous pairing in common wheat. Can J Genet Cytol 19:585–593

Sears ER, Okamoto M (1958) Intergenomic chromosome relationships in hexaploid wheat. Proc Int Congr Genetics 10:258–259

Sherman JD, Stack SM (1995) Two-dimensional spreads of synaptonemal complex from solanaceous plants. VI High resolution recombination nodule map for tomato (*Lycopersicum esculentum*). Genetics 141:683–708

Shibata F, Hizume M, Kuroki Y (1999) Chromosome painting of the Y chromosomes and isolation of a Y chromosome-specific repetitive sequence in the dioecious plant *Rumex acetosa*. Chromosoma 108:266–270

Shibata F, Hizume M, Kuroki Y (2000) Differentiation and the polymorphic nature of the Y chromosomes revealed by repetitive sequences in the dioecious plant, *Rumex acetosa*. Chromosome Res 8:229–236

Simchen G, Stamberg D (1969) Fine and coarse controls of genetic recombination. Nature 222:329–332

Smiryaev AV, Nechaev IM, Rogozhina IV (2000) The possibilities of selection for higher frequency of meiotic recombination: simulation and experiment. Russ J Genetics 36:331–339

Suzuki K, Yamagiwa Y, Matsui T, Yoshida K (1994) Restriction enzyme-resistant high molecular weight telomeric DNA fragments in tobacco. DNA Res 1:129–138

Timmermans MCP, Das OP, Bradeen JM, Messing J (1997) Region-specific *cis*- and *trans*-acting factors contribute to genetic variability in meiotic recombination in maize. Genetics 146:1101–1113

Tulsieram L, Compton WA, Morris R, Thomas-Compton M, Eskridge K (1992) Analysis of genetic recombination in maize populations using molecular markers. Theor Appl Genet 84:65–72

U N (1935) Genome-analysis in *Brassica* with special reference to the experimental formation of *B. napus* and peculiar mode of fertilization. Jpn J Bot 7:389–452

Vyskot B, Hobza R (2004) Gender in plants: sex chromosomes are emerging from the fog. Trends Genet 20:432–438

Wall AM, Riley R, Chapman V (1971) Wheat mutant permitting meiotic chromosome pairing. Genet Res 18:311–328

Wang G, Hyne V, Chao S, Henry Y, De Buyser J, Gale MD, Snape JW (1995) A comparison of male and female recombination frequency in wheat using RFLP maps of homoeologous group 6 and 7 chromosomes. Theor Appl Genet 91:744–746

Westergaard M (1958) The mechanism of sex determination in dioecious flowering plants. Adv Genet 9:217–281

Williams CG, Goodman MM, Stuber CW (1995) Comparative recombination distances among *Zea mays* L. inbreds, wide crosses and interspecific hybrids. Genetics 141:1573–1581

Wilson JY (1959a) Chiasma frequency in relation to temperature. Genetica 29:290–303

Wilson JY (1959b) Temperature effect on chiasma frequency in the bluebell *Endymion nonscriptus*. Chromosoma 10:337–354

Woods DL, Capcara JJ, Downey RK (1991) The potential of mustard (*Brassica juncea* (L.) Coss.) as an edible oil crop on the Canadian prairies. Can J Plant Sci 71:195–198

Wu J, Mizuno H, Hayashi-Tsugane M, Ito Y, Chiden Y, Fujisawa M, Kategiri S, Saji S, Yoshiki S, Karasawa W, Yoshihara R, Hayashi A, Kobayashi H, Ito K, Hamada M, Okamoto M, Ikeno M, Ichikawa Y, Katayose Y, Yano M, Matsumoto T, Sasaki T (2003) Physical maps and recombination frequency of six rice chromosomes. Plant J 36:720–730

Yandeau-Nelson MD, Nikolau BJ, Schnable PS (2006) Effects of *trans*-acting genetic modifiers on meiotic recombination across the *a1-sh2* interval of maize. Genetics 174:101–112

Yao H, Schnable PS (2005) *Cis*-effects on meiotic recombination across distinct *a1-sh2* intervals in a common zea genetic background. Genetics 170:1929–1944

Yu HG, Hiatt EN, Chan A, Sweeney M, Dawe RK (1997) Neocentromere-mediated chromosome movement in maize. J Cell Biol 139:831–840

Zhong XB, Fransz PF, Wenneskes van Eden J, Ramanna MS, Kammen A van, Zabel P, Jong JH de (1998) FISH studies reveal the molecular and chromosomal organization of individual telomere domains in tomato. Plant J 13:507–517

Functional Markers in Resistance Breeding

Christina Roenn Ingvardsen(✉), Britt Schejbel, and Thomas Lübberstedt

Contents

Abstract Severe yield loss in plant crops caused by pathogens constantly demands plant breeders to improve disease resistance in crop species. To date, more than 90 plant resistance (R) genes have been isolated, many of them belonging to the NBS-LRR class of R genes. These genes are often found in clusters along the chromosomes and frequently lack genetic synteny between species, which makes them difficult to identify based on sequence homology to previously isolated R genes. However, once identified, R genes provide the basis for the application of functional markers (FMs), which efficiently assist phenotypic selection in all phases of resistance breeding. FMs are derived from polymorphic sites within genes causally involved in phenotypic trait variation. The major advantage of FMs is that they are in complete linkage disequilibrium with causative genes. Consequently, they can be used in a number of different genetic backgrounds. In fact, only a few FMs for R genes have been published in crop species up to now, although several alleles have been identified for a number of cloned R genes. But the rapidly increasing number of cloned R genes, decreasing costs for allele sequencing and the increasing availability of TILLING populations (*t*argeting-*i*nduced *l*ocal *l*esions *i*n *g*enomes) for most crop species will significantly contribute to the establishment of FMs for most major disease R genes within the next decade. Thus, this review summarises the current status of FM development in R genes and discusses implications of the availability of FMs for resistance breeding.

Christina Roenn Ingvardsen
Department of Genetics and Biotechnology, Research Centre Flakkebjerg, University of Aarhus, Forsoegsvej 1, DK-4200 Slagelse, Denmark
e-mail: Christina.Ingvardsen@agrsci.dk

U. Lüttge et al. (eds.), *Progress in Botany 69*.

Abbreviations *Avr*-gene: avriulence gene; BC: backcross; CAPS: cleaved amplified polymorphic sequence; CC: coiled coil; DUS: distinctness, uniformity and stability; *eIF4E*: eukaryotic translation initiation factor 4E; FM: functional marker; GUS: β-glucuronidase; HR: hypersensitive response; LD: linkage disequilibrium; LRR: leucine-rich repeat; MAS: marker-assisted selection; NBS: nucleotide binding site; NIL: near isogenic line; PCR: polymerase chain reaction; QTL: quantitative trait locus; RFLP: restriction fragment length polymorphism; RGA: resistance gene analogue; R gene: resistance gene; SNP: single nucleotide polymorphism; SSR: simple sequence repeat (microsatellite); TILLING: targeting-induced local lesions in genomes; TIR: *Drosophila* *T*oll and mammalian *I*nterleukin (IL)-1 *R*eceptors.

1 Introduction

Plant diseases are estimated to reduce global food production by more than 10% (James 1998). However, in some years total yield loss can occur, the most famous example being the potato late blight epidemic in Europe in 1845. Therefore, plant breeders have through all times aimed to improve disease resistance in crop species. Traditionally, resistance breeding is been performed by conventional breeding based on phenotypic selection. Conventional resistance breeding is demanding, as plants can only be assayed for resistance in the presence of respective pathogens at natural locations with high levels of disease incidence, or after artificial infection in the field or under controlled conditions. The success of field trials under natural conditions is strongly influenced by environment, interference with other pathogens or abiotic stress factors and the genetic composition of the pathogen population. Artificial inoculation experiments, which overcome these problems, are more laborious and often raise the question of transferability to field conditions. Molecular markers have been suggested to assist in overcoming these problems.

Marker-assisted selection (MAS) is based on markers closely linked to genes of interest and is used to assist phenotypic selection in variety (parent) development (Lande and Thompson 1990; Francia et al. 2005). In contrast to phenotypic selection, DNA marker-based selection is not affected by environment. Selection can be performed at any stage of plant development and from any tissue, reducing the time required to develop a new cultivar (Francia et al. 2005). In order to use molecular markers for MAS, it is a basic prerequisite: (a) to develop genetic maps to establish linkage between mapped molecular markers and major genes or quantitative trait loci (QTL) of interest and (b) to have the facilities to analyse large numbers of individuals in a time- and cost-effective manner (Lande and Thompson 1990). In addition, MAS is used in backcross (BC) programmes to recover the recurrent parent. This is particularly useful when the trait of interest is difficult to phenotype, for instance in the case of a recessive allele introgressed into a recurrent parent line carrying the dominant allele. Here MAS markedly reduces the time spent on BC procedures because the use of markers overcomes additional selfing or testcross generations after each BC. MAS can also be used in BC to select progeny with minimum amount of donor parent germplasm outside the region of interest (Francia et al. 2005). The overall cost compared with conventional BC might be significantly reduced (Kuchel et al. 2005). Another use of MAS is in pyramiding genes of interest in a breeding scheme by combining two or more genes in a breeding line, such as disease resistance genes (Servin et al. 2004). However, for any marker application there is a risk of recombination between the marker and the gene of interest leading to a false selection where the gene of interest is lost. This risk can be overcome by the use of functional markers (FMs; Andersen and Lübberstedt 2003).

The number of isolated plant resistance (R) genes has increased tremendously during the past 10 years, especially from *Arabidopsis* and rice, both of which have been fully sequenced. The number of isolated R genes can be expected to increase substantially in the near future, especially for those species with extensive genomics programs such as maize and cereals (http://www.ncbi.nlm.nih.gov/Genomes). Based on this accumulating information on R genes, the development of FMs is becoming reality. The objective of this review is to summarise the current status of FM development in R genes and to discuss the implications of the availability of FMs for resistance breeding.

2 Development of Functional Markers

FMs are derived from polymorphic sites within genes causally involved in phenotypic trait variation. The major advantages of FMs compared with conventional markers are: (a) no recombination between marker and the gene of interest and thus no loss of information over time, (b) transfer of information to unknown materials resulting in a high predictive value, and (c) targeted exploitation of genetic diversity (Andersen and Lübberstedt 2003). FMs benefit all phases in plant breeding. Genetic resources can be studied by FMs to determine the degree of within and between population genetic diversity. In the first phase of plant breeding, i.e the generation of genetic variation, FMs are useful in selecting complementary parents. During the development of parents of varieties (such as inbred lines in hybrid breeding), FMs are useful in pyramiding genes. In hybrid breeding, prediction of the best hybrid is another potential long-term benefit of FMs. In DUS testing (*d*istinctness, *u*niformity, *s*tability) of new cultivars, FMs are useful in reducing the amount of field tests required. Finally FMs can be used to accurately describe the properties of different available cultivars, depending on the number of FMs available. The development of FMs requires a set of steps to be fulfilled. First of all the gene of interest has to be identified, sequenced and functionally characterised. Gene identification is achieved by map-based gene isolation, expression profiling, use of sequence homology to characterised genes from other species, or other methods such as transposon tagging. Methods used to determine gene function include transformation, RNA interference and mutant characterisation. The second step towards FM development involves the study of allelic variation within characterised genes. Therefore, allele sequencing has to be performed between genotypes contrasting in the phenotype affected by the respective gene. The next crucial step is to identify those polymorphisms that are causative for the phenotypic trait variation, e.g. by association studies (Thornsberry et al. 2001). A more direct approach is the use of isogenic genotypes produced by homologous recombination or induced mutation assay screened by targeting-induced local lesions in genomes (TILLING; Table 1). EcoTILLING, a non-transgenic variant of TILLING has proven to be an efficient method to reveal polymorphisms between different alleles in the resistance genes *Mla* and *mlo* (Mejlhede et al. 2006). Finally, on the basis of the allele sequencing, co-dominant markers such as RFLP, SNP, SSR and CAPS markers can be developed, to distinguish the polymorphic sites causally involved in the phenotypic trait variation.

To date, only a few FMs have been published in crop species. In wheat, FMs have been developed for quality traits such as content of glutenin, puroindolines and granule-bound starch synthase activity (Bagge et al. 2007). The development of these markers was slightly different from the process outlined above. Biochemical markers previously characterised different phenotypes. Using public sequence databases, the allelic sequences of quality trait genes were obtained and, based on these sequences, functional genetic markers were developed (Gale 2005). One functional and two RFLP markers were used to improve cooking quality in rice (Zhou et al. 2003). In maize, several genomic resources are available such as public EST libraries, BAC libraries and publicly available microarrays of more than 10 000 different ESTs, rendering the identification of genes of interest practicable (Lübberstedt et al. 2005). Recently, an example of FM development of a maize forage quality gene was published where, based on association studies, insertion/deletion (INDEL) polymorphisms were identified in the *Bm3* gene associated with digestibility. These polymorphisms were converted into simple PCR-based marker assays (Lübberstedt et al. 2005).

3 Genetics of Disease Resistance

Plants contain and employ hundreds of different defence genes, which act upon pathogen attack (Meyers et al. 2003; Zhou et al. 2004). Since plants and pathogens co-evolve, every time a pathogen changes its mode of attack, the plant has to overcome this new challenge by developing a new mode of defence. This arms race led to the gene-for-gene theory, which was developed by Flor in the 1940s. He discovered that for resistance to occur, complementary pairs of genes, one in the host (resistance, R gene) and one in the pathogen (avirulence, *Avr*-gene), are required (reviewed by Flor 1971). It would be quite resource-intensive for the plant to express defence pathways all the time. Therefore plants need to detect when the pathogen attempts to attack in order to turn on the defence mechanisms only when appropriate.

More than 90 R genes have been isolated so far by the use of methods such as map-based cloning, transposon tagging and homology-based DNA library screening (Table 2). The first R gene isolated was *Hm1* from maize, which confers resistance

Table 1 Steps in FM development. The basis of functional marker development is a candidate gene. Once allelic sequences of the gene are obtained, the functional motif needs to be found. Finally, based on this motif(s), a marker is developed

Gene identification	Candidate gene approach	Expression profiling	RNA interference	QTL mapping	Trans-formation	Map-based cloning	Transposon gene tagging
Gene polymorphism				Allele sequencing			
Motif identification	Association studies			Homologous recombination			TILLING
Marker development				Co-dominant markers to distinguish alleles			

Table 2 Cloned and functionally characterised R genes

Plant	R gene	Pathogen[a]	Method[b]	Class[c]	Reference
Apple					
	Vf	*Venturia inaequalis* (F)	MB	2. LRR	Vinatzer et al. (2001)
Arabidopsis					
	HRT	TCV (V)	MB	1. CC-NBS-LRR	Cooley et al. (2000)
	Lsp1	TuMV (V)	MB	*eIF4E*	Lellis et al. (2002)
	PBS1	*Pseudomonas syringae* (B)	MB	3. Ser/Thr protein kinase	Swiderski and Innes (2001)
	RAC1	*Albugo candida* (F)	MB	1. TIR-NBS-LRR	Borhan et al. (2004)
	RCY1	*CMV-Y* (V)	MB	1. CC-NBS-LLR	Takahashi et al. (2002)
	RLM1 locus At1g64070 At1g63880	*Leptosphaeria maculans* (F)	MB/genome sequence	1. TIR-NBS-LRR	Staal et al. (2006)
	RPM1	*Pseudomonas syringae* (B)	MB	1. CC-NBS-LRR	Grant et al. (1995)
	RPP1	*Peronospora parasitica* (F)	MB	1. TIR-NBS-LRR	Botella et al. (1998)
	RPP13	*Peronospora parasitica* (F)	MB	1. NBS-LRR	Bittner-Eddy et al. (2000)
	RPP2	*Peronospora parasitica* (F)	MB	1. TIR-NBS-LRR	Sinapidou et al. (2004)
	RPP27	*Peronospora parasitica* (F)	MB	2. LRR	Tor et al. (2004)
	RPP4	*Peronospora parasitica* (F)	MB	1. TIR-NBS-LRR	van der Biezen et al. (2002)
	RPP5	*Peronospora parasitica* (F)	MB	1. TIR-NBS-LRR	Parker et al. (1997)
	RPP8	*Peronospora parasitica* (F)	MB	1. CC-NBS-LRR	McDowell et al. (1998)
	RPS2	*Pseudomonas syringae pv. tomato* (B)	MB	1. NBS-LRR	Bent et al. (1994)
	RPS4	*Pseudomonas syringae* pv. *tomato* (B)	MB	1. TIR-NBS-LRR	Gassmann et al. (1999)
	RPS5	*Pseudomonas syringae* pv. *tomato* (B)	MB	1. NBS-LRR	Warren et al. (1998)
	RPW8.1	*Erysiphe cruciferarum* (F)	MB	5. Transmembrane/ signal peptide/CC	Xiao et al. (2001)
	RPW8.2	*Erysiphe cruciferarum* (F)	MB	5. Transmembrane/ signal peptide/CC	Xiao et al. (2001)
	RRS1	*Ralstonia solanacearum*	MB	TIR-NBS-WRKY	Deslandes et al. (2002)

	RTM1	TEV (V)	MB	Jacalin like	Chisholm et al. (2000)
	RTM2	TEV (V)	MB	Small HSP-like	Whitham et al. (2000)
Barley					
	Hv-eIF4E	BYMV (V)	MB	*eIF4E*	Stein et al. (2005)
	Mla1	*Blumeria graminis* f.sp. *hordei* (F)	TE/hom/syn	1. CC-NBS-LRR	Zhou et al. (2001)
	Mla10	*Blumeria graminis* f.sp. *hordei* (F)	RT-PCR/hom/syn	1. NBS-LRR	Halterman and Wise (2004)
	Mla12	*Blumeria graminis* f.sp. *hordei* (F)	DNA lib/hom/syn	1. CC-NBS-LRR	Shen et al. (2003)
	Mla13	*Blumeria graminis* f.sp. *hordei* (F)	DNA lib/hom/syn	1. CC-NBS-LRR	Halterman et al. (2003)
	Mla6	*Blumeria graminis* f.sp. *hordei* (F)	DNA lib/hom/syn	1. CC-NBS-LRR	Halterman et al. (2001)
	Mla7	*Blumeria graminis* f.sp. *hordei* (F)	RT-PCR/hom/syn	1. NBS-LRR	Halterman and Wise (2004)
	Mlo	*Blumeria graminis* f.sp. *hordei* (F)	MB	Membrane-associated	Buschges et al. (1997)
	Rpg1	*Puccinia graminis* (F)	MB	Receptor kinase	Brueggeman et al. (2002)
Bean					
	RT4-4	Virus	RT-PCR	1. TIR-NBS-LRR	Seo et al. (2006)
Capsicum					
	Bs2	*Xanthomonas campestris* pv. *Vesictoria* (B)	MB	1. NBS-LRR	Tai et al. (1999)
	Pvr1/Pvr2	PepMoV (V) PVY (V) TEV (V)	Can gen/hom/syn	*eIF4E*	Ruffel et al. (2002), Kang et al. (2005)
Flax					
	Fis1	*Melampsora lini* (F)	Sub hyb/hom/syn	Dehydrogenase	Roberts and Pryor (1995)
	L6	*Melampsora lini* (F)	TT	1. NBS-LRR	Lawrence et al. (1995)
	M	*Melampsora lini* (F)	Mutation/TT	1. NBS-LRR	Anderson et al. (1997)
	P2	*Melampsora lini* (F)	TT/homology	1. TIR-NBS-LRR	Dodds et al. (2001)
Lettuce					
	Dm3	*Bremia lactucae* (F)	Mutation	1. NBS-LRR	Shen et al. (2002)
	mo1/Ls-eIF4E	*LMV* (V)	Homology	eIF4E	Nicaise et al. (2003)

(continued)

Table 2 (continued)

Plant	R gene	Pathogen[a]	Method[b]	Class[c]	Reference
Maize					
	Hm1	*Cochliobolus carbonum* (F)	TT	Reductase activity	Johal and Briggs (1992)
	Hm2	*Cochliobolus carbonum* (F)	TT	Reductase activity	Multani et al. (1998)
	Rp1-D	*Puccinia sorghi* (F)	TT	1. TIR-NBS-LRR	Collins et al. (1999)
	Rxo1/Rba1	*Xanthomonas oryzae* *Burkholderia andropogonis*	BAC library/ homology	1. CC-NBS-LRR	Zhao et al. (2005)
Melon					
	Cm-eIF4E	*MNSV* (V)	MB/synteny	eIF4E	Nieto et al. (2006)
Pea					
	eIF4E	*PSbMV* (V)	MB/homology	eIF4E	Gao et al. (2004)
Potato					
	Gpa2	*Globodera pallida* (N)	MB	1. CC-NBS-LRR	van der Vossen et al. (2000)
	Gro1–4	*Globodera rostochiensis* (N)	DNA lib	1. TIR-NBS-LRR	Paal et al. (2004)
	R1	*Phytophthora infestans* (F)	MB	1. CC-NBS-LRR	Ballvora et al. (2002)
	R3a	*Phytophthora infestans* (F)	Synteny	1. CC-NBS-LRR	Huang et al. (2005)
	RB	*Phytophthora infestans* (F)	MB	1. CC-NBS-LRR	Song et al. (2003)
	Rx1	PVX (V)	MB	1. NBS-LRR	Bendahmane et al. (1999)
	Rx2	PVX (V)	TE	1. CC-NBS-LRR	Bendahmane et al. (2000)
Rice					
	Nbs4-Pi2	*Magnaporthe grisea* (F)	MB	1. NBS-LRR	Zhou et al. (2006)
	Nbs2-Piz-t	*Magnaporthe grisea* (F)	MB	1. NBS-LRR	Zhou et al. (2006)
	Pi9	*Magnaporthe grisea* (F)	MB	1. NBS-LRR	Qu et al. (2006)
	Pib	*Magnaporthe grisea* (F)	MB	1. NBS-LRR	Wang et al. (1999)
	Pi-d2	*Magnaporthe grisea* (F)	MB	3. Ser/Thr protein kinase	Chen et al. (2006)
	Pi-k(h)	*Magnaporthe grisea* (F)	MB	1. NBS-LRR	Sharma et al. (2005)
	Pi-ta	*Magnaporthe grisea* (F)	MB	1. NBS-LRR	Bryan et al. (2000)

	R8	*Magnaporthe grisea* (F)	DNA lib/ hom/syn	Calmodulin-binding	Zheng et al. (2004)
	Rymv1/eIF (iso)4G	*RYMV* (V)	MB	eIF(iso)4G	Albar et al. (2006)
	Xa1	*Xanthomonas oryzae* pv. *Oryzae* (B)	MB	1. NBS-LRR	Yoshimura et al. (1998)
	Xa21	*Xanthomonas oryzae* pv. *Oryzae* (B)	MB	4. LRR/Ser/Thr Protein kinase	Song et al. (1995)
	Xa26/Xa3	*Xanthomonas oryzae* pv. *Oryzae* (B)	MB	4. LRR/Ser/Thr protein kinase	Sun et al. (2004), Xiang et al. (2006)
	Xa5	*Xanthomonas oryzae* pv. *Oryzae* (B)	MB	Transcription factor IIAγ	Iyer and McCouch (2004)
Solanum bulbocastnum					
	RB	*Phytophthora infestans* (F)	MB	1. CC-NBS-LRR	Song et al. (2003)
Soybean					
	Rhg1 Rhg4	*Heterodera glycines* (N)	Unknown	4. LRR/Ser/Thr protein kinase	Lightfoot and Meksem (2000), Hauge et al. (2001), Ruben et al. (2006)
Sugar beet					
	Hs1	*Heterodera schachtii* (N)	MB	2. LRR	Cai et al. (1997)
Tobacco					
	N	TMV (V)	TT	1. TIR-NBS-LRR	Whitham et al. (1994)
Tomato					
	Bs4	*Xanthomonas campestris* pv. *Vesictoria* (B)	MB	1. TIR-NBS-LRR	Schornack et al. (2004)
	Hero	*Globodera rostochiensis*	MB	1. NBS-LRR	Ernst et al. (2002)
	Hcr2 family	*Cladosporium fulvum* (F)	MB	2. LRR	Dixon et al. (1996, 1998)
	Cf-2				
	Cf-5				
	Hcr9 family	*Cladosporium fulvum* (F)		2. LRR	
	Cf-4		MB		Thomas et al. (1997)

(continued)

Table 2 (continued)

Plant	R gene	Pathogen[a]	Method[b]	Class[c]	Reference
	Cf-4A		TT		Takken et al. (1998)
	Cf-9		MB		Jones et al. (1994)
	Fen	*Pseudomonas syringae pv. tomato* (B)	MB	3. Ser/Thr protein kinase	Martin (1994)
	I2	*Fusarium oxysporum* (F)	MB	1. CC-NBS-LRR	Ori et al. (1997)
	Mi	*Meloidogyne incognita* (N)	MB	1. CC-NBS-LRR	Milligan et al. (1998)
	Prf	*Pseudomonas syringae* pv. *tomato* (B)	MB	1. CC-NBS-LRR	Salmeron et al. (1996)
	Pti1	*Pseudomonas syringae* pv. *tomato* (B)	Two-hybrid	3. Ser/Thr protein kinase	Zhou et al. (1995)
	Pto	*Pseudomonas syringae* pv. *tomato* (B)	MB	3. Ser/Thr protein kinase	Martin (1994)
	Sw-5	Tospovirus (V)	MB	1. NBS-LRR	Brommonschenkel et al. (2000)
	Tm-2	ToMV (V)	Hom	1. CC-NBS-LRR	Lanfermeijer et al. (2005)
	Tm2-2	ToMV (V)	TT	1. CC-NBS-LRR	Lanfermeijer et al. (2003)
	Ve1	*Verticillium albo-atrum*	MB	1. CC-NBS-LRR	Kawchuk et al. (2001)
	Ve2				
Tomato (*L. pimpinellifolium*)					
	Hcr9 family	*Clodosporium fulvum* (F)	Phage library	2. LRR	van der Hoorn et al. (2001), Kruijt et al. (2004)
	9DC2				
Wheat					
	Cre3	*Heterodera avenae* (N)	MB	1. NBS-LRR	Lagudah et al. (1997)
	Lr10	*Puccinia triticina* (F)	MB	1. CC-NBS-LRR	Feuillet et al. (2003)
	Lr21	*Puccinia triticina* (F)	MB	1. NBS-LRR	Huang et al. (2003)
	Pm3b	*Blumeria graminis* f.sp. *tritici* (F)	MB	1. CC-NBS-LRR	Yahiaoui et al. (2004)

[a]Pathogen: B = bacteria, F = fungus, V = virus and N = nematode

[b]Method abbreviations: TE: Transient expression, hom: homology, syn: synteny, DNA lib: DNA library screening, MB: map-based cloning, TT: transposon tagging, sub hyb: subtraction hybridisation, can gen: candidate gene

[c]Class refers to the five classes described by Dangl and Jones (2001) o r otherwise to the structure/function of the isolated R gene

to the necrotrophic fungus *Cochliobolus carbonum. Hm1* functions by reductase activity detoxifying the *C. carbonum* toxin (Johal and Briggs 1992).

With the combined use of molecular tools and genetic maps, several R genes have been cloned and their structure deduced (Table 2). In spite of the vast amount of R genes found against such diverse pathogens as fungi, bacteria, nematodes and viruses, most cloned R genes can be divided into five major classes based on structural characteristics (Dangl and Jones 2001). The largest of the R gene classes (class 1) consists of genes with a nucleotide-binding site (NBS) and leucine rich repeats (LRR) domain. The LRRs have an average consensus length of 23–24 amino acids, they are involved in ligand binding and the variable number of LRR repeat units is believed to determine R gene specificity (Hammond-Kosack and Jones 1997; Jones and Jones 1997). The NBS fragment of the genes contains several conserved domains such as P-loop, resistance nucleotide binding site A (RNBS-A), kinase 2, RNBS-B, RNBS-C, GLPL (amino acid motif), RNBS-D and MHDV (amino acid motif) and is involved in signal transduction (Meyers et al. 2003). The NBS-LRR genes can be subdivided into two groups according to the N-terminal structural features. One group has a domain (TIR) with homology to the domains of *Drosophila T*oll and the mammalian *i*nterleukin (IL)-1 *r*eceptors. This group is therefore called TIR-NBS-LRR. The other group have a coiled-coil domain (the CC-NBS-LRR group). In dicot species, both the TIR-NBS-LRR and the CC-NBS-LRR groups have been identified, whereas in monocot species, only the CC-NBS-LRR group is present (Pan et al. 2000). However, resent findings suggest that the TIR-NBS-LRR type might not be completely absent in monocots, as a few TIR-NBS-LRR type R genes have been found in barley and rice (Zhou et al. 2004; Mammadov et al. 2006). The other classes of R genes are characterised by: (class 2) transmembrane domains with extracellular LRR, (class 3) cytoplasmic Ser/Thr kinase, (class 4) extracellular LRR, a transmembrane domain and an intracellular kinase, or (class 5) a CC domain with a signal anchor at the N-terminus.

Since Dangl and Jones (2001) defined the five classes of R genes, many more R genes belonging to these classes have been cloned. In particular, class 1 (NBS-LRR) has proven to be a widely distributed class with so far 42 of the 69 cloned R genes belonging to this class (Table 2). R genes are known to be located in clusters in the genomes and the clusters of NBS-LRR genes are believed to be a result of microscale chromosomal duplication and deletion events (Meyers et al. 2003). This class contains R genes for diverse pathogens such as viruses, bacteria, fungi and nematodes, whereas classes 3 and 4 so far only represent bacterial resistance (*Pseudomonas syringae* and *Xanthomonas oryzae* respectively) and class 5 only represents fungal resistance genes (*E. cruciferarum*). Class 2 contains both fungal and nematode resistance genes (Table 2).

A few resistance genes described could not be assigned to any of the five classes defined. The *mlo* gene, which confers broad-spectrum resistance towards powdery mildew, is membrane-associated with six membrane-spanning helices (Buschges et al. 1997). A new class of R genes towards viruses was identified more recently: three different R genes for virus resistance show homology to the eukaryotic translation initiation factor 4E (*eIF4E*; Table 2).

Table 2 contains R genes for which the full sequence has been obtained. Many studies have been performed in which fragments of R genes have been isolated, mainly by the use of degenerate primers. Thus, R gene candidates have been identified in pea (Timmerman-Vaughan et al. 2000), barley (Madsen et al. 2003; Mammadov et al. 2006), apple (Lee et al. 2003; Calenge et al. 2005), potato (van der Linden et al. 2004), wheat (Dilbirligi et al. 2004), apricot (Soriano et al. 2005), *Avena* species (Irigoyen et al. 2006), poplar (Zhang et al. 2006) and ryegrass (Li et al. 2006).

4 Organisation of R Genes Within and Between Genomes

Plants contain large numbers of R genes. In the Columbia ecotype of *Arabidopsis*, 149 NBS-LRR-encoding genes have been identified, whereas as many as 480 have been found in rice (Meyers et al. 2003; Zhou et al. 2004). These R genes are non-randomly distributed among chromosomes both in rice (e.g. 49 on chromosome 1, two on chromosome 2) and in *Arabidopsis* (e.g. 133 on chromosome 1, 20 on chromosome 9; *Arabidopsis* Genome Initiative 2000; Dangl and Jones 2001; Zhou et al. 2004). Some R genes are found in clusters or even superclusters whereas others are singletons (Madsen et al. 2003; Mammadov et al. 2006; Yang et al. 2006). In species with lower gene densities, R genes, although clustered, are usually farther apart than is seen in species with a smaller genome (Sun et al. 2001; Meyers et al. 1998). Although R genes generally have a higher degree of nucleotide diversity than other gene families (Xing et al. 2007), the level of nucleotide diversity varies among NBS-LRR genes. In rice, it seems that singleton genes are conserved, whereas the genes in multi-gene families or gene clusters have a higher average diversity (Yang et al. 2006). It can be speculated that the rather conserved singleton genes might be involved in the detection of general elicitors or other conserved avirulence proteins, whereas the R genes having a high degree of variation are involved in the guarding of virulence targets (Jones and Dangl 2006). As the pathogen might be able to change these products, plants try to cope with it by constantly having a variety of potentially new R genes in their toolbox.

Genetic analyses of R genes are often hampered by the fact that R genes are members of a gene family. In many studies, fragments of R genes are isolated by PCR using degenerative primers placed in conserved motifs of the NBS-LRR gene (Soriano et al. 2005; Irigoyen et al. 2006; Mammadov et al. 2006). Even though several genes can be cloned by this technique, not all R genes present in the genome are found, as some might be very similar in the conserved region (and still have very different functions). Genes with minor differences are often excluded from further analysis and are not mapped (Irogoyen et al. 2006). Zhang et al. (2006) developed a technique, whereby it is possible to concentrate on one chromosome at the time. Using this technique, R genes from one location are not mixed up with (nearly) identical R genes from other chromosomes. This strategy is advantageous when the same family has members in different loci, as seen for pic19 in maize (Quint et al. 2003). There seems to be a large difference between the number of gene sequences found and the number of ESTs, indicating that a significant proportion of the NBS-encoding genes might be pseudogenes (Dilbirligi et al. 2004). However, it must be kept in mind that the low number of ESTs for R genes might be due to low transcript levels resulting in an under-representation of these genes in EST libraries.

When working with gene families, it is often difficult to determine whether genes mapped in related species are orthologues or paralogues (Ilag et al. 2000). Clustered genes are usually members of multigene families, belonging to the same phylogenetic clade (Leister 2004), although the sequence diversity between the family members is high. These clusters are usually the result of tandem duplications of paralogous sequences (Richly et al. 2002; Baumgarten et al. 2003, Meyers et al. 2003). Examples of such multigene families are the *Rp3* and *Rp1* clusters in maize (Webb et al. 2002; Smith et al. 2004) and the *RGC2* cluster in lettuce (Kuang et al. 2004). It seems that intergenic exchange is dramatically lower between NBS-LRR sequences located in different chromosome regions than exchange within the same chromosome region (Baumgarten et al. 2003). However, some resistance gene clusters carry different resistance genes that are clearly not derived from recent duplication events (Richly et al. 2002). These clusters might have formed by chance due to random rearrangements and might thus be relatively rare (Meyers et al. 2003). Furthermore, some clusters of R genes contain a set of genes that control very closely related pathogens, whereas others contain genes that control unrelated pathogens such as viruses, fungi and nematodes (Cooley et al. 2000). Since at least some R genes do not recognise the pathogen itself, but rather pathogen activity, we need to know a lot more about the specific compound detected by R genes to predict the R gene specificity by looking at the genetic sequence or the active sites within R proteins. The same holds true for a better understanding of *Avr* genes in pathogens, which have in several cases not been identified so far.

Although there is a high co-linearity between grasses at the macro level (Wilson et al. 1999), this might not always be the case at the micro co-linearity level (Paterson et al. 2000; Brunner et al. 2003, 2005). Interspecific analyses of R genes frequently reveal non-syntenic map locations between rice, barley and foxtail millet (Leister et al. 1998). This might to a great extent complicate the use of synteny studies to uncover the specificity of mapped and/or cloned R genes. Even among varieties within species, clusters in one variety will not always be co-linear to clusters in another (Yang et al. 2006). A remarkable number of unpaired genes were found when two rice genomes were compared (Shen et al. 2006; Yang et al. 2006). Another complication might be that some clusters clearly contain members of non-functional alleles or other types of pseudo-genes (Dangl and Jones 2001; Backes et al. 2003). However, some families, such as the Solanaceae, have a higher level of microcolinearity (Paterson et al. 2000), making it possible to use genomic information from the model plant tomato to clone a gene from potato (Huang et al. 2005). Several pathogens are species-specific, which also impairs the usefulness of synteny-based gene isolation of R genes.

The number of family members can be highly variable within a cluster and within a species. Five different haplotypes of a resistance gene cluster were found in wild diploid wheat (*Aegilops taushii*), ranging from all three genes being present to none (Brooks et al. 2006). It was concluded that deletion haplotypes were occurring frequently and independently in the genome. However, a gene present in one species missing from its homologous position in another species does not necessarily mean that it is not found in the genome, as the matching gene homologue might be found in a non-orthologous location (Song et al. 2002). When comparing the *Pph7* locus for leaf rust resistance in rice and barley, it was found that a major rearrangement had occurred. In barley, two HGA genes flank this locus. In the orthologous locus in rice, five HGA genes were found, four of which were also found at the barley locus. However, in the barley locus, six additional genes were inserted between the HGA genes. These six genes have homologues on eight different rice chromosomes (Brunner et al. 2003).

Direct jumping into novel R genes based on previously isolated R genes will probably be difficult due to lack of synteny on the sequence level. However, the growing body of physical maps and even complete genome sequences for crop species will be major resources for efficient isolation of novel R genes in the future. A new PCR-based approach (NBS profiling) combines the use of restriction enzymes and degenerate primers targeting the NBS-encoding region (van der Linden et al. 2004). NBS profiling can rapidly provide numerous polymorphic markers tightly linked to R genes and R gene clusters, and might even be used to identify diversity in resistance loci (van der Linden et al. 2004; Calenge et al. 2005). If a close sequence homologue is found and/or a cluster is identified, this information can be used to efficiently isolate additional candidates for particular resistance genes. The model plant sequences will still be an important resource even if there is a lack of synteny at the R gene level. If the regions around a mapped gene/OTL are conserved, then this can be used to develop new markers closer to the mapped gene (Kota et al. 2006; Mammadov et al. 2005).

5 Allelic Diversity in Resistance Genes

Bakker et al. (2006) investigated the degree of polymorphism in R genes in *Arabidopsis*. It was found that, in general, R genes have higher nucleotide diversity, a greater number of non-synonymous mutations and more recombination than the background genome. However, R genes were divided into three groups, showing high, medium and low levels of polymorphism, respectively. This emphasises that R genes are a very diverse and complex group of genes (Bakker et al, 2006). A similar grouping was found for R genes in rice (Yang et al. 2006). NBS-LRR genes in *Lolium perenne* showed a very high SNP density, with one SNP every 22 bp between two randomly sampled allele sequences. In contrast, non-NBS-LRR resistance gene candidates showed a lower degree of nucleotide diversity, with one SNP every 112 bp (Xing et al. 2007). The NBS domain is the most conserved region among the domains encoded by R genes, whereas the LRR region shows a very high diversity (Zhou et al. 2004). In *L. perenne*, one SNP every 10 bp between two sequences was found in the LRR region in the most divergent haplotypes (Xing et al. 2007).

Gene duplication is assumed to be a major evolutionary mechanism, acting to expand R genes in plant genomes (Michelmore and Meyers 1998). Moreover, several genetic mechanisms such as point mutation, recombination and unequal crossing-over have been suggested as potential causes for genetic variation within R genes. Li et al. (2006) found that gene mutation probably is the primary source of variation for the majority of R genes in *Lolium* spp as well as in other monocots, even though recombination events are also present. Furthermore, different kinds of selection were found in different types of R genes. Purifying selection was found in *Lolium* sp. (Li et al. 2006), whereas at least some R genes in rice seem to be under positive selection (Yang et al. 2006). Due to the great number of R genes in most species, however, it is challenging to perform such analyses on all R genes. At least in some species, there seem to be two types of R genes: a slow- and a fast-developing type (Kuang et al. 2004). Thus, the available data suggest that it might not be possible to formulate universally applicable models explaining R gene evolution.

In order to be able to develop a FM, the functional motifs causative for the phenotypic trait variation have to be identified, once alleles have been sequenced from susceptible as well as resistant genotypes. A low intragenic LD as found in the majority of R genes will be useful to pinpoint causative polymorphisms by association studies, which can subsequently be converted into FMs. In the case of several functional motifs per R gene, a low LD increases the chance to identify the optimal allele combining all favourable resistance motifs, but also to identify novel potentially beneficial alleles in the plant–pathogen arms race. If the starting point for a candidate gene-based association study is an R gene cluster, application of association genetics requires a high number of markers in the case of a low LD.

6 Functional Markers in Disease Resistance – Current Status

With knowledge gained from cloned R genes it is now becoming possible to develop diagnostic FMs for particular disease resistances. The first step in FM development, gene identification (Table 1), is well on the way for several disease resistances, with more than 90 different R genes cloned (Table 2). There is, however, a long way from gene identification to FMs. The next step is to identify and discriminate different alleles of the gene. Subsequently mutant or association studies are required to determine which polymorphisms are causative for the resistance phenotype. Finally, development of allele-specific markers for these polymorphisms is needed.

Several alleles have been identified for some of the cloned resistance genes (Table 3). In general, susceptibility can be explained by two different scenarios: (a) single base pair or amino acid changes explain resistance/susceptibility, (b) larger rearrangements lead to loss/gain of function. In some cases, as with *Pi-d2*, *Cm-eIF4E*, *RB* and *Xa26/Xa3*, only two alleles, a resistant and a susceptible allele, have been reported. Where the susceptibility is due to a single mutation, a FM, such as a SNP marker, might be developed based on this difference. However, the usefulness of

Table 3 Resistance genes where alleles have been identified

Pathogen	Plant	R gene	Alleles	Reference
Virus				
BYM-complex	Barley	*Hv-eIF4E*	7	Stein et al. (2005)
LMV	Lettuce	*Ls-eIF4E*	3	Nicaise et al. (2003)
MNSV	Melon	*Cm-eIF4E*	2	Nieto et al. (2006)
PepMoV, PVY, TEV	*Capsicum*	*Pvr1/eIF4E*	4	Kang et al. (2005)
PSbMV	Pea	*eIF4E*	3	Gao et al. (2004)
RYMV	Rice	*Rymv1/eIF (iso) 4G*	5	Albar et al. (2006)
TMV	Tomato	*Tm-2*^{2}	4	Lanfermeijer et al. (2005)
Bacteria				
Pseudomonas syringae	*Arabidopsis*	*RPS2*	7	Caicedo et al. (1999)
Pseudomonas syringae	Tomato	*Pto*	16	Rose et al. (2005)
Ralstonia solanacearum	*Arabidopsis*	*RRS1*	2	Deslandes et al. (2002)
Xanthomonas oryzae	Rice	*Xa26/Xa3*	2	Xiang et al. (2006)
Fungi				
Blumeria gaminis	Barley	*mlo*	11	Buschges et al. (1997), Srichumpa et al. (2005),
Blumeria graminis	Wheat	*Pm3*	8	Yahiaoui et al. (2006)
Cocliobolus carbonum	*Zea* sp.	*Hm2*	4–5	Tiffin et al. (2004)
Magnaporthe grisea	Rice	*Pi-ta*	4	Bryan et al. (2000)
Magnaporthe grisea	Rice	*Pi-d2*	2	Chen et al. (2006)
Melampsore lini	Flax	*L*	13	Ellis et al. (1999)
Peronospora parasitica	*Arabidopsis*	*RPP13*	24	Rose et al. (2004)
Phytophthora infestans	*Solanum bulbocastanum*	*RB*	2	Song et al. (2003)
Nematodes				
Heterodera glycines	Soybean	*rhg1*	9	Ruben et al. (2006)

such a marker might be limited until more alleles are sequenced, as also other mutations might give susceptibility. The development of FMs in NBS-LRR genes might be hampered by the fact that these genes often belong to gene families or are part of gene clusters built of closely related genes. The situation in the *Rp1* locus shows that even if several alleles are sequenced, it might be difficult to differentiate between true alleles and paralogues (Smith et al. 2004). Even for single gene loci, the development of PCR-based FMs can be a complex task if many alleles are identified, as in the L locus of flax (Hausner et al. 1999). However, newer SNP techniques, such as the Illumina technique with 1536-plexes, allow the establishment of high-level multiplexes which should facilitate discrimination of both multiple alleles and paralogous sequences (http://www.illumina.com/products/snp/default.ilmn).

R genes might have only one functional motif, implying that a single SNP in this motif mutates the R gene and switches the resistance from on to off (or vice versa, as in the case with eIF4E). Many R genes, however, have more than one important functional motif (e.g. one for pathogen reception, another for signal transduction) and several different mutations could make the R protein non-functional. However, some mutations might not turn the resistance off completely, but rather make it less efficient. This might result in a gradient of more or less susceptible genotypes or give resistance to different pathotypes. However, a complete 'library' of all possible SNPs and their potential role in resistance is so far not available for any of the cloned R genes.

Even though loss/gain of function types of resistance might seem simpler than a multi-allele situation, no FM marker assays have so far been developed based on larger rearrangements or complete loss of a resistance gene. This might be due to the fact that a co-dominant analysis is needed in order to be sure that a missing PCR band is due to a missing gene and not to a mismatch between primers and genomic sequence. This often requires sequence availability not only for the gene itself but also for the regions surrounding the gap.

Several markers have been developed to distinguish between the resistant *Pi-ta* allele and the susceptible *pi-ta* allele in rice based on different motifs in coding as well as non-coding regions of this gene (Jia et al. 2002, 2004). Already when it was first cloned, a single amino acid difference within the LRR domain of the gene was observed between the *pi-ta* allele of susceptible and resistant rice varieties, respectively (Bryan et al. 2000). It was later found that this mutation is not in all cases the causative mutation (Jia et al. 2002). Moreover, susceptibility in some varieties is caused by a complete lack of the *Pi-ta* gene. In conclusion, as a first step, the presence or absence of the *Pi-ta* gene needs to be determined. Subsequently, different causative intragenic polymorphisms yet to be identified are required to develop a set of FMs in order to discriminate all haplotypes.

The strategy of having a specific marker for each resistance allele was successfully used by Yeam et al. (2005). They developed a co-dominant marker system for the selection of three recessive alleles for eIF4E-based virus resistance in *Capsicum* using CAPS markers based on the SNPs causal for the phenotype of interest (Kang et al. 2005; Yeam et al. 2005). It should be straightforward to use similar systems to develop FMs in other crops in which eIF4E is involved in virus resistance, e.g. barley, lettuce, pea and rice (Table 3).

FMs specific for seven *Pm3* resistance alleles in wheat have been developed (Tommasini et al. 2006). Although highly conserved (98.5% nucleotide sequence identity), the authors were able to identify nucleotide polymorphisms that uniquely characterised each allele and to convert these into marker assays. Validation in a collection of 93 cultivars or breeding lines showed the usefulness of these markers and confirmed that *Pm3a-Pm3g* forms a true allelic series (Tommasini et al. 2006). In conclusion, the first FMs for R genes in plants have been established and should be useful tools for plant breeders. Given the rapidly increasing number of cloned R genes, decreasing costs for allele sequencing and the increasing availability of TILLING populations for most crop species, FMs will probably be established for most major disease R genes within the next decade.

7 The Impact of Functional Markers on Resistance Breeding

Compared with random markers that are generated from an anonymous region of the genome, FMs have some additional advantages. The major advantages of FMs are that they allow reliable application of markers in populations without prior mapping and that the markers can be used in mapping populations without the risk of information loss due to recombination. The *Pm3* allele specific markers have been used in 93 accessions of winter wheat and spelt, which is a large part of the global wheat material carrying the *Pm3* allele. This illustrates the broad usefulness of FMs, once they are established (Tommasini et al. 2006).

Elite varieties are often susceptible to major diseases and resistance genes have to be introgressed from wild relatives (Backes at al. 2003). MAS using FMs will be useful to efficiently reduce the amount of linkage drag in combination with closely linked markers. FMs allow for efficient selection of recombinants between the target gene and closely linked markers in large seedling populations. This might significantly reduce the number of BC generations (Frisch 2005).

In clusters of several different R genes eventually conferring resistance to a broad range of pathogens, the optimal multigenic allele combination will occur very rarely and will be difficult to obtain without markers or with markers linked at some distance to such a complex locus. The availability of FMs for different R genes would give plant breeders the possibility to select rare recombinants without the need for screening large numbers of plants with several pathogens. More generally, correct prediction of the resistance genotype for multiple diseases at the single-plant level is a particular strength of FMs, especially over phenotypic selection but also compared with traditional MAS. This is also true for situations where several redundant resistance genes confer resistance to the same pathogen. Moreover, FMs should be a useful tool to describe the pathovar or isolate specificity of R gene alleles, to discriminate functionally different alleles and to search for novel alleles for known R genes.

It has been discussed that alleles conferring resistance may confer a fitness disadvantage in the absence of a pathogen carrying the corresponding *Avr* gene. Although the major part of these disadvantages is now believed to be due to linkage drag of closely linked genes with unwanted effects on, e.g. yield, direct negative effects of resistance alleles on seed yield cannot be ruled out completely. As resistance on the cost of grain yield would not be acceptable in most cases, the general performance will still need to be evaluated in the field. Thus the full strength of FMs will first materialise when FMs become available for both resistance and other agronomic traits, especially in the light of continued progress in sequencing and marker technology. Until that time, beneficial FMs for a particular resistance have to be critically inspected for negative pleitropic effects, but also at later time-points for their effectiveness against novel pathogen isolates.

Acknowledgements Research in the authors' laboratory on resistance and resistance genes is supported by The Danish Ministry of Food, Agriculture and Fisheries. We thank Bruno Studer for critically reading and improving the manuscript.

References

Albar L, Bangratz-Reyser M, Hébrard E, Ndjiondjop M-N, Jones M, Ghesquière A (2006) Mutations in the eIF(iso)4G translation initiation factor confer high resistance of rice to *Rice yellow mottle virus*. Plant J 47:417–426

Andersen JR, Lübberstedt T (2003) Functional markers in plants. Trends Plant Sci 8:554–560

Anderson PA, Lawrence GJ, Morrish BC, Ayliffe MA, Finnegan EJ, Ellis JG (1997) Inactivation of the flax rust resistance gene *M* associated with loss of a repeated unit within the leucine-rich repeat coding region. Plant Cell 9:641–651

Arabidopsis Genome Initiative (2000) Analysis of the genome sequence of the flowering plant *Arabidopsis thaliana*. Nature 308:796–815

Backes G, Madsen LH, Jaiser H, Stougaard J, Herz M, Mohler V, Jahoor A (2003) Localisation of genes for resistance against *Blumeris graminis* f.sp. *hordei* and *Puccinia graminis* in a cross between a barley cultivar and a wild barley (*Hordeum vulgare* ssp. *spontaneum*) line. Theor Appl Genet 106:353–362

Bagge M, Xia X, Lübberstedt T (2007) Functional markers in wheat. Curr Opin Plant Biol 10:211–216

Bakker EG, Toomajian C, Kreitman M, Bergelson J (2006) A genome-wide survey of R gene polymorphisms in *Arabidopsis*. Plant Cell 18:1803–1818

Ballvora A, Ercolano MR, Weiss J, Meksem K, Bormann CA, Oberhagemann P, Salamini F, Gebhardt C (2002) The *R1* gene for potato resistance to late blight (*Phytophthora infestans*) belongs to the leucine zipper/NBS/LRR class of plant resistance genes. Plant J 30:361–371

Baumgarten A, Cannon S, Spangler R, May G (2003) Genome-level evolution of resistance genes in *Arabidopsis thaliana*. Genetics 165:309–319

Bendahmane A, Kanyuka K, Baulcombe DC (1999) The *Rx* gene from potato controls separate virus resistance and cell death responses. Plant Cell 11:781–792

Bendahmane A, Querci M, Kanyuka K, Baulcombe DC (2000) *Agrobacterium* transient expression system as a tool for the isolation of disease resistance genes: application to the *Rx2* locus in potato. Plant J 21:73–81

Bent AF, Kunkel BN, Dahlbeck D, Brown KL, Schmidt R, Giraudat J, Leung J, Staskawicz BJ (1994) *Rps2* of *Arabidopsis thaliana* – a leucine rich repeat class of plant disease resistance genes. Science 265:1856–1860

Biezen EA van der, Freddie CT, Kahn K, Parker JE, Jones JD (2002) *Arabidopsis RPP4* is a member of the *RPP5* multigene family of TIR-NB-LRR genes and confers downy mildew resistance through multiple signalling components. Plant J 29:439–451

Bittner-Eddy PD, Crute IR, Holub EB, Beynon JL (2000) *RPP13* is a simple locus in *Arabidopsis thaliana* for alleles that specify downy mildew resistance to different avirulence determinants in *Peronospora parasitica*. Plant J 21:177–188

Borhan MH, Holub EB, Beynon JL, Rozwadowski K, Rimmer SR (2004) The *Arabidopsis* TIR-NB-LRR gene *RAC1* confers resistance to *Albugo candida* (white rust) and is dependent on *EDS1* but not *PAD4*. Mol Plant Microbe Interact 17:711–719

Botella MA, Parker JE, Frost LN, Bittner-Eddy PD, Beynon JL, Daniels MJ, Holub EB, Jones JDG (1998) Three genes of the *Arabidopsis RPP1* complex resistance locus recognize distinct *Peronospora parasitica* avirulence determinants. Plant Cell 10:1847–1860

Brommonschenkel SH, Frary A, Frary A, Tanksley SD (2000) The broad-spectrum tospovirus resistance gene *Sw-5* of tomato is a homolog of the root-knot nematode resistance gene *Mi*. Mol Plant Microbe Interact 13:1130–1138

Brooks SA, Hunag L, Herbel MN, Gill BS, Brown-Guedira G, Fellers JP (2006) Structural variation and evolution of a defense-gene cluster in natural populations of *Aegilops tauschii*. Theor Appl Genet 112:618–626

Brueggeman R, Rostoks N, Kudrna D, Kilian A, Han F, Chen J, Druka A, Steffenson B, Kleinhofs A (2002) The barley stem rust-resistance gene *Rpg1* is a novel disease-resistance gene with homology to receptor kinases. Proc Natl Acad Sci USA 99:9328–9333

Brunner S, Keller B, Feuillet C (2003) A large rearrangement involving genes and low-copy DNA interrupts the microcollinearity between rice and barley at the *Rph7* locus. Genetics 164:673–683
Brunner S, Fengler K, Morgante M, Tingey S, Rafalski A (2005) Evolution of DNA sequences nonhomologies among maize inbreds. Plant Cell 17:343–360
Bryan GT, Wu K-S, Farral L, Jia Y, Hershey HP, McAdams SA, Faulk KN, Donaldson GK, Tarchini R, Valent B (2000) A single amino acid difference distinguishes resistant and susceptiple alleles of the rice blast resistance gene *Pi-ta*. Plant Cell 12:2033–2045
Buschges R, Hollricher K, Panstruga R, Simons G, Wolter M, Frijters A, Daelen R van, Lee T van der, Diergaarde P, Groenendijk J, Topsch S, Vos P, Salamini F, Schulze-Lefert P (1997) The barley *mlo* gene: a novel control element of plant pathogen resistance. Cell 88:695–705
Cai DG, Kleine M, Kifle S, Harloff HJ, Sandal NN, Marcker KA, KleinLankhorst RM, Salentijn EMJ, Lange W, Stiekema WJ, Wyss U, Grundler FMW, Jung C (1997) Positional cloning of a gene for nematode resistance in sugar beet. Science 275:832–834
Caicedo AL, Schaal BA, Kunkel BN (1999) Diversity and molecular evolution of the *RPS2* resistance gene in *Arabidopsis thaliana*. Proc Natl Acad Sci USA 96:302–306
Calenge F, Linden CG van der, Weg E van der, Schouten HJ, Arkel G van, Denancé C, Durel C-E (2005) Resistance gene analogues identified through the NBS-profiling method map close to major genes and QTL for disease resistance in apple. Theor Appl Genet 110:660–668
Chen X, Shang J, Chen D, Lei C, Zou Y, Zhai W, Liu G, Xu J, Ling Z, Cao G, Ma B, Wang Y, Zhao X, Li S, Zhu L (2006) A B-lectin receptor kinase gene conferring rice blast resistance. Plant J 46:794–804
Chisholm ST, Mahajan SK, Whitham SA, Yamamoto ML, Carrington JC (2000) Cloning of the *Arabidopsis RTM1* gene, which controls restriction of long-distance movement of tobacco etch virus. Proc Natl Acad Sci USA 97:489–494
Collins N, Drake J, Ayliffe M, Sun Q, Ellis J, Hulbert S, Pryor T (1999) Molecular characterization of the maize *Rp1-D* rust resistance haplotype and its mutants. Plant Cell 11:1365–1376
Cooley MB, Pathirana S, Wu HJ, Kachroo P, Klessig DF (2000) Members of the *Arabidopsis HRT/RPP8* family of resistance genes confer resistance to both viral and oomycete pathogens. Plant Cell 12:663–676
Dangl JL, Jones JDG (2001) Plant pathogens and integrated defence responses to infection. Nature 411:826–833
Deslandes L, Olivier J, Theulières F, Hirsch J, Feng DX, Bittner-Eddy P, Beynon J, Marco Y (2002) Resistance to *Ralstonia solanacearum* in *Arabidopsis thaliana* is conferred by the recessive *RRS1-R* gene, a member of a novel family of resistance genes. Proc Natl Acad Sci USA 99:2404–2409
Dilbirligi M, Erayman M, Sandhu D, Sidhu D, Gill KS (2004) Identification of wheat chromosomal regions containing expressed resistance genes. Genetics 166:461–481
Dixon MS, Jones DA, Keddie JS, Thomas CM, Harrison K, Jones JDG (1996) The tomato *Cf-2* disease resistance locus comprises two functional genes encoding leucine-rich repeat proteins. Cell 84:451–459
Dixon MS, Hatzixanthis K, Jones DA, Harrison K, Jones JDG (1998) The tomato *Cf-5* disease resistance gene and six homologs show pronounced allelic variation in leucine-rich repeat copy number. Plant Cell 10:1915–1925
Dodds PN, Lawrence GJ, Ellis JG (2001) Six amino acid changes confined to the leucine-rich repeat β-strand/β-turn motif determine the difference between the *P* and *P2* rust resistance specificities in flax. Plant Cell 13:163–178
Ellis JG, Lawrence GJ, Luck JE, Dodds PN (1999) Identification of regions in alleles of the flax rust resistance gene *L* that determine differences in gene-for-gene specificity. Plant Cell 11:495–506
Ernst K, Kumar A, Kriseleit D, Kloos D-U, Phillips MS, Ganal MW (2002) The broad-sprectrum potato cyst nematode resistance gene (*Hero*) from tomato is the only member of a large gene family of NBS-LRR genes with an unusual amino acid repeat in the LRR region. Plant J 31:127–136

Feuillet C, Travella S, Stein N, Albar L, Nublat A, Keller B (2003) Map-based isolation of the leaf rust disease resistance gene *Lr10* from the hexaploid wheat (*Triticum aestivum* L.) genome. Proc Natl Acad Sci USA 100:15253–15258

Flor HH (1971) Current status of gene-for-gene concept. Annu Rev Phytopathol 9:275–296

Francia E, Tacconi G, Crosatti C, Barabaschi D, Bulgarelli D, Dall'Aglio E, Vale G (2005) Marker assisted selection in crop plants. Plant Cell Tissue Organ Cult 82:317–342

Frisch M (2005) Optimum design of marker-assisted backcross programs. In: Loerz H, Wenzel G (eds) Biotechnology in agriculture and forestry, vol 55. Springer, Berlin Heidelberg New York. pp 319–334

Gale KR (2005) Diagnostic DNA markers for quality traits in wheat. J Cereal Sci 41:181–192

Gao Z, Johansen E, Eyers S, Thomas CL, Ellis THN, Maule AJ (2004) The potyvirus recessive resistance gene, *sbm1*, identifies a novel role for translation initiation factor eIF4E in cell-to-cell trafficking. Plant J 40:376–385

Gassmann W, Hinsch ME, Staskawicz BJ (1999) The *Arabidopsis RPS4* bacterial-resistance gene is a member of the TIR-NBS-LRR family of disease-resistance genes. Plant J 20:265–277

Grant MR, Godiard L, Straube E, Ashfield T, Lewald J, Sattler A, Innes RW, Dangl JL (1995) Structure of the *Arabidopsis Rpm1* gene enabling dual-specificity disease resistance. Science 269:843–846

Halterman DA, Wise RP (2004) A single amino acid substitution in the sixth leucine-rich repeat of barley *Mla6* and *Mla13* alleviates dependence on *RAR1* for disease resistance signaling. Plant J 38:215–226

Halterman D, Zhou FS, Wei FS, Wise RP, Schulze-Lefert P (2001) The *MLA6* coiled-coil, NBS-LRR protein confers *AvrMla6*-dependent resistance specificity to *Blumeria graminis* f. sp. *hordei* in barley and wheat. Plant J 25:335–348

Halterman DA, Wei FS, Wise RP (2003) Powdery mildew-induced *Mla* mRNAs are alternatively spliced and contain multiple upstream open reading frames. Plant Physiol 131:558–567

Hammond-Kosack KE, Jones JDG (1997) Plant disease resistance genes. Annu Rev Plant Physiol Plant Mol Biol 48:575–607

Hauge BM, Wang ML, Parsons JD, Parnell LD (2001) Monsanto. Nucleic acid molecules and other molecules associated with soybean cyst nematode resistance. Patent US 20030005491

Hausner G, Rashid KY, Kenaschuk EO, Procunier JD (1999) The development of codominant PCR/RFLP based markers for the flax rust-resistance alleles at the L locus. Genome 42:1–8

Hoorn RAL van der, Kruijt M, Roth R, Brandwagt BF, Joosten MHAJ, Witt PJGM de (2001) Intragenic recombination generated two distinct *Cf* genes that mediate AVR9 recognition in the natural population of *Lycopersicon pimpinellifolium*. Proc Natl Acad Sci USA 98:10493–10498

Huang L, Brooks SA, Li W, Fellers JP, Trick HN, Gill BS (2003) Map-based cloning of leaf rust resistance gene *Lr21* from the large and polyploid genome of bread wheat. Genetics 164:655–664

Huang S, Vossen EAG van der, Kuang H, Vleeshouwers VGAA, Zhang N, Borm TJA, Eck HJ van, Baker B, Jacobsen E, Visser RGF (2005) Comparative genomics enables the isolation of the *R3a* late blight resistance gene in potato. Plant J 42:251–261

Ilag LL, Yadav RC, Huang N, Ronald PC, Ausubel FM (2000) Isolation and characterization of disease resistance gene homologues from rice cultivar IR64. Gene 255:245–255

Irigoyen ML, Ferrer E, Loarce Y (2006) Cloning and characterization of resistance gene analogs from *Avena* species. Genome 49:54–63

Iyer AS, McCouch SR (2004) The rice bacterial blight resistance gene *xa5* encodes a novel form of disease resistance. Mol Plant Microbe Interact 17:1348–1354

James C (1998) Global food security. Abstr Int Congr Plant Pathol 7:4.1GF. Available at: www.bspp.org.uk/icpp98/4/1GF.html

Jia Y, Wang Z, Singh P (2002) Development of dominant rice blast *Pi-ta* resistance gene markers. Crop Sci 42:2145–2149

Jia YL, Redus M, Wang ZH, Rutger JN (2004) Development of a SNLP marker from the *Pi-ta* blast resistance gene by tri-primer PCR. Euphytica 138:97–105

Johal GS, Briggs SP (1992) Reductase activity encoded by the *Hm1* disease resistance gene in maize. Science 258:985–987

Jones DA, Jones DG (1997) The role of leucine-rich repeat proteins in plant defences. Advances in botanical research 24:89–167

Jones DA, Thomas CM, Hammondkosack KE, Balintkurti PJ, Jones JDG (1994) Isolation of the tomato *Cf-9* gene for resistance to *Cladosporium fulvum* by transposon tagging. Science 266:789–793

Jones JDG, Dangl JL (2006) The plant immune system. Nature 444:323–329

Kang B-C, Yeam I, Frantz JD, Murphy JF, Jahn MM (2005) The *pvr1* locus in *Capsicum* encodes a translation initiation factor eIF4E that interacts with *Tobacco etch virus* Vpg. Plant J 42:392–405

Kawchuk LM, Hachey J, Lynch DR, Kulcsar F, Rooijen G van, Waterer DR, Robertson A, Kokko E, Byers R, Howard RJ, Fischer R, Prüfer D (2001) Tomato *Ve* disease resistance genes encode cell surface-like receptors. Proc Natl Acad Sci USA 98:6511–6515

Kota R, Spielmeyer W, McIntish RA, Lagudah ES (2006) Fine genetic mapping fails to dissociate durable stem rust resistance gene *Sr2* from pseudo-black chaff in common wheat (*Triticum aestivum* L.). Theor Appl Genet 112:492–499

Kruijt M, Brandwagt BF, Wit PJGM de (2004) Rearrangements in the *Cf-9* disease resistance gene cluster of wild tomato have resulted in three genes that mediate Avr9 responsiveness. Genetics 168:1655–1663

Kuang H, Woo S-S, Meyers BC, Nevo E, Michelmore RW (2004) Multiple genetic processes result in heterogeneous rates of evolution within major cluster disease resistance genes in lettuce. Plant Cell 16:2870–2894

Kuchel H, Ye G, Fox R, Jefferies S (2005) Genetic and economic analysis of a targeted marker-assisted wheat breeding strategy. Mol Breed 16:67–78

Lagudah ES, Moullet O, Appels R (1997) Map-based cloning of a gene sequence encoding a nucleotide binding domain and a leucine-rich region at the *Cre3* nematode resistance locus of wheat. Genome 40:659–665

Lande R, Thompson R (1990) Efficiency of marker assisted selection in the improvement of quantitative traits. Genetics 124:743–756

Lanfermeijer FC, Dijkhuis J, Sturre MJ, Haan P de, Hille J (2003) Cloning and characterization of the durable tomato mosaic virus resistance gene *Tm-2*^{2} from *Lycopersicon esculentum*. Plant Mol Biol 52:1037–1049

Lanfermeijer FC, Warmink J, Hille J (2005) The products of the broken *Tm-2* and the durable *Tm-2*^{2} resistance genes from tomato differ in four amino acids. J Exp Bot 56:2925–2933

Lawrence GJ, Finnegan EJ, Ayliffe MA, Ellis JG (1995) The *L6* gene for flax rust resistance is related to the *Arabidopsis* bacterial resistance gene *Rps2* and the tobacco viral resistance gene *N*. Plant Cell 7:1195–1206

Lee S-Y, Seo J-S, Rodriguez-Lanetty M, Lee D-H (2003) Comparative analysis of superfamilies of NBS-encoding disease resistance gene analogs in cultivated and wild apple species. Mol Genet Genomics 269:101–108

Leister D (2004) Tandem and segmental gene duplication and recombination in the evolution of plant disease resistance genes. Trends in Genetics 20:116–122

Leister D, Kurth J, Laurie DA, Yano M, Sasaki T, Devos K, Graner A, Schulze-Lefert P (1998) Rapid reorganization of resistance gene homologues in cereal genomes. Proc Natl Acad USA 95:370–375

Lellis AD, Kasschau KD, Whitham SA, Carrington JC (2002) Loss-of-susceptibility mutants of *Arabidopsis thaliana* reveal an essential role for eIF(iso)4E during potyvirus infection. Curr Biol 12:1046–1051

Li J, Xu Y, Fei S, Li L (2006) Isolation, characterization and evolutionary analysis of resistance gene analogs in annual ryegrass, perennial ryegrass and their hybrid. Physiol Plant 126:627–638

Lightfoot DA, Meksem K (2000) Novel polynucleotides and polypeptides relating to loci underlying resistance to soybean cyst nematode and methods of use thereof (patent pending 09/772,134). Patent filing date 01-29–2000

Linden CG van der, Wouters DCAE, Mihalka V, Kochieva EZ, Smulders MJM, Vosman B (2004) Efficient targeting of plant disease resistance loci using NBS profiling. Theor Appl Genet 109:384–393

Lübberstedt T, Zein I, Andersen J, Wenzel G, Krutzfeldt B, Eder J, Ouzunova M, Chun S (2005) Development and application of functional markers in maize. Euphytica 146:101–108

Madsen LH, Collins NC, Rakwalska M, Backes G, Sandal N, Krusell L, Jensen J, Waterman EH, Jahoor A, Ayliffe M, Pryor AJ, Langridge P, Schulze-Lefert P, Stougaard J (2003) Barley disease resistance gene analogs of the NBS-LRR class: identification and mapping. Mol Genet Genomics 269:150–161

Mammadov JA, Steffenson BJ, Saghai Maroof MA (2005) High-resolution mapping of the barley leaf rust resistance gene *Rph5* using barley expressed sequence tags (ESTs) and synteny with rice. Theor Appl Genet 111:1651–1660

Mammadov JA, Liu Z, Biyashev RM, Muehlbauer GJ, Saghai Maroof (2006) Cloning, genetic and physical mapping of resistance gene analogs in barley (*Hordeum vulgare* L.). Plant Breed 125:32–42

Martin GB (1994) Analysis of the molecular basis of *Pseudomonas syringae* pv. *tomato* resistance in tomato. Euphytica 79:187–193

McDowell JM, Dhandaydham M, Long TA, Aarts MGM, Goff S, Holub EB, Dangl JL (1998) Intragenic recombination and diversifying selection contribute to the evolution of downy mildew resistance at the *RPP8* locus of *Arabidopsis*. Plant Cell 10:1861–1874

Mejlhede N, Kyjovska Z, Backes G, Burhenne K, Rasmussen SK, Jahoor A (2006) EcoTILLING for the identification of allelic variation in the powdery mildew resistance genes *mlo* and *Mla* in barley. Plant Breed 125:461–467

Meyers BC, Chin DB, Shen KA, Sivaramakrishnan S, Lavelle DO, Zhang Z, Michelmore RW (1998) The major resistance gene cluster in lettuce is highly duplicated and spans several megabases. Plant Cell 10:1817–1832

Meyers BC, Kozik A, Griego A, Kuang H, Michelmore RW (2003) Genome-wide analysis of NBS-LRR-encoding genes in *Arabidopsis*. Plant Cell 15:809–834

Michelmore RW, Meyers BC (1998) Clusters of resistance genes in plants evolve by divergent selection and a birth-and-death progress. Genome Research 8:1113–1130

Milligan SB, Bodeau J, Yaghoobi J, Kaloshian I, Zabel P, Williamson VM (1998) The root knot nematode resistance gene *Mi* from tomato is a member of the leucine zipper, nucleotide binding, leucine-rich repeat family of plant genes. Plant Cell 10:1307–1319

Multani DS, Meeley RB, Paterson AH, Gray J, Briggs SP, Johal GS (1998) Plant–pathogen microevolution: Molecular basis for the origin of a fungal disease in maize. Proc Natl Acad Sci USA 95:1686–1691

Nicaise V, German-Retana S, Sanjuán R, Dubrana M-P, Mazier M, Maisonneuve B, Candresse T, Caranta C, LeGall O (2003) The eukaryotic translation initiation factor 4E controls lettuce susceptibility to the potyvirus *Lettuce mosaic virus*. Plant Physiol 132:1272–1282

Nieto C, Morales M, Orjeda G, Clepet C, Monfort A, Sturbois B, Puigdomènech P, Pitrat M, Caboche M, Dogimont C, Garcia-Mas J, Aranda MA, Bendahmane A (2006) An *eIF4E* allele confers resistance to an uncapped and non-polydenylated RNA virus in melon. Plant J 48:452–462

Ori N, Eshed Y, Paran I, Presting G, Aviv D, Tanksley S, Zamir D, Fluhr R (1997) The *I2C* family from the wilt disease resistance locus *I2* belongs to the nucleotide binding, leucine-rich repeat superfamily of plant resistance genes. Plant Cell 9:521–532

Paal J, Henselewski H, Muth J, Meksem K, Menéndez CM, Salamini F, Ballvora A, Gebhardt C (2004) Molecular cloning of the potato *Gro1-4* gene conferring resistance to pathotype Ro1 of the root cyst nematode *Globodera rostochiensis*, based on a candidate gene approach. Plant J 38:285–297

Pan Q, Wendel J, Fluhr R (2000) Divergent evolution of plant NBS-LRR resistance gene homologues in dicot and cereal genomes. J Mol Evol 50:203–213

Parker JE, Coleman MJ, Szabo V, Frost LN, Schmidt R, Biezen EA van der, Moores T, Dean C, Daniels MJ, Jones JDG (1997) The *Arabidopsis* downy mildew resistance gene *RPP5* shares similarity to the toll and interleukin-1 receptors with *N* and *L6*. Plant Cell 9:879–894

Paterson AH, Bowers JE, Burow MD, Draye X, Elsik CG, Jiang C-X, Katsar CS, Lan T-H, Lin Y-R, Ming R, Wright RJ (2000) Comparative genomics of plant chromosomes. Plant Cell 12:1623–1539

Qu S, Liu G, Zhou B, Bellizzi M, Zeng L, Dai L, Han B, Wang GL (2006) The broad-spectrum blast resistance gene *Pi9* encodes a nucleotide-binding site leucine-rich repeat protein and is a member of a multigene family in rice. Genetics 172:1901–1914

Quint M, Dussle CM, Melchinger AE, Lübberstedt T (2003) Identification of genetically linked RGAs by BAC screening in maize and implications for gene cloning, mapping and MAS. Theor Appl Genet 106:1171–1177

Richly E, Kurth J, Leister D (2002) Mode of amplification and reorganization of resistance genes during recent *Arabidopsis thaliana* evolution. Mol Biol Evol 19:76–84

Roberts JK, Pryor A (1995) Isolation of a flax (*Linum usitatissimum*) gene induced during susceptible infection by flax rust (*Melampsora lini*). Plant J 8:1–8

Rose LE, Bittner-Eddy PD, Langley CH, Holub EB, Michelmore RW, Beynon JL (2004) The maintenance of extreme amino acid diversity at the disease resistance gene, *RPP13*, in *Arabidopsis thaliana*. Genetics 166:1517–1527

Rose LE, Langley CH, Bernal AJ, Michelmore RW (2005) Natural variation in the *Pto* pathogen resistance gene within species of wild tomato (*Lycopersicon*). I. Functional analysis of *Pto* alleles. Genetics 171:345–357

Ruben E, Jamai A, Afzal J, Njiti VN, Triwitayakorn K, Iqbal MJ, Yaegashi S, Bashir R, Kazi S, Arelli P, Town CD, Ishihara H, Meksem K, Lightfoot DA (2006) Genomic analysis of the *rhg1* locus: candidate genes that underlie soybean resistance to the cyst nematode. Mol Genet Genomics 276:503–516

Ruffel S, Dussault MH, Palloix A, Moury B, Bendahmane A, Robaglia C, Caranta C (2002) A natural recessive resistance gene against potato virus Y in pepper corresponds to the eukaryotic initiation factor *4E (eIF4E)*. Plant J 32:1067–1075

Salmeron JM, Oldroyd GED, Rommens CMT, Scofield SR, Kim HS, Lavelle DT, Dahlbeck D, Staskawicz BJ (1996) Tomato *Prf* is a member of the leucine-rich repeat class of plant disease resistance genes and lies embedded within the *Pto* kinase gene cluster. Cell 86:123–133

Schornack S, Ballvora A, Gurlebeck D, Peart J, Baulcombe D, Ganal M, Baker B, Bonas U, Lahaye T (2004) The tomato resistance protein *Bs4* is a predicted non-nuclear TIR-NB-LRR protein that mediates defense responses to severly truncated derivatives of *AvrBs4* and overexpressed *AvrBs3* (vol 37, pg 46, 2004). Plant J 37:787–787

Seo Y-S, Rojas MR, Lee J-Y, Lee S-W, Jeon J-S, Ronald P, Lucas WJ, Gilbertson RL (2006) A viral resistance gene from common bean functions across plant families and is up-regulated in a non-virus-specific manner. Proc Natl Acad Sci USA 103:11856–11861

Servin B, Martin OC, Mezard M, Hospital F (2004) Toward a theory of marker-assisted gene pyramiding. Genetics 168:513–523

Sharma TR, Madhav MS, Singh BK, Shanker P, Jana TK, Dalal V, Pandit A, Singh A, Gaikwad K, Upreti HC, Singh NK (2005) High-resolution mapping, cloning and molecular characterization of the *Pi-k^h^* gene of rice, which confers resistance to *Magnaporthe grisea*. Mol Genet Genomics 274:569–578

Shen J, Araki H, Chen L, Chen J-Q, Tian D (2006) Unique evolutionary mechanism in *R*-genes under the presence/absence polymorphism in *Arabidopsis thaliana*. Genetics 172:1243–1250

Shen KA, Chin DB, Arroyo-Garcia R, Ochoa OE, Lavelle DO, Wroblewski T, Meyers BC, Michelmore RW (2002) *Dm3* is one member of a large constitutively expressed family of nucleotide binding site-leucine-rich repeat encoding genes. Mol Plant Microbe Interact 15:251–261

Shen QH, Zhou FS, Bieri S, Haizel T, Shirasu K, Schulze-Lefert P (2003) Recognition specificity and *RAR1/SGT1* dependence in barley *Mla* disease resistance genes to the powdery mildew fungus. Plant Cell 15:732–744

Sinapidou E, Williams K, Nott L, Bahkt S, Tor M, Crute I, Bittner-Eddy P, Beynon J (2004) Two TIR:NB:LRR genes are required to specify resistance to *Peronospora parasitica* isolate Cala2 in *Arabidopsis*. Plant J 38:898–909

Smith SM, Pryor AJ, Hulbert SH (2004) Allelic and haplotypic diversity at the *Rp1* rust resistance locus of maize. Genetics 167:1939–1947

Song J, Bradeen JM, Naess SK, Raasch JA, Wielgus SM, Haberlach GT, Liu J, Kuang H, Austin-Phillips S, Buell CR, Helgeson JP, Jiang J (2003) Gene *RB* cloned from *Solanum bulbocastanum* confers broad spectrum resistance to potato late blight. Proc Natl Acad Sci USA 100:9128–9133

Song R, Llaca V, Messing J (2002) Mosaic organization of orthologous sequences in grass genomes. Genome Res 12:1549–1555

Song WY, Wang GL, Chen LL, Kim HS, Pi LY, Holsten T, Gardner J, Wang B, Zhai WX, Zhu LH, Fauquet C, Ronald P (1995) A receptor kinase-like protein encoded by the rice disease resistance gene, *Xa21*. Science 270:1804–1806

Soriano JM, Vilanova S, Romero C, Llácer G, Badenes ML (2005) Characterization and mapping of NBS-LRR resistance gene analogs in apricot (*Prunus armeniaca* L.). Theor Appl Genet 110:980–989

Srichumpa P, Brunner S, Keller B, Yahiaoui N (2005) Allelic series of four powdery mildew resistance genes at the *Pm3* locus in hexaploid bread wheat. Plant Physiol 139:885–895

Staal J, Kaliff M, Bohman S, Dixelius C (2006) Transgenic segregation reveals two Arabidopsis TIR-NB-LRR resistance genes effective against *Leptospaeria maculans*, causal agent of blackleg disease. Plant J 46:218–230

Stein N, Perovic D, Kumlehn J, Pellio B, Stracke S, Streng S, Ordon F, Graner A (2005) The eukaryotic translation initiation factor 4E confers multiallelic recessive Bymovirus resistance in *Hordeum vulgare* (L.). Plant J 42:912–922

Sun Q, Collins NC, Ayliffe M, Smith SM, Drake J, Pryor T, Hulbert SH (2001) Recombination between paralogues at the *rp1* rust resistance locus in maize. Genetics 158:423–438

Sun X, Cao Y, Yang Z, Xu C, Li X, Wang S, Zhang Q (2004) *Xa26*, a gene conferring resistance to Xanthomonas oryzae pv. oryzae in rice, encodes an LRR receptor kinase-like protein. Plant J 37:517–527

Swiderski MR, Innes RW (2001) The Arabidopsis *PBS1* resistance gene encodes a member of a novel protein kinase subfamily. Plant J 26:101–112

Tai TH, Dahlbeck D, Clark ET, Gajiwala P, Pasion R, Whalen MC, Stall RE, Staskawicz BJ (1999) Expression of the *Bs2* pepper gene confers resistance to bacterial spot disease in tomato. Proc Natl Acad Sci USA 96:14153–14158

Takahashi H, Miller J, Nozaki Y, Sukamto, Takeda M, Shah J, Hase S, Ikegami M, Ehara Y, Dinesh-Kumar SP (2002) *RCY1*, an *Arabidopsis thalinana RPP8/HRT* family resistance gene, conferring resistance to cucumber mosaic virus requires salicylic acid, ethylene and a novel signal transduction mechanism. Plant J 32:655–667

Takken FLW, Schipper D, Nijkamp HJJ, Hille J (1998) Identification and *Ds*-tagged isolation of a new gene at the *Cf-4* locus of tomato involved in disease resistance to *Cladosporium fulvum* race 5. Plant J 14:401–411

Thomas CM, Jones DA, Parniske M, Harrison K, Balint-Kurti PJ, Hatzixanthis K, Jones JDG (1997) Characterisation of the tomato *Cf-4* gene for resistance to *Cladosporium fulvum* identifies sequences that determine recognitional specificity in *Cf-4* and *Cf-9*. Plant Cell 9:2209–2224

Thornsberry JM, Goodman MM, Doebley J, Kresovich S, Nielsen D, Buckler IV ES (2001) *Dwarf8* polymorphisms associate with variation in flowering time. Nature Gen 28:286–289

Tiffin P, Hacker R, Gaut BS (2004) Population genetic evidence for rapid changes in intraspecific diversity and allelic cycling of a specialist defense gene in *Zea*. Genetics 168:425–434

Timmerman-Vaughan GM, Frew TJ, Weeden NF (2000) Characterization and linkage mapping of R-gene analogous DNA sequences in pea (*Pisum sativum* L.). Theor Appl Genet 101:241–247

Tommasini L, Yahiaoui N, Srichumpa P, Keller B (2006) Development of functional markers specific for seven *Pm3* resistance alleles and their validation in the bread wheat gene pool. Theor Appl Genet 114:165–175

Tor M, Brown D, Cooper A, Woods-Tor A, Sjolander K, Jones JDG, Holub EB (2004) *Arabidopsis* downy mildew resistance gene *RPP27* encodes a receptor-like protein similar to *CLAVATA2* and tomato *Cf-9*. Plant Physiol 135:1100–1112

Vinatzer BA, Patocchi A, Gianfranceschi L, Tartarini S, Zhang HB, Gessler C, Sansavini S (2001) Apple contains receptor-like genes homologous to the *Cladosporium fulvum* resistance gene family of tomato with a cluster of genes cosegregating with *Vf* apple scab resistance. Mol Plant Microbe Interact 14:508–515

Vossen EA van der, Voort JN van der, Kanyuka K, Bendahmane A, Sandbrink H, Baulcombe DC, Bakker J, Stiekema WJ, Klein-Lankhorst RM (2000) Homologues of a single resistance-gene cluster in potato confer resistance to distinct pathogens: a virus and a nematode. Plant J 23:567–576

Wang ZX, Yano M, Yamanouchi U, Iwamoto M, Monna L, Hayasaka H, Katayose Y, Sasaki T (1999) The *Pib* gene for rice blast resistance belongs to the nucleotide binding and leucine-rich repeat class of plant disease resistance genes. Plant J 19:55–64

Warren RF, Henk A, Mowery P, Holub E, Innes RW (1998) A mutation within the leucine-rich repeat domain of the *Arabidopsis* disease resistance gene *RPS5* partially suppresses multiple bacterial and downy mildew resistance genes. Plant Cell 10:1439–1452

Webb CA, Richter TE, Collins NC, Nicolas M, Trick HN, Pryor T, Hulbert SH (2002) Genetic and molecular characterization of the maize *rp3* rust resistance locus. Genetics 162:381–394

Whitham S, Dinesh-Kumar SP, Choi D, Hehl R, Corr C, Baker B (1994) The product of the tobacco mosaic virus resistance gene *N*: similarity to toll and the interleukin-1 receptor. Cell 78:1101–1115

Whitham SA, Anderberg RJ, Chisholm ST, Carrington JC (2000) *Arabidopsis RTM2* gene is necessary for specific restriction of tobacco etch virus and encodes an unusual small heat shock-like protein. Plant Cell 12:569–582

Wilson WA, Harrington SE, Woodman WL, Lee M, Sorrells ME, McCouch SR (1999) Inferences on the genome structure of progenitor maize through comparative analysis of rice, maize and the domesticated panicoids. Genetics 153:453–473

Xiang Y, Cao Y, Xu C, Li X, Wang S (2006) *Xa3*, conferring resistance for rice bacterial blight and encoding a receptor kinase-like protein, is the same as *Xa26*. Theor Appl Genet 113:1347–1355

Xiao SY, Ellwood S, Calis O, Patrick E, Li TX, Coleman M, Turner JG (2001) Broad-spectrum mildew resistance in *Arabidopsis thaliana* mediated by *RPW8*. Science 291:118–120

Xing Y, Andreasen BS, Frei U, Lübberstedt T (2007) Development of ryegrass allele-specific markers (GRASP) for rust resistance in *Lolium perenne*. In: Perugia D, Rosselini D (eds) Proceedings of the 26th EUCARPIA fodder crops and amenity grasses section and 16th *Medicago* spp group joint meeting. Università degli Studi di Perugia/Facoltà di Agraria, Perugia, pp 373–377

Yahiaoui N, Srichumpa P, Dudler R, Keller B (2004) Genome analysis at different ploidy levels allows cloning of the powdery mildew resistance gene *Pm3b* from hexaploid wheat. Plant J 37:528–538

Yahiaoui N, Brunner S, Keller B (2006) Rapid generation of new powdery mildew resistance genes after wheat domestication. Plant J 47:85–98

Yang S, Feng Z, Zhang X, Jiang K, Jin X, Hang Y, Chen J-Q, Tian D (2006) Genome-wide investigation on the genetic variations of rice disease resistance genes. Plant Mol Biol 62:181–193

Yeam I, Kang B-C, Lindeman W, Frantz JD, Faber N, Jahn MM (2005) Allele specific CPAS markers based on point mutations in resistance alleles at the *pvr1* locus encoding eIF4E in *Capsicim*. Theor Appl Genet 112:178–186

Yoshimura S, Yamanouchi U, Katayose Y, Toki S, Wang ZX, Kono I, Kurata N, Yano M, Iwata N, Sasaki T (1998) Expression of *Xa1*, a bacterial blight-resistance gene in rice, is induced by bacterial inoculation. Proc Natl Acad Sci USA 95:1663–1668

Zhang Y, Zhang S, Qi L, Liu B, Xiong B, Gao J, Chen X, Chen C, Li X, Song W (2006) Cloning and Characterization of disease resistance gene analogs from poplar (*Populus tremula*) chromosome 1. Int J Plant Sci 167:403–412

Zhao B, Lin X, Poland J, Trick H, Leach J, Hulbert S (2005) A maize resistance gene functions against bacterial streak disease in rice. Proc Natl Acad Sci USA 102:15383–15388

Zheng X, Chen X, Zhang X, Lin Z, Shang J, Xu J, Zhai W, Zhu L (2004) Isolation and identification of a gene in response to rice blast disease in rice. Plant Mol Biol 54:99–109

Zhou B, Qu S, Liu G, Dolan M, Sakai H, Lu G, Bellizzi, Wang G-L (2006) The eight amino-acid differences within three leucine-rich repeats between Pi2 and Piz-t resistance proteins determine the resistance specificity to *Magnaporthe grisea*. Mol Plant Microbe Interact 19:1216–1228

Zhou FS, Kurth JC, Wei FS, Elliott C, Vale G, Yahiaoui N, Keller B, Somerville S, Wise R, Schulze-Lefert P (2001) Cell-autonomous expression of barley *Mla1* confers race-specific resistance to the powdery mildew fungus via a *Rar1*-independent signaling pathway. Plant Cell 13:337–350

Zhou JM, Loh YT, Bressan RA, Martin GB (1995) The tomato gene *Pti1* encodes a serine/threonine kinase that is phosphorylated by Pto and is involved in the hypersensitive response. Cell 83:925–935

Zhou PH, Tan YF, He YQ, Xu CG, Zhang Q (2003) Simultaneous improvement for four quality traits of Zhenshan 97, an elite parent of hybrid rice, by molecular markers-assisted selection. Theor Appl Genet 106:326–331

Zhou T, Wang Y, Chen JQ, Araki H, Jing Z, Jiang K, Shen J, Tian D (2004) Genome-wide identification of NBS genes in japonica rice reveals significant expansion of divergent non-TIR NBS-LRR genes. Mol Genet Genomics 271:402–415

Extranuclear Inheritance: Plastid–Nuclear Cooperation in Photosystem I Assembly in Photosynthetic Eukaryotes

Mark Aurel Schöttler and Ralph Bock(✉)

Contents

Abstract Photosystem I (PSI), the final complex of the photosynthetic electron transport chain, is composed of at least 15 protein subunits. PSI accumulation is tightly regulated with approximately constant PSI amounts being present under all environmental and developmental conditions. Only about one-third of the PSI subunits is encoded in the plastid genome, the other two-thirds are nucleus-encoded, made in the cytosol and post-translationally imported into the chloroplast. In higher plants, the nucleus-encoded subunits must be distributed to dozens or hundreds of chloroplasts per

Ralph Bock
Max-Planck-Institut für Molekulare Pflanzenphysiologie, Am Mühlenberg 1, D-14476 Potsdam-Golm, Germany
e-mail: rbock@mpimp-golm.mpg.de

U. Lüttge et al. (eds.), *Progress in Botany 69.*

cell and, as the demand for nucleus-encoded PSI proteins is likely to vary between different chloroplasts in a cell, this may require control of PSI biogenesis at the level of the individual plastid. In recent years, genetic work in *Chlamydomonas*, *Arabidopsis* and tobacco has identified protein factors specifically involved in the assembly of PSI complexes. Also, first insights have been gained into the regulatory mechanisms underlying PSI biogenesis in photosynthetic eukaryotes. The picture that emerges from these studies is that the availability of the plastid-encoded reaction center subunits PsaA, PsaB and PsaC limits PSI biogenesis. All plastid-encoded subunits are predominantly regulated at the translational level, while transcriptional control is of limited relevance. Besides translation initiation, also cofactor synthesis and/or insertion into the assembling complex and the activity of assembly chaperones may contribute to the regulation of PSI accumulation. The majority of the proteins involved in the regulation of translation initiation, cofactor provision and PSI assembly are nucleus-encoded, thus allowing the nucleus to exert coarse control over PSI biogenesis.

1 Introduction

Plant cells arose through the uptake of a free-living cyanobacterium by endosymbiosis and the gradual conversion of this photosynthetic prokaryote into DNA-containing organelles: the plastids (chloroplasts). Capture of the cyanobacterial endosymbiont was followed by a dramatic restructuring of the genomes of both the host and the symbiont. This involved the loss of redundant and dispensable genetic information and, most significantly, the massive translocation of genes from the prokaryotic genome of the endosymbiont to the eukaryotic genome in the nucleus of the host cell. As a consequence, contemporary chloroplast genomes are drastically reduced and contain only a small proportion of the genes that their free-living cyanobacterial ancestors once encoded. This gene transfer from the plastid to the nucleus took place over a time-scale of hundreds of millions of years; and both phylogenetic and experimental evidence suggests that it is still an ongoing process (Millen et al. 2001; Huang et al. 2003; Stegemann et al. 2003; Stegemann and Bock 2006; for reviews see Timmis et al. 2004; Bock 2006).

The intimate division of labor between two genetic compartments as implemented with endosymbiosis and gene transfer poses new challenges for regulation and coordination of gene expression. For example, only about one-third of the chloroplast ribosomal proteins are encoded in the plastid genome (plastome) of higher plants, the other two-thirds are encoded in the nuclear genome, synthesized on cytosolic ribosomes and post-translationally imported into the chloroplast. Likewise, none of the protein complexes involved in photosynthesis is composed entirely of plastid-encoded subunits. The two-subunit enzyme ribulose 1,5-bisphosphate carboxylase/oxygenase (Rubisco) provides the classic example for this intimate plastid–nuclear cooperation. In all flowering plants, the large subunit of Rubisco is encoded in the plastid genome, whereas the small subunit is encoded in the nucleus, typically by a small gene family. The situation is far more complicated for the multiprotein complexes involved in the light reactions of photosynthesis. They contain many more subunits, a large number of cofactors and, in addition, require sophisticated assembly systems to be put together in the thylakoid membrane. Moreover, the presence of their components in uncoordinated stoichiometries would pose a severe threat to the integrity of photosynthetic membranes (by provoking photooxidative damage and production of reactive oxygen species). Here, we review the current knowledge about plastid–nuclear cooperation in the synthesis and assembly of one of the largest pigment–protein complexes in nature: photosystem I.

2 Structure and Function of Higher Plant Photosystem I

2.1 Photosystem I Structure

Photosystem I (PSI) catalyzes the final step of photosynthetic electron transport. The photoactivated chlorophyll-a dimer in the PSI reaction center, P_{700}^{*}, donates an electron via phylloquinone and the three iron–sulfur clusters on the PSI acceptor side to ferredoxin. After its photooxidation, P_{700}^{+} functions as a strong oxidant, abstracting electrons from the components of the high-potential chain (i.e., cytochrome-f and the diffusible redox carrier protein plastocyanin or cytochrome c_6), which functions as a direct electron donor to PSI.

In angiosperms, the PSI reaction center complex consists of 15 subunits (Khrouchtchova et al. 2005), five of which (PsaA–PsaC, PsaI, PsaJ) are chloroplast-encoded (Jensen et al. 2003; Fig. 1). In addition to the 15 PSI subunits, there are five specific LHCI antenna proteins (Lhca1–Lhca5; Ganeteg et al. 2005; Lucinski et al. 2006) which bind to the complex and, in this way, increase the antenna cross-section of PSI. Under certain light conditions, the binding of additional LHCII antenna proteins normally attached to photosystem II (PSII) can further increase the PSI antenna cross-section, a process commonly referred to as "state transitions" (Lunde et al. 2000). Among the five chloroplast-encoded PSI subunits, there are three crucial subunits (PsaA, PsaB, PsaC) which bind all redox-active cofactors for electron flux through PSI. In addition to the chlorophyll-a dimer P_{700}, these include the chlorophyll acceptor A_0, the phylloquinone A_1 functioning as the first non-chlorophyll acceptor, and the three 4Fe-4S clusters F_X, F_A and F_B. The function of the other two plastome-encoded PSI subunits, PsaI and PsaJ, is less well understood. PsaI and PsaJ, both located at the periphery of the PSI complex, are not essential for PSI biogenesis and function, as evidenced by the lack of a strong phenotype in the corresponding knock-out mutants under standard growth conditions (Schöttler et al. 2007a: authors' unpublished data). Whereas the J subunit is involved in the binding and stabilization of the PSI antenna system (Schöttler et al. 2007a), the role of the I subunit is still totally unclear.

Likewise, the majority of the nucleus-encoded subunits seem to play non-essential roles for PSI function: T-DNA insertion lines and antisense or RNAi mutants often display only relatively mild phenotypes, usually comprising slight changes in PSI accumulation, stability and/or antenna function (Jensen et al. 2003). The sole exception is PsaD, the subunit forming the ferredoxin binding site, whose gene

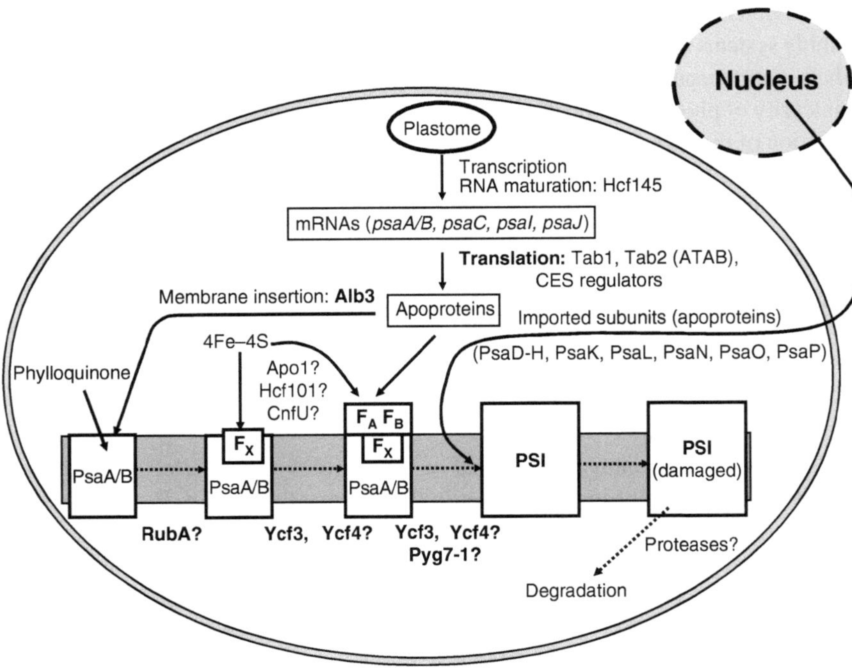

Fig. 1 Hypothetical model of PSI biogenesis. The factors that most likely determine rate-limiting steps in PSI biogenesis are indicated in *bold*. *Question marks* denote uncertain involvement or unclear mode of action (see text for details)

inactivation results in a strong PSI-deficient phenotype that is comparable with deletion of the genes for the A, B and C subunits (Haldrup et al. 2003; Ihnatowicz et al. 2004).

The chimeric nature of PSI, with some subunits being nucleus- and others plastid-encoded, poses the challenge of tightly coordinating gene expression and subunit synthesis between the chloroplasts and the nucleus. This represents a highly complicated problem because, depending on plant species and cell type, hundreds of chloroplasts can exist per cell. The turn-over of photosynthetic complexes can differ even between the chloroplasts in one and the same cell, so that the demand for nucleus-encoded subunits may be chloroplast-specific and variable. Recent evidence obtained for both the unicellular green alga *Chlamydomonas reinhardtii* and higher plants, especially *Arabidopsis thaliana* and tobacco (*Nicotiana tabacum*), strongly indicated that PSI biogenesis is controlled at the translational and post-translational levels in the individual chloroplasts. Therefore, this review predominantly focuses on these post-transcriptional steps and their contribution to the control of PSI biogenesis. An excellent recent review covering the transcription of PSI subunits and the post-transcriptional processing of plastid-encoded mRNAs was published by Rochaix (2006) and the interested reader is referred to it.

2.2 Photosystem I, Photosynthetic Flux Control and Environmental Acclimation

The role of PSI for photosynthetic flux control and the environmental acclimation of the photosynthetic apparatus is not yet fully resolved. Some reports suggest strong changes in PSI accumulation in response to changing light qualities and altered redox poise of the electron transport chain (Pfannschmidt et al. 1999; Allen and Pfannschmidt 2000; see Section 3 for a more detailed discussion). However, the vast majority of investigations on environmental acclimation of the photosynthetic apparatus revealed no significant alterations in PSI contents in response to changing environmental conditions. Neither light intensity or quality (Anderson et al. 1988; Murchie and Horton 1998; Bailey et al. 2001; Table 1), nor leaf age or sink to source balance (Schöttler et al. 2004, 2007b) had a pronounced effect on PSI contents. Interestingly, some differences between species were observed, with PSI contents varying across 1.5–2.5 mmol PSI mol^{-1} chlorophyll (chl.; Table 1). However, part of the reported interspecific variation in PSI contents may be attributable to the methods of PSI quantification applied: PSI quantitation by light-induced difference absorption signals of P_{700} (according to Hiyama and Ke 1972) tends to result in higher estimates of PSI contents than quantification based on difference absorption measurements by chemical oxidation and reduction of P_{700} (according to Whitmarsh and Ort 1984; Schöttler, unpublished data).

When determined in parallel to PSI, the other complexes of the photosynthetic apparatus display strong changes in contents in response to alterations in environmental or metabolic conditions. Especially for the cytochrome-bf complex, more than 5-fold changes in its abundance have been reported in response to changed light conditions (Table 1) or an altered metabolic state of the leaf (Schöttler et al. 2004). Cytochrome-bf complex contents varied across 0.25–1.5 mmol cytochrome-bf complex mol^{-1} chl. Stoichiometric adjustments of PSII and ATP synthase contents can be almost as pronounced as those of the cytochrome-bf complex. Therefore, the relative constancy of PSI contents under a variety of experimental conditions is a rather exceptional feature of the photosynthetic apparatus (Table 1).

Generally, PSI seems to accumulate to much higher levels than actually required to support electron flow. It is well established that linear electron flux is predominantly controlled at the level of the cytochrome-bf complex (Anderson 1992; Anderson et al. 1997), which catalyzes the slowest reaction of linear electron transport, the reoxidation of plastoquinole. This reaction has a half-time in the range of at least 2–3 ms (Haehnel 1984; Hope 2000). Although the amount of cytochrome-bf complex is the most important determinant of electron transport rates, plastocyanin diffusion between cytochrome-f and PSI may also contribute to flux control (Kirchhoff et al. 2004; Schöttler et al. 2004). In addition to accumulating to much higher amounts than the cytochrome-bf complex (with cytochrome-bf complex:PSI ratios typically varying between 1.0 and 0.1; Table 1), PSI also catalyzes reactions that are much faster than those catalyzed by the cytochrome-bf complex. Oxidation of plastocyanin by PSI requires less than 200 µs (Haehnel 1984; Hope 2000;

Table 1 Contents of photosynthetic complexes in different plant species and under different growth light conditions. Whereas the contents of PSII and cytochrome-bf complex (cyt-bf) display large changes depending on growth light intensities, the PSI contents (as determined by the amounts of the PSI reaction center chlorophyll P_{700}) show only small variations with light intensity. Between species, P_{700} contents can vary over 1.5–3.0 mmol mol^{-1} chlorophyll (chl.). Note that always, P_{700} contents are higher or at least as high as cyt-bf contents. *ND* No data

Plant species	Growth condition	PSII (mmol mol^{-1} chl.)	Cyt-bf (mmol mol^{-1} chl.)	P_{700} (mmol mol^{-1} chl.)	Cyt-bf/P_{700}	Reference
Arabidopsis thaliana	100 µE white light	3.0	ND	1.7	–	Bailey et al. (2001)
	600 µE white light	8.5	ND	1.9	–	
Pisum sativum	50 µE white light	1.5	0.7	1.9	0.27	Anderson et al. (1988)
	800 µE white light	5.1	2.1	2.1	0.72	
Nicotiana tabacum	50 µE low light	1.7	0.4	2.4	0.17	Schöttler et al. (2007b, unpublished data)
	500 µE high light	3.0	1.3	2.3	0.57	
Hordeum vulgare	50 µE white light	1.8	1.6	2.3	0.70	De la Torre and Burkey (1990)
	500 µE white light	3.0	2.3	2.3	1.00	
Plantago lanceolata	50 µE white light	2.0	0.4	1.3	0.31	Murchie and Horton (1998)
	50 µE FR-enriched	2.8	0.4	1.5	0.27	
	300 µE white light	3.0	1.1	1.5	0.73	
Digitalis purpurea	50 µE white light	1.3	0.4	1.5	0.27	
	50 µE FR-enriched	2.7	0.5	1.5	0.33	
	300 µE white light	2.7	1.5	1.5	1.00	
Brachypodium sylvaticum	50 µE white light	2.3	0.5	1.9	0.26	
	50 µE FR-enriched	2.4	0.6	1.5	0.40	
	300 µE white light	2.0	0.5	2.2	0.23	

Finazzi et al. 2005) and electron transfer through the PSI acceptor side to ferredoxin takes less than 1 ms (Setif 2001). Therefore, due to both its presence in higher contents and its much faster reaction halftimes, a considerable overcapacity of PSI, relative to linear electron flux, should exist in higher plants. These considerations render a contribution of PSI to photosynthetic flux control highly unlikely. Recent data on mutants which dispay reduced PSI accumulation, but yet have unaltered or even increased assimilation capacities, lend support to this conclusion (Schöttler et al. 2007a; Schöttler, Piepenburg, Flügel, Thiele and Bock, unpublished data).

In contrast, a strict adjustment of PSI activity to the metabolic demand for ATP and especially NADPH is absolutely essential for plant viability, as otherwise electron accumulation at the PSI acceptor side could result in high rates of production of reactive oxygen species, especially superoxide, hydroxyl radicals and hydrogen peroxide (H_2O_2). These reactive oxygen species are well known to initiate cell death programs so that, in addition to oxidative damage to the photosynthetic apparatus, the viability of the mesophyll tissue would be compromised by imbalances between PSI electron donation and carbon metabolism (Desikan et al. 2001; Laloi et al. 2006). Therefore, we suggest that, in higher plants, PSI capacity is not controlled by its contents, but rather by adjustments of electron flux towards PSI through changes in the contents of the cytochrome-bf complex and plastocyanin.

When electron flux towards PSI is restricted, P_{700} accumulates in its photochemically inactive oxidized state. Due to the relatively low redox potential of P_{700}^+ (+470 mV), this chlorophyll-a dimer radical is highly stable. It does not function as a strong and non-specific oxidant, as the oxidized PSII reaction center chlorophyll-a dimer P_{680}^+ does, with its far more positive redox potential (+1100 mV). P_{700}^+ dissipates excitation by internal conversion, thus functioning as a strong quencher of excitation energy harvested by the PSI antenna bed (Ort 2001).

The advantage of regulating electron flux towards PSI rather than PSI contents themselves may be that the cytochrome-bf complex comprises only eight subunits and far fewer cofactors than PSI, so that tuning at the level of this much smaller and simpler-structured complex might be favored. This is even more true for plastocyanin, which is only a single-subunit nucleus-encoded protein, and therefore would facilitate nuclear control of photosynthetic electron flux in a much more straightforward and economic way than adjustment of the PSI content.

While providing a rational for photosynthetic flux control not involving dynamic changes of PSI amounts, this model does not explain why PSI accumulates to higher amounts than theoretically required to sustain electron transport, and why the PSI contents are about constant under a variety of growth conditions. A potential explanation for the existence of these "surplus" amounts of PSI might lie in the requirement for a threshold level of PSI to initiate the lateral segregation of the thylakoid membrane as a prerequisite for grana stacking (Borodich et al. 2003; Chow et al. 2005). In this way, PSI contents may influence membrane structure in the chloroplast. Alternatively, high PSI contents could optimize photosynthetic electron flux by compensating for low-affinity plastocyanin binding.

To address the functional implications of the seemingly unnecessarily large PSI pool and its constancy under a variety of environmental conditions, a detailed understanding of the assembly pathway of PSI and the regulatory mechanisms governing PSI accumulation is of pivotal importance. For example, the identification of limiting steps in PSI biogenesis would facilitate the construction of mutants with deregulated PSI accumulation, which in turn would allow to study the functional consequences of increased or reduced PSI amounts without altering subunit composition of the complex.

3 Regulation of Genes for Photosystem I Subunits: Transcription and mRNA Maturation

Most nucleus-encoded genes for photosystem I proteins are believed to be predominantly or exclusively regulated at the level of transcription, although at least for some genes, translational control mechanisms also operate (Sherameti et al. 2002). Conflicting data exist on the significance of transcriptional regulation of the plastid-encoded PSI genes for the control of PSI biogenesis: For higher plants, Pfannschmidt et al. (1999) and Allen and Pfannschmidt (2000) reported pronounced changes in *psaA/psaB* transcript accumulation in response to changes in the redox poise of the electron transport chain. Using inhibitors of either plastoquinone reduction (DCMU) or plastoquinol oxidation (DBMIB), or light sources preferentially exciting either PSII or PSI, *psaA/psaB* transcription rates were correlated with the redox state of the plastoquinone pool. When the plastoquinone pool is highly reduced, transcription of the plastid-encoded PSI reaction center subunits is increased, whereas in response to reoxidation of the pool, transcription goes down again.

While a transcriptional upregulation of plastid PSI gene expression in response to overreduction of the plastoquinone pool would make physiological sense, the significance of the observed transcriptional changes is still not quite clear: Whereas Pfannschmidt et al. (1999) reported up to 2-fold changes in PSI contents paralleling the changes in *psaA/psaB* transcript abundance, other data do not support such a close correlation between transcription rates and PSI complex biogenesis in photosynthetic eukaryotes. In *Chlamydomonas*, no redox-related changes in PSI mRNA abundance and PSI contents have been observed at all (Stampacchia et al. 1997). Also, higher plant photosynthetic mutants with altered accumulation of the thylakoid protein complexes (which suffer from a strong disturbance of the chloroplast redox poise) do not display any significant changes in PSI accumulation (Anderson et al. 1997; Baena-Gonzalez et al. 2003; Allahverdiyeva et al. 2005; Schöttler et al. 2007b; authors' unpublished data).

A potential explanation for these discrepancies might be that Pfannschmidt et al. (1999) used young mustard seedlings early during photomorphogenesis. The capacity for acclimation could be lost once the photosynthetic apparatus is fully assembled, so that differences in the developmental state of the plants investigated might account for the discrepancy between different studies. However, in several of the photosynthetic mutants investigated, perturbations of chloroplast redox poise occurred already during the early steps of chloroplast development (Baena-Gonzalez et al. 2003; Allahverdiyeva et al. 2005; authors' unpublished data) and thus, the significance of changes in *psaA/psaB* transcript accumulation for PSI biogenesis remains questionable. In this context, it seems noteworthy that, in cyanobacteria, a strong correlation between PSI subunit transcription and PSI reaction center accumulation has been demonstrated to occur in response to changing environmental conditions or alterations in the cell's metabolic state (Li and Sherman 2001; Muramatsu and Hihara 2003; Herranen et al. 2005). However, different from higher plants, cyanobacteria have no choice but to regulate photosynthetic electron transport at the level of the photosystems. This is because all components of the intersystem chain, including the cytochrome-bf complex, plastocyanin and cytochrome c, are also used for respiratory electron transfer processes, which in cyanobacteria are not spatially separated from photosynthetic electron transport (Scherer 1990; Duran et al. 2004). Therefore, any adjustment of photosynthetic fluxes via regulation at the level of the mobile redox carriers or the cytochrome-bf complex would unavoidably also affect respiration. In contrast, compartmentalization of respiratory (mitochondria) and photosynthetic (chloroplast) electron transport processes in higher plants obviates the necessity to avoid a regulation at the level of the intersystem electron transport chain. This has allowed the evolution of alternative mechanisms of photosynthetic flux control in chloroplasts, which perhaps are more efficient than adjustments of photosystem stoichiometry. Thus, the occurrence of transcriptional responses to alterations in the redox poise of the chloroplast in higher plants (Pfannschmidt et al. 1999) might be an evolutionary relic of limited functional significance. In conclusion, there is currently no strong evidence for a significant contribution of transcriptional regulation of plastid PSI genes to the control of PSI biogenesis in photosynthetic eukaryotes.

Plastid primary transcripts undergo a series of RNA maturation steps, including 5′ and 3′ processing, intron splicing and, in most lineages of land plants, also RNA editing. Potentially, these mRNA processing mechanisms could also be employed to exert nuclear control over the expression of plastid genes. However, as far as plastid PSI genes are concerned, none of the RNA processing steps seems to be highly regulated or contributes to the control of PSI biogenesis. In some respects, mRNA maturation in higher plants is even less complex than in *Chlamydomonas*. For example, the *Chlamydomonas psaA* mRNA is transcribed as three separate exons, resulting in three distinct *psaA* precursor mRNAs. These three mRNAs are then joined together by two trans-splicing reactions, for which, in addition to a small chloroplast-encoded RNA, at least 14 different nucleus-encoded gene products are required (reviewed by Rochaix 2006). In contrast, *psaA* forms a contiguous reading frame in plastid genomes of higher plants and is cotranscribed with *psaB* (and the ribosomal protein gene *rps14*) to form a polycistronic mRNA.

Until now, mutant screens in higher plants have revealed only a single RNA maturation/RNA stability mutant (*hcf145*) that exhibits a specific deficiency in functional PSI. The nucleus-encoded HCF145 protein appears to be involved in stabilization and possibly also processing of the polycistronic *psaA-psaB-rps14* mRNA (Lezhneva and Meurer 2004). In its absence, the stability of the *psaA* region of the transcript is drastically reduced, resulting in reduction of *psaA/psaB* transcript abundance to approximately 10% of the wild-type level. Interestingly, the regions closer to the 3′-end of the polycistronic mRNA were less severely depleted, with *rps14* still accumulating to 25% of wild-type mRNA levels. As an immediate consequence of the strong reduction in mRNAs encoding the two PSI reaction center subunits PsaA and PsaB, PSI accumulation is also severely reduced in the *hcf145* mutant. The trace amounts of PSI that are still produced in the mutant do not allow for a sufficient reoxidation of the electron transport chain, which therefore suffers from severe over-reduction (Lezhneva and Meurer 2004).

4 Translational Control of PSI Biogenesis

In *Chlamydomonas reinhardtii*, the rate of translation initiation of several chloroplast-encoded subunits of the cytochrome-bf complex (Choquet et al. 2001) and PSII (Minai et al. 2006) turned out to be controlled by a mechanism of translational autoregulation termed "control by epistasy of synthesis" (CES). Recently, this CES mechanism was also demonstrated to control translation of several PSI subunits (Wostrikoff et al. 2004). Generally, the assembly of protein complexes in the thylakoid membrane is dependent on one or two master subunits (also referred to as "dominant subunits"), which integrate into the thylakoid membrane early in the assembly process and, in whose absence, other subunits downstream in the assembly process cannot assemble. While these unassembled subunits are often unstable and rapidly turned over, some of them can accumulate to low amounts in their unbound form. For several chloroplast-encoded subunits, it was shown that the presence of the unassembled subunit suppresses further translation of its own mRNA by negative feed-back regulation. This negative feed-back is mediated by a normally unexposed domain of the subunit, which if incorporated into the complex, is shielded and hidden inside the thylakoid membrane. However, if the protein is present in its unassembled from, the domain becomes exposed to the stroma and exerts its negative regulatory function by sequestering ternary factors essential for translation initiation of its own mRNA. It is important to note that, in theory, it is sufficient to regulate one key subunit of the complex very tightly, namely the master subunit (dominant subunit) which functions as the starter of the assembly process. If this dominant subunit is not available, all subunits downstream in the assembly process cannot accumulate, either because their translation is shut off (if they are regulated by CES) or because the unassembled proteins are highly unstable and condemned to immediate degradation.

The CES process has been elucidated in great detail for *petA*, encoding cytochrome-f, an essential subunit of the cytochrome-bf complex. Translation of *petA* is blocked in the absence of its assembly partners, subunit IV and cytochrome-b_6, the two dominant subunits of the complex. This is because, in the absence of subunit IV and cytochrome-b_6, the C-terminus of cytochrome-f is exposed to the stroma and inhibits translation of the *petA* transcript. The free C-terminus does not interact directly with the *petA* mRNA, but rather binds the ternary effector TCA1, a nucleus-encoded specific activator of *petA* translation initiation (Wostrikoff et al. 2001).

TCA1 acts as an essential translational activator protein by binding to the 5′ untranslated region (5′ UTR) of the *petA* mRNA and recruiting the ribosomes to the start codon. If TCA1 is sequestered away by the cytochrome-f C-terminal domain, no translation initiation can occur (Choquet and Vallon 2000). Control of translation initiation by such mRNA-specific nucleus-encoded ternary factors seems to be a general feature of plastid gene expression (Zerges 2000).

In the regulation of PSI biogenesis in *Chlamydomonas*, PsaB acts as the master subunit. In its absence, no significant translation of *psaA* mRNA occurs. The absence of PsaA protein, in turn, results in suppression of *psaC* translation (Wostrikoff et al. 2004). In terms of the CES mechanism, this means that, in the absence of PsaB, a ternary activator of *psaA* translation, which has not yet been identified, is bound to the unassembled PsaA proteins and thus is no longer available to bind to the 5′ UTR of the *psaA* mRNA. Consequently, translation initiation cannot occur and the *psaA* mRNA remains untranslated. In this way, translation of the *psaA* mRNA is autoregulated by unassembled PsaA protein in *Chlamydomonas* chloroplasts. An analogous CES mechanism controls translation initiation of the *psaC* mRNA: in the absence of PsaA protein, unassembled PsaC autoregulates its own translation, again by decreasing translation initiation (Wostrikoff et al. 2004). Whereas this CES mechanism is well established for PSI biogenesis in *Chlamydomonas*, experimental evidence for its existence in higher plants is still lacking.

In view of the crucial role of PsaB synthesis and membrane insertion for the overall rate of PSI biogenesis, it is of pivotal importance to identify mechanisms, by which the accumulation of PsaB in the thylakoids is controlled. The (most likely nucleus-encoded) protein that regulates the rate of PsaB translation initiation might represent some kind of master switch for the regulation of PSI biogenesis.

In *Chlamydomonas*, *psaB* translation initiation is controlled by the Tab1 and Tab2 proteins. Loss-of-function mutations in the nuclear genes for Tab1 and Tab2 are characterized by normal *psaB* transcript accumulation, but lack of *psaB* translation. The precise function of the *Tab1* gene has not yet been fully resolved. It encodes a large protein with five putative transmembrane helices at its N-terminus and a lipase-like domain at its C-terminus, which is essential for Tab1–mediated translation initiation (Rochaix 2006). Tab2 is a soluble, stromal RNA-binding protein, which binds to the 5′ UTR of the *psaB* transcript. As it does not contain any recognizable RNA-binding domain, the mechanism is not yet understood. The binding site for Tab2 appears to be formed by A/U-rich stretches in the *psaB* 5′ UTR. In addition to the *psaB* mRNA, also the 5′ UTRs of other plastid mRNAs with a low GC content are recognized by the Tab2 protein in vitro. For the molecular function of Tab2, two alternative models are discussed: Tab2 might either be involved in targeting of the *psaB* transcript to ribosomes associated with the thylakoid membrane to facilitate cotranslational membrane insertion of the PsaB protein, or act as a genuine and mRNA-specific translation initiation factor (Dauvillée et al. 2003).

Tab2 is ubiquitously present in all photosynthetic organisms, suggesting that it has evolved already in photosynthetic prokaryotes. Its orthologue in *A. thaliana*, ATAB2, was recently shown to be essential for *psaB* translation initiation in higher plant chloroplasts (Barneche et al. 2006). However, different from Tab2 in *Chlamydomonas*,

ATAB2 not only affects PSI biogenesis, but also seems to be involved in PSII biogenesis. In addition to complete loss of PSI, PSII contents were found to be strongly reduced in *atab2* T-DNA insertion lines, something not observed for other PSI mutants. Interestingly, *Atab2* gene expression is induced during photomorphogenesis by blue light in a cryptochrome-dependent manner. This finding links PSI biogenesis to light-dependent control of plant development and may indicate that the nuclear control of plastid mRNA translation initiation through the availability of translation initiation factors represents indeed the key regulatory step in photosystem biogenesis.

Another translational regulator of PSI accumulation could be the pentatricopeptide repeat (PPR) protein CRP1, which has been described as a translational activator essential for accumulation of the cytochrome-bf complex in maize. In its absence, the *petA* mRNA encoding the cytochrome-f subunit of the complex is not translated. Furthermore, in the *crp1* mutant, translation of the *psaC* mRNA is also severely reduced, resulting in a strongly diminished PSI accumulation (Barkan et al. 1994). CRP1 functions through specific interaction with the 5′ UTRs of both *petA* and *psaC* and binds in close proximity to the start codons of both mRNAs (Schmitz-Linneweber et al. 2005).

There is also evidence for translational regulation of several nucleus-encoded PSI subunits, at least in some higher plant species: Association of the *PsaD*, *PsaF* and *PsaL* mRNAs to cytosolic poly-ribosomes (polysomes) is strictly light-dependent in spinach seedlings (Sherameti et al. 2002). This light-dependent ribosome binding to mRNAs for nucleus-encoded PSI subunits can be suppressed by the application of inhibitors of photosynthetic electron transport, indicating that the light signal is transmitted via changes in the cellular redox poise after activation of photosynthetic electron transport. For *PsaD*, a light- and redox-responsive element could be identified in the first 30 nucleotides of the 5′ UTR of the *PsaD* mRNA. The mechanistic details of this light-regulated translation of PSI mRNAs in the cytosol have not yet been elucidated. Potentially, the cellular redox poise could modulate the binding of a translational activator to the 5′ UTR under reducing conditions or, alternatively, could result in dissociation of a translational repressor (Sherameti et al. 2002). Interestingly, in another higher plant, tobacco, no such regulation of *PsaD* mRNA translation by light and redox could be observed and therefore, the exact contribution of redox-regulated translation initiation in the cytosol to the synthesis of nucleus-encoded PSI subunits and thus to the regulation of PSI accumulation remains to be clarified.

5 Assembly Factors for Photosystem I

Following translation, the next step in PSI biogenesis is the insertion of the different membrane-intrinsic subunits into the thylakoids (Fig. 1). In the case of the two large reaction center proteins PsaA and PsaB, membrane insertion occurs cotranslationally. Further steps, occurring either in parallel to membrane insertion or afterwards, are the attachment of cofactors to the protein subunits and the assembly of the complex through specific subunit interactions. These steps of PSI biogenesis are supported by helper proteins, membrane insertases and chaperones, which mediate distinct steps of

PSI biogenesis (Fig. 1). The majority of these assembly factors were first identified in the unicellular green alga *Chlamydomonas* or in the cyanobacterium *Synechocystis*, but many of these auxiliary factors seem to be highly conserved between cyanobacteria, eukaryotic algae and higher plants. Normally, they are present at substoichiometric amounts relative to PSI, as expected for proteins required only to support assembly processes. A notable exception is Ycf4 (see below). The low accumulation levels and the transient nature of the interaction between the assembly factors and the PSI subunits and/or PSI assembly intermediates has greatly impaired the precise functional characterization of these helper proteins. The vast majority of putative assembly chaperones were identified only indirectly due to a severely reduced or even abolished accumulation of functional PSI complexes, while mRNAs for all PSI genes still accumulated and were faithfully translated (as evidenced, for example, by polysome-loading analyses). This is usually taken as a strong indication that a post-translational step in PSI biogenesis is impaired (i.e. complex assembly and/or stability). However, the same criteria are often also fulfilled by proteins involved in PSI cofactor synthesis or cofactor transfer to the assembling PSI reaction center proteins. This is because lack of an essential cofactor can also result in abolished PSI complex formation (see Section 6). Therefore, it can be notoriously difficult to faithfully distinguish between assembly chaperones and proteins involved in the provision of essential PSI cofactors. An additional and more stringent criterion for chaperones would be copurification or coimmunoprecipitation with PSI assembly intermediates, but this again is often hampered by the low binding affinities between PSI subunits and their chaperones, and until now, has only be successful for a single assembly factor (Ycf3; see below).

5.1 Albino3

Recently, Albino3 (Alb3) was identified as an interaction partner of the PSI-A subunit (PsaA). Alb3 is a plastid-localized homologue of the mitochondrial OxaI and the bacterial YidC proteins, which function as membrane insertases. Members of the Oxa1/Alb3/YidC protein family mediate both co- and post-translational insertion of transmembrane proteins or protein domains into bioenergetic membranes of prokaryotes, mitochondria and chloroplasts. In addition to membrane insertion, they may also block non-specific interactions of the newly inserted proteins with the hydrophobic transmembrane segments of other proteins. Oxa1/Alb3/YidC-like proteins are thus thought to function as chaperones for the very first assembly steps of large membrane-intrinsic protein complexes, including the ATP synthase and the respiratory complexes in bacteria and mitochondria (van der Laan et al. 2005). The thylakoid-intrinsic protein Alb3 functions in both cyanobacteria and plant thylakoids as a membrane insertase by mediating the cotranslational thylakoid insertion of plastid-encoded proteins. In addition, a function of Alb3 as a chaperone in the assembly of reaction center proteins of PSII has been reported (Ossenbühl et al. 2004). Whether or not Alb3 also plays a role in PSI biogenesis is still uncertain, but protein–protein interaction assays using the yeast split-ubiquitin system have identified PsaA as an interaction partner of Alb3 (Pasch et al. 2005). In *Chlamydomonas reinhardtii*, RNAi-mediated depletion of Alb3.2, one of two Alb3 isoforms in *Chlamydomonas* thylakoids, resulted in a pronounced reduction in the accumulation of both photosystems (Göhre et al. 2006), thus also providing further evidence for a function of Alb3 in PSI biogenesis, potentially as a mediator of membrane insertion of PsaA and PsaB. If confirmed, Alb3 would function in the very early steps of PSI assembly.

5.2 *Ycf3*

Ycf3 is, arguably, the most thoroughly analyzed and best understood chaperone-like factor in the PSI assembly process. Together with Ycf4, another candidate assembly factor for PSI (see below), it is plastome-encoded and therefore a bit more difficult to manipulate genetically. Before its function in the PSI assembly process was uncovered (Ruf et al. 1997; Boudreau et al. 1997), the *ycf3* gene was designated hypothetical *c*hloroplast reading *f*rame number *3*, in accordance with the nomenclature conventions for plastid genome-encoded open reading frames (ORFs). The Ycf3 protein turned out to be absolutely essential for PSI accumulation in cyanobacteria (Schwabe and Kruip 2000), *Chlamydomonas* (Boudreau et al. 1997) and higher plants (Ruf et al. 1997). The protein is characterized by the presence of three conserved tetratrico-peptide repeat (TPR) domains. A TPR domain is formed by two short amphipathic α-helices of altogether 34 amino acids length, and is known to mediate protein–protein interactions. In tobacco, analysis of chloroplast gene expression and translation confirmed that Ycf3 is involved neither in transcription, mRNA maturation nor translation of any of the plastid-encoded PSI subunits. Rather, the assembly of functional PSI reaction centers is blocked. A detailed analysis of Ycf3 function was conducted in *Chlamydomonas* (Naver et al. 2001). Biochemical studies revealed that Ycf3 is a stromal protein which is only weakly associated with thylakoids. Nonetheless, it can be coimmunoprecipitated after thylakoid solubilization with both the PsaA and PsaD subunits. In *Synechocystis*, evidence for homo-oligomerization of Ycf3 has been obtained, resulting in the formation of complexes as large as 260 kDa, which would be equivalent to an oligomer of approximately 13 subunits (Schwabe et al. 2003). Therefore, the TPR domains of Ycf3 could function either in Ycf3 oligomerization or in Ycf3 interaction with its assembly partners. In conclusion, Ycf3 seems to form, or at least be part of, an oligomeric protein complex that is loosely attached to the stromal side of the thylakoids and is likely involved in the assembly of the reducing side of PSI, especially the attachment of PsaD to the PsaA/B reaction center core.

5.3 *Ycf4*

The plastome-encoded Ycf4 protein has thus far been characterized only in *Chlamydomonas*, where a knock-out of the *ycf4* gene results in loss of functional PSI (Boudreau et al. 1997). Comparable with Ycf3, Ycf4 is localized on the stromal side of the thylakoid membrane, but it appears to be more stably associated with the membrane than Ycf3. In contrast to all other assembly factors/chaperones, Ycf4 is suggested to accumulate to almost equimolar amounts relative to PSI. Nonetheless, Ycf4 is not a bona fide PSI subunit, as it accumulates in PSI knock-out mutants. This circumstantially argues in favor of a chaperone function, as all true PSI subunits are highly unstable in the absence of their assembly partners and are present at best in very low amounts. Interestingly, in the cyanobacterium *Synechocystis*, Ycf4 is not essential for PSI biogenesis, as in its absence, low amounts of PSI can still accumulate (Schwabe et al. 2003).

5.4 *Pyg7-1*

Pyg 7-1 (pale yellow green 7) is a recently identified nucleus-encoded higher plant homologue of the cyanobacterial PSI assembly factor Ycf37 (Stöckel et al. 2006). In cyanobacteria, Ycf37 is not essential for PSI assembly, but instead, seems to have an accessory function by supporting the more efficient accumulation of PSI (Wilde et al. 2001). In contrast, Pyg7-1 is absolutely essential for PSI assembly in higher plants. In its absence, no PSI accumulates and the subunits of the complex, while being faithfully synthesized, are condemned to rapid degradation. The phenomenon that an assembly factor for PSI biogenesis is non-essential in cyanobacteria, but adopts an essential function in eukaryotic algae and higher plants is not unusual and, in addition to Ycf37/Pyg7-1, has also been observed for Ycf4 (Wilde et al. 1995; Boudreau et al. 1997) and Hcf101 (Stöckel et al. 2006).

The molecular function of Ycf37/Pyg7-1 in PSI biogenesis is not yet understood. In wild-type plants, the 27-kDa Pyg7-1 protein copurifies with the thylakoid membrane and, upon thylakoid fractionation, the highest amount is found in the PSI enriched fractions, indicating a relatively stable association of Pyg7-1 with PSI (Stöckel et al. 2006). Both Pyg7-1 and Ycf37 harbor three TPR domains, which were suggested to mediate the binding to PSI subunits (Wilde et al. 2001; Stöckel et al. 2006). In cyanobacteria, Ycf37 is additionally involved in PSI trimerization, as the accumulation of PSI trimers is reduced to 70% of wild-type levels in Δ*ycf37* mutants. Monomeric PSI complexes were observed exclusively in the mutant and do not normally occur in cyanobacteria (Düring et al. 2006). PSI trimerization does not occur in photosynthetic eukaryotes and therefore, the essentiality of Pyg7-1 for PSI assembly in *Arabidopsis* makes it clear that the protein must fulfill a different, trimerization-independent obligatory function in higher plants. As Ycf37 participates in a late step of cyanobacterial PSI biogenesis, it seems conceivable that Pyg7-1 also functions in one of the later steps of PSI biogenesis, after the reaction center core complex has been formed. An obvious possibility for such a late function in PSI assembly would be, for example, incorporation of the peripheral subunits into the complex. It is known that the loss of single peripheral subunits in higher plants does not result in strong PSI phenotypes. Only deletion of PsaA, PsaB, PsaC or PsaD leads to a loss of PSI accumulation that is comparable with the phenotype of the *pyg7-1* mutant (see Section 2.1). However, if Pyg7-1 was involved in the coordinated binding of several peripheral subunits, its absence might have a severe effect on PSI biogenesis.

6 Cofactor Biogenesis and Cofactor Attachment to Photosystem I

PSI biogenesis could also be controlled at the level of cofactor biosynthesis and insertion into PSI. PSI contains one unique cofactor, phylloquinone, and three 4Fe-4S clusters termed F_X, F_A and F_B (Fig. 1). As all known enzymes involved in cofactor biogenesis and incorporation into PSI are nucleus-encoded and PSI is highly unstable and rapidly degraded in the absence of either phylloquinone or the 4Fe-4S cofactors, a highly regulated production or insertion of cofactors could represent another mechanism of nuclear control over PSI biogenesis. In theory, regulation at the level of cofactor synthesis should be most efficient for the unique cofactor phylloquinone, whereas the 4Fe-4S clusters, which PSI shares with other plastid enzymes, could only be specifically regulated at the level of their insertion into the complex.

Recently, phylloquinone biosynthesis was elucidated in higher plants (Gross et al. 2006; Lohmann et al. 2006). Except for the final step of phylloquinone biosynthesis, all mutations in the biosynthetic pathway result in a PSI-deficient phenotype. Only the *menG* mutant, which is defective in the final step of phylloquinone synthesis, the methylation of 2-phythyl-1,4-naphtoquinone to phylloquinone, does not display a strong phenotype. In this mutant, 2-phytyl-1,4-naphtoquinone is incorporated into PSI, resulting in small changes of the cofactor's redox potential and steric orientation, which in turn increases the rate of oxidative damage to PSI under stress conditions, such as high light or chilling (Lohmann et al. 2006). However, there is no indication that phylloquinone availability could limit PSI biogenesis, as a considerable fraction of the phylloquinone pool (more than 30% of total phylloquinone) accumulates in plastoglobules, small lipoprotein vesicles attached to the thylakoids, and in the inner envelope membrane of the chloroplast. Only a maximum of 60% of phylloquinone seems to localize to the thylakoids, where it is almost exclusively associated with PSI (Lohmann et al. 2006). It thus seems reasonable to assume that the phylloquinone subpools in the plastoglobules and the inner envelope membrane provide a storage from which phylloquinone can be rapidly mobilized to PSI upon demand.

In contrast to phylloquinone biosynthesis, the biogenesis of the 4Fe-4S clusters in plastids is far less understood. In addition to PSI, at least seven other plastid enzymes, such as ferredoxin–thioredoxin oxidoreductase, nitrite reductase, sulfite reductase and adenosine 5′-phosphosulfate reductase, also require 4Fe-4S clusters (Balk and Lobréaux 2005). Furthermore, the early steps in the biogenesis of 2Fe-2S clusters share the same enzymatic machinery with 4Fe-4S biosynthesis. The biogenesis of Fe-S clusters takes place inside the chloroplast and recent evidence indicates that the plastid Fe-S assembly machinery is related to the bacterial *suf* system (standing for *s*ulfur *u*tilization *f*actor) and not to the *isc*-type regulon. The *isc* regulon (*i*ron *s*ulfur *c*luster assembly) encodes the predominantly active enzymatic machinery of Fe-S cluster synthesis in both prokaryotes and mitochondria. However, under conditions of oxidative stress or iron limitation, the *suf* regulon is additionally induced in bacteria. The Suf enzyme system appears to be more stable under oxidative conditions, which may explain why it evolved into the preferred system for Fe-S cluster synthesis in plastids, whereas in the less aerobic environment of the mitochondria, the Isc-type enzyme system is active (Balk and Lobréaux 2005).

The Suf system of chloroplasts utilizes the cysteine desulfurase SufS (also named NFS2, CpSufS or CpNifS in higher plants) to generate the organic sulfide required for Fe-S cluster synthesis by cleavage of cysteine (Ye et al. 2005; Xu and Moller 2006). The SufE (AtSufE in *Arabidopsis*) protein functions as an essential stromal activator of SufS and increases its substrate affinity to cysteine (Xu and Moller 2006). Interestingly, SufE over-expression results in deregulation of Fe-S cluster synthesis and a general increase in Fe-S cluster-containing proteins, including PsaC and the Rieske protein. This indicates that provision of the Fe-S clusters may be a highly regulated and limiting factor of photosynthetic complex biogenesis. It is noteworthy that, in addition to having elevated levels of cytochrome-bf complex and PSI, the *SufE* overexpressors also suffer from severe growth retardation and chlorosis, potentially due to elevated sulfur levels (Xu and Moller 2006).

The function of the other genes of the Suf system is less well understood. The sulfur mobilized from cysteine by the action of SufS and SufE could be transferred to the scaffold protein SufA, where the further steps of Fe-S cluster biogenesis might take place. The SufB homologue AtNAP1 (Xu et al. 2005) and the SufC homologue AtNAP7 (Xu and Moller 2004) seem to function together with SufD as part of a large multiprotein complex that provides the Fe^{3+} ions for Fe-S cluster biogenesis. SufC belongs to the ATP-binding cassette (ABC) proteins and seems to energize the whole Suf system via its ATPase activity, presumably by providing the free energy to release Fe^{3+} from the plastid iron storage protein ferritin. Whether or not synthesis of the complete 4Fe-4S clusters occurs outside the target apoproteins (e.g., the PSI reaction center subunits) is currently unknown. Alternatively, the final steps of cluster assembly could take place at the apoprotein. Solid experimental evidence supports the idea that the transfer of the nascent or the fully assembled clusters to some target proteins, including the PSI reaction center heterodimer, requires additional transfer proteins.

Studies conducted in the cyanobacterium *Synechocystis* PCC 6803 indicate that binding of F_X to the PsaA-PsaB heterodimer is a prerequisite for subsequent binding of PsaC to the reaction center core complex. However, PsaC also cannot bind as an apoprotein, but must bind its two 4Fe-4S clusters F_A and F_B prior to attachment as mature holoprotein to PsaA-PsaB (Jung et al. 1997; Yu et al. 1997). In the following section, proteins identified as candidates for the 4Fe-4S cluster insertion process into PSI are described in more detail. These proteins are not components of the standard bacterial Suf system and their proposed functions range from additional scaffolds for cluster assembly and transfer to roles in the adjustment of the redox state of the PSI acceptor side to facilitate cluster incorporation. Such additional proteins might be necessary to adjust the machinery for cluster synthesis to both the highly aerobic environment and the extreme redox potentials inside the chloroplast.

6.1 RubA

RubA belongs to the family of rubredoxins, small redox-active proteins involved in electron transport processes in sulfur-metabolizing bacteria and archaea. A stromal rubredoxin protein is essential for the biogenesis of the F_X cluster in *Synechocystis*, and, as it is highly conserved in eukaryotic algae and higher plants, it may serve a comparable function there (Shen et al. 2002a, b). In *rubA* deletion strains of *Synechocystis*, no functional PSI can assemble. Nevertheless, transient formation of an unstable PsaA-PsaB heterodimer binding both the chlorophylls and phylloquinone and being capable of charge separations in the reaction center, is possible in the mutants. However, all further assembly steps at the reducing side of PSI are blocked, and F_X is absent from the complex. As binding of F_X to the reaction center heterodimer is a prerequisite for the subsequent assembly of PsaC and PsaD into the PSI acceptor side, a function of RubA for F_X insertion into the heterodimer was suggested (Shen and Golbeck 2006). The current hypothesis for RubA function is that it keeps the strongly reducing PSI acceptor side, especially the phylloquinone A_1, oxidized. Otherwise, A_1 might reduce the assembling F_X and thereby impair the insertion into the reaction center.

6.2 CnfU

AtCnfU-V and AtCnfU-IVb are plastid-localized *Arabidopsis* homologues of the NifU protein from *Azotobacter vinelandii*, which is involved in the biogenesis of the 4Fe-4S cluster of the nitrogenase, but does not seem to be involved in the synthesis of other bacterial Fe-S clusters. NifU functions as a scaffold protein for the assembly and transfer of the Fe-S cluster. In higher plants, CnfU seems to be involved in the synthesis of both 2Fe-2S and 4Fe-4S clusters, as in a T-DNA insertion line lacking AtCnfU-V, the biogenesis of ferredoxin is impaired and PSI contents are also strongly reduced (Yabe et al. 2004). Interestingly, the biogenesis of other proteins containing Fe-S clusters, such as the Rieske protein, glutamate synthase and sulfite reductase, is not impaired, possibly indicating that, for the synthesis and transfer of Fe-S clusters to these proteins, different scaffolds such as SufA could be used (Touraine et al. 2004; Yabe et al. 2004). However, it is known that a severe antisense repression of ferredoxin to levels comparable with those observed in the CnfU insertion mutants can result in quantitative loss of PSI (Holtgreve et al. 2003), most likely due to electron accumulation at the PSI acceptor side and oxidative damage (PSI photoinhibition; see Section 7). Therefore, the reduced PSI contents in the AtCnfU-V mutants need not be a direct consequence of an impaired synthesis or insertion of the 4Fe-4S clusters, but could also arise as a secondary consequence of an impaired ferredoxin biogenesis. In conclusion, further analyses are required to resolve whether or not the CnfU proteins contribute directly to PSI biogenesis.

6.3 *Apo1*

Novel plant-specific proteins involved in Fe-S cluster synthesis have been identified in screens for photosynthetic mutants. One of them is the *a*ccumulation of *p*hotosystem *o*ne 1 mutant (Apo1; Amann et al. 2004). In contrast to all other proteins involved in Fe-S cluster biogenesis in chloroplasts, which have homologues also in cyanobacteria, Apo1 is one out of four members of a small gene family exclusively found in vascular plants. The other members of the Apo family either also localize to the chloroplast stroma (Apo2) or to mitochondria (Apo3, Apo4). Apo1 is essential for the biogenesis of 4Fe-4S clusters, whereas the accumulation of proteins containing 2Fe-2S clusters is not affected in the corresponding T-DNA insertion lines.

6.4 *Hcf101*

Hcf101 was identified by Stöckel and Oelmüller (2004) and Lezhneva et al. (2004) as a nucleus-encoded protein, whose knock-out results in a high chlorophyll-a fluorescence (HCF) phenotype. Electron transport in *hcf101* mutants is blocked due to the absence of functional PSI, resulting in elevated PSII fluorescence due to over-reduction of the electron transport chain. In *hcf101* knock-out mutants, PSI subunits are synthesized, but they do not assemble into a stable complex, indicating a function of the Hcf101 protein in a post-translational step of PSI biogenesis. In addition to PSI biogenesis, the accumulation of ferredoxin-thioredoxin reductase (FTR) is also strongly and specifically impaired, whereas accumulation of proteins containing 2Fe-2S clusters, such as the Rieske protein or ferredoxin, is not affected. This suggests an involvement of the stromal Hcf101 protein in the biogenesis of 4Fe-4S clusters (Lezhneva et al. 2004). In other PSI mutants, accumulation of FTR is not affected, ruling out that this effect could be a direct consequence of the loss of functional PSI. Hcf101 belongs to the ubiquitous family of P-loop ATPases. As is the case for Apo1, the exact function of Hcf101 in 4Fe-4S cluster biogenesis and its relationship to the Suf system is still unknown.

7 PSI Stability, Turnover and Degradation

In addition to the regulation of PSI biogenesis, the level of PSI accumulation could also be determined by complex stability and rate of degradation. Unfortunately, next to nothing is known about PSI stability and turnover in higher plants. Once the photosynthetic apparatus is assembled, the turnover rates of PSI seem to be very low, clearly differing from PSII, where the D1 protein of the reaction center has a high turnover rate and can be replaced under light stress conditions at least once per hour. Whereas, due to this high D1 turnover, PSII assembly and repair can be easily studied by pulse-chase experiments in algae and higher plants (Plücken et al. 2002; Ossenbühl et al. 2004), no such studies have yet been performed with PSI. This is because the rates of radiolabel incorporation are extremely low, with PsaA/PsaB normally being the only detectable PSI subunits, suggesting that PSI as a whole is highly stable and has a much lower turnover than PSII.

This scenario is further supported by the observation that, upon PSI photoinhibition by chilling stress, damaged PSI complexes disassemble and cannot efficiently be replaced by de novo synthesized PSI. Chilling stress conditions combined with only moderate light intensities damage the 4Fe-4S clusters at the PSI acceptor side, as low temperatures decrease both the rate of NADPH consumption by the Calvin cycle and the capacity of the antioxidative systems, whereas photosynthetic electron transport is less severely affected. This results in an over-reduction of the PSI acceptor side and increased generation of reactive oxygen species, which then can no longer be detoxified by the Mehler–Asada cycle. Depending on the capacity of the antioxidative system, even a few hours under chilling conditions can be sufficient to damage more than 50% of the PSI complexes. Damaged PSI complexes appear to disassemble and the reaction center subunits are subsequently degraded. Thus, PSI can only be replaced by de novo biogenesis of PSI (reviewed by Scheller and Haldrup 2005). In some plant species, such as *Arabidopsis* (Zhang and Scheller 2004), recovery of PSI requires several days, whereas in other species, such as cucumber, loss of PSI is irreversible and therefore recovery from PSI photoinhibition results in a readjustment of the whole photosynthetic apparatus to the remaining PSI contents, including a proportional degradation of PSII and antenna proteins (Kuduh and Sonoike 2002). In addition, we have observed leaf age-dependent differences in the capacity of PSI to recover from chilling stress in tobacco: Whereas young tobacco leaves can completely recover within a week after the end of chilling stress, mature leaves have lost the capacity for sufficient de novo synthesis of PSI and therefore suffer from strongly accelerated senescence (C. Flügel, M. Schöttler and R. Bock, unpublished data). Obviously, once the photosynthetic apparatus is assembled in young leaves, the capacity for PSI biogenesis declines sharply or may be even almost completely lost. The molecular basis for these ontogenetic and/or developmental changes in the capacity for PSI biogenesis is not yet understood.

7.1 BtpA

In *Synechocystis* PCC 6803, a screen for PSI mutants resulted in the identification of BtpA (Bartsevich and Pakrasi 1997). BtpA (*b*iogenesis of *t*hylakoid *p*roteins *A*) is an extrinsic thylakoid protein, facing the cytoplasm (Zak et al. 1999). A *btpA* knock-out strain suffers from severely reduced PSI contents, accumulating less than 20% of the wild-type level of PSI. As the transcript accumulation of PSI subunits was normal, Bartsevich and Pakrasi concluded that BtpA functions at a post-transcriptional level of PSI biogenesis. However, subsequent experiments by Zak and Pakrasi (2000) revealed a role for BtpA rather in the stabilization of existing

PSI than in PSI biogenesis, as a block of translation by chloramphenicol application resulted in a dramatically accelerated loss of PSI in the mutants, which could not be explained by a direct effect on PSI biogenesis. This effect was most pronounced under chilling conditions, resulting in the complete loss of PSI in Δ*btpA* strains within hours, whereas the wild-type *Synechocystis* strain was not affected. Another indication for BtpA acting as a PSI-stabilizing factor is the strong induction of this protein under stress conditions (Zak and Pakrasi 2000). A possible explanation of this protective function of BtpA for assembled PSI complexes might be its involvement in the activation or stabilization of antioxidative enzymes or their sequestration towards PSI under specific stress conditions, thereby reducing the oxidative damage that induces PSI degradation. Alternatively, BtpA could counteract specific PSI proteases and block their access to the complex.

7.2 PSI Proteases

Thus far, it has not been possible to identify proteases involved in PSI degradation. Therefore, nothing is known about PSI proteolysis and its regulation. Likewise, the conditions and mechanisms by which PSI may become prone to degradation are largely unknown. However, the appearance of distinct degradation intermediates for both PsaA and PsaB after chilling stress was observed in barley, possibly indicating a complex series of different proteolytic processing steps. For PsaB, as many as seven distinct degradation fragments could be resolved by immunoblotting (Tjus et al. 1999).

8 Outlook

Due to recent progress, we now have gained a rough overview over the different steps of PSI biogenesis (summarized in Fig. 1). The presently available evidence suggests that PSI accumulation is predominantly regulated at the level of translation initation of the plastid-encoded reaction center subunits PsaA, PsaB and PsaC, with PsaB synthesis representing the most important control point (at least in *Chlamydomonas*, where PsaB constitutes the dominant subunit of the CES process). In contrast, transcriptional regulation, the predominant level of regulation of gene expression in the nucleus, does not seem to contribute significantly to PSI gene regulation in plastids. Regulation at the level of cofactor synthesis and especially cofactor insertion into the assembling PSI complex cannot be ruled out yet. However, several important aspects of PSI biogenesis and turn-over still remain enigmatic and therefore represent important challenges for future research efforts. As the exact molecular functions of the assembly factors involved in PSI biogenesis are not yet understood, protein–protein interaction studies will be of utmost importance to: (a) reveal the mechanistic details of the assembly process and (b) identify further assembly chaperones and accessory factors for PSI biogenesis. Also, PSI stability and turnover rates in mature leaves, whose photosynthetic apparatus is fully established, warrant further analysis, including attempts to identify possible PSI-specific proteases and their regulation. A better understanding of these processes will be crucial to our picture of the integration of PSI biogenesis into the developmental program of the leaf, and the elucidation of the molecular basis of the leaf age-dependent differences in the capacity for PSI biogenesis (as suggested by the different capacities of PSI to recover from chilling stress-induced photoinhibition).

Furthermore, it will be necessary to dissect the influence of key endogenous and exogenous factors, like phytohormones, light intensity gradients and the metabolic state of the leaf, on the developmental changes in PSI biogenesis capacity during leaf ontogenesis. In this context, the interaction between chloroplasts and mitochondria will also be of high relevance, as revealed by the recent identification of a maize mutant with a defect in mitochondrial electron transport. The NCS6 mutant is defective in the mitochondrial cytochrome-c oxidase, which as a secondary consequence also results in strongly impaired PSI accumulation, whereas other photosynthetic complexes, such as PSII and the ATP synthase, are not affected by the mitochondrial mutation underlying the NCS6 phenotype (a cytochrome oxidase subunit 2 deletion; Jiao et al. 2005). The loss of PSI was attributed to a severe reduction in PsaC and the nucleus-encoded PsaD and PsaE subunits, whereas the PsaA-PsaB reaction center dimer seemed to accumulate at least transiently. Although the underlying mechanisms are not understood at all, these observations strongly indicate that the signaling network controlling the different steps of PSI biogenesis might be much more complex than currently recognized. In addition to intimate plastid–nuclear cooperations, plastid–mitochondrial cross-talk can also strongly affect PSI biogenesis, a discovery opening up a new direction for future research.

Acknowledgements Our work on photosystem I is supported the Max-Planck Society and by a grant from the Deutsche Forschungsgemeinschaft (SFB 429/Project A12).

References

Allahverdiyeva Y, Mamedov F, Mäenpää P, Vass I, Aro EM (2005) Modulation of photosynthetic electron transport in the absence of terminal electron acceptors: characterization of the rbcL deletion mutant of tobacco. Biochim Biophys Acta 1709:69–83

Allen JF, Pfannschmidt T (2000) Balancing the two photosystems: photosynthetic electron transport governs transcription of reaction center genes in chloroplasts. Philos Trans R Soc Lond B Biol Sci 355:1351–1359

Amann K, Lezhneva L, Wanner G, Hermann RG, Meurer J (2004) Accumulation of photosystem 1, a member of a novel gene family, is required for accumulation of [4Fe-4S] clsuer-containing chloroplast complexes and antenna proteins. Plant Cell 16:3084–3097

Anderson JM (1992) Cytochrome-bf complex: dynamic molecular organization, function and acclimation. Photosynth Res 34:341–357

Anderson JM, Chow WS, Goodchild DJ (1988) Thylakoid membrane organization in sun/shade acclimation. Aust J Plant Physiol 15:11–26

Anderson JM, Price GD, Chow WS, Hope AB, Badger MR (1997) Reduced levels of cytochrome bf complex in transgenic tobacco leads to a marked photochemical reduction of the plastoquinone pool, without significant change in acclimation to irradiance. Photosynth Res 53:215–227

Baena-Gonzalez E, Allahverdiyeva Y, Svab Z, Maliga P, Josse EM, Kuntz M, Mäenpää P, Aro EM (2003) Deletion of the tobacco plastid psbA gene triggers an upregulation of the thylakoid-associated NAD(P)H dehydrogenase complex and the plastid terminal oxidase (PTOX). Plant J 35:704–716

Bailey S, Walters RG, Jansson S, Horton P (2001) Acclimation of *Arabidopsis thaliana* to the light environment: the existence of separate low light and high light responses. Planta 213:794–801

Balk J, Lobréaux S (2005) Biogenesis of iron–sulfur proteins in plants. Trends Plant Sci 10:324–331

Barkan A, Walker M, Nolasco M, Johnson D (1994) A nuclear mutation in maize blocks the processing and translation of several chloroplast mRNAs and provides evidence for the differential translation of alternative mRNAs forms. EMBO J 13:3170–3181

Barneche F, Winter V, Crevecoeur M, Rochaix JD (2006) ATAB2 is a novel factor in the signalling pathway of light-controlled synthesis of photosystem proteins. EMBO J 25:5907–5918

Bartsevich VV, Pakrasi HB (1997) Molecular identification of a novel protein that regulates biogenesis of photosystem I, a membrane protein complex. J Biol Chem 272:6382–6387

Bock R (2006) Extranuclear inheritance: Gene transfer out of plastids. Prog Bot 67:75–98

Borodich A, Rojdestvenski I, Cottam M (2003) Lateral heterogeneity of photosystems in thylakoid membranes studied by Brownian dynamics simulations. Biophys J 85:774–789

Boudreau E, Takahashi Y, Lemieux C, Turmel M, Rochaix JD (1997) The chloroplast ycf3 and ycf4 open reading frames of *Chlamydomonas reinhardtii* are required for the accumulation of the photosystem I complex. EMBO J 16:6095–6104

Choquet Y, Vallon O (2000) Synthesis, assembly and degradation of thylakoid membrane proteins. Biochimie 82:615–634

Choquet Y, Wostrikoff K, Rimbault B, Zito F, Girard-Bascou J, Drapier D, Wollman FA (2001) Assembly-controlled regulation of chloroplast gene translation. Biochem Soc Trans 29:421–426

Chow WS, Hope AB (2004) Kinetics of reactions around the cytochrome bf complex studied in intact leaf discs. Photosynth Res 81:153–163

Chow WS, Kim EH, Horton P, Anderson JM (2005) Grana stacking of higher plant thylakoid membranes in higher plant chloroplasts: the physicochemical forces at work and the functional consequences that ensue. Photochem Photobiol Sci 4:1081–1090

Dauvillée D, Stampacchia O, Girard-Bascou J, Rochaix JD (2003) Tab2 is a novel conserved RNA binding protein required for translation of the chloroplast psaB mRNA. EMBO J 23:6378–6388

De la Torre W, Burkey KO (1990) Acclimation of barley to changes in light intensity: photosynthetic electron transport activity and components. Photosynth Res 24:127–136

Desikan R, Mackerness SAH, Hancock JT, Neill SJ (2001) Regulation of the *Arabidopsis* transcriptome by oxidative stress. Plant Phys 127:159–172

Duran RV, Hervas M, De la Rosa MA, Navarro JA (2004) The efficient functioning of photosynthesis and respiration in *Synechocystis* sp. PCCC 6803 strictly requires the presence of either cytochrom c_6 or plastocyanin. J Biol Chem 279:7229–7233

Düring U, Irrgang KD, Lünser K, Kehr J, Wilde A (2006) Analysis of photosynthetic complexes from a cyanobacterial cyf37 mutant. Biochim Biophys Acta 1757:3–11

Finazzi G, Sommer F, Hippler M (2005) Release of oxidized plastocyanin from photosystem I limits electron transfer between photosystem and cytochrome-b6f complex in vivo. Proc Natl Acad Sci USA 102:7031–7036

Ganeteg U, Klimmek F, Jansson S (2005) Lhca5 – an LHC-type protein associated with photosystem I. Plant Mol Biol 54:641–651

Göhre V, Ossenbühl F, Crevecoeur M, Eichacker LA, Rochaix JD (2006) One of two alb3 proteins is essential for the assembly of the photosystems and for cell survival in *Chlamydomonas*. Plant Cell 18:1454–1466

Gross J, Cho W, Lezhneva L, Falk J, Krupinska K, Shinozaki K, Seki M, Herrmann RG, Meurer J (2006) A plant locus essential for phylloquinone (vitamin K1) biosynthesis originated from a fusion of four eubacterial genes. J Biol Chem 281:17189–17196

Haehnel W (1984) Photosynthetic electron transport in higher plants. Annu Rev Plant Physiol 35:659–693

Haldrup A, Lunde C, Scheller HV (2003) Arabidopsis thaliana plants lacking the PSI-D subunit of photosystem I suffer severe photoinhibition, have unstable photosystem I complexes, and altered redox homeostasis in the chloroplast stroma. J Biol Chem 278:33276–33283

Herranen M, Tyystjarvi T, Aro EM (2005) Regulation of photosystem I reaction center genes in *Synechocystis* sp strain PCC 6803 during light acclimation. Plant Cell Physiol 46:1484–1493

Hiyama T, Ke B (1972) Difference spectra and extinction coefficients of P-700. Biochim Biophys Acta 267:14003–14010

Holtgrefe S, Bader KM, Horton P, Scheibe R, Schaewen A von, Backhausen JE (2003) Decreased content of leaf ferredoxin changes electron distribution and limits photosynthesis in transgenic potato plants. Plant Physol 133:1768–1778

Hope AB (2000) Electron transfers amongst cytochrome f, plastocyanin and photosystem I: kinetics and mechanisms. Biochim Biophys Acta 1456:5–26

Huang CY, Ayliffe MA, Timmis JN (2003) Direct measurement of the transfer rate of chloroplast DNA into the nucleus. Nature 422:72–76

Ihnatowicz A, Pesaresi P, Varotto C, Richly E, Schneider A, Jahns P, Salamini F, Leister D (2004) Mutants of photosystem I subunit D of *Arabidopsis thaliana*: effects on photosynthesis, photosystem I stability and expression of nuclear genes of chloroplast functions. Plant J 37:839–852

Jensen PE, Haldrup A, Rosgaard L, Scheller HV (2003) Molecular dissection of photosystem I in higher plants: topology, structure and function. Physiol Plant 119:313–321

Jiao S, Thornsberry JM, Elthon TE, Newton KJ (2005) Biochemical and molecular characterization of photosystem I deficiency in the NCS6 mutant of maize. Plant Mol Biol 57:303–313

Jung YS, Vassiliev IR, Yu J, McIntosh L Goldbeck JH (1997) Strains of Synechocystis sp. PCC 6803 with altered PsaC. Part II. EPR and optical spectroscopic properties of FA and FB in aspartate, serine and alanine replacements of cysteines 14 and 51. J Biol Chem 272:8040–8049

Khrouchtchova A, Hanson M, Paakkarinen V, Vainonen JP, Zhang S, Jensen PE, Scheller HV, Vener AV, Aro EM, Haldrup A (2005) A previously found thylakoid membrane protein of 14 kDa (TMP14) is a novel subunit of plant photosystem I and is designated PSI-P. FEBS Lett 579:4805–4812

Kirchhoff H, Schöttler MA, Maurer J, Weis E (2004) Plastocyanin redox kinetics in spinach chloroplasts: evidence for disequilibrium in the high potential chain. Biochim Biophys Acta 1659:63–72

Kuduh H, Sonoike K (2002) Irreversible damage to photosystem I by chilling in the light: cause of the degradation of chlorophyll after returning to normal growth temperature. Planta 215:541–548

Laloi C, Przybyla D, Apel K (2006) A genetic approach towards elucidating the biological activity of different reactive oxygen species in *Arabidopsis thaliana*. J Exp Bot 567:1719–1724

Lezhneva L, Meurer J (2004) The nuclear factor HCF145 affects chloroplast psaA-psaB-rps14 transcript abundance in *Arabidopsis thaliana*. Plant J 38:740–753

Lezhneva L, Amann K, Meurer J (2004) The universally conserved HCF101 protein is involved in the assembly of [4Fe-4S]-cluster containing complexes in *Arabidopsis thaliana* chloroplasts. Plant J 37:174–185

Li H, Sherman LA (2001) A redox-responsive regulator of photosynthesis gene expression in the cyanobacterium *Synechocystis* sp. strain PCC 6803. J Bacteriology 182:4268–4277

Lohmann A, Schöttler MA, Bréhélin C, Kessler F, Bock R, Cahoon EB, Dörmann P (2006) Deficiency in phylloquinone (vitamin K1) methylation affects prenyl quinine distribution, photosystem I abundance, and anthocyanin accumulation in the *Arabidopsis* menG mutant. J Biol Chem 281:40461–40472

Lucinski R, Schmid VH, Jansson S, Klimmek F (2006) Lhca5 interaction with plant photosystem I. FEBS Lett 580:6485–6488
Lunde C, Jensen PE, Haldrup A, Knoetzel J, Scheller HV (2000) The PSI-H subunit of photosystem I is essential for state transitions in plant photosynthesis. Nature 408:6153–6158
Millen RS, Olmstead RG, Adams KL, Palmer JD, Lao NT, Heggie L, Kavanagh TA, Hibberd JM, Gray JC, Morden CW, Calie PJ, Jermiin LS, Wolfe KH (2001) Many parallel losses of infA from chloroplast DNA during angiosperm evolution with multiple independent transfers to the nucleus. Plant Cell 13:645–658
Minai L, Wostrikoff K, Wollman FA, Choquet Y (2006) Chloroplast biogenesis of photosystem II cores involves a series of assembly-controlled steps that regulate translation. Plant Cell 18:159–175
Muramatsu M, Hihara Y (2003) Transcriptional regulation of genes encoding subunits of photosystem I during acclimation to high light conditions in *Synechocystis* sp. PC6803. Planta 216: 446–453
Murchie EH, Horton P (1998) Contrasting patterns of photosynthetic acclimation to the light environment are dependent on the differential expression of the responses to altered irradiance and spectral quality. Plant Cell Environ 21:139–148
Naver H, Boudreau E, Rochaix JD (2001) Functional studies of Ycf3: its role in assembly of photosystem I and interactions with some of its subunits. Plant Cell 13:2731–2745
Ort DR (2001) When there is too much light. Plant Physiol 125:29–32
Ossenbühl F, Göhre V, Meurer J, Krieger-Liszkay A, Rochaix JD, Eichacker LA (2004) Efficient assembly of photosystem II in *Chlamydomonas reinhardtii* requires Alb 3.1p, a homolog of *Arabidopsis* Albino3. Plant Cell 16:1790–1800
Pasch JC, Nickelsen J, Schünemann D (2005) The yeast split-ubiquitin system to study chloroplast membrane protein interactions. Appl Microbiol Biotechnol 69:440–447
Pfannschmidt T, Nillson A, Allen JF (1999) Photosynthetic control of chloroplast gene expression. Nature 397:625–628
Pilon M, Abdel-Ghany SE, Hoewyk D van, Ye H, Pilon-Smits EA (2006) Biogenesis of iro-sulfur cluster proteins in plastids. Genet Eng 27:101–117
Plücken H, Müller B, Grohmann D, Westhoff P, Eichacker LA (2002) The HCF136 protein is essential for assembly of the photosystem II reaction center in *Arabidopsis thaliana*. FEBS Lett 532:85–90
Rochaix JD (2006) Genetic dissection of photosystem I assembly and turnover in eukaryotes. In: Golbeck JH (ed) Photosystem I. The light-driven plastocyanin:ferredoxin oxidoreductase. (Advances in photosynthesis and respiration, vol 24), Springer, Berlin Heidelberg New York, pp 515–527
Ruf S, Kössel H, Bock R (1997) Targeted inactivation of a tobacco intro-containing open reading frame reveals a novel chloroplast-encoded photosystem I-related gene. J Cell Biol 139:95–102
Savitch LV, Barker-Astrom J, Ivanov AG, Hurry V, Öquist V, Huner NPA, Gardestrom P (2001) Cold acclimation of *Arabidopsis thaliana* results in incomplete recovery of photosynthetic capacity, associated with an increased reduction of the chloroplast stroma. Planta 214:295–303
Scheller HV, Haldrup A (2005) Photoinhibition of photosystem I. Planta 221:5–8
Scherer S (1990) Do photosynthetic and respiratory electron transport chains share redox proteins? Trends Biol Sci 15:458–462
Schmitz-Linneweber C, Williams-Carrier R, Barkan A (2005) RNA immunoprecipitation and microarray analysis show a chloroplast pentatricopeptide repeat protein to be associated with the 5′-region of mRNAs whose translation it activates. Plant Cell 17:2791–2804
Schöttler MA, Kirchhoff H, Weis E (2004) The role of plastocyanin in the adjustment of the photosynthetic electron transport to the carbon metabolism in tobacco. Plant Physiol 136:4265–4274
Schöttler MA, Flügel C, Thiele W, Stegemann S, Bock R (2007a) The plastome-encoded PsaJ subunit is required for efficient photosystem I excitation, but not for plastocyanin oxidation in tobacco. Biochem J (in press)

Schöttler MA, Flügel C, Thiele W, Bock R (2007b) Knock-out of the plastid-encoded PetL subunit results in reduced stability and accelerated leaf-age dependent loss of the cytochrome b6f complex. J Biol Chem 282:976–985
Schwabe T, Kruip J (2000) Biogenesis and assembly of photosystem I. Indian J Biochem Biol 37:351–359
Schwabe TM, Gloddek K, Schluesener D, Kruip J (2003) Purification of recombinant BtpA and Ycf3, proteins involved in membrane protein biogenesis in *Synechocystis* PCC 6803. J Chromatogr B Anal Technol Biomed Life Sci 786:45–59
Setif P (2001) Ferredoxin and flavoredoxin reduction by photosystem I. Biochim Biophys Acta 1507:161–179
Shen G, Golbeck JH (2006) Assembly of the bound iron-sulfur clusters in photosystem I. In: Golbeck JH (ed) Photosystem I. The light-driven plastocyanin:ferredoxin oxidoreductase. (Advances in photosynthesis and respiration, vol 24) Springer, Berlin Heidelberg New York, pp 529–547
Shen G, Zhao J, Reimer SK, Antonkine ML, Cai Q, Weiland SM, Golbeck JH, Bryant DA (2002a) Assembly of photosystem I. I. Inactivation of the rubA gene encoding a membrane-associated rubredoxin in the cyanobacterium Synechococcus sp. PCC7002 causes a loss of photosystem I activity. J Biol Chem 277:20343–20354
Shen G, Antonkine ML, van der Est A, Vassiliev IR, Brettel K, Bittl R, Zech SG, Zhao J, Stehlik D, Bryant DA, Golbeck JH (2002b) Assembly of photosystem I. II. Rubredoxin is required for the in vivo assembly of F(X) in *Synechococcus* sp. PCC7002 as shown by optical and EPR spectroscopy. J Biol Chem 277:20355–20366
Sherameti I, Nakamura M, Yamamoto YY, Pfannschmidt T, Obokata J, Oelmüller R (2002) Polyribosome loading of spinach mRNAs for photosystem I subunits is controlled by photosynthetic electron transport. Plant J 32:631–639
Stampacchia O, Girard-Bascou J, Zanasco JL, Zerges W, Bennoun P, Rochaix JD (1997) A nuclear-encoded function essential for translation of the chloroplast *psaB* mRNA in *Chlamydomonas*. Plant Cell 9:773–782
Stegemann S, Bock R (2006) Experimental reconstruction of functional gene transfer from the tobacco plastid genome to the nucleus. Plant Cell 18:2869–2878
Stegemann S, Hartmann S, Ruf S, Bock R (2003) High-frequency gene transfer from the chloroplast genome to the nucleus. Proc Natl Acad Sci USA 100:8828–8833
Stöckel J, Oelmüller (2004) A novel protein for photosystem I biogenesis. J Biol Chem 279:10243–10251
Stöckel J, Bennewitz S, Hein P, Oelmüller R (2006) The evolutionary conserved tetratrico peptide repeat protein Pale Yellow Green7 is required for photosystem I accumulation in *Arabidopsis* and copurifies with the complex. Plant Physiol 141:870–878
Timmis JN, Ayliffe MA, Huang CY, Martin W (2004) Endosymbiotic gene transfer: organelle genomes forge eukaryotic chromosomes. Nat Rev Genet 5:123–136
Tjus SE, Lindberg Moller B, Scheller HV (1999) Photoinhibition of photosystem I damages both reaction centre proteins PSI-A and PSI-B and acceptor-side located small photosystem I polypeptides. Photosynth Res 60:75–86
Touraine B, Boutin JP, Marion-Poll A, Briat JF, Peltier G, Lobréaux S (2004) Nfu2: a scaffold protein required for [4Fe-4S] and ferredoxin iron–sulphur cluster assembly in *Arabidopsis thaliana*. Plant J 40:101–111
Van der Laan M, Nouven NP, Driessen AJM (2005) YidC – an evolutionary conserved device for the assembly of energy-transducing membrane protein complexes. Curr Opin Microbiol 8:182–187
Whitmarsh J, Ort DR (1984) Stoichiometries of electron transport complexes in spinach chloroplasts. Arch Biochem Biophysiol 231:378–389
Wilde A, Hartel H, Hübschmann T, Hoffmann P, Shestakov SV, Börner T (1995) Inactivation of Synechocystis sp strain PCC-6803 gene with homology to conserved chloroplast open reading frame-184 increases the photosystem-II to photosystem-I ratio. Plant Cell 7:649–658

Wilde A, Lunser K, Ossenbühl F, Nickelsen J, Börner T (2001) Characterization of cyanobacterial Ycf37: mutation decreases the photosystem I content. Biochem J 357:211–216

Wostrikoff K, Choquet Y, Wollman FA, Girard-Bascou J (2001) TCA1, a single nuclear-encoded translational activator specific for *petA* mRNA in *Chlamydomonas reinhartii* chloroplast. Genetics 159:119–132

Wostrikoff K, Girard-Bascou J, Wollman FA, Choquet Y (2004) Biogenesis of PSI involves a cascade of translational autoregulation in the chloroplast of *Chlamydomonas*. EMBO J 23:2696–2705

Xu XM, Moller SG (2004) AtNAP7 is a plastidic SufC-like ATP-binding cassette/ATPase essential for *Arabidopsis* embryogenesis. Proc Natl Acad Sci USA 24:9143–9148

Xu XM, Moller SG (2006) AtSufE is an essential activator of plastidic and mitochondrial desulfurases in *Arabidopsis*. EMBO J 25:900–909

Xu XM, Adams S, Chua NH, Moller SG (2005) At NAP1 represents an atypical SufB protein in *Arabidopsis* plastids. J Biol Chem 280:6648–6654

Yabe T, Morimoto K, Kikuchi S, Nishio K, Terashima I, Nakai M (2004) The Arabidopsis chloroplastic NifU-like protein CnfU, which can act as an iron–sulfur cluster scaffold protein, is required for biogenesis of ferredoxin and photosystem I. Plant Cell 16:993–1007

Ye H, Garifullina GF, Abdel-Ghany SE, Zhang L, Pilon-Smits EAH, Pilon M (2005) The chloroplast NifS-like protein of *Arabidopsis thaliana* is required for iron–sulfur cluster formation in ferredoxin. Planta 220:602–608

Yu J, Smart LB, Young YS, Golbeck J, McIntosh L (1995) Absence of PsaC subunit allows assembly of photosystem I core but prevents the binding of PsaD and PsaE in Synechocystis sp. PCC 6803. Plant Mol Biol 29:331–342

Yu J, Vassiliev IR, Jung YS, Golbeck JH, McIntosh L (1997) Strains of Synechocystis sp. PCC 6803 with altered PsaC. Part I. Mutations incorporated in the cysteine ligands of the two (4Fe-4S)-clusters F_A and F_B of photosystem I. J Biol Chem 272:8032–8039

Zak A, Pakrasi HB (2000) The BtpA protein stabilizes the reaction center proteins of photosystem I in the cyanobacterium *Synechocystis* sp. PCC 6803 at low temperatures. Plant Physiol 123:215–222

Zak E, Norling B, Andersson B, Pakrasi HB (1999) Subcellular localisation of the BtpA protein in the cyanobacterium *Synechocystis* sp. PCC 6803. Eur J Biochem 261:311–316

Zerges W (2000) Translation in chloroplasts. Biochemie 82:583–601

Zhang S, Scheller HV (2004) Photoinhibition of photosystem I at chilling temperature and subsequent recovery in *Arabidopsis thaliana*. Plant Cell Physiol 45:1595–1602

Molecular Cell Biology: Are Reactive Oxygen Species Regulators of Leaf Senescence?

Ulrike Zentgraf(✉) and Vera Hemleben

Contents

Abstract Senescence processes can influence many important agricultural traits; however, our knowledge concerning regulatory mechanisms controlling senescence is still limited. Free radicals are thought to play an essential role in senescence, especially those derived from oxygen. In addition to their deleterious functions, they might serve as signalling molecules. The critical balance between production and scavenging of reactive oxygen species (ROS), which normally is very tightly regulated, appears to be specifically disrupted during the progression of senescence in different cellular compartments either by depletion of antioxidants or excess production of ROS. Hydrogen peroxide (H_2O_2) is very likely the most important ROS. In contrast to other ROS, it has a relatively long half-life and can also pass membranes; therefore, it can be assumed that it executes signalling functions. Hydrogen peroxide is produced in different cell compartments but can also be released into the cytosol or vice versa. The role of ROS originating from different cellular compartments like chloroplasts, peroxisomes or mitochondria is discussed here with respect to senescence.

Ulrike Zentgraf
ZMBP, General Genetics, University of Tübingen, Auf der Morgenstelle 28, D-72076 Tübingen, Germany
e-mail: ulrike.zentgraf@zmbp.uni-tuebingen.de

1 Introduction

With regard to global climate changes, one of our future challenges will be to develop crop plants that cope better with changing environmental conditions. Abiotic stress is estimated to be the primary cause of crop loss worldwide, with the potential to cause a reduction of more than 50% in the average yield of the main crops. Climatic extremes are known to trigger senescence processes. Many different agriculturally important traits are affected by senescence, like number and quality of seeds, timing of seed set, fruit ripening, etc.. Despite the importance of the senescence processes, our knowledge on the regulatory mechanisms of senescence is still poor. However, senescence is not a chaotic breakdown, but an orderly loss of normal cell functions. In contrast to aging processes which have a passive and non-regulated degenerative character (for a review, see Krupinska et al. 2003), senescence is an active and highly regulated process. Senescence can be initiated by exogenous and endogenous triggers. The most important endogenous factors inducing senescence are the age of the leaves and the age and developmental stage of the plant. The leaves of annual plants show a continuous decrease in their photosynthesis rate after full expansion (Batt and Woolhause 1975; Hensel et al. 1993). In fast-aging plants like *Arabidopsis*, photosynthetic capacity of the leaves decreases by 50% within 4–6 days of full leaf expansion under continuous light conditions (Hensel et al. 1993). The endogenous signals that are involved in the onset of leaf senescence are still under debate. A decline in photosynthetic activity under a certain threshold may act as a senescence-inducing signal (Matile et al. 1992; Smart 1994). However, autumn senescence in free-growing aspen (*Populus tremula*) is apparently initiated solely by the photoperiod, and progresses steadily without any obvious influence of other environmental signals (Keskitalo et al. 2005).

It is also discussed whether leaf senescence is induced by sugar starvation or by sugar accumulation. In contrast to a possible role of sugar starvation in the senescence of petals (van Doorn 2004), an external supply of sugars has been shown to induce symptoms of senescence, such as leaf yellowing. In tobacco, sugars accumulate during early senescence, but decline during later stages (Masclaux et al. 2000). Leaf sugar contents have been shown to increase during senescence in several plant species, e.g. tobacco, soybean, wheat, maize, castor bean (Noodén et al. 1997; Wingler et al. 1998; Jongebloed et al. 2004), and *Arabidopsis* (Quirino et al. 2001; Stessman et al. 2002; Diaz et al. 2005). In contrast to sugars, total amino acid content declines throughout leaf development. Changes in the sugar relative to nitrogen content during the sink/source transition of leaves could thus play a role in the induction of leaf senescence and thereby stimulate nitrogen remobilisation from the old leaves (Masclaux et al. 2000; Pourtau et al. 2004; Wingler et al. 2004). Glucose and fructose accumulate strongly during senescence in *Arabidopsis* leaves, indicating that senescing leaves are not sugar-starved. The observed hexose accumulation could result from phloem blockage (Jongebloed et al. 2004) or from nitrogen starvation (Pourtau et al. 2004). Recent microarray analyses revealed that the senescence-associated gene *SAG12*, which was previously thought to be sugar-repressible, was induced over 900-fold by glucose. Induction of *SAG12*, which is expressed during late senescence, demonstrates that processes characteristic for late stages are sugar-inducible (Pourtau et al. 2006).

However, besides sugar many other signals trigger developmental senescence, including more or less all phytohormones. Cytokines are generally regarded as inhibitors of senescence (Van Staden et al. 1988; Gan and Amasino 1995). Ethylene has been shown to play an essential role in the timing of leaf senescence in *Arabidopsis* rosette leaves but it is neither necessary nor sufficient as a senescence-promoting signal. The ethylene-insensitive mutant *etr1-1* of *Arabidopsis* and antisense 1-amino-cyclopropane-1-carboxylic acid (ACC)-oxidase tomato plants indicate that mainly age-dependent factors are involved in the regulation of leaf senescence (Grbic and Bleecker 1995; John et al. 1995). Recently, expression profiles of senescing leaves of *Arabidopsis* lines defective in signalling pathways involving salicylic acid (SA), jasmonic acid (JA) and ethylene have shown that all three pathways are required for expression of many genes during developmental senescence (Buchanan-Wollaston et al. 2005).

Brassinosteroids also have a leaf senescence-promoting role (He et al. 2001), as has JA. It was found that several genes involved in JA biosynthesis are upregulated during leaf senescence in *Arabidopsis* and that the JA level in senescing *Arabidopsis* leaves increases (He et al. 2001). Since JA induces the expression of several senescence-enhanced genes (He et al. 2002; Buchanan-Wollaston et al. 2005), JA appears to be an important senescence-promoting factor. Salicylic acid, which is mainly involved in pathogen defence, has also been shown to be required for expression of some SAGs (He et al. 2002; Buchanan-Wollaston et al. 2005). Abscisic acid (ABA) is implicated in the regulation of stress-induced senescence and the expression of several SAGs can be induced upon treatment with ABA (BuchananWollaston et al. 2005). By contrast, auxin and gibberellic acid have a negative effect on leaf senescence. Recently, the involvement of auxin responsive transcription factors in senescence regulation was reported (Ellis et al. 2005). However, overall senescence is controlled most likely by the interaction of all hormones in specific concentrations acting in synergistic or antagonistic ways and in concert with other signals like sugar, calcium and free radicals.

2 Oxygen Free Radicals and Life Span

There are several lines of evidence that reactive oxygen species (ROS) trigger leaf senescence. Oxidative stress arises from an imbalance between generation and elimination of ROS, often leading to cell death. Oxidative stress occurs when this critical balance is disrupted because of depletion of antioxidants or excess accumulation of ROS. Regardless of how or where they are generated, an increase in intracellular oxidants results in two very important effects: damage to various cell components and activation of specific signalling pathways, both of which influence numerous cellular processes (Fig. 1). However, oxidative damage in plant tissues is

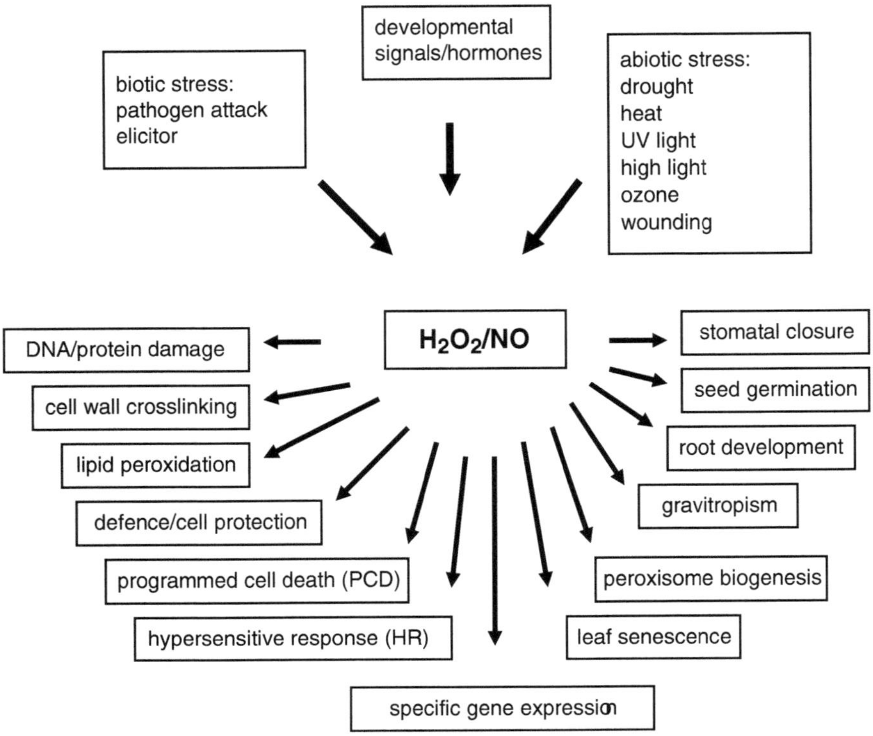

Fig. 1 Involvement of H_2O_2 (and NO) in cellular responses to various stimuli, modified after Desikan et al. (2004) and Neill et al. (2002)

especially important during senescence and is characterised by a notable increase in the metabolism of activated oxygen species (Kar and Feierabend 1984; Thompson et al. 1987; Halliwell and Gutteridge 1989). For the mediterranean plant *Cistus clusii*, it could be shown that oxidative stress associated with the aging in plants accumulates progressively in chloroplasts (Munne-Bosch and Alegre 2002).

Obviously, oxidative stress resistance and potential life span seem to be correlated in many organisms ranging from *Caenorhabditis* to mammals (Harman 1956, 1998; Orr and Sohal 1994; Martin et al. 1996; Sohal and Weindruch 1996). The loss of antioxidative capacity during the progression of senescence has also been reported for many different plants (Chia et al. 1981; Dhinsda et al. 1981; McRae and Thompson 1983; Pauls and Thompson 1984; Pastori and del Rio 1994; Jimenez et al. 1998; Panavas and Rubinstein 1998; Orendi et al. 2001), implying that this may be a more general phenomenon for many aerobic organisms. However, the relationship between life span and oxidative stress tolerance in plants is poorly understood.

Analyses of different late-flowering/extended-longevity mutants suggested that control of longevity and oxidative stress tolerance are tightly linked in *Arabidopsis* (Kurepa et al. 1998). The most extended longevity mutant was *gigantea3* which

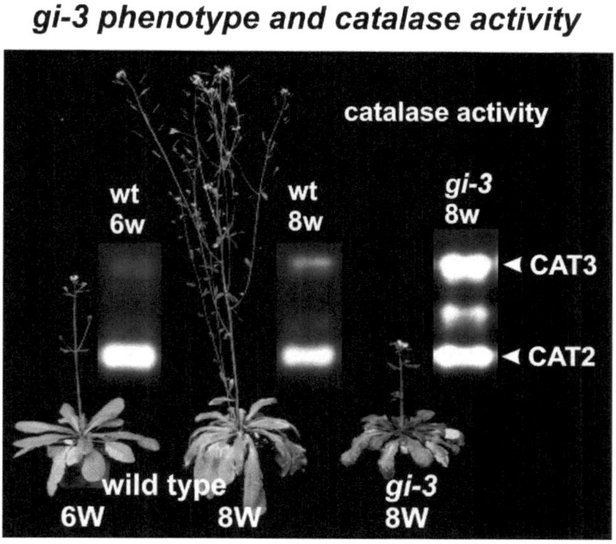

Fig. 2 Phenotype and catalase activity profiles of wild-type and *gigantea3* mutant plants (*gi3*). When plants were grown under the same conditions, *gi3* flowering time and leaf senescence was delayed. Catalase staining of native protein gels revealed that *gi3* showed higher CATALASE3 activity than wild-type plants of the same age (8-week-old plants, *8w*) or the same developmental stage (6-week-old plants, *6w*). *CAT3* CATALASE3 activity, *CAT2* CATALASE2 activity

also showed the highest tolerance against paraquat treatment. GIGANTEA was previously reported to function in red light signalling, central clock function, and flowering time regulation. Recent results indicate that GIGANTEA acts in blue light signalling and has biochemically separable roles in circadian clock and flowering-time regulation (Martin-Tryon et al. 2006). However, the link between this nuclear localised protein and resistance to oxidative stress is still unclear. Clock-controlled genes like CATALASE2 (CAT2) and CATALASE3 (CAT3) might be candidates for this link. Analyses of catalase activities in the *gigantea3* mutant revealed a much higher activity of the stress inducible catalase3 isoform (Fig. 2) which might be responsible for the elevated oxidative stress tolerance. The delayed leaf senescence mutants of *Arabidopsis thaliana ore1*, *ore3*, and *ore9* also exhibit increased tolerance to various types of oxidative stress. However, the activities of antioxidant enzymes were similar or lower in these mutants, as compared with the wild type, providing evidence that oxidative stress tolerance is also genetically linked to control of leaf longevity in plants (Woo et al. 2004). Conversely, elevated levels of ROS influence the expression of SAGs. Eight of 12 SAGs characterised for *Arabidopsis* are induced by ozone (Miller et al. 1999) and the expression of many other SAGs was also enhanced by increased levels of ROS (Navabpour et al. 2003), indicating that elevated levels of oxygen free radicals might be used as a signal to promote senescence.

3 Oxygen Free Radical Sources and Antioxidants

Aerobic organisms use molecular oxygen as a terminal oxidant during respiration as it is relatively harmless and not very reactive. However, it has the potential to be reduced incompletely to toxic intermediates like singlet oxygen ($^1O_2$), the superoxide radical ($O_2^{\bullet-}$), the hydroperoxyl radical ($HO_2^{\bullet-}$), hydrogen peroxide (H_2O_2), and the hydroxyl radical ($^{\bullet}OH$). These radicals are extremely reactive and are able to oxidise all kinds of macromolecules, but have different specificities (for a review, see Dat et al. 2000). The most reactive radical is the hydroxyl radical, which has no preference for specific macromolecules, often leading to irreparable metabolic misfunction and cell death (Knox and Dodge 1985). Superoxide radicals preferentially oxidise amino acids like histidine, methionine and tryptophane. Hydrogen peroxide oxidises SH groups; it differs from the other ROS through its higher stability and its diffusibility; and it is most likely the most important ROS. It can be produced in almost all cell compartments but it can also pass membranes. The endogenous H_2O_2 contents of plant cells can be much higher than those found in animals and bacteria; plant cells happily survive H_2O_2 levels that would kill animal cells. This high tolerance is due to an extensive antioxidant system composed of enzymatic and non-enzymatic components (Noctor and Foyer 1998). Ascorbate peroxidase (APX) is the most important enzyme scavenging H_2O_2 produced in the chloroplast and uses ascorbate to reduce H_2O_2 to water. To maintain ascorbate levels, it has to be regenerated via the ascorbate–glutathione cycle (Bowler et al. 1992). The respective enzymes driving these reactions are also present in peroxisomes and mitochondria; and the participation of this cycle in the control of H_2O_2 concentration in both cell organelles has been proposed (Jiménez et al. 1997, 1998). The capacity of leaves to produce ascorbate declines with leaf age (Foyer 2004). The content of reduced glutathione (GSH) also significantly decreases in pea leaf mitochondria during senescence, which is probably due to the decrease in glutathione reductase (GR) activity under these conditions (Jiménez et al. 1998). Decreases in total glutathione have also been linked to nodule senescence (Dalton et al. 1993; Evans et al. 1999; Puppo et al. 2005). Exposure to stress can result in changes in antioxidants levels, particularly in glutathione and ascorbate. In many plant organs, altered levels of these compounds and the ratio of their reduced to oxidised forms act as a signal to trigger specific cellular responses (Noctor and Foyer 1998; Pastori et al. 2003). However, ascorbate and glutathione are multifunctional compounds with functions that extend beyond the antioxidative system (May et al. 1998; Noctor and Foyer 1998), e.g. ascorbate and glutathione are also required for the operation of the cell cycle (Potters et al. 2004). The degree of coupling between the ascorbate and glutathione redox couples varies greatly between different cellular compartments, and this flexibility of coupling between these antioxidant pools is crucial to differential signalling by ascorbate and glutathione (Noctor et al. 2002; Foyer et al. 2005).

Catalases (CAT) detoxify H_2O_2 produced in the peroxisomes. Whereas APX has a high affinity for H_2O_2 and is able to detoxify low concentrations of H_2O_2, CAT has a high reaction rate, but a low affinity for H_2O_2. However, besides its role in the elimination of peroxisomal H_2O_2, catalase plays a critical role in maintaining the redox balance during oxidative stress and appears to be indispensable for stress defence in some C3 plants (Willekens et al. 1997).

In addition, many different peroxidases are involved in many physiological processes during the plant life cycle. Their activity and expression is modulated by internal and external stimuli and is probably regulated by a fine-tuning that has yet to be elucidated and that meets the demands of plants during stress conditions and senescence (Passardi et al. 2005). Plant metallothioneins are involved in the reduction of metal-induced oxidative stress (Thomas et al. 2005). Peroxiredoxins (Prx) constitute the most recently identified group of H_2O_2-decomposing antioxidant enzymes. In addition to the reduction of H_2O_2, Prx proteins also detoxify alkyl hydroperoxides and peroxinitrite, despite the fact that significant differences exist in substrate specificity and kinetic properties (Dietz et al. 2006).

Furthermore, additional non-enzymatic low-molecular-weight antioxidants are involved in redox balancing. Nitric oxide (NO) is also a bioactive molecule which is able to scavenge ROS and can strongly counteract many ROS-mediated cytotoxic processes in plants (Beligni and Lamattina 1999). In *Arabidopsis*, it was shown that NO can also act as a negative regulator of leaf senescence (Mishina et al. 2007). Even though considered a non-essential element in plants, selenium (Se) might also act as an antioxidant and was shown to play a role as an antioxidative protectant in soybean during senescence (Djanaguiraman et al. 2005). Tocopherols are also well known low-molecular-weight antioxidants (Munne-Bosch and Falk 2004). The content of alpha-tocopherol as well as that of gamma-tocopherol increased significantly in leaves of aging *Arabidopsis* plants (Hollander-Czytko et al. 2005). In conclusion, the various antioxidative systems appear to be tightly regulated during senescence. Transgenic tobacco plants with enhanced glutathione reductase activity or with autoregulated senescence-induced production of cytokinins show that the antioxidative capacity is sufficiently balanced with the plant metabolism and that its decline with increasing age is not a cause but a consequence of senescence and aging in plants (Dertinger et al. 2003). However, there is increasing evidence that ROS are also involved in the regulation of senescence as signalling molecules.

4 The Role of Different Cellular Compartments

More or less all cellular compartments produce ROS and have their own scavenging systems. However, the kind of ROS and the amounts produced can highly differ between compartments, and as already mentioned H_2O_2 can also pass membranes and might be released from different compartments into the cytosol (Fig. 3).

4.1 Chloroplasts

In contrast to animal systems, chloroplasts are the main source of ROS in plants. During photosynthesis, light energy is absorbed by a series of redox reactions and transferred to the reaction centres of the photosystems. Finally, the electrons are

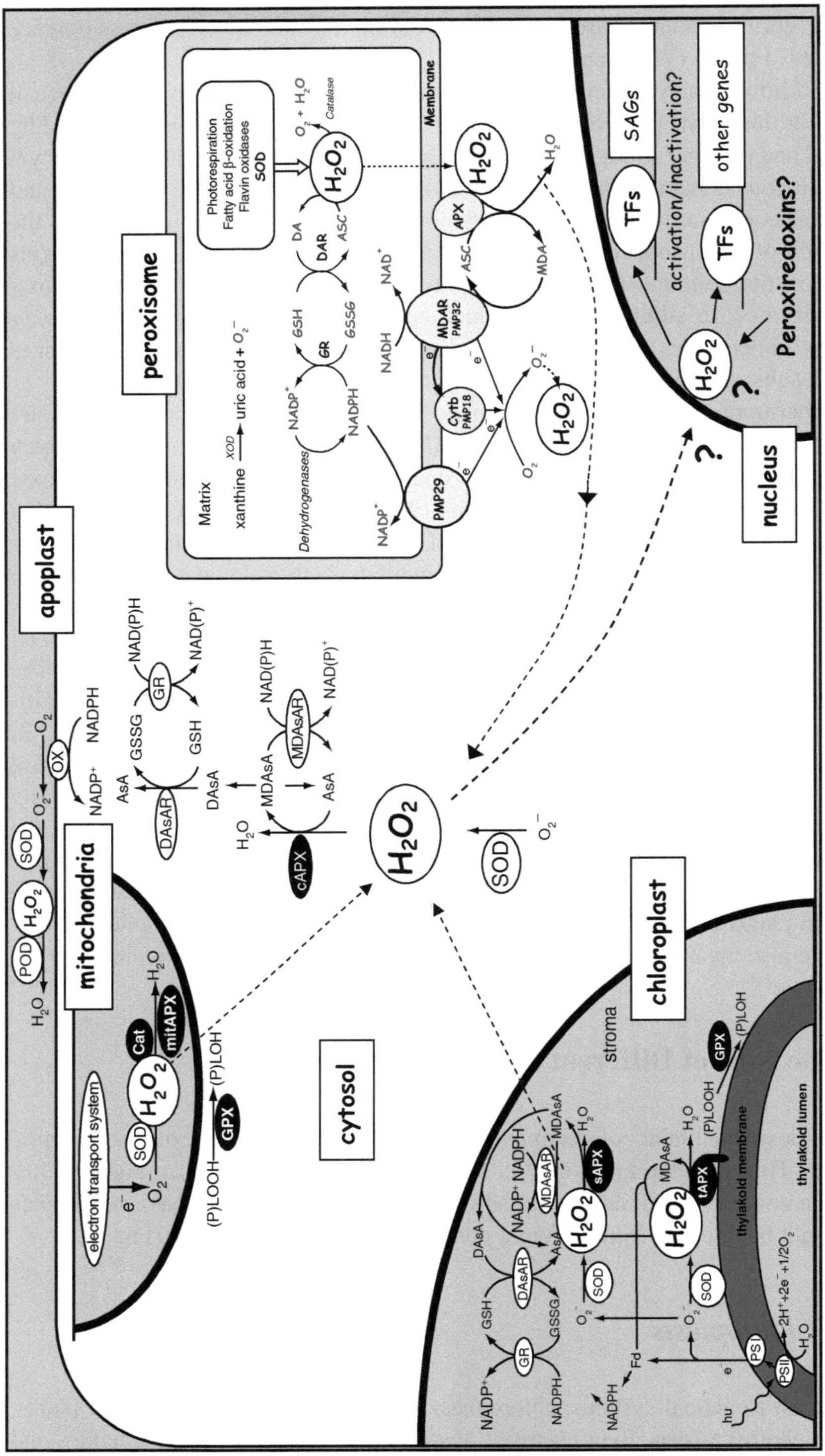
apoplast
peroxisome
Matrix
xanthine
XOD
uric acid + O_2^-
Dehydrogenases
$NADP^+$
NADPH
GR
GSH
GSSG
DA
DAR
ASC
Photorespiration
Fatty acid β-oxidation
Flavin oxidases
SOD
H_2O_2
$O_2 + H_2O$
Catalase
Membrane
PMP29
Cytb PMP18
MDAR PMP32
NADH
NAD^+
APX
MDA
O_2
O_2^-
H_2O
nucleus
SAGs
other genes
TFs
activation/inactivation?
Peroxiredoxins?
?
mitochondria
electron transport system
SOD
Cat
mitAPX
(P)LOOH
(P)LOH
GPX
POD
OX
NADPH
cytosol
cAPX
DAsAR
MDAsAR
AsA
DAsA
MDAsA
NAD(P)H
$NAD(P)^+$
chloroplast
stroma
sAPX
tAPX
thylakoid membrane
thylakoid lumen
PSI
PSII
Fd
hν
$2H^+ + 2e^- + 1/2O_2$

transmitted to CO_2. Since in most plants the rate of CO_2 fixation is not high enough to convert more than 50% of the light energy (Baker 1991), alternative electron acceptors like molecular oxygen are used, leading to the formation of superoxide radicals ($O_2^{\bullet-}$). Furthermore, the chloroplasts can form significant amounts of singlet oxygen ($^1O_2$). Normally, the excited singlet status of the chlorophyll serves the transfer of energy or electrons. To emit energy, chlorophyll uses either fluorescence or conversion to the triplet status, which can, in combination with oxygen, lead to the formation of singlet oxygen (Arora et al. 2002). During senescence, chloroplast are converted into "gerontoplasts". This term was established to describe the organelle of a senescing, formerly green tissue (Parthier 1988). Electron microscopy revealed that the chloroplasts of senescing leaves show an increased number of enlarged plastoglobuli, a disorientation of the grana stacks, and a swelling of the thylacoids. It is assumed that the formation of plastoglobuli is associated with the degradation of the thylacoids (Smart 1994). In cucumber (*Cucumis sativus*), the involvement of the light-harvesting complex of photosystem II (LHC II) in the structural reorganisation of the thylakoid membranes during cotelydon senescence was shown (Prakash et al. 2001); and recently the protease FtsH6 degrading LHC II during leaf senescence was isolated from *Arabidopsis* (Zelisko et al. 2005). It is likely that FtsH6 is a general LHC II protease and that FtsH6-dependent LHC II proteolysis is a feature of all higher plants (Zelisko et al. 2005).

This loss of chloroplast integrity can be observed in the very early stages of senescence; it is associated with the breakdown of the chlorophylls, and the degradation products are transported into the vacuole (Thomas and Stoddart 1980; Matile et al. 1996; Gan and Amasino 1997). The composition and fluidity of the thylacoid membrane appears not to be changed. The conversion of chloroplasts to gerontoplasts in leaves is reversible in some, possibly all higher plants (Zavaleta-Mancera et al. 1999a, b; Thomas et al. 2003). However, the chloroplasts may play a regulatory role during leaf senescence reminiscent to that of mitochondria during animal programmed cell death (PCD). Transgenic tobacco with a knock-out of the plastid *ndhF* gene (Δ*ndhF*) shows low levels of the plastid Ndh complex (plastidic NAD(P)H dehydrogenase analogous to complex I of mitochondria) and a >30-day delay in leaf senescence with respect to wild-type tobacco plants (Zapater et al. 2005). The level and activity of the Ndh complex increase during leaf senescence (Zapater et al. 2005), which strikingly parallels the increased transcription of the mitochondrial complex I during human aging (Pich et al. 2004). Chloroplastic control of leaf senescence provides an unexpected role of the plastid *ndh* genes that are present in almost all higher plants.

The regulation of leaf senescence by chloroplasts opens up the question whether chloroplasts are targets for proapoptotic and antiapoptotic proteins

Fig. 3 Production and scavenging of H_2O_2 in the different cellular compartments, modified after Shigeoka et al. (2002) and del Rio et al. (2006). Whether H_2O_2 is also formed in the nucleus or exclusive imported and whether there are nucleus specific scavenging systems like those proposed for the peroxiredoxins is still an open question. Most likely the redox state of different transcription factors (TFs) and thereby their activity and/or DNA-binding can be influenced by the concentration of H_2O_2 in the nucleus

which affect mitochondria in animal PCD. There, mitochondria integrate signals of proapoptotic and antiapoptotic proteins regulating the release of cytochrome *c* and the production of ROS that direct subsequent apoptotic processes (Green and Reed 1998; Jones 2000; Ferri and Kroemer 2001; Dufour and Larsson 2004). In this case, complex I is the main site for the production of superoxide; complexes II and III are involved to a lesser extent (Dufour and Larsson 2004). In chloroplasts, the Ndh complex regulates the redox level of cyclic electron transporters by providing electrons that are removed by the Mehler reaction, and the coordinated action of superoxide dismutase (SOD) and peroxidase when transporters become over-reduced. Cytochrome *c* release from mitochondria and the decrease of Calvin cycle activity in chloroplasts both lead to an increased generation of ROS in the respective organelle. In addition, the decrease of SOD activity (Casano et al. 1994; Orr and Sohal 1994; Jimenez et al. 1998) in both organelles would amplify the levels of ROS. Moreover, the chloroplast NADPH thioredoxin reductase (NTRC), with a thioredoxin domain, uses NADPH to reduce the chloroplast 2-Cys peroxiredoxin BAS1 and might be a key detoxification system during darkness, with NADPH produced by the oxidative pentose phosphate pathway as the source of reducing power. Prolonged darkness followed by light/dark incubation produced an increase in H_2O_2 and lipid peroxidation in leaves and an accelerated senescence of NTRC-deficient plants, also indicating a regulatory role for ROS produced in the chloroplast during plant senescence (Perez-Ruiz et al. 2006).

4.2 *Mitochondria*

Mitochondria are also an important source of reactive oxygen in plants. The mitochondrial electron transport chain consists of several dehydrogenase complexes which reduce a common pool of ubiquinone (Millenar and Lambers 2003). Cytochrome *c* oxidase or alternative oxidases serve as terminal electron acceptors. Here, the superoxide radical is mainly produced by ubiquinone and the NADH dehydrogenases, namely by autooxidation of the reduced components of the respiration chain (Richter and Schweizer 1997). In contrast to chloroplasts and due to the persisting energy demand, the function of mitochondria is maintained during the gradual breakdown of the cell up to a late time-point in senescence. However, electron microscopy studies showed that senescence in pea leaves induced deterioration in the mitochondrial membrane structure and a slight disorganisation in the matrix and cristae (Pastori and del Río 1994). A decrease in mitochondrial membrane integrity could allow the leakage of H_2O_2 from mitochondria into the cytosol during senescence. This extrusion of H_2O_2 could be favoured by the decrease of APX and monodehydroascorbate reductase (MDHAR) activities in mitochondrial membranes (Jiménez et al. 1998).

In parallel, the alternative respiration pathway is activated during senescence (Hiser and McIntosh 1990; Svensson and Rasmusson 2001; Maxwell et al. 2002).

An important function of the alternative oxidase (AOX) is to prevent the formation of excess free oxygen radicals. AOX ensures a low reduction status of the ubiquinone pool by oxidising ubiquinol. Thus, the electron flow is guaranteed (Millenaar and Lambers 2003). This reaction is necessary, if the cytochrome *c* dependent pathway is restricted by naturally occurring cyanide, NO, sulfide, high concentrations of CO_2, low temperatures, or phosphorus deprivation (Millenaar and Lambers 2003), as well as wounding, drought, osmotic stress, ripening, and pathogen infection (McIntosh 1994; Moore et al. 2002).

There is some evidence that alternative respiration is also correlated with senescence. Aging potato slides showed a decline in the capacity of cytochrome *c*-dependent respiration whereas both the alternative respiration and the protein content of AOX increased (Hiser and McIntosh 1990). In the fungus *Podospora anserina*, inactivation of subunit V of the cytochrome *c* oxidase complex led to the exclusive use of the alternative respiration pathway and to a decline in ROS formation in these mutants. This inactivation of the cytochrome *c* oxidase resulted in an extraordinary longevity of this fungus (Dufour et al. 2000). Long-term treatment of *Arabidopsis* plants with antimycin A also lead to a reduction of H_2O_2 levels but not to a delay of senescence (P. Zimmermann and U. Zentgraf, unpublished data). However, a microarray study using an AOX anti-sense line of *Arabidopsis* showed AOX influences outside mitochondria, particularly in chloroplasts and on several carbon metabolism pathways (Umbach et al. 2005).

In *Podospora anserina*, copper depletion also leads to the induction of an alternative respiratory pathway which appears to be induced by specific impairments of the copper-dependent cytochrome *c* oxidase. During senescence of the wild-type strain, copper is released from mitochondria. The mechanism involved is still unknown. However, it is striking that the permeability of mitochondrial membranes in animal systems changes during apoptosis and that mitochondrial proteins with an important impact on this type of cellular death are released (Borghouts et al. 2001). In *Arabidopsis*, a copper chaperone (CCH) is upregulated during leaf senescence, suggesting that certain metal ions are mobilised in leaves and are transported to other growing parts of the plants. The CCHs are also involved in defence mechanisms against oxidative stress in *Arabidopsis*, tomato, and poplar. Emerging data suggest that the mechanisms regulating plant copper homeostasis could be implicated in stress and senescence signal transduction pathways (Himmelblau et al. 1998; Mira et al. 2002; Lee et al. 2005).

Mitochondrial NO production is also involved in senescence regulation, since dark-induced senescence of detached leaves and intact plants progressed more rapidly in the *nos1*-mutant compared with the wild type (Guo and Crawford 2005). However, there is some doubt whether this protein really encodes an plant nitric oxide synthase (NOS; Zemojtel et al. 2006). Although it has been confirmed that basal NO production is impaired or that *nos1* mutants have a reduced capacity to mount a NO burst, no direct NOS activity can be confirmed so far (Crawford et al. 2006). In contrast, it was suggested that AtNOS1 is a GTPase involved in mitochondrial ribosome biogenesis and/or processes of translation.

In this scenario, AtNOS1 deletion would lead to defects in mitochondrial biogenesis, resulting in an observed decrease in NO production (Zemojtel et al. 2006). Whether the At*NOS* gene might encode a new kind of NOS with low activity which uses N-ώ-hydroxyarginine (NOHA) as substrate is still under investigation (Guo 2006).

The copy number of mitochondrial genomes is kept constant during leaf development, indicating that organellar gene expression in higher plants is not significantly regulated at the level of genome copy number (Li et al. 2006).

4.3 Peroxisomes

Another source for ROS formation, especially for H_2O_2, is photorespiration in the peroxisomes. During CO_2 fixation, ribulose-1,5-bisphosphate-carboxylase (RubisCO) uses CO_2 to carboxylate ribulose-1,5-bisphosphate. This enzyme can also use molecular oxygen to oxygenate ribulose-1,5-bisphosphate (Foyer 1996). During this reaction, glycolate is formed and transported from the chloroplasts into the peroxisomes and then oxidised there, with H_2O_2 formed as a byproduct. After the reverse transition of leaf peroxisomes into glyoxysomes, which occurs during senescence, additional H_2O_2 is produced by increased beta-oxidation of fatty acids. Moreover, it is very likely that the peroxisomal NADH-dependent production of $O_2^{\bullet-}$ radicals is intensified by this transition (Landolt and Matile 1990; Pistelli et al. 1996; Pastori and del Río 1997), since more NADH would also be available as a result of the induction of fatty acid beta-oxidation and the glyoxylate cycle (Jiménez et al. 1998). Peroxisomes and ROS generated in these organelles were shown to play a central role in natural and dark-induced senescence in pea (del Rio et al. 1998). Here, peroxisomes have a ROS-mediated cellular function in leaf senescence and in stress situations induced by xenobiotics and heavy metals. Moreover, peroxisomes appear to play an important role as a supplier of signal molecules like $NO^{\bullet}$ (nitric oxide), $O_2^{\bullet-}$, H_2O_2, and possibly S-nitrosoglutathione (Pastori and del Rio 1997; del Rio et al. 1998, 2002, 2003). Whereas the superoxide- and H_2O_2-generating enzymes (e.g. xanthine oxidase, urate oxidase, MnSOD) and the NADPH-dependent generation of superoxide on the membranes of these organelles increased during leaf senescence (Pastori and del Rio 1997; del Rio et al. 1998), CAT activity was almost completely lost (Pastori and del Rio 1994, 1997). Therefore, H_2O_2 levels and lipid peroxidation rate significantly increased in these organelles during senescence. Remarkable differences in redox components and progression of senescence can be observed when pea plants grow under different nitrogen supply conditions (Vanacker et al. 2006). In contrast to ROS, NO production is clearly down-regulated during leaf senescence of pea plants. Confocal laser microscopy analyses with 4,5-diaminofluorescein diacetate in pea leaf sections revealed that endogenous NO was predominantly detected in the vascular tissues, suggesting that it could be involved in long-distance communication (Corpas et al. 2004). Taken together, these data suggest that peroxisomes could act as subcellular sensors of plant stress and senescence by releasing NO, superoxide, and H_2O_2 as signalling molecules to the cytosol and thereby also triggering specific gene expression (Corpas et al. 2001, 2004; del Rio et al. 2002). If the mitochondrial and peroxisomal ascorbate–glutathione cycles are compared during the progression of senescence, it can be speculated that peroxisomes may participate in the cellular oxidative mechanism of leaf senescence longer than mitochondria, since mitochondria appear to be affected by oxidative damage earlier than peroxisomes (Jiménez et al. 1998; del Rio et al. 2003). Moreover, stresses that generate H_2O_2 as a signalling molecule can result in peroxisome proliferation via up-regulation of components (*PEX* genes) required for the biogenesis of these organelles and import of proteins into these organelles (Lopez-Huertas et al. 2000). However, whether the number of peroxisomes generally increases during leaf senescence still has to be elucidated; in senescent pea leaves the number of peroxisomes increases up to five-fold (Pastori and del Rio 1994; del Rio et al. 1998).

4.4 The Nucleus

In contrast to the numerous studies on antioxidants in the cytoplasm, the nuclear antioxidant system has not been studied in much detail. However, nuclear redox states influence the activities of several transcription factors; and oxidative injury or DNA replication errors caused by ROS are serious problems for aerobic organisms. Therefore, it is reasonable to speculate that eukaryotic cells evolved nuclear antioxidant systems distinct from their cytosolic ones. In barley, a dormancy-related peroxiredoxin anti-oxidant, PER1, was localised to the nucleus of barley embryo and aleurone cells and protective roles for PER1 in seeds were discussed (Stacy et al. 1999). Recently, the localisation of at least one family member of the peroxiredoxins was reported in the nucleoplasm of rice and *Arabidopsis* (Dietz et al. 2006).

Oxidative stress can lead to DNA damage by the formation of 8-oxo-7,8-dihydro-2′-deoxyguanosine (8-oxo-dG) in cucumber, most likely induced by H_2O_2 (Watanabe et al. 2006). In general, the nucleotide pool is also a significant target for the formation of extranuclear 8-oxo-dG (Haghdoost et al. 2006). In mammalian cells, it was observed that H_2O_2 plus Cu(II) induces 8-oxo-dG formation in telomere sequences more efficiently than in non-telomere sequences. These oxidative damages are repaired less efficiently in telomeric DNA than elsewhere in the chromosome. Moreover, oxidative stress accelerates telomere loss whereas antioxidants delays it (Von Zglincki 2002). During development from immature embryos to mature barley plants with senescent leaves, telomeres shorten dramatically. This telomere shortening cannot be explained by the continuous loss of telomeric repeats during replication, indicating that a different mechanism for teleomere shortening exists, which is most likely involved in developmental regulation (Kilian et al. 1995). In other plants, telomere length is kept constant during development and post-replicative senescence (Fajkus and Zentgraf 2000).

However, the intranuclear localisation of the telomeres in *Arabidopsis* at the nucleolar boundary changes with the age of the cells (Fajkus and Zentgraf 2000) and the protein structure of the *Arabidopsis* telomeres is modified at an early stage in leaf senescence when a small protein is recruited to the telomeres by protein–protein interaction (Zentgraf et al. 2000). It still has to be elucidated whether this has any implication for telomere function or nuclear architecture and, as a consequence, active chromatin domains and gene expression patterns. In humans, oxidative DNA damage may exert deleterious effects on telomeres by disrupting the association of the telomere-maintenance proteins TRF1 and TRF2 (Opresko et al. 2005).

5 Oxygen Free Radical Signalling

ROS were originally considered to be exclusively detrimental to cells, but it is clear now that redox regulation involving ROS plays a key role in cell signalling. ROS likely participate far more in cellular activities than anticipated (see also Fig. 1). In yeast, transcriptome analyses revealed that the response to oxidative stress involves about one-third of the genome (Gasch et al. 2000); and a number of transcription factors that regulate the expression of antioxidant genes are well characterised, e.g. ACE1, MAC1, YAP1, YAP2, HAP1, and HAP2/3/4 of yeast (for a review, see Scandalios 2002). In yeast, the glutathione peroxidase (GPX)-like enzyme GPX3 was identified as a second component of the pathway, serving the role of sensor and transducer of the H_2O_2 signal to the YAP1 transcription factor (Delaunay et al. 2002). In higher eukaryotes, oxidative stress responses are more complex and are modulated by many regulators.

Despite some limitations, it became clear through the use of microarrays that in plants there are far more genes and gene clusters responding to ROS and/or senescence than previously thought and that there is a remarkable overlap (Desikan et al. 2001; Scandalios 2002; Navabpour et al. 2003; Buchanan-Wollaston et al. 2005). Many of the H_2O_2-responsive genes have no obvious direct role in oxidative stress, but may be linked to stress or developmental signalling functions, explaining their sensitivity to H_2O_2 (Desikan et al. 2001). Indeed, global expression analysis revealed that H_2O_2 is involved in the regulation of a large number of genes in tobacco (Vandenabeele et al. 2003) and *Arabidopsis* (Vanderauwera et al. 2005). In plant cells, little is known about how the signals are perceived and transduced.

In alfalfa, an oxidative stress-activated MAP triple kinase 1 (OMTK1) was identified which is exclusively activated by H_2O_2 and which directly interacts with a MAPK (MMK3) and induces cell death (Nakagami et al. 2004). H_2O_2 can also activate a specific *Arabidopsis* MAP triple kinase, ANP1, which initiates a phosphorylation cascade involving two stress MAPKs, AtMPK3 and AtMPK6 (Kovtun et al. 2000). Expression of the MAP triple kinase1 (MEKK1) of *Arabidopsis* can also be induced by H_2O_2 and shows its expression maximum during onset of leaf senescence (Miao et al. 2007; http://www.genevestigator.ethz.ch). This triple MAP kinase can directly phophorylate the senescence-associated transcription factor WRKY53 and thereby increase its DNA binding activity (Miao et al. 2007). WRKY53 was shown to be involved in senescence regulation in an upstream position (Miao et al. 2004). A coordinated regulation of the H_2O_2 scavenging enzymes CAT and APX on the transcriptional and posttranscriptional level most likely creates a distinct increase of the hydrogen peroxide right at that time-point when plants start to bolt and a coordinated senescence process of all rosette leaves should be induced (Ye et al. 2000; Zimmermann et al. 2006). Most likely the expression of MEKK1 is induced by this increase in H_2O_2 levels; and *WRKY53*, which can also be induced by H_2O_2 treatment (Miao et al. 2004), is switched from a leaf age to a plant age-dependent expression (Hinderhofer and Zentgraf 2001) at that time-point.

Whether H_2O_2-induced expression of SAGs is transduced by MAPK signalling or directly by redox-sensitive transcription factors still has to be elucidated. However, the redox-sensitive zinc-finger DNA-binding domain of the WRKY proteins in which two cysteines together with two histidines interact electrostatically with a zinc atom to form a "zinc finger" makes it an excellent candidate for direct redox regulation (Arrigo 1999). For WRKY53 there is some evidence that this could be the case. It could be shown that WRKY53 can negatively regulate its own expression (Miao et al. 2004). Hydrogen peroxide can induce reporter gene expression which was driven by the *WRKY53* promoter in protoplasts prepared from wild-type plants. In contrast, this induction was abolished in protoplast prepared from *WRKY53* knock-out lines, indicating that the WRKY53 protein is directly involved in the transduction of the hydrogen peroxide signal to the *WRKY53* promoter (Miao et al. 2007). Further hints for the involvement of ROS and antioxidants came from the ascorbate-deficient *Arabidopsis vtc1* mutant which shows altered timing of senescence (Barth et al. 2004). ROS, ascorbate, and glutathione decline in a regulated manner during nodule development and senescence. This does not necessarily cause oxidative stress but all components might be involved in signalling processes or in a development-related shift in redox-linked metabolite cross-talk during nodule senescence (Groten et al. 2005).

6 Conclusions and Perspectives

Many different agriculturally important traits are affected by plant senescence, such that understanding senescence processes might contribute to solve the problems approaching with global climatic changes. As described above, ROS play an important role during leaf senescence in two different aspects: in signalling and in molecule degradation. Despite extensive investigation of the role of H_2O_2, to date no complete ROS signal transduction pathways have been described (Foyer et al. 2005). Obviously, plants have developed a very fine-tuned network of enzymatic and low-molecular-weight antioxidative components in different cell compartments; and different plants have different strategies to balance their redox potential and regulate their ROS status. The network managing the ROS balance is highly dynamic and redundant, and includes ROS-scavenging and ROS-producing proteins. Although recent studies have unravelled some of the key players in the network, many unanswered questions remain, related to its mode of regulation, its protective roles, and its modulation of signalling networks that control growth, development, and stress response (Mittler et al. 2004). A very comprehensive overview is also given in the special issue of the *Journal of Experimental Botany* published in May 2006: "*Oxygen metabolism, ROS, and redox signalling in plants*". New approaches like metabolic modelling are being developed as tools to analyse network regulations, like e.g. redox reactions composing the SOD–ascorbate–glutathione pathway (Polle 2001). These kinds of modelling approaches cannot yet make exact predictions but can contribute to the theoretical understanding of the functioning of antioxidant systems by pointing out questions that need to be validated (Polle 2001). Developing in vivo imaging systems for different ROS to visualise local changes in ROS levels in different compartments will help to integrate our current data in a holistic view of the cells.

References

Arora A, Sairam RK, Srivastava GC (2002) Oxidative stress and antioxidative system in plants. Curr Sci 82:1227–1238Arrigo AP (1999) Gene expression and the thiol redox state. Free Radical Biol Med 27: 936–944

Baker NR (1991) A possible role for photosystem II in environmental perturbations of photosynthesis. Physiol Plant 81:563–570

Barth C, Moeder W, Klessig DF, Conklin PL (2004) The timing of senescence and response to pathogens is altered in the ascorbate-deficient *Arabidopsis* mutant vitamin c-1. Plant Physiol 134:1784–1792

Batt T, Woolhouse HW (1975) Changing activities during senescence and sites of synthesis of photosynthetic enzymes in leaves of labiate, *Perilla frutenscens* (L.). British J Exp Bot 26:569–579

Beligni MV, Lamattina L (1999) Nitric oxide counteracts cytotoxic processes mediated by reactive oxygen species in plant tissues. Planta 208:337–344

Borghouts C, Werner A, Elthon T, Osiewacz HD (2001) Copper-modulated gene expression and senescence in the filamentous fungus *Podospora anserine*. Mol Cell Biol 21:390–399

Bowler C, Van Montagu M, Inzé D (1992) Superoxide dismutase and stress tolerance. Annu Rev Plant Physiol Plant Mol Biol 43:83–116

Buchanan-Wollaston V, Page T, Harrison E, Breeze E, Lim PO, Nam HG, Lin JF, Wu SH, Swidzinski J, Ishizaki K, Leaver CJ (2005) Comparative transcriptome analysis reveals significant differences in gene expression and signalling pathways between developmental and dark/starvation-induced senescence in *Arabidopsis*. Plant J 42:567–585

Casano LM, Martín M, Sabater B (1994) Sensitivity of superoxide dismutase transcript levels and activities to oxidative stress is lower in mature-senescent than in young barley leaves. Plant Physiol 106:1033–1039

Chia LS, Thompson JE, Dumbroff EB (1981) Simulation of the effects of leaf senescence on membranes by treatment with paraquat. Plant Physiol 67:415–420

Corpas FJ, Barroso JB, Río LA del (2001) Peroxisomes as a source of reactive oxygen species and nitric oxide signal molecules in plant cells. Trends Plant Sci 6:145–150

Corpas FJ, Barroso JB, Carreras A, Quirós M, León AM, Romero-Puertas MC, Esteban FJ, Valderrama R, Palma JM, Sandalio LM, Gómez M, del Río LA (2004) Cellular and subcellular

localization of endogenous nitric oxide in young and senescent pea plants. Plant Physiol 136:2722–2733
Crawford NM, Galli M, Tischner R, Heimer YM, Okamoto M, Mack A (2006) Response to Zemojtel et al. Plant nitric oxide synthase: back to square one. Trends Plant Sci 11:526–527
Dalton DA, Langeberg L, Treneman NC (1993) Correlations between the ascorbate–glutathione pathway and effectiveness in legume root nodules. Physiol Plant 87:365–370
Dat JF, Vandenabeele S, Vranová E, Van Montagu M, Inzé D, Van Breusegem F (2000) Dual action of the active oxygen species during plant stress responses. Cell Mol Life Sci 57:779–795
Delaunay A, Pflieger D, Barrault MB, Vinh J, Toledano MB (2002) A thiol peroxidase is an H_2O_2 receptor and redox-transducer in gene activation. Cell 111:471–481
Dertinger U, Schaz U, Schulze ED (2003) Age-dependence of the antioxidative system in tobacco with enhanced glutathione reductase activity or senescence-induced production of cytokinins. Physiol Plant 119:19–29
Desikan, R, Mackerness AHS, Hancock JT, Neill SJ (2001) Regulation of the Arabidopsis transcriptome by oxidative stress. Plant Physiol 127:159–172
Desikan R, Cheung MK, Bright J, Henson D, Hancock JT, Neill SJ (2004) ABA, hydrogen peroxide and nitric oxide signalling in stomatal guard cells. J Exp Bot 55: 205–212
Dhinsda RJ, Dhinsda PP, Thorpe TA (1981) Leaf senescence: correlation with increased levels of membrane permeability and lipid peroxidation, and decreased levels of superoxide dismutase and catalase. J Exp Bot 32:93–101
Diaz C, Purdy S, Christ A, Morot-Gaudry J-F, Wingler A, Masclaux-Daubresse C (2005) Characterization of markers to determine the extent and variability of leaf senescence in *Arabidopsis thaliana*: a metabolic profiling approach. Plant Physiol 138:898–908
Dietz KJ, Jacob S, Oelze ML, Laxa M, Tognetti V, Nunes de Miranda SM, Baier M, Finkemeier I (2006) The function of peroxiredoxins in plant organelle redox metabolism. J Exp Bot 57:1697–1709
Djanaguiraman M, Devi DD, Shanker AK, Sheeba JA, Bangarusamy U (2005) Selenium – an antioxidative protectant in soybean during senescence. Plant Soil 272:77–86
Dufour E, Larsson NG (2004) Understanding aging: revealing order out of chaos. Biochim Biophys Acta 1658:122–132
Dufour E, Boulay J, Rincheval V, Sainsard-Chanet A (2000) A causal link betweeen respiration and senescence in *Podospora anserina*. Proc Natl Acad Sci USA 97:4138–4143
Ellis CM, Nagpal P, Young JC, Hagen G, Guilfoyle TJ, Reed JW (2005) *AUXIN RESPONSE FACTOR1* and *AUXIN RESPONSE FACTOR2* regulate senescence and floral organ abscission in *Arabidopsis thaliana*. Development 132:4563–4574
Evans PJ, Gallesi D, Mathieu C, Hernandez MJ, Felip M de, Halliwell B, Puppo A (1999) Oxidative stress occurs during soybean nodule senescence. Planta 208:73–79
Fajkus J, Zentgraf U (2002) Structure and maintenance of chromosome ends. In: Krupp G, Parwaresch R (eds) Telomeres and telomerases: cancer and biology. Bioscience/Gen-Expression, available at: www.eurekah.com
Ferri KF, Kroemer G (2001) Organelle-specific initiation of cell death pathways. Nat Cell Biol 3: E255–E263
Foyer CH (1996) Oxygen processing in photosynthesis. Biochem Soc Trans 24:427–433
Foyer CH (2004) The role of ascorbic acid in defense networks and signaling in plants. In: Asard H, May JM, Smirnoff S (eds) Vitamin C. Functions and biochemistry in animals and plants. Bios, Oxford, pp 65–82
Foyer CH, Trebstm A, Noctor G (2005) Protective and signalling functions of ascorbate, glutathione and tocopherol in chloroplasts. In: Demmig-Adams B, Adams WW (eds) Advances in photosynthesis and respiration: photoprotection, photoinhibition, gene regulation, and environment, vol 19. Kluwer, Dordrecht, pp 241–268
Gan S, Amasino RM (1995) Inhibition of leaf senescence by autoregulated production of cytokinin. Science 270:1986–1988

Gan S, Amasino RM (1997) Making sense of senescence: molecular genetic regulation and manipulation of leaf senescence. Plant Physiol 113:313–319

Gasch A, Spellman P, Kao C, Harel O, Eisen M, Storz G, Botstein D, Brown P (2000) Genomic expression programs in the response of yeast cells to environmental changes. Mol Biol Cell 11:4241–4257

Grbic V, Bleecker AB (1995) Ethylene regulates the timing of leaf senescence in *Arabidopsis*. Plant J 8:595–602

Green DR, Reed JC (1998) Mitochondria and apoptosis. Science 281:1309–1312

Groten K Vanacke, H, Dutilleul C, Bastian F, Bernhard S, Carzaniga R, Foyer CH (2005) The roles of redox processes in pea nodule development and senescence. Plant Cell Environ 28:1293–1304

Guo FQ (2006) Response to Zemojtel et al. Plant nitric oxide synthase: AtNOS1 is just the beginning. Trends Plant Sci 11:527–528

Guo FQ, Crawford NM (2005) *Arabidopsis* Nitric oxide synthase1 is targeted to mitochondria and protects against oxidative damage and dark-induced senescence. Plant Cell 17:3436–3450

Haghdoost S, Sjolander L, Czene S, Hanns-Ringdahl M (2006) The nucleotide pool is a significant target for oxidative stress. Free Radical Biol Med 41:620–626

Halliwell B, Gutteridge JMC (1989) Free radicals in biology and medicine, 2nd edn. Oxford University Press, Oxford

Harman D (1956) Aging: a theory based on free radical and radiation chemistry. J Geront 11:298–300

Harman D (1998) Extending functional life span. Exp Geront 33:95–112

He Y, Tang W, Swain JD, Green AL, Jack TP, Gan S (2001) Networking senescence-regulating pathways by using *Arabidopsis* enhancer trap lines. Plant Physiol 126:707–716

He Y, Fukushige H, Hildebrande DF, Gan S (2002) Evidence supporting a role for jasmonic acid in *Arabidopsis* leaf senescence. Plant Physiol 128:876–884

Hensel LL, Grbić V, Baumgarten D, Bleecker AB (1993) Developmental and age-related processes that influence the longevity and senescence of photosynthetic tissues in *Arabidopsis*. Plant Cell 5:553–564

Himelblau E, Mira H, Lin SJ, Culotta VC, Penarrubia L, Amasino RM (1998) Identification of a functional homolog of the yeast copper homeostasis gene ATX1 from *Arabidopsis*. Plant Physiol 117:1227–1234

Hinderhofer K, Zentgraf U (2001) Identification of a transcription factor specifically expressed at the onset of leaf senescence. Planta 213:469–473

Hiser C, McIntosh L (1990) Alternative oxidase of potato is an integral membrane protein synthesized de novo during aging of tuber slices. Plant Physiol 93:312–318

Hollander-Czytko H, Grabowski J, Sandorf I, Weckermann K, Weiler EW (2005) Tocopherol content and activities of tyrosine aminotransferase and cystine lyase in *Arabidopsis* under stress conditions. J Plant Physiol 162: 767–770

Jiménez A, Hernández JA, Río LA del, Sevilla F (1997) Evidence for the presence of the ascorbate–glutathione cycle in mitochondria and peroxisomes of pea (*Pisum sativum* L.) leaves. Plant Physiol 114:275–284

Jiménez A, Hernandez JA, Pastori G, Rio LA del, Sevilla F(1998) Role of the ascorbate–glutathione cycle of mitochondria and peroxisomes in the senescence of pea leaves. Plant Physiol 118:1327–1335

John I, Drake R, Farrell A, Cooper W, Lee P, Horton P, Grierson D (1995) Delayed leaf senescence in ethylene-deficient ACC-oxidase antisense tomato plants: molecular and physiological analysis. Plant J 7:483–490

Jones A (2000) Does the plant mitochondrion integrate cellular stress and regulate programmed cell death? Trends Plant Sci 5:225–230

Jongebloed U, Szederkényi J, Hartig K, Schobert C, Komor E (2004) Sequence of morphological and physiological events during natural ageing of a castor bean leaf: sieve tube occlusions and carbohydrate back-up precede chlorophyll degradation. Physiol Plant 120:338–346

Kar M, Feierabend J (1984) Metabolism of activated oxygen in detached wheat and rye leaves and its relevance to the initiation of senescence. Planta 160:385–391

Keskitalo J, Bergquist G, Gardestrom P, Jansson S (2005) A cellular timetable of autumn senescence. Plant Physiol 139:1635–1648

Kilian A, Stiff C, Kleinhofs A (1995) Barley telomeres shorten during differentiation but grow in callus culture. Proc Natl Acad Sci USA 92:9555–9559

Knox JP, Dodge AD (1985) Singlet oxygen and plants. Phytochemistry 24:889–896

Kovtun Y, Chiu WL, Tena G, Sheen J (2000) Functional analysis of oxidative stress-activated mitogen-activated protein kinases cascade in plants. Proc Natl Acad Sci USA 97:2940–2945

Krupinska K, Falk J, Humbeck K (2003) Genetic, metabolic and environmental factors associated with ageing in plants. In: Osiewacz HD (ed) Aging of organisms. Kluwer, Dordrecht, pp 55–78

Kurepa J, Smalle J, Van Montagu M, Inzé D (1998) Oxidative stress tolerance and longevity in *Arabidopsis*: the late flowering mutant *gigantea* is tolerant to paraquat. Plant J 14:759–764

Landolt R, Matile P (1990) Glyoxysome-like microbodies in senescent spinach leaves. Plant Sci 72:159–163

Lee H, Lee JS, Bae EK, Choi YI, Noh EW (2005) Differential expression of a poplar copper chaperone gene in response to various abiotic stresses. Tree Physiol 25:395–401

Li W, Ruf S, Bock R (2006) Constancy of organellar genome copy numbers during leaf development and senescence in higher plants. Mol Genet Genomics 275:185–192

Lopez-Huertas E, Charlton WL, Johnson B, Graham IA, Baker A (2000) Stress induces peroxisome biogenesis genes. EMBO J 19:6770–6777

Martin GM, Austad SN, Johnson TE (1996) Genetic analysis of aging: role of oxidative damage and environmental stresses. Nat Genet 13:25–34

Martin-Tryon EL, Kreps JA, Harmer SL (2006) GIGANTEA acts in blue light signaling and has biochemically separable roles in circadian clock and flowering time regulation. Plant Physiol 143:473–486

Masclaux C, Valadier MH, Brugière N, Morot-Gaudry JF, Hirel B (2000) Characterization of the sink/source transition in tobacco (*Nicotiana tabacum* L) shoots in relation to nitrogen management and leaf senescence. Planta 211:510–518

Matile P, Schellenberg M, Peisker C (1992) Production and release of a chlorophyll catabolite in isolated senescent chloroplasts. Planta 187:230–235

Matile P, Hörtensteiner S, Thomas H, Kräutler B (1996) Chlorophyll breakdown in senescent leaves. Plant Physiol 112:1403–1409

Maxwell DP, Nickels R, McIntosh L (2002) Evidence of mitochondrial involvement in the transduction of signals required for the induction of genes associated with pathogen attack and senescence. Plant J 29:269–279

May JM, Qu ZC, Mendiratta S (1998) Protection and recycling of alpha-tocopherol in human erythrocytes by intracellular ascorbic acid. Arch Biochem Biophys 349:281–289

McIntosh L (1994) Molecular biology of the alternative oxidase. Plant Physiol 105:781–786

McRae DG, Thompson JE (1983) Senescence dependent changes in superoxide anion production by illuminated chloroplast from bean leaves. Planta 158:185–193

Miao Y, Laun T, Zimmermann P, Zentgraf U (2004) Targets of the WRKY53 transcription factor and its role during leaf senescence in *Arabidopsis*. Plant Mol Biol 55:853–867

Miao Y, Laun T, Smykowski A, Zentgraf U (2007) Arabidopsis MEKK1 can take a short cut: it can directly interact with the senescence-related WRKY53 transcription factor on the protein level and it can bind to its promoter. Plant Mol Biol, DOI 10.1007/s11103-007-9198-z

Millenaar FF, Lambers H (2003) The alternative oxidase: in vivo regulation and function. Plant Biol 5:2–15

Miller JD, Arteca RN, Pell EJ (1999) Senescence-associated gene expression during ozone-induced leaf senescence. Plant Physiol 120:1015–1023

Mira H, Martinez N, Penarrubia L (2002) Expression of a vegetative-storage-protein gene from *Arabidopsis* is regulated by copper, senescence and ozone. Planta 214:939–946

Mishina TE, Lamb C, Zeier J (2007) Expression of a nitric oxide degrading enzyme induces a senescence programme in *Arabidopsis*. Plant Cell Environ 30:39–52
Mittler R, Vanderauwera S, Gollery M, Van Breusegem F (2004) Reactive oxygen gene network of plants. Trends Plant Sci 9:490–498
Moore AL, Albury MS, Crichton PG, Affourtit C (2002) Function of the alternative oxidase: is it still a scavenger? Trends Plant Sci 7:478–481
Munne-Bosch S, Alegre L (2002) Plant aging increases oxidative stress in chloroplasts. Planta 214:608–615
Munne-Bosch S, Falk J (2004) New insights into the function of tocopherols in plants. Planta 218:323–326
Nakagami H, Kiegerl S, Hirt H (2004) OMTK1, a novel MAPKKK, channels oxidative stress signalling through direct MAPK interaction. J Biol Chem 279:26959–26966
Navabpour S, Morris K, Allen R, Harrison E, Mackerness SAH, Buchanan-Wollaston V (2003) Expression of senescence-enhanced genes in response to oxidative stress. J Exp Bot 54:2285–2292
Neill SJ, Desikan R, Clarke A, Hurst RD, Hancock JT (2002) Hydrogen peroxide and nitric oxide as signalling molecules in plants. J Exp Bot 53:1237–1247
Noctor G, Foyer CH (1998) Ascorbate and glutathione: keeping active oxygen under control. Annu Rev Plant Physiol Plant Mol Biol 49:249–279
Noctor G, Veljovic-Jovanovic S, Driscol S, Novitskaya L, Foyer CH (2002) Drought and oxidative load in wheat leaves: a predominant role for photorespiration? Ann Bot 89:841–850
Noodén LD, Guiamét JJ, John I (1997) Senescence mechanisms. Physiol Plant 101:746–753
Orendi G, Zimmermann P, Baar C, Zentgraf U (2001) Loss of stress-induced expression of catalase3 during leaf senescence in *Arabidopsis thaliana* is restricted to oxidative stress. Plant Sci 161:301–314
Opresko PL, Fan JH, Danzy S, Wilson DM, Bohr VA (2005) Oxidative damage in telomeric DNA disrupts recognition by TRF1 and TRF2. Nucleic Acid Res 33:1230–1239
Orr WC, Sohal RS (1994) Extension of life-span by overexpression of superoxide dismutase and catalase in *Drosophila melanogaster*. Science 263:1128–1130
Panavas T, Rubinstein B (1998) Oxidative events during programmed cell death of daylily (*Hemerocallis* hybrid) petals. Plant Sci 133:125–138
Parthier B (1988) Gerontoplasts – the yellow end in the ontogenesis of chloroplasts. Endocytobiosis Cell Res 5:163–190
Passardi F, Cosio C, Penel C, Dunand C (2005) Peroxidases have more functions than a Swiss army knife. Plant Cell Rep 24:255–265
Pastori GM, Rio LA del (1994) An activated-oxygen-mediated role for peroxisomes in the mechanism of senescence of *Pisum sativum* L. leaves. Planta 193:385–391
Pastori GM, Río LA del (1997) Natural senescence of pea leaves: an activated oxygen-mediated function for peroxisomes. Plant Physiol 113:411–418
Pastori GM, Kiddle G, Antoniw J, Bernard S, Veljovic-Jovanovi, S, Verrier PJ, Noctor G, Foyer CH (2003) Leaf vitamin C contents modulate plant defense transcripts and regulate genes that control development through hormone signaling. Plant Cell 15:939–951
Pauls KP, Thompson JE (1984) Evidence for the accumulation of peroxidized lipids in membranes of senescing cotyledons. Plant Physiol 75:1152–1157
Pérez-Ruiz JM, Spínola MC, Kirchsteiger K, Moreno J, Sahrawy M, Cejudo FJ (2006) Rice NTRC is a high-efficiency redox system for chloroplast protection against oxidative damage. Plant Cell 18:356–2368 http://www.plantcell.org/egi/content/full/18/9/2356 - FNI
Pich MM, Raule N, Catani L, Fagioli ME, Faenza I, CoccoL, Lenaz G (2004) Increased transcription of mitochondrial genes for complex I in human platelets during ageing. FEBS Lett 558:19–22
Pistelli L, Nieri B, Smith SM, Alpi A, De Bellis L (1996) Glyoxylate cycle enzyme activities are induced in senescent pumpkin fruits. Plant Sci 119:23–29
Polle A (2001) Dissecting the superoxide dismutase–ascorbate–glutathione pathway in chloroplasts by metabolic modeling. Computer simulations as a step towards flux analysis. Plant Physiol 126:445–462

Potters G, Horemans N, Bellone S, Caubergs RJ, Trost P, Guisez Y, Asard H (2004) Dehydroascorbate influences the plant cell cycle through a glutathione-independent reduction mechanism. Plant Physiol 134:1479–1487
Pourtau N, Marès M, Purdy S, Quentin N, Ruël A, Wingler A (2004) Interactions of abscisic acid and sugar signalling in the regulation of leaf senescence. Planta 219:765–772
Pourtau N, Jennings R, Pelze, E, Pallas J, Wingler A (2006) Effect of sugar-induced senescence on gene expression and implications for the regulation of senescence in *Arabidopsis*. Planta 224:556–568
Prakash JS, Baig MA, Mohanty P (2001) Senescence induced structural reorganization of thylakoid membranes in Cucumis sativus cotyledons; LHC II involvement in reorganization of thylakoid membranes. Photosynth Res 68:153–161
Puppo A, Groten K, Bastian F, Carzaniga R, Soussi M, Lucas MM, Felipe MR de, Harrison J, Vanacker H, Foyer CH (2005) Legume nodule senescence: roles for redox and hormone signalling in the orchestration of the natural aging process. New Phytol 165:683–701
Quirino BF, Reiter WD, Amasino RM (2001) One of two tandem *Arabidopsis* genes homologous to monosaccharide transporters is senescence-associated. Plant Mol Biol 46:447–457
Richter C, Schweizer M (1997) Oxidative stress in mitochondria. In: Scandalios JG (ed) Oxidative stress and the molecular biology of antioxidant defences. Cold Spring Harbor Laboratory Press, New York, pp 169–200
Río LA del, Pastori, GM, Palma JM, Sandalio LM, Sevilla F, Corpas FJ, Jiménez A, López-Huertas E, Hernández JA (1998) The activated oxygen role of peroxisomes in senescence. Plant Physiol 116:1195–1200
Río LA del, Corpas FJ, Sandalio LM, Palma JM, Gómez M, Barroso JB (2002) Reactive oxygen species, antioxidant systems and nitric oxide in peroxisomes. J Exp Bot 53:1255–1272
Rio LA del, Sandalio LM, Altomare DA, Zilinskas BA (2003) Mitochondrial and peroxisomal manganese superoxide dismutase: differential expression during leaf senescence. J Exp Bot 54:923–933
Rio LA del, Sandalio LM, Corpas FJ, Palma JM, Barroso JB (2006) Reactive oxygen species and reactive nitrogen species in peroxisomes. Production, scavenging, and role in cell signaling. Plant Physiol 141:330–335
Scandalios JG (2002) Oxidative stress responses – what have genome-scale studies taught us? Genome Biol 3:R1019.1–R1019.6
Shigeoka S, Ishikawa T, Tamoi M, Miyagawa Y, Takeda T, Yabuta Y, Yoshimura K (2002) Regulation and function of ascorbate peroxidase isoenzymes. J Exp Bot 53:1305–1319
Smart CM (1994) Gene expression during leaf senescence. New Phytol 126:419–448
Sohal RS, Weindruch R (1996) Oxidative stress, caloric restriction, and aging. Science 273:59–63
Stacy RAP, Nordeng TW, Culiáñez-Macià FA, Aelen RB (1999) The dormancy-related peroxiredoxin anti-oxidant, PER1, is localized to the nucleus of barley embryo and aleurone cells. Plant J 19:1–8
Stessman D, Miller A, Spalding M, Rodermel S (2002) Regulation of photosynthesis during *Arabidopsis* leaf development in continuous light. Photosynth Res 72:27–37
Svensson AS, Rasmusson AG (2001) Light-dependent gene expression for proteins in the respiratory chain of potato leaves. Plant J 28:73–82
Thomas H, Stoddart JL (1980) Leaf senescence. Annu Rev Plant Physiol 31:83–111
Thomas H, Ougham H, Wagstaff C, Stead AD (2003) Defining senescence and death. J Exp Bot 54:1127–1132
Thomas JC, Perron M, LaRosa PC, Smigocki AC (2005) Cytokinin and the regulation of a tobacco metallothionein-like gene during copper stress. Physiol Plant 123:262–271
Thompson JE, Ledg RL, Barber RF (1987) The role of free radicals in senescence and wounding. New Phytol 105:317–344
Umbach AL, Fiorani F, Siedow JN (2005) Characterization of transformed *Arabidopsis* with altered alternative oxidase levels and analysis of effects on reactive oxygen species in tissue. Plant Physiol 139:1806–1820

Van Doorn WG (2004) Is petal senescence due to sugar starvation? Plant Physiol 134:35–42
Van Staden J, Cook EL, Noodén LD (1988) Cytokinins and senescence. In: Noodén LD, Leopold AC (eds) Senescence and aging in plants. Academic, San Diego, pp 282–328
Vanacker H, Sandalio L, Jimenez A, Palma JM, Corpas FJ, Meseguer V, Gomez M, Sevilla F, Leterrier M, Foyer CH, Rio LA del (2006) Roles for redox regulation in leaf senescence of pea plants grown on different sources of nitrogen nutrition. J Exp Bot 57:1735–1745
Vandenabeele S, Van der Kelen K, Dat J, Gadjev I, Bonefaes T, Morsa S, Rottiers P, Sloten L, Van Montagu M, Zabeau M, Inzé D, Van Breusegem F (2003) A comprehensive analysis of hydrogen peroxide-induced gene expresión in tobacco. Proc Natl Acad Sci USA 100:16113–16118
Vanderauwera S, Zimmermann P, Rombauts S, Vandenabeele S, Langebartels C, Gruissem W, Inzé D, Van Breusegem F (2005). Genome-wide analysis of hydrogen peroxide-regulated gene expression in Arabidopsis reveals a high light-induced transcriptional cluster involved in anthocyanin biosynthesis. Plant Physiol 139:806–821
Von Zglincki T (2002) Oxidative stress shortens telomeres. Trends Biochem Sci 27:339–344
Watanabe K, Yamada N, Takeuchi Y (2006) Oxidative DNA damage in cucumber cotyledons irradiated with ultraviolet light. J Plant Res 119:239–246
Willekens H, Chamnongpol S, Davey M, Schraudner M, Langebartels C, Van Montagu M, Inzé D, Van Camp W (1997) Catalase is a sink for H_2O_2 and is indispensable for stress defence in C_3 plants. EMBO J 16:4806–4816
Wingler A, Schaewen A von, Leegood RC, Lea PJ, Quick WP (1998) Regulation of leaf senescence by cytokinin, sugars, and light effects on NADH-dependent hydroxypyruvate reductase. Plant Physiol 116:329–335
Wingler A, Marès M, Pourtau N (2004) Spatial patterns and metabolic regulation of photosynthetic parameters during leaf senescence. New Phytol 161:781–789
Woo HR, Kim JH, Nam HG, Lim PO (2004) The delayed leaf senescence mutants of *Arabidopsis*, ore1, ore3, and ore9 are tolerant to oxidative stress. Plant Cell Physiol 45:923–932
Ye ZZ, Rodriguez R, Tran A, Hoang H, Santos D de los, Brown S, Vellanoweth RL (2000) The developmental transition to flowering represses ascorbate peroxidase activity and induces enzymatic lipid peroxidation in leaf tissue in *Arabidopsis thaliana*. Plant Sci 158:115–127
Zapater JM, Guera A, Esteban-Carrasco A, Martin M, Sabater B (2005) Chloroplasts regulate leaf senescence: delayed senescence in transgenic ndhF-defective tobacco. Cell Death Diff 12:1277–1284
Zavaleta-Mancera HA, Franklin KA, Ougham HJ, Thomas H, Schott IM (1999a) Regreening of senescent *Nicotiana* leaves. I. Reappearance of NADPH-protophyllide oxidoreductase and light harvesting chlorophyll a/b-binding protein. J Exp Bot 50:1677–1682
Zavaleta-Mancera HA, Thomas BJ, Thomas H, Scott IM (1999b) Regreening of *Nicotiana* leaves. II. Redifferentiation of plastids. J Exp Bot 50:1683–1689
Zelisko A, Garcia-Lorenzo M, Jackowski G, Jansson S, Funk C (2005) AtFtsH6 is involved in the degradation of the light-harvesting complex II during high-light acclimation and senescence. Proc Natl Acad Sci USA 102:13699–13704
Zemojtel T, Frohlich A, Palmieri MC, Kolanczyk M, Mikula I, Wyrwicz LS, Wanker EE, Mundlos S, Vingron M, Martasek P, Durner J (2006) Plant nitric oxide synthase: a never-ending story? Trends Plant Sci 11:524–525
Zentgraf U, Hinderhofer K, Kolb D (2000) Specific association of a small protein with the telomeric DNA-protein complex during the onset of leaf senescence in *Arabidopsis thaliana*. Plant Mol Biol 42:429–438
Zimmermann P, Heinlein C, Orendi G, Zentgraf U (2006) Senescence specific regulation of catalases in *Arabidopsis thaliana* (L.) Heynh. Plant Cell Environ 29:1049–1060

Physiology

Application of Laser-Assisted Microdissection for Tissue and Cell-Specific Analysis of RNA, Proteins, and Metabolites

Dirk Hölscher and Bernd Schneider(✉)

Contents

Abstract The present state of different laser-assisted microdissection methods and their application in various disciplines of plant science is reviewed. The first part of the present review describes the basic effects of a laser beam on biological tissue, and following sections dealing with practical aspects of sample preparation and the technologies implemented in laser-assisted microdissection procedures. An attempt is made to distinguish laser-capture microdissection from laser cutting, to distinguish different varieties of laser cutting, and to discuss the advantages and drawbacks of the various methods for specific applications. The second part of the review covers applications of laser-assisted microdissection in various areas of plant science. Using lasers to conduct microsurgery on plant tissue and to dissect chromosome parts are two highly specialized areas that are discussed. RNA isolation and cell-specific gene expression analysis are the most frequent reasons for using laser-assisted microdissection methods, so a section comprising recent applications is included. Although proteomic methods for analyzing the contents of specific cell populations are available, proteins have been analyzed infrequently in laser-microdissected plant tissue. Therefore, the corresponding section also includes examples from animal samples.

Bernd Schneider
Max-Planck-Institute for Chemical Ecology, Beutenberg Campus, Hans-Knöll-Str. 8, D-07745 Jena, Germany
e-mail: schneider@ice.mpg.de

The final section is dedicated to the emerging field of cell type-specific metabolite profiling, including secondary metabolites, in extracts of laser-microdissected plant samples by mass spectrometry and nuclear magnetic resonance spectroscopy.

Abbreviations LAM: laser-assisted microdissection; generic term; LCM: laser capture microdissection; target cells are bonded by laser to a plastic film and captured by lifting; LMD: laser microdissection; laser cutting method, sample collected by gravity; LMPC: laser microdissection and pressure catapulting; laser cutting method, sample transferred to the collecting device by pressure catapulting.

1 Introduction

Detailed knowledge of non-uniform expression of RNA transcripts and cell type-specific biosynthesis of proteins and metabolites is fundamental for understanding the biochemical and physiological processes of organisms and their ecological interaction. Hence, microanalytical approaches suitable for miniaturized analysis of biological tissue and single cells have been developed, based on different methodological platforms and intended to achieve various objectives (Kehr 2003; Brandt 2005; Day et al. 2005). In the scope of the present review, the chemical composition, i.e. the identification and quantification of biomacromolecules and low-molecular-weight compounds occurring within specific cell types or single cells is of particular interest. Depending on the underlying physical principles, information about cell ingredients can be obtained in a minimally destructive manner or after disrupting the tissue or cells of interest. Methods such as highly sensitive UV and fluorescence detection, Fourier-transform infrared, and Raman microscopy are suitable for gaining spectroscopic information non-invasively. Moreover, widely used immunolocalization, histological staining, and in situ hybridization can be categorized as minimally invasive. In contrast, amplification of nucleic acid and methods of proteomic and metabolomic analysis require tissue to be disrupted and the contents processed. Therefore, although the results obtained by invasive methods from isolated plant cells might not always reflect the in vivo situation, the analysis of contents of isolated cells or cell populations represents considerable progress.

Several methods for harvesting single cells or specimens of microscopic size have been developed. Methods based on cell sorting by flow cytometry and micromanipulation of cultured cells, protoplasts, or homogenized tissue are not within the scope of this review. In a first step toward identifying cell ingredients, we focus here on the physical selective collection of specialized cells from the matrix of intact tissue or organs. Manual microdissection has been used, for example, to harvest specific plant cells and analyze their contents for metabolites (Outlaw and Lowry 1977). Glass microcapillaries are common tools for harvesting the contents of single cells (Karrer et al. 1995; Lochmann et al. 1998; Brandt et al. 1999). They are controlled by a micromanipulator and can be used in combination with a microchisel to dissect tissue areas of interest. A protocol for chisel-assisted mechanical microdissection of vascular bundles from *A. thaliana* and cucumber leaves has been reported (Brandt et al. 2003). These mechanical methods, though time-consuming, are useful for selecting and isolating groups of cells rather than individual cells. However, laser-assisted technologies seem to be more efficient than manual methods and currently are the first choice for harvesting single cells or other small-scale samples from intact tissue. After the effects of focused light energy on matter (see Section 2) were discovered, a variety of such laser-assisted microdissection techniques was developed (Meier-Ruge et al. 1976; Greulich and Weber 1992; Kubo et al. 1995; Emmert-Buck et al. 1996; Schütze and Lahr 1998). All these otherwise relatively different methods (see Section 4) directly visualize the cells of interest using a microscope and select them on the basis of morphological or other distinguishing characteristics. Laser-assisted dissection methods were primarily used for microsurgery and the microautopsy of mammalian tissue, especially neoplastic cells, in order to characterize disease morphology. After being used successfully for mammalian cancer tissue, their application extended to other life science disciplines, including plant physiology and plant biochemistry (Asano et al. 2002; Kerk et al. 2003; Nakazono et al. 2003).

Although this review is focused mainly on the application of laser-assisted microdissection (LAM) to plants, it starts with a description of the major basic processes that occurs when a laser impinges on biological tissue (see Section 2). Generally, the effects of the laser are not plant-specific but operate in almost the same way on all living matter. Moreover, LAM has been combined with various downstream analytical methods, which, in contrast to some steps of sample preparation (see Section 3), a priori are not plant-specific. Therefore, although our focus in this review is on plants, examples have also been selected from animal tissue, if they appeared feasible. Sections 5 and 6 deal with laser microsurgery of plant tissue and chromosome dissection, respectively. Among the analytical downstream methods, various techniques for gene expression analysis (see Section 7) have been combined with LAM and are used frequently for both plant and animal tissue. The situation is different for protein analysis of laser-microdissected tissue (see Section 8), which is still the domain of animal researchers. Nevertheless, a number such studies are reviewed here because the techniques are generally transferable to plants. To our knowledge, in contrast to proteins, metabolite analyses have been reported for plants only. Hence, Section 8 is dedicated to the emerging field of analysis of low-molecular-weight compounds in laser-microdissected plant samples.

2 Relationships Between Laser and Biological Systems

The development of the laser technique can be regarded as a technical revolution. Since the first demonstration of light amplification by stimulated emission of radiation by Maiman (1960), lasers are used in an almost uncountable number of applications in many fields of science, technology, and engineering, in addition to medical and biological sciences (Solon et al. 1961; Krasnov 1973; Srinivasan 1986; Gitomer and Jones 1991).

In contrast to metals and other solids such as synthetic polymers, biological tissues are low-absorbing and low-scattering materials. Biological samples show interaction phenomena not only on their surface but also in their interior (Lubatschowski and Heisterkamp 2004). The processes involved are complex because of the tissues' large water content and structural and chemical heterogeneity. Depending on the intensity and duration of the laser radiation, different processes of interaction with biological tissues have been described: photochemical reactions, coagulation, vaporization, photoablation, and photodisruption (Boulnois 1986). The laser parameters for photochemical reactions, coagulations, and vaporizations include low laser intensity and longer pulse duration times. According to the literature, photodisruption can be distinguished from the other processes because of the different energy absorption mechanisms involved. Details of these processes have been reviewed (Boulnois 1986) and are not discussed here.

In the early 1980s, a first description of a UV-laser treatment of organic polymers was published by Srinivasan and Mayne-Banton (1982). Using pulsed energy-rich UV radiation, a precise and gentle removal of organic material was achieved. A few years later, pulsed IR lasers were developed for removing biological tissue. Because the two different laser systems (UV, pulsed IR) lead to very similar phenomenological results, the term "photoablation" is used to describe the effect on biological material for both of them. However, depending on the laser system used and the chosen parameters, different mechanisms are thought to be responsible for removing biological tissue (Kawamura et al. 1982; Srinivasan and Mayne-Barton 1982; Garrison and Srinivasan 1985). In the case of UV lasers, photochemical processes seem to dominate the removal of the bioorganic matrix (Srinivasan and Leigh 1982; Garrison and Srinivasan 1985). Explosion-like vaporizations of water caused by IR laser radiation are suggested for removing biological material. The reasons for the differences between pulsed IR and UV laser ablation are still under discussion (Schmidt et al. 1998; Vogel and Venugopalan 2003).

Generally speaking, the term "ablation" can be used for any process of tissue removal, regardless of the photochemical or photophysical mechanisms involved. But not only the laser parameters determine the ablation process: the elasticity and mechanical strength of the biological tissue, its optical properties, and the amount, distribution, and organization of different compounds occurring in the target material also influence the process of tissue removal by pulsed laser ablation. Depending on the laser wavelength, tissue type, and optical penetration depth of laser activity in the biological material, the tissue area around the direct site of laser treatment can be seriously damaged. In practical applications, this necrotic area of biological tissue caused by the laser radiation should be kept as small as possible and the energy deposition should be confined to a small volume. Thermal damage to the biological tissue and heat transfer into neighboring tissue areas can be minimized by choosing laser duration times shorter than the so-called thermal relaxation time (Boulnois 1986). In unfavorable cases, however, the mechanical damage of the biological material induced by cavitation can be recognized even as far as 0.5 mm from the ablation site of the laser application. To avoid large thermal and mechanical side-effects and to achieve as precise a pulsed laser ablation as possible, the laser wavelength chosen should have a very small optical penetration depth and an ultrashort pulse duration.

The thermal denaturation of biopolymers can start with the weakening of hydrogen bonds and van der Waals interactions, which in intact chemical structures are a stabilizing configuration (Cantor and Schimmel 1980). In the case of UV laser ablation especially, the fission of covalent chemical bonds offers the possibility of removing biological tissue. The main polymer chain was found to be degraded after the photochemically induced breaking of side-chain bonds of biomacromolecules and the generation of volatile fragments (Kitai et al. 1991). But UV laser ablation seems to be an interplay of photochemical and thermochemical mechanisms. The process of ablation is driven by the formation of cavitation bubbles, the collapse of these bubbles, and the development of a jet formation (Oraevsky et al. 1995; Paltauf and Schmidt-Kloiber 1995, 1996). The process of ablation starts with the development of a vapour plume, and the ejection of material is the next step in the ablation process, coupled with photoacoustic phenomena.

The formation of shock waves and (under certain conditions) the development of a shock front are caused by explosive heating, the formation of bubbles, and the subsequent expansion of tissue after laser treatment (Emmony et al. 1976; Walsh and Deutsch 1991; Brinkmann et al. 1996; Vogel et al. 1996, 1999; Frenz et al. 1998). The recoil of the ablation fragments leaving the sample with extremely high velocity can cause very high pressure amplitudes in the samples (Zweig 1991; Ventzek et al. 1992; Doukas et al. 1995; Oraevsky et al. 1995). Although considerable progress has been made in recent years, much work remains to be done for the further understanding of pulsed laser ablation processes (Vogel and Venugopalan 2003; Vogel et al. 2005).

3 Sample Preparation

One of the most critical steps in LAM is the development and application of optimized specimen preparation methods. Preparation and (if necessary) staining techniques should both avoid causing structural disturbance within the cells and distorting the arrangement of the cells of interest in the plant tissue, and exclude the harvest of non-target neighboring cell groups. Furthermore, tissue preparation for localization studies requires preserving the chemical integrity of the target molecules, i.e. mRNA, proteins, or metabolites. Hence, the method of sample preparation must be balanced with the requirements for downstream analysis of the probed molecule type (Fig. 1).

Goldsworthy et al. (1999) reported protocols for the fixation and laser microdissection-assisted harvesting of frozen and paraffin-embedded animal material. One of the advantages of using frozen material, compared with paraffin embedding, is the higher yield of RT-PCR amplification products obtained from cell-specific RNA. As a consequence of specific structures and the compositions of plant and animal cells, different histological methods have to be used. The existence of cells walls and vacuoles in plant cells requires plant-specific tissue preparation. Vacuolated cells can be easily damaged after freezing and thawing because of the

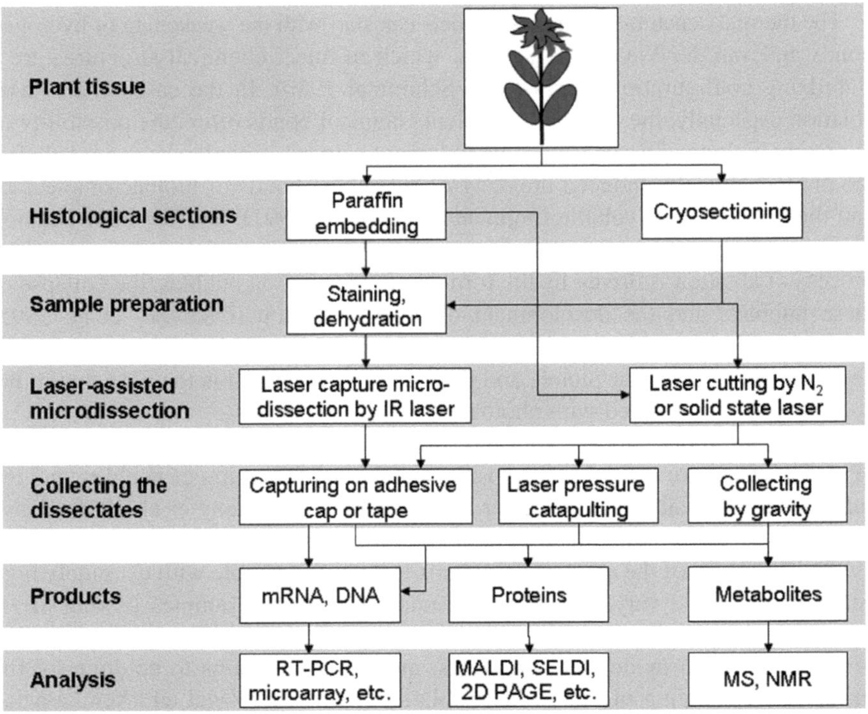

Fig. 1 Workflow of laser-assisted microdissection (LAM) and analysis of dissected cells or tissue. *RT-PCR* Reverse transcriptase polymerase chain reaction, *MALDI* matrix-assisted laser desorption/ionization, *SELDI* surface-enhanced laser desorption/ionization, *2D PAGE* two-dimensional polyacrylamide gel electrophoresis, *MS* mass spectrometry, *NMR* nuclear magnetic resonance spectroscopy

formation of large ice crystals in the vacuoles and intercellular spaces. Plant cell walls can tightly link cells of neighboring tissue, which should be separated from each other during LAM. To ensure cellular resolution of LAM and stability of plant tissue material, Kerk et al. (2003) and Nakazono et al. (2003) developed protocols to use paraffin-embedded plant material to harvest cell-specific RNA for amplification and expression analysis. Although different LAM technologies (see Section 4 for details) such as laser capture microdissection (LCM; Liu et al. 2005; Murata and De Luca 2005; Cai and Lashbrook 2006) and laser cutting methods, like laser microdissection (LMD; Klink et al. 2005; Sanders et al. 2005; Nakada et al. 2006) and laser microdissection and pressure catapulting (LMPC; Ivashikina et al. 2003; Ramsay et al. 2004), have been used for collecting specific cell types or tissue from plant material, many experiments are based on the protocol developed by Kerk et al. (2003) for fixation and sectioning. Other studies have introduced slight modifications to optimize the sample preparation procedure. Kerk et al. (2003) reported the differences in RNA yield when starting the fixation of plant cells with a coagulating (precipitative) fixative like ethanol–acetic acid or a cross-linking (non-coagulating) fixative like formaldehyde–acetic acid–ethanol. Two to three times more RNA was recovered from plant material fixed with the coagulating fixative. However, this fixative was frequently found to be suboptimal in preserving cell structures (Kerk et al. 2003; Ramsay et al. 2004). If necessary, the fixation step was coupled with vacuum treatment of the plant material to improve the sinking and extent to which the fixative infiltrated the tissue. Prior to paraffin embedding, Kerk et al. (2003) dehydrated the material with a graded series of ethanol followed by treatment with an ethanol:xylene series. After sectioning, the samples were deparaffinized in xylene and then warmed to remove traces of xylene. After that procedure, sections were ready to be cut.

In an additional preparation step, Nakazono et al. (2003) transferred maize kernel sections into 10% and 15% sucrose solutions to minimize the disastrous effect of ice crystal formation (Woll et al. 2005; Wu et al. 2006). The density of embryonic tissue of *Arabidopsis thaliana* made it possible to use freezing in liquid nitrogen-cooled isopentane instead of paraffin embedding prior to LCM to harvest intact cells for RNA isolation (Casson et al. 2005).

As demonstrated by Kerk et al. (2003) and other authors, treatment of plant tissue with solvents for fixation and dehydration is an important prerequisite for extracting intact RNA. However, this approach is not useful for tissue dedicated to the analysis of small molecules, which are soluble in the solvent used. For example, the fixative-free procedure of cryosectioning offered the only chance to exclude the total extraction of secondary metabolites from *Dilatris* species during preparation of the specimen. The development of a glass metal frame system made it possible to directly cut the secretory cavities from the leaves of herbarium material of *Dilatris* using a laser (Hölscher and Schneider 2007) without any sectioning method; and this seems to be the method of choice for desiccated plant material. Additionally, this method is also promising for other objects of limited thickness (Hölscher and Schneider, unpublished data). In conclusion, it has to be noted that the procedure for sample preparation depends on the type of target molecule and that it is necessary to optimize conditions for each cell type.

4 Laser-Assisted Microdissection Technologies

After the tissue has been prepared for LAM, the microtome section, which usually is mounted on a supporting membrane or a glass slide, has to be placed on the microscope table. The next critical step is the proper identification of the cells or areas of interest by means of the microscope. In favorable cases, this is possible on the basis of the shape, color, or autofluorescence of the target cells. Discrimination of originally non-recognizable target cells from surrounding non-target tissue frequently requires pretreatment such as staining, immunolocalization, or reporter gene expression. The potential impact of the pretreatment on the integrity of the material to be analyzed has to be taken into consideration early in the LAM procedure (see Section 3). Automatic cell recognition algorithms are integrated into most commercial laser-microdissection instruments. They are capable of automatically detecting and discriminating different cell types, if the target cells exhibit features which can be distinguished by the recognition software.

For the purposes of this review, the central step in harvesting cells and tissue of interest is laser-assisted microdissection. As outlined in Section 2 on the effect of laser light on biological tissue, the criterion for denominating a method as LAM is the ability to remove or to support the removal of tissue pieces, cells, or cell compartments by the action of a laser beam. LAM is the most inclusive term for this group of methods. According to the specific approaches and instrumentation, different terms and abbreviations were used to name special LAM methods. However, all these slightly different approaches can be attributed to two basic variations: LCM and laser cutting. The LCM methodology (Emmert-Buck et al. 1996) makes use of a low-energy near-infrared laser to melt a thermoplastic polymer film onto the region of interest of a tissue section, which is mounted on a glass slide. The film is attached to a transparent transfer cap through which the laser beam is targeted to the tissue. The cap moves downwards until the film is close to or touches the tissue, then the IR laser starts pulsing to expand the thermoplastic polymer and make it adhere to the target cells. When the cap is lifted, the fused cells are selectively removed while the surrounding tissue remains on the glass slide. Thus LCM does not ablate the tissue but mechanically separates it, assisted by the laser beam which attaches the cells of interest to the target.

Unlike LCM, laser cutting (comprising "laser microdissection", LMD, LMPC), makes use of a pulsed UV laser microbeam (nitrogen gas or solid state) that is focused on the tissue to be dissected. The energy of the laser beam at the point of the focus is sufficiently high to disintegrate biomolecules by photofragmentation (see Section 2). The tissue outside the focus remains almost unaffected. For cutting a regular or irregular area of interest, the laser beam moves, following an imaginary track predefined by the software (Fig. 2A). When, after one round of cutting around the delineated area (or several rounds in the case of thick sections or dense material), the laser beam reaches the starting point again, the desired specimen is separated from the matrix (Fig. 2B). At this point, the separated cells or tissue piece has to be transferred into a collection device for further treatment. In an upright microscope

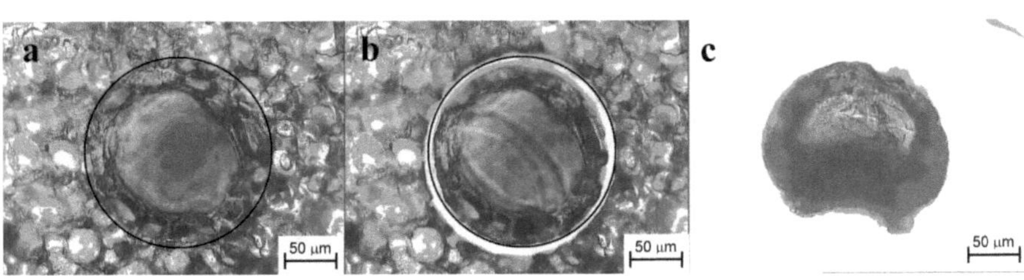

Fig. 2 Microscopic images illustrating how a secretory cavity of *Dilatris pillansii* (Haemodoraceae) is cut by a laser beam (Hölscher and Schneider 2007). **a** Track for microdissection marked by LMD software. **b** Secretory cavity immediately after microdissection. **c** Secretory cavity detected by the microscope in the collection device

configuration, the laser comes from above and the transfer happens simply by gravity. The laser-microdissectate falls into a microcentrifuge cap placed directly beneath the membrane slide on which the section is mounted. In the case of an inverted microscope configuration, the laser comes from below and collection by gravity is not possible. However, a laser from the same source, but in a defocused mode, can be used for catapulting the separated specimen upward where it is captured in a collection device. Alternatively, capturing the laser-dissected specimen onto an adhesive film is possible. Manual transfer using a sharp dissecting needle under a stereomicroscope (Hölscher and Schneider 2007), though time-consuming, is feasible for large cells or multicellular specimens.

From the many applications reported in the literature, it is obvious that LAM methods are extremely useful for harvesting specific tissue or cells from various organisms, including plants, and that intact biomacromolecules (see Sections 7, 8) and small metabolites (see Section 9) can be isolated from the dissectates. Each of the LAM technologies, in addition to sample preparation (see Section 3), has its specific advantages and drawbacks (Nelson et al. 2006). Negative or undesired effects may occur due to enhanced temperature and laser irradiation. Contamination may originate from adjacent tissue, polymer foil, adhesive material, and other external sources. The pluses and minuses of a LAM workflow may differently affect the downstream analysis of DNA, RNA, proteins, and small molecules. For example, contact with adhesive material is not a problem for nucleic acid amplification and protein analysis but has to be taken into account for the identification of low-molecular-weight metabolites. Contamination especially is problematic for analysis, e.g. by ^{1}H NMR spectroscopy, of crude extracts without separating individual components (Hölscher and Schneider 2007; Li et al. 2007). The use of nitrogen or solid state lasers does not influence the structural integrity of DNA and RNA in the target cells for two reasons. First, the energy is high only in the narrow focus of the laser. Second, the irradiation wavelength of the nitrogen and the solid state laser are outside the absorption maxima of nucleic acids and most proteins. In the case of low-molecular-weight metabolites, photochemical reactions cannot be excluded. However, no photochemical degradation of cell-specific metabolites has been reported yet upon laser cutting (Schad et al. 2005b; Hölscher and Schneider 2007; Li et al. 2007).

During LAM, the samples are exposed to room temperature or, due to the action of the IR laser in LCM, even elevated temperature. This is also the case for cryosections because the dwell time on the microscope table almost exceeds the thawing time for frozen tissue. Hence, sensitive proteins may denature during the procedure.

During laser-cutting, the motion of dissectates to the collecting vial, forced either by gravity (during LMD) or laser catapulting (during LMPC), may by disturbed by hydrodynamic instability, density changes, and convection, resulting in deviation from the desired straight trajectory line. Thus, although most of the microdissected specimens properly approach the target vial, a certain fraction does not, but instead drifts laterally away. The dynamic behavior of dissectates depends on their particle size, shape, and mass:size ratio (Elvers et al. 2005). The yield, i.e. the fraction of successfully collected dissectates depends also on the velocity of the specimen and the distance of the collecting vial from the dissection site. It is plausible that larger dissectates having higher mass and compact shape fall down faster and are less influenced by convection in a gravity-driven collection system. Items catapulted against gravity have an upper limit of mass and the efficiency seems to depend on the shape of the dissectate. Such issues do not account for LCM and LMD variants which capture the dissectate by adhesion to a transfer cap.

In summary, the availability of various commercial LAM instruments making use of different methodology allows the costumer to select the most appropriate one for user-specific applications. Further information about LAM technology is available by personnel communication with the producers, at their websites (http://www.arctur.com/; http://www.leica-microsystems.com; http://www.molecular-machines.com/; http://www.palm-microlaser.com/; Burgemeister 2005), and in partly or fully independent studies and reviews (Elvers et al. 2005; Kehr 2003; Nelson et al. 2006; Day et al. 2007).

5 Laser-Ablative Microsurgery of Plant Tissue

Laser ablation for removing or damaging specific microscopic plant tissue or organs in order to observe the effect of this manipulation offers the opportunity to revisit and extend concepts of classic microsurgical plant investigations carried out between the 1920s and the 1970s. The combination of modern microscopy techniques and LAM technologies provides access to smaller and better-defined pieces of plant tissues, with minimal destruction of adjacent tissue, as was possible in classic microsurgical plant experiments. Depending on the optical penetration depth, only a small zone of plant tissue at the ablation site is damaged by laser activities.

A series of high-precision laser ablation and microsurgical tissue removal experiments to test the functions of different parts of the meristem of *Lycopersicon esculentum* and to reveal their interactions have been reported (Reinhardt et al. 2003). An infrared Er:YAG laser with high ablation efficiency and precision was used to ablate the central zone and different cell layers (e.g. the external L1 layer) of the tomato meristem. The results indicated that, while the zones of the meristem show a remarkable capacity to regenerate after interference, L1 cells were never observed to regenerate after ablation. Instead, terminal differentiation occurred. Elegant control experiments excluded wound effects from the lesion or secondary stress signals as causes for the ectopic induction of the *WUSCHEL* gene, which plays a key role in specifying and maintaining stem cells.

Repeating classic studies (Snow and Snow 1931) on phyllotaxis and the leaf polarity of *Lycopersicon esculentum*, Reinhardt et al. (2005) showed that laser ablation techniques yielded different results, compared with classic incision procedures. The precise removal of the incipient primordium by laser ablation led to the formation of a new primordium either close to the lesion or in an ectopic position. Previous surgical incisions had not shown such a high degree of flexibility in the meristem (Snow and Snow 1931). Unlike laser ablations, mechanical incisions caused major damage to the peripheral zone and reduced the space available for a new primordium. Two examples (Reinhardt et al. 2003, 2005) demonstrate the technical superiority of laser ablation compared with the conventional surgical techniques used in classic studies on phyllotaxis and leaf polarity.

6 Laser-Assisted Microdissection of Chromosomes

The preciseness of LAM not only allows tissue pieces and individual cells to be cut, but also subcellular-sized particles. Experimental laser devices or commercial laser microdissection instruments have been used since the early 1980s for collecting chromosomes or chromosome fragments of human and animal material (Scalenghe et al. 1981; Monajembashi et al. 1986), and only a decade later botanists were starting to use this technique. Fukui et al. (1992) used a cell work station equipped with an argon-ion laser to cut the chromosomes of *Oryza sativa* cv. Nipponbare and *Hordeum vulgare* cv. New Golden. These authors determined the optimal intensity needed for a laser beam to dissect chromosomes and were able to prepare chromosomal samples of the two species on a polyester membrane and on ordinary glass slides. The laser proved useful for obtaining specific regions of the centromere, satellites, and short or long arms of the barley and rice chromosomes. A successful transfer of the microdissected chromosome material into Eppendorf tubes completed this first laser-assisted separation of chromosomes and chromosome fragments from plant material.

Sex chromosomes have been studied in the dioecious plant *Silene latifolia*. Scutt et al. (1997) isolated X and Y chromosomes from mitotic metaphase preparations of *S. latifolia* on polyester membranes. Autosomes were ablated using an argon-ion laser and the DNA from isolated sex chromosomes was subjected to degenerate oligonucleotide-primed (DOP)-PCR. The results were used to get more information about the repeat sequences and genomic organization of the sex chromosomes.

After a hole was drilled in a pollen grain from the anthers of a male flower of *S. latifolia* with a UV laser microbeam, the pollen grain was catapulted by LMPC into the cap of a PCR tube (Matsunaga et al. 1999a). The entire genome of the single pollen grain was amplified with primer-extension preamplification (PEP)-PCR. The authors detected a number of genes in a single pollen grain. In the same year, Matsunaga et al. (1999b) published a UV laser application approach for isolating Y chromsomes to analyze sex-determining genes located on the male chromosome of the heteromorphic chromosomes of *S. latifolia*.

For further investigations into the sex chromosomes of the diploid genome of *S. latifolia*, Hobza et al. (2004) developed an improved FAST-FISH protocol. The X and Y chromosomes, which can be easily distinguished during metaphase from each other and from the autosomes, were isolated by LMPC. Ten microdissected chromosomes were subjected to DOP-PCR. The FISH experiments revealed a strong signal of the DOP-X probe on the entire X chromosome and signals of low intensity on all the other chromosomes.

7 Analysis of RNA in Laser-Microdissected Samples

Numerous examples exist in the biomedical literature of how LAM, especially LCM, has been used to isolate RNA and to analyze gene expression. Although launched later than LMD for mammalian systems, RNA isolation combined with gene expression profiling is by far the most frequent downstream application for plant dissectates (Kehr 2003; Day et al. 2005; Lange et al. 2005; Nelson et al. 2006; Ramsay et al. 2006).

Asano et al. (2002) were the first to publish the application of LAM for plant tissue. A modified version of LMPC was used to dissect phloem cells of rice (*Oryza sativa*) leaf tissue and subsequently to analyze gene expression profiles. The morphology of the phloem cells is apparently different from that of the surrounding cells and therefore they are easily detectable with the microscope. Cryosections of 20 μm thickness were used to harvest approximately 150 individual phloem cells. From this material, total RNA was isolated, amplified using T7 RNA polymerase, and a cDNA library was constructed. Sequencing the library identified a variety of genes localized in the phloem, indicating that a limited number of microdissected phloem cells were sufficient for gene expression analysis. The efficiency of the method was proved by in situ hybridization-based detection of a putative amino acid permease, which is specifically expressed in phloem cells.

Kerk et al. (2003) optimized the protocols of sample preparation for a variety of plant tissues and species, and evaluated the suitability of LCM for gaining cellular information on the genome and proteome levels. The authors suggested the precipitative fixation with ethanol–acetic acid as a method of choice for successful RNA recovery. The advantages of paraffin embedding were reported, different RNA extraction methods tested, and the specificity and quality of RNA from neighboring mesophyll and bundle sheath cells evaluated from paraffin-embedded sections of *Zea mays* leaves. The successful detection of the presence of C4-specific NADP-malic enzyme showed the possibility of RNA recovery from laser-microdissected individual plant cells.

Nakazono et al. (2003) used ethanol–acetic acid-fixed coleoptiles of *Z. mays* to obtain more than 10 000 laser-microdissected epidermal cells and a complex of vascular bundles and bundle sheath cells. Approximately 40 ng of RNA were isolated and used for T7 RNA amplification. Hybridization with a *Z. mays* cDNA microarray showed, for example, the high expression of genes of the phenylpropanoid and phenylpropanoid–acetate pathway in epidermal cells. In total, expression data specific to cell type were reported for more than 120 novel maize genes. LCM and cDNA microarray techniques were demonstrated to be a feasible combination for conducting high-resolution global gene expression analyses of plants.

Root-knot nematodes (*Meloidogyne* spp) induce the formation of specific feeding cells called giant cells in host roots. Knowledge about the molecular processes involved in the induction and differentiation of giant cells is limited because it is difficult to obtain pure cytoplasm specifically from the highly specialized cells. LMPC was used by Ramsay et al. (2004) to overcome this problem. Paraffin-embedded root material of *Lycopersicon esculentum* inoculated with *Meloidogyne* was used four days post-inoculation to get pure cytoplasmic contents from individual giant cells. Acridine orange staining indicated that RNA is retained through the various stages of sample preparation. Total RNA was isolated from the dissected giant cells and used in RT-PCR to investigate gene expression. The expression of cell cycle-related cyclin genes was higher in giant cells than in cells of the surrounding tissue. Finally, the authors reported the construction of a cDNA library for studying further host–pathogen interactions.

Conditions for LAM are not generally useful for each plant tissue, but it is necessary to optimize tissue preparation and the dissection procedure for each target cell type. For example, due to the specific morphology, isolating certain cells or tissue for cell-specific transcriptome and proteome analysis from the model plant *A. thaliana* can be difficult. The leaf vasculature, especially, is a difficult tissue to access (Pommerrenig et al. 2006; Hölscher and Schneider, unpublished data). Ivashikina et al. (2003) applied LMPC to the isolation of *Arabidopsis* vascular mRNA from flower stalks. One hundred and fifty phloem regions were pooled and probed for the vascular-specific expression of novel phloem-specific transcripts, like the *AtSUC2* Suc transporter gene, the *AKT2* K^+-channel gene, and *AHA3*, the gene of the plasma membrane H^+-ATPase. In the same article, 700 expressed sequence tags (ESTs) were reportedly derived from *Arabidopsis* companion cell (CC) protoplasts and stress-induced *AtSUC3* transcripts detected in the protoplasted mesophyll cells.

Casson et al. (2005) applied LCM to analyze embryogenesis of *A. thaliana*. A special cryosectioning procedure for the dense embryonic cells was developed to avoid formation of ice crystals, which often appear in vacuoles and intracellular spaces of mature tissue. Using nitrogen-cooled isopentane and cryosectioning at −22 °C were essential steps for successful RNA extraction. The RNA quality from the microdissected cells was sufficient for RNA amplification and for hybridization to Affymetrix microarrays. The combination of LCM with DNA microarray analysis identified strongly differentially expressed genes, which are regulated both spatially and temporally. Approximately 65% of the genome of *A. thaliana* was shown to be expressed in the developing embryo. The limits of combining LCM and DNA microarray analysis on genes that encode for low levels of transcripts were also discussed.

A topical research involves cell-specific localization of the biosynthesis of the dimeric indole alkaloids vinblastine and vincristine, which are important anticancer drugs. Recent data obtained by in situ hybridization and immunolocalization of *Catharanthus roseus* suggested that specific leaf cells are involved in vindoline biosynthesis. Murata and De Luca (2005) demonstrated that LCM is a useful technique for harvesting different cell types from leaves of *C. roseus*. They adapted the methodology of tissue fixation, embedding, and LCM reported by Kerk et al. (2003) to capture specific cell types from epidermis cells, palisade mesophyll cells, palisade-assisted idioblast cells, cross-connected laticifer cells, and vascular cells from the base part of young leaves of *C. roseus*. RNA was extracted from 2000–5000 pooled cells from each cell type. Nevertheless, the amount of RNA obtained from LCM-dissected cells was insufficient for direct RT-PCR analysis. Therefore the RNA was subjected to T7-based amplification. The majority of key enzymes of the vindoline biosynthesis, e.g. tabersonine 16-hydroxylase, were shown to be detectable in the epidermal cells; and it was shown that at least the last three steps of the pathway are probably expressed in separate cells (mesophyll/idioblast/laticifer) within the leaf of *C. roseus*. A complementary non-laser-assisted technology, a modified carborundum abrasion technique, was applied to obtain epidermis-enriched leaf extracts to measure alkaloid metabolite levels, enzyme activity, and gene expression.

The adventitious rootless 1 (*ARL*1) gene, a key regulator of adventitious root development in *Oryza sativa*, was identified by Liu et al. (2005) from a tissue culture-derived rice mutant. *Arl*1 mutants showed impaired formation of the adventitious primordium. The investigation of genes expressed at the initiation of periclinal division revealed that the rice *WUSCHEL*-type homeobox gene QHB and the *OsSCR* gene, compared with wild-type plants, were not expressed in *arl*1 mutant plants. Tissue-specific RNA was obtained from paraffin-embedded and laser capture-microdissected stem base tissues and used for RT-PCR analysis. The quality of the LCM-based tissue separation was validated by strong expression of the marker gene *OsNAS1* exclusively in pericycle cells of both the wild-type and *arl*1 mutant plants.

The plant cells selected by nematodes as a site for feeding and accommodation during certain phases of their life cycle undergo a variety of well studied cytological changes. Ongoing molecular changes during the formation of syncytia caused by the soybean (*Glycine max)* cyst nematode *Heterodera glycines* were studied by Klink et al. (2005). Root samples enriched in syncytial cells were collected at a defined time-point (day 8) by LCM. RNA from LCM-derived tissue enriched for syncytia was used to create a cDNA library for cloning and gene ontology analysis, to perform RT-PCR experiments, and to generate probes for in situ hybridization. Using this strategy, a subset of genes was detected which was differently expressed in LCM-derived syncytia-enriched tissue and whole roots.

In the case of *Nicotiana tabacum*, dehiscence, which is the terminal step in anther development that releases pollen grains, requires two groups of cells: the stomium and the circular cell cluster. Sanders et al. (2005) isolated approximately 400–500 stomium cells (a specialized set of epidermal cells that degenerate and break at flower opening to allow pollen grains to be released) from the notch region of the anthers. The paraffin-embedded anther material was microdissected using the LMD method. Real-time quantitative reverse transcription (qRT)-PCR revealed the presence of TA56 and TA20 mRNAs, which encoded for a thiol endopeptidase and a protein of unknown function, respectively, and confirmed earlier in situ hybridization procedures (Cox and Goldberg 1988). The authors estimated the presence of 7000 TA20 mRNA and 2000 TA56 mRNA molecules per stomium cell, representing approximately 1.4% and 0.4% of the mRNA transcripts at a certain stage (+6) of stomium development. The results indicated that temporally regulated cellular processes involving specific gene sets are required for anther dehiscence.

Cai and Lashbrock (2006) used LCM to harvest cells from ovules, replums, and stamen abscission zones of *A. thaliana* to extract structurally intact RNA. Paraffin embedding and an ethanol–acetic acid mixture for fixation were used for preparing samples in these experiments. The authors checked the RNA yield and integrity throughout the entire procedure and optimized the protocol. Mounting the sections onto slides was identified as the most critical part of the procedure. A tape-based paraffin section-mounting system (which makes use of adhesive-coated slides and completely avoids exposure to both water and heat) resulted in maximum RNA yield and structural integrity. Integrity was assessed directly by electrophoretically separating picogram and nanogram levels of total RNA isolated from multiple cell types. The system was shown to have the sensitivity to reliably detect transcripts below 0.002% mRNA abundance. Nevertheless, the expression of other extremely rare transcripts likely falls below the sensitivity threshold of the experimental system. RNA amplifications were hybridized to GeneChips and 37 significantly regulated genes were identified from gene families involved in modifying the structure of cell walls, signal transduction, cellular metabolism, protein transport, and protein storage.

Recently Corpas et al. (2006) used LMPC on cryosections of different cell types of *Olea europaea* cells. They reported using 14–16 µm thick microdissected leaf sections of spongy, mesophyll, palisade mesophyll, xylem, and phloem tissue. The region-specific RNA was isolated and subjected to qRT-PCR, and the differences of distribution of superoxide dismutase (SOD) were measured. Gene expression of the different SOD isozymes (Fe-SOD, Mn-SOD, CuZn-SOD) varied depending on the tested cell types of the leaf.

Jiang et al. (2006) used LCM on paraffin sections of different cell types from the root tip of *Zea mays* to verify results obtained from qRT-PCR experiments and Affymetrix GeneChip arrays after manually microdissecting *Z. mays* root cap tissue. The RNA was obtained from laser-microdissected material from the proximal meristem, from the quiescent center, and from cells from two regions of the root cap: the central columella and the lateral root cap (LRC). The upregulation of GDP-mannose pyrophosphorylase, especially, in laser-microdissected LRC could easily explain the high fucose concentration in the root cap of *Z. mays*.

Since flower buds of *Muscari armeniacum* are too small to be manually separated into floral organs, LMD was used to collect the individual organs, and from these, total RNA was then prepared for RT-PCR (Nakada et al. 2006). In this study, *MaDEF*, a DEF-like class B gene (classified into *MADS*-box genes) from *M. armeniacum*, was isolated and its expression pattern assessed by Northern blot analysis and RT-PCR. *MaDEF* expression was searched for in 20 sections of the outer tepal, inner tepal and stamens, and in 21 sections of the carpels from *M. armeniacum*; and it was found in the inner and outer tepals and stamens. The authors recognized the non-similarity of the expression pattern of the *DEF*-like genes compared with other members of the asparagales and intend to start further investigations of the expression pattern of class B genes from Convallariaceae and Asparagaceae.

Cryosections of the ovary of *Gossypium hirsutum* were subjected to LMPC in order to probe for spatial and temporal expression patterns of genes involved in early cotton fiber development (Wu et al. 2006). The extracted RNA was subjected to RT-PCR. The homeodomain gene and a Myb transcription factor (GhMyb25) were shown to be specifically expressed in ovules and to coincide with the time and location of cotton fiber initiation. GhMyb25 overexpression in transgenic tobacco revealed similarities between cotton fiber initiation and leaf trichome formation, depending on DNA endoreduplication.

The regulation of the *GAMYB* gene, a component of gibberellin signaling in cereal aleurone cells which plays an important role in flower development, was examined in *Oryza sativa* (Tsuji et al. 2006). When co-expressed, *GAMYB* and *GAMYB*-like genes showed a negative correlation with microRNA (miRNA159) levels during anther development. After in situ hybridization experiments failed, LMD was used to obtain the tapetum layer as well as the epidermis with endodermis and vascular tissues from anthers, and to confirm the regulation of the expression of *GAMYB* genes by miR159. RNA was extracted from the microdissected material and amplified by qRT-PCR. Tissue-specific expression of *GAMYB* and *GAMYB*-like genes was found and the level of the product derived from miR159 was similar in all investigated tissues, confirming the miR159 regulation of the expression of *GAMYBs* in their natural context. Microarray analysis revealed the regulation of different sets of genes by *GAMYB* in anthers and aleurone cells.

Woll et al. (2005) studied cell-type-specific transcriptome profiling of the primary root pericycle of maize (*Z. mays*), where lateral roots are initiated. The authors compared pericycle cells from *Z. mays* wild type and the *rum1* (*rootless with undetectable meristems 1*) mutant, which were affected in lateral and seminal root initiation. Using cryosections and a tape-transfer system, LCM resulted in the isolation of over 13 000 cells of primary roots. Subsequent microarray analyses of 12k maize microarray chips revealed 90 genes preferentially expressed in the wild-type pericycle and 73 genes preferentially expressed in the *rum1* pericycle. A detailed discussion of the differences of auxin transport in the *rum1* mutant in comparison with the wild type of *Z. mays* showed the importance of LCM as one of the tools for elucidating putative checkpoints in the molecular network of *Z. mays* root system formation.

8 Analysis of Proteins in Laser-Microdissected Samples

The total protein level has been reported to be as low as 2% in potato tuber cells and 37% in lupin seed cells (Pühler et al. 2000). One single cell may contain up to several thousand different proteins with molecule numbers ranging from 10 to 10^{12}. In contrast to RNA and DNA, amplification techniques for proteins are not available. Clearly protein analysis suffers from low concentrations and is dependent on tissue and sample. Taking the limited availability of laser-microdissected material into account, only a small number of the most abundant and sufficiently stable proteins are available for analysis, and these may in addition have only housekeeping functions rather than a role in specific questions of interest. Despite recent developments in protein isolation and separation, identifying proteins in some cases is still a challenge. Nevertheless a number of investigations have identified proteins from laser-microdissected cell populations, most of which were obtained from human tissue; their aim was to detect differences in protein patterns between healthy and diseased cells. The methodology, especially when combined with techniques of protein analysis developed in human and animal studies, may be useful for plants as well.

Details of tissue preparation and the suitability of various human pathological samples and their impact on the potential of LCM in combination with 2D-PAGE have been evaluated (Craven et al. 2002). In this study, the success of this combination was judged by protein yield as well as the degree of enrichment relative to the original material, protein modifications or degradation, and the time taken for dissection.

Immunoassays are another group of techniques with great potential for detecting a subset of proteins in a specific tissue. For example, a quantitative chemoluminescent immunoassay has been used to determine a prostate marker protein from stained human prostate tissue cells selected and procured by LCM (Shekouh et al. 2000). For further examples of combining antibody technologies and LCM, see the review of Craven and Banks (2001).

LCM has been coupled to reverse phase protein arrays, a technology involving proteomics microarrays with which the fluctuating state of the proteome in minute quantities of cells can be studied. The method has been used, for example, to investigate signaling changes during the progression of prostate cancer (Paweletz et al. 2001; Grubb et al. 2003).

A protein mixture obtained by LCM from dehydrated breast cancer cells and solubilized with a denaturing buffer was digested with trypsin; the resulting peptide mixture was fractionated using reversed-phase HPLC coupled to mass spectrometry (LC-MS/MS). The identity of the proteins was determined by comparing fragments with genomic and proteomic databases (Wu et al. 2003). A LC-MS/MS approach was also employed by Zang et al. (2004).

LCM was used to generate sufficient proteins of suitable quality for 2D electrophoresis-based protein expression profiling (Shekouh et al. 2003). Coomassie-stained protein spots of interest were excised from the SDS PAGE gel and trypsin-digested. Differences in protein expression between normal and malignant cells were reproducibly detected and one protein was identified by matrix-assisted laser desorption/ionization–time of flight (MALDI-TOF) mass spectrometric analysis.

Direct acquisition of MALDI-TOF spectra without elaborate sample preparation and purification steps was used to obtain abridged protein expression profiles from microdissected cells of human samples (Palmer-Toy et al. 2000; Bhattacharya et al. 2003). An improved method for preparing cells obtained by LCM and subsequent analysis with MALDI MS was developed and used to study various human and animal samples (Xu et al. 2002). The authors demonstrated that histological staining of tissue prior to LCM reduced the quality of the spectra, while dehydration of tissue using organic solvent washes did not result in significant protein loss from the samples.

SELDI-TOF-MS represents an alternative protein expression profiling technique that makes use of ProteinChip technology to selectively analyze subsets of proteins retained on affinity surfaces due to their physico-chemical characteristics, followed by direct TOF-MS analysis. In combination with 2D gel electrophoresis, tandem mass spectrometry, and immunohistochemistry, for example, the method was applied to identify a marker protein from malignant human cell populations (Melle et al. 2003) and other human carcinoma cells (Melle et al. 2004). In contrast to other human and animal protein studies cited in this review, all of which employed LCM to dissect the cells, the two latter investigations used the LMPC technique. The method did not influence the protein profile. In these studies, the minimum quantity of tissue that was required to generate satisfactory results by SELDI TOF-MS analysis was about 700 cells. It was reported that 500–2000 laser-microdissected cells were sufficient for recording complex protein profiles (Kwapiszewska et al. 2004).

The possibilities and limitations of LCM in proteomic analysis have been reviewed (Craven and Banks 2001; Ball and Hunt 2004). The authors emphasized that established protein detection methods such as 2D PAGE and Western blotting are compatible with LCM methodology. Though mass spectrometry-based methods, especially MALDI MS and SELDI MS, are more sensitive, they are considered complementary to PAGE since they are more useful for smaller proteins between 2 kDa and 70 kDa (Caldwell and Caprioli 2005). Most of the sample preparation procedures originally developed for human and animal samples can be easily adapted to plant applications.

Schad et al. (2005a) evaluated 2D electrophoresis and LC-MS/MS for the tissue-specific protein profiling of laser-microdissected material of *A. thaliana*. Preparation of samples from 40-day-old plants was carefully optimized. Cryopreparation retained the morphology of plant tissue sections and was efficient in protein extraction because it did not lead to protein degradation or modification. Vascular bundles and sections without vascular bundles were dissected from cryosections using LMPC (nitrogen laser method). Protein analysis was conducted either by classic 2D gel electrophoresis or by LC-MS/MS. The latter method required less plant material and was selected as the more suitable approach. The optimized nanoLC-MS/MS method resulted in the identification of 33 proteins in vascular bundles and 131 proteins from sections without vascular bundles.

9 Laser-Microdissection for Analysis of Cell Type-Specific Metabolites

The spatial distribution of small molecules in plant tissue is of great interest to researchers in plant physiology, biochemistry, and chemical ecology. Various approaches have been used to identify the distribution of primary and secondary metabolites in or from plant tissue. These methods can be divided into in vivo methods, which non-invasively detect the metabolites within the intact cell of interest, and methods based on sampling the cell contents followed by physical and chemical analytical methods. In situ measurements of UV absorption (microspectral photometry) and confocal laser scanning microscopy-based fluorescence detection are suitable tools for detecting metabolites or groups of related metabolites by means of their absorption or emission properties in intact cells. Raman spectroscopy is increasingly employed to analyze metabolite composition on the plant surface with cellular resolution (Petry et al. 2003). One possible way to sample the contents of an individual cell is to use a microcapillary mounted on a micromanipulator. Such techniques, combined with analytical methods which are sufficiently sensitive to separate and/or identify metabolites obtained from an individual cell, were reviewed by Tomos and Sharrock (2001). Though an established method, sampling from individual cells using a microcapillary is very time-consuming and almost restricted to cells at the surface of a tissue or section. A suitable technique for separating the contents of a single cell is micellar electrokinetic chromatography, which has been used for amino acid analysis in nanoliter samples from plants (Zhu et al. 2005).

Metabolic profiling of vascular bundles obtained by laser microdissection of *Arabidopsis* cross-sections has been carried out (Schad et al. 2005b). Cryosectioned stem material (cross-sections of 30 μm thickness) was subjected to LMPC. Vascular bundles were dissected by a focused laser beam (UV 337 nitrogen laser) and catapulted by a defocused laser pulse into the cap of a 0.5-ml reaction tube which was filled with ethanol to denature metabolic enzymes and protect the cell contents from unwanted modification. About 100 vascular bundles, corresponding to approximately 5000 cells, were harvested, extracted with ethanol and, after derivatization with MSTFA, subjected to GC-TOF MS analysis. The results were compared with those from the remaining material (sections without vascular bundles). Sixty-eight metabolites were identified in vascular bundles and 65 in the sections without vascular bundles. Principal component analysis was used for statistical evaluation. About half of the identified metabolites were shown to be either enriched or depleted in vascular bundles compared with the sections without vascular bundles. Since only about the half of the metabolites identified from *Arabidopsis* vascular bundles and samples without vascular bundles were specified (Schad et al. 2005b), it is unclear whether secondary metabolites were among these compounds. Some of the metabolites found in this study proved sensitive to the sample preparation procedure. For example, the ratio of dehydroascorbic acid found in fresh cryosections versus that found in LMPC sections was 8.5. Moreover, the procedures used to prepare tissue for protein and RNA analysis, especially dehydration by organic solvent (see Section 3), would have resulted in the loss of small molecules. However, it is essential for metabolite analysis to avoid using staining, cryoprotecting, and dehydrating agents because the materials used for sample preparation are the sources of many low-molecular-weight contaminants such as plastifiers, adhesives, and lipids. In contrast to metabolite identification, proteomic and nucleic acid analysis is less susceptible to low-molecular-weight impurities. Contamination originating from adhesive tape were found in our preliminary experiments to laser-dissected plant samples. Chemically different contaminants were detected by NMR when adhesive tape from different suppliers was used, even if the tape did not directly touch the dissected cells.

Quantitative microanalysis of hydrolysed cell wall polysaccharides of lignified and unlignified parenchyma cells and xylem fibres obtained by LMPC of *Urtica dioica* has been reported (Angeles et al. 2006). Purification, cell wall hydrolysis, and other downstream steps for analyzing the micro-quantity material of each cell

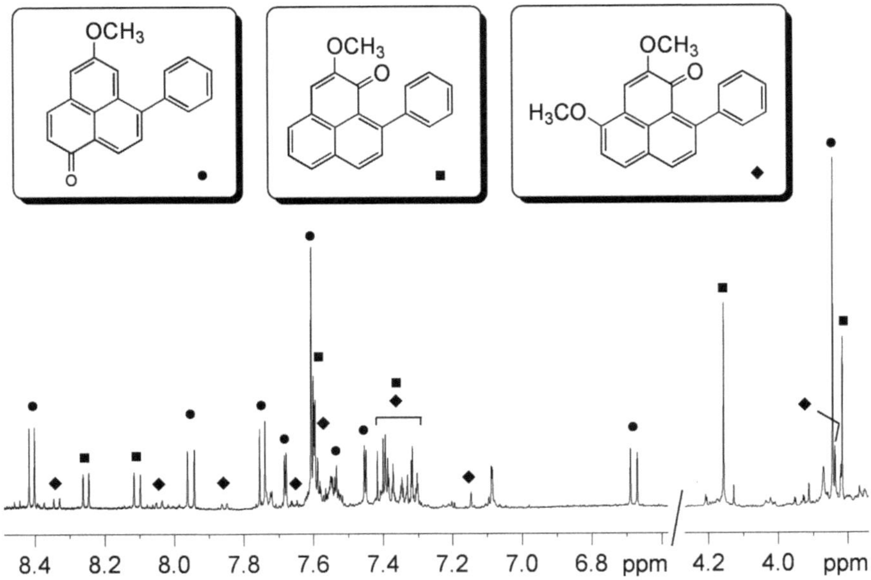

Fig. 3 ^{1}H NMR spectrum (500 MHz) of an extract of laser-microdissected secretory cavities of *Dilatris pillansii* (Hölscher and Schneider 2007). The compounds of the phenylphenalenone type shown above the spectrum were identified by means of authentic standards. The symbols indicate assignment of the ^{1}H NMR signals to hydrogen atoms of the three structures

type were performed in a one-pot sequential procedure. After derivatization, the monomeric sugar components were analyzed by GC-MS. Contamination originating from paraffin embedding was distinguished by analyzing blanks.

LMD (nitrogen laser beam) was used as a tool to harvest secretory cavities from the leaves of *Dilatris pillansii* Barker (Haemodoraceae; Hölscher and Schneider 2007; Fig. 3). The storage cavities, relatively large lipophilic cells (75–120 µm) with low water content, are well suited for laser dissection. After cryosectioning the leaf, slices were mounted on a simple glass slide and cut by laser beam. An extremely sharp dissecting needle was used to pick up the laser-dissected cells under a stereo microscope. The cavities were transferred directly to a microcentrifuge tube containing extraction solvent. The extract was analyzed by reversed-phase HPLC and cryogenic ^{1}H NMR spectroscopy. Comparing the UV traces and ^{1}H NMR spectra with those of leaf total extracts identified several phenylphenalenones, which are typical secondary metabolites of the Haemodoraceae. A signal-to-noise ratio of 310 was obtained, for example, from only ten dissected cavities when the sample was measured by ^{1}H NMR spectroscopy for 35 min. Integrating the corresponding ^{1}H NMR signals of the three major metabolites allowed their relative amounts in the secretory cavities of *D. pillansii* to be determined. Moreover, secretory cells were laser-dissected from leaves and flowers of herbarium specimens (more than 180 years old) of *D. corymbosa* Berg. and *D. viscosa* L (Hölscher and Schneider 2007). The material was simply fixed between a thin glass slide and a specially manufactured metal frame. Secretory cavities dissected from herbarium material were collected in the cap of an Eppendorf tube by gravity. Again, ^{1}H NMR spectroscopy using a cryogenically cooled probe head was used to identify phenylphenalenones in these cells.

LMD was also applied to ablate tissue from woody samples, namely so-called stone cells from the bark of *Picea abies* (Norway spruce; Li et al. 2007). The stone cells were detected in cryosections by means of their characteristic fluorescence and microdissected using a nitrogen laser beam. Non-fluorescing phloem tissue, which did not contain stone cells, was microdissected from the same cryosections. The dissected cells were collected by gravity and, without any further treatment, transferred to a NMR tube to which deuterated methanol was added for extraction and subsequent NMR analysis. ^{1}H NMR, 2D COSY, and ^{1}H-^{13}C heterocorrelation (HSQC) spectra were measured using a cryogenically cooled probe head. Comparison with the spectra of authentic reference compounds, together with mass spectrometric analyses, identified astringin and dihydroxyquercetin 3'-*O*-*β*-D-glucopyranoside as the two major components in both the stone cells and stone-cell-free phloem tissue. An internal standard (rutin) was added and the samples measured again to quantitatively determine the two major compounds. Surprisingly, the stone cells contained lower concentrations of astringin and dihydroxyquercetin glucoside than did surrounding phloem cells, indicating that they are not only involved in mechanical defense but also contribute to chemical defense against pathogens.

In light of the intrinsic low sensitivity of NMR spectroscopy, the unambiguous identification of secondary metabolites in special laser-microdissected plant cells and the determination of their ratios (in secretory cavities of *Dilatris*) and contents (in stone cells and phloem tissue of Norway spruce) are astonishing. The recently introduced cryogenically cooled probe heads (Kovacs et al. 2005) and other technical improvements have considerably increased NMR sensitivity and made this progress possible. Though only the major components were accessible in the two studies by Li et al. (2007) and Hölscher and Schneider (2007), it was shown that the combination of LMD and cryogenic NMR spectroscopy is suitable for analyzing secondary metabolites from specialized cell populations. Some of the advantages of NMR spectroscopy for metabolite analysis are, for example, that: (a) almost all biomolecules contain protons, i.e. they are detectable by ^{1}H NMR, (b) ^{1}H NMR enables metabolite ratios to be determined simply by integrating signals in the spectrum, (c) derivatization is not required, and (d) analysis of mixtures containing only a limited number of components is possible without separation. Furthermore, coupling methods such as HPLC-SPE-NMR and HPLC-NMR-MS, applied to extracts of laser-microdissected plant cells, are anticipated to become important microphytochemical tools for identifying low-molecular-weight compounds in complex mixtures.

A study on the association of UV-protecting flavonols with the nuclei of various conifer species (Polster et al. 2006) used LMPC methodology. For example, seed wings of *Tsuga canadensis* were treated with "Nuclear Fast Red" in order to improve absorption of the laser beam for catapulting the nuclei out of a cell. The excised nuclei were then catapulted into a drop of *p*-dimethylamino-cinnamaldehyde on a cover slide for staining. The blue coloration of the nuclei indicated the presence of flavanols. More specifically, titration experiments found associations with phenolics such as rutin or quercetin but not with DNA.

10 Conclusions and Perspectives

Although known for many years and used in biomedical investigations for more than a decade, laser-assisted microdissection technologies have been used only recently for harvesting single plant cells and cell populations. Protocols describe how to prepare plant tissue for laser capture and laser cutting, and outline which dissection methods are most useful for the specific requirements of the plant and the molecules being studied; results from these studies have been impressive. Clearly the focus is still on collecting cells and tissues for RNA extraction, because the methodology available for gene expression analysis can be applied without restrictions to such microsamples. Protein analysis, though rarely performed with laser-microdissected plant samples, is an emerging area of research. Further developments in instrumental analysis – microcapillary electrophoresis and mass-spectrometry-based methods for protein and proteomic analysis, and NMR and hyphenated techniques for metabolite analysis – will open the doors for further applications. Microphytochemical metabolite profiling studies are a reality and it may be that even de novo structural elucidation of natural products from specific cell types will soon be possible by combining NMR and mass spectroscopic techniques.

Acknowledgement Emily Wheeler (Jena) is gratefully acknowledged for editorial assistance.

References

Angeles G, Berrio-Sierra J, Joseleau J-P, Lorimier P, Lefèbvre A, Ruel K (2006) Preparative laser capture microdissection and single-pot cell wall material preparation: a novel method for tissue-specific analysis. Planta 224:228–232

Asano T, Masumura T, Kusano H, Kikuchi S, Kurita A, Shimada H, Kadowaki KI (2002) Construction of a specialized cDNA library from plant cells isolated by laser capture microdissection: toward comprehensive analysis of the genes expressed in the rice phloem. Plant J 32:401–408

Ball HJ, Hunt NH (2004) Needle in a haystack: microdissecting the proteome of a tissue. Amino Acids 27:1–7

Bhattacharya SH, Gal AA, Murray KK (2003) Laser capture microdissection MALDI for direct analysis of archival tissue. J Proteome Res 2:95–98

Boulnois JL (1986) Photophyiscal processes in recent medical laser developments: a review. Lasers Med Sci 1:47–66

Brandt S (2005) Microgenomics: gene expression analysis at the tissue-specific and single-cell levels. J Exp Bot 56:495–505

Brandt S, Kehr J, Walz C, Imlau A, Willmitzer L, Fisahn J (1999) A rapid method for detection of plant gene transcripts from single epidermal, mesophyll and companion cell of intacts leaves. Plant J 20:245–250

Brandt S, Walz C, Schad M, Pavlovic N, Kehr J (2003) A simple, chisel-assisted mechanical microdissection system for harvesting homogenous plant tissue suitable for the analysis of nucleic acids and proteins. Plant Mol Biol Rep 21:417–427

Brinkmann R, Hansen C, Mohrenstecher D, Scheu M, Birngruber R (1996) Analysis of cavitation dynamics during pulsed laser tissue ablation by optical on-line monitoring. IEEE J Sel Top Quantum Electron 2:826–835

Burgemeister R (2005) New aspects of laser microdissection in research and routine. J Histochem Cytochem 53:409–412

Cai S, Lashbrook CC (2006) Laser capture microdissection of plant cells from tape-transferred paraffin sections promotes recovery of structurally intact RNA for global gene profiling. Plant J 48:628–637

Caldwell RL, Caprioli RM (2005) Tissue profiling by mass spectrometry. Mol Cell Proteomics 4:394–401

Cantor CR, Schimmel PR (1980) Biophysical chemistry (I). Freeman, New York

Casson S, Spencer M, Walker K, Lindsey K (2005) Laser capture microdissection for the analysis of gene expression during embryogenesis of Arabidopsis. Plant J 42:111–123

Corpas FJ, Fernández-Ocaña A, Carreras A, Valderrama R, Luque F, Esteban FJ, Rodríguez-Serrano M, Chaki M, Pedrajas JR, Sandalio LM, Río LA del, Barroso JB (2006) The expression of different superoxide dismutase forms is cell-type dependent in olive (*Olea europaea* L.) leaves. Plant Cell Physiol 47:984–994

Cox KH, Goldberg RB (1988) Analysis of gene expression. In: Shaw CH (ed) Plant molecular biology: a practical approach. IRL, Oxford, pp 1–34

Craven RA, Banks RE (2001) Laser capture microdissection and proteomics: possibilities and limitation. Proteomics 1:1200–1204

Craven RA, Totty N, Harnden P, Selby PJ, Banks RE (2002) Laser capture microdissection and two-dimensional polyacrylamide gel electrophoresis. Am J Pathol 160:815–822

Day RC, Grossniklaus U, Macknight RC (2005) Be more specific! Laser-assisted microdissection of plant cells. Trends Plant Sci 10:398–406

Day RC, McNoeb LA, Macknight RC (2007) Transcript analysis of laser microdissected plant cells. Physiol Plant 129:267–282

Doukas AG, McAucliffe DJ, Lee S, Venugopalan V, Flotte TJ (1995) Physical factors involved in stress-wave-induced cell injury – the effect of stress gradient. Ultrasound Med Biol 21:961–975

Elvers D, Remer L, Arnold N, Bäuerle D (2005) Laser microdissection of biological tissues: process optimization. Appl Phys A 80:55–59

Emmert-Buck MR, Bonner RF, Smith PD, Chuaqui RF, Zhuang Z, Goldstein SR, Weiss RA, Liotta LA (1996) Laser capture microdissection. Science 274:998–1001

Emmony DC, Geerken BM, Straaijer A (1976) Interaction of 10.6 mm laser radiation with liquids. Infrared Phys 16:87–92

Frenz M, Konz F, Pratisto H, Weber HP, Silenok AS, Konov VI (1998) Starting mechanisms and dynamics of bubble formation induced by a Ho:Yttrium aluminium garnet laser in water. J Appl Phys 84:5905–5912

Fukui K, Minezawa M, Kamisugi Y, Ishikawa M, Ohmido N, Yanagisawa T, Fujishita M, Sakai E (1992) Microdissection of plant chromosomes by argon-ion laser beam. Theor Appl Genet (1992) 84:787–791

Garrison BJ, Srinivasan R (1985) Laser ablation of organic polymers – microscopic models for photochemical and thermal-processes. J Appl Phys 57:2909–2914

Gitomer SJ, Jones RD (1991) Laser-produced plasmas in medicine. IEEE Trans Plasma Sci 19:1209–1219

Goldsworthy SM, Stockton PS, Trempus CS, Foley JF, Maronpot RR (1999) Effects of fixation on RNA extraction and amplification from laser capture microdissected tissue. Mol Carcinogen 25:86–91

Greulich KO, Weber G (1992) The light microscope on its way from an analytical to a preparative tool. J Microsc 167:127–151

Grubb RL, Calvert VS, Wulkuhle JD, Paweletz CP, Linehan WM, Phillips JL, Chuaqui R, Valasco A, Gillespie J, Emmert-Buck M, Liotta LA, Petricoin EF (2003) Signal pathway profiling of prostate cancer using reverse phase protein arrays. Proteomics 3:2142–2146
Hobza R, Lengerova M, Cernohorska M, Rubes J, Vyskot B (2004) FAST-FISH with laser beam microdissected DOP-PCR probe distinguishes the sex chromosomes of *Silene latifolia*. Chromosome Res 12:245–250
Hölscher D, Schneider B (2007) Laser microdissection and cryogenic nuclear magnetic resonance spectroscopy, an alliance for cell type-specific metabolite profiling. Planta 225:763–770
Ivashikina N, Deeken R, Ache P, Kranz E, Pommerrenig B, Sauer N, Hedrich R (2003) Isolation of AtSUC2promoter-GFP-marker companion cells for patch-clamp studies and expression profiling. Plant J 36:931–945
Jiang K, Zhang S, Lee S, Tsai G, Kim K, Huang H, Chilcott C, Zhu T, Feldman LJ (2006) Transcription profile analyses identify genes and pathways central to root cap functions in maize. Plant Mol Biol 60:343–363
Karrer EE, Lincoln JE, Hogenhout S, Bennett AB, Bostock RM, Martineau B, Lucas WJ, Gilchrist DG, Alexander D (1995) *In situ* isolation of messenger-RNA from individual plant-cells-creation of cell-specific cDNA libraries. Proc Natl Acad Sci USA 92:3814–3818
Kawamura Y, Toyoda K, Namba S (1982) Effective deep ultraviolet photoetching of polymethyl methacrylate by an excimer laser. Appl Phys Lett 40:374–375
Kehr J (2003) Single cell technology. Curr Opin Plant Biol 6:1–5
Kerk N, Ceserani T, Tausta SL, Sussex IM, Nelson TM (2003) Laser capture microdissection of cells from plant tissues. Plant Physiol 132:27–35
Kitai MS, Popkov VL, Semchishen VA, Kharizov AA (1991) The physics of uv laser cornea ablation. IEEE J Quantum Electron 27:302–307
Klink VP, MacDonald M, Alkharouf N, Matthews BF (2005) Laser capture microdissection (LCM) and expression analyses of *Glycine max* (soybean) syncytium containing root regions formed by the plant pathogen *Heterodera glycines* (soybean cyst nematode). Plant Mol Biol 59:969–983
Kovacs H, Moskau D, Spraul M (2005) Cryogenically cooled probes – a leap in NMR technology. Progr Nucl Mag Res Spec 46:131–155
Krasnov MM (1973) Laseropuncture of anterior chamber angle in glaucoma. Am J Ophthalmol 75:674–678
Kubo Y, Klimek F, Kikuchi Y, Bannasch P, Hino O (1995) Early detection of knudson two-hits in preneoplastic renal cells of the eker rat model by the laser microdissection procedure. Cancer Res 55:989–990
Kwapiszewska G, Meyer M, Bogumil R, Bohle RM, Seeger W, Weissmann N, Fink L (2004) Identification of proteins in laser-microdissected small cell numbers by SELDI-TOF and Tandem MS. BMC Biotechnol 4:30
Lange BM (2005) Single-cell genomics. Curr Opin Plant Biol 8:236–241
Li S-H, Schneider B, Gershenzon J (2007) Microchemical analysis of laser-microdissected stone cells of Norway spruce by cryogenic nuclear magnetic resonance spectroscopy. Planta 225:771–779
Liu H, Wang S, Yu X, Yu J, He X, Zhang S, Shou H, Wu P (2005) ARL1, a LOB-domain protein required for adventitious root formation in rice. Plant J 43:47–56
Lochmann H, Bazzanella A, Bächmann K (1998) Analysis of solutes and metabolites in single plant cell vacuoles by capillary electrophoresis. J Chromatogr 817:337–343
Lubatschowski H, Heisterkamp A (2004) Interaction with biological tissue. In: Dausinger F, Lichtner F, Lubatschowski H (eds) Topics in applied physics, vol 96. Springer, Berlin Heidelberg New York, pp 91–104
Maiman TH (1960) Stimulated optical rotation in ruby. Nature 187:493–494
Matsunaga S, Kawano S, Michimoto T, Higashiyama T, Nakao S, Sakai A, Kuroiwa T (1999a) Semi-automatic laser beam microdissection of the Y chromosome and analysis of Y chromosome DNA in a dioecious plant, *Silene latifolia*. Plant Cell Physiol 40:60–68

Matsunaga S, Schütze K, Donnison IS, Grant SR, Kuroiwa T, Kawano S (1999b) Single pollen typing combined with laser-mediated manipulation. Plant J 20:371–378

Meier-Ruge W, Bielser W, Remy E, Hillenkamp F, Nitsche R, Unsöld R (1976) The laser in the Lowry technique for microdissection of freeze-dried tissue slices. Histochem J 8:387–401

Melle C, Ernst G, Schimmel B, Bleul A, Koscielny S, Wiesner A, Bogumil R, Möller U, Osterloh D, Halbhuber KJ, Eggeling F von (2003) Biomarker discovery and identification in laser microdissected head and neck squamous cell carcinoma with ProteinChip technology, two-dimensional gel electrophoresis, tandem mass spectrometry, and immunohistochemistry. Mol Cell Proteomics 2:443–452

Melle C, Kaufmann R, Hommann M, Bleul A, Driesch D, Ernst G, Eggeling F von (2004) Proteomic profiling in microdissected hepatocellular carcinoma tissue using ProteinChip technology. Int J Oncol 24:885–891

Monajembashi S, Cremer C, Cremer T, Wolfrum J, Greulich KO (1986) Microdissection of human-chromosomes by a laser microbeam. Exp Cell Res 167:262–265

Murata J, De Luca V (2005) Localization of tabersonine 16-hydroxylase and 16-OH tabersonine-16-O-methyltransferase to leaf epidermal cells defines them as a major site of precursor biosynthesis in the vindoline pathway in *Catharanthus roseus*. Plant J 44:581–594

Nakada M, Komatsu M, Ochiai T, Ohtsu K, Nakazono M, Nishizawa NK, Nitta K, Nishiyama R, Kameya T, Kanno A (2006) Isolation of MaDEF from *Muscari armeniacum* and analysis of its expression using laser microdissection. Plant Sci 170:143–150

Nakazono M, Qiu F, Borsuk LA, Schnable PS (2003) Laser-capture microdissection, a tool for the global analysis of gene expression in specific plant cell types: identification of genes expressed differentially in epidermal cells or vascular tissues of maize. Plant Cell 15:583–596

Nelson T, Tausta SL, Gandotra N, Liu T (2006) Laser microdissection of plant tissue: what you see is what you get. Annu Rev Plant Biol 57:181–201

Oraevsky AA, Jacques SL, Tittel FK (1995) Mechanism of laser-ablation for aqueous-media irradiated under confined-stress conditions. J Appl Phys 78:1281–1290

Outlaw WH, Lowry OH (1977) Organic-acid and potassium accumulation in guard cells during stomatal opening. Proc Natl Acad Sci USA 74:4434–4438

Palmer-Toy DE, Sarracino DA, Sgroi D, LeVangie R, Leopold PE (2000) Direct acquisition of matrix-assisted laser desorption/ionization time-of-flight mass spectra from laser capture microdissected tissues. Clin Chem 46:1513–1516

Paltauf G, Schmidt-Kloiber H (1995) Model study to investigate the contribution of spallation to pulsed-laser ablation of tissue. Lasers Surg Med 16:277–287

Paltauf G, Schmidt-Kloiber H (1996). Microcavity dynamics during laser-induced spalltion of liquids and gels. Appl Phys A 62:303–311

Paweletz CP, Charboneau L, Bichsel VE, Simone NL, Chen T, Gillespie JW, Emmert-Buck MR, Roth MJ, Petricoin EF, Liotta LA (2001) Reverse phase protein microarrays which capture disease progression show activation of pro-survival pathways at the cancer invasion front. Oncogene 20:1981–1989

Petry R, Schmitt M, Popp J (2003) Raman spectroscopy – a prospective tool in the life sciences. ChemPhysChem 4:14–30

Polster J, Dithmar H, Burgemeister R, Friedemann G, Feucht W (2006) Flavonoids in plant nuclei: detection by laser microdissection and pressure catapulting (LMPC), *in vivo* staining, and uv–visible spectroscopic titration. Physiol Plant 128:163–174

Pommerrenig B, Barth I, Niedermeier M, Kopp S, Schmid J, Dwyer RA, McNair RJ, Klebl F, Sauer N (2006) Common plantain. A collection of expressed sequence tags from vascular tissue and a simple and efficient transformation method. Plant Physiol 142:1427–1441

Pühler A, Regitz M, Schmid RD (eds) (2000) Römpp Lexikon Biochemie und Molekularbiologie. Thieme, Stuttgart

Ramsay K, Wang Z, Jones MGK (2004) Using laser capture microdissection to study gene expression in early stages of giant cells induced by root-knot nematodes. Mol Plant Pathol 5:587–592

Ramsay K, Jones MGK, Wang Z (2006) Laser capture microdissection: a novel approach to microanalysis of plant–microbe interactions. Mol Plant Pathol 7:429–435

Reinhardt D, Frenz M, Mandel T, Kuhlemeier C (2003) Microsurgical and laser ablation analysis of interactions between the zones and layers of the tomato shoot apical meristem. Development 130:4073–4083

Reinhardt D, Frenz M, Mandel T, Kuhlemeier C (2005) Microsurgical and laser ablation analysis of leaf positioning and dorsoventral patterning in tomato. Development 132:15–26

Sanders PM, Bui AQ, Le BH, Goldberg RB (2005) Differentiation and degeneration of cells that play a major role in tobacco anther dehiscence. Sex Plant Reprod 17:219–241

Scalenghe F, Turco E, Edstrom JE, Pirrotta V, Melli M (1981) Micro-dissection and cloning of DNA from a specific region of *Drosophila melanogaster* polytene chromosomes. Chromosoma 82:205–216

Schad M, Lipton MS, Giavalisco P, Smith RD, Kehr J (2005a) Evaluation of two dimensional electrophoresis and liquid chromatography–tandem mass spectrometry for tissue-specific protein profiling of laser-microdissected plant samples. Electrophoresis 26:2729–2738

Schad M, Mungur R, Fiehn O, Kehr J (2005b) Metabolic profiling of laser microdissected vascular bundles of *Arabidopsis thaliana*. Plant Methods 1:2

Schmidt H, Ihlemann J, Wolff-Rottke B, Luther K (1998) Ultraviolet laser ablation of polymers: spot size, pulse duration, and plume attenuation effects explained. J Appl Phys 83, 5458–5468

Schütze K, Lahr G (1998) Identification of expressed genes by laser-mediated manipulation of single cells. Nat Biotechnol 16:737–742

Scutt CP, Kamisugi Y, Sakai F, Gilmartin PM (1997) Laser isolation of plant sex chromosomes: studies on the DNA composition of the X and Y sex chromosomes of *Silene latifolia*. Genome 40:705–715

Shekouh AR, Thompson CC, Prime W, Campbell F, Hamlett J, Herrington CS, Lemoine NR, Crnogorac-Jurcevic T, Buechler MW, Friess H, Neoptolemos JP, Pennington SR, Simone NL, Remaley AT, Charboneau L, Petricoin EF III, Glickman JW, Emmert-Buck MR, Fleisher TA, Liotta LA (2000) Sensitive immunoassay of tissue cell proteins procured by laser capture microdissection. Am J Pathol 156:445–452

Snow M, Snow R (1931) Experiments on phyllotaxis. I. The effect of isolating a primordium. Philos Trans R Soc Lond B Biol Sci 221:1–43

Solon AR, Aronson R, Gould G (1961) Physiological implications of laser beams. Science 134:1506–1508

Srinivasan R (1986) Ablation of polymers and biological tissue by ultraviolet-lasers. Science 234:559–565

Srinivasan R, Leigh W (1982) Ablative photodecomposition: action of far-ultraviolet laser radiation (193 nm) on poly(ethylene terephthalate) films. J Am Chem Soc 104:6784–6785

Srinivasan R, Mayne-Banton V (1982) Self developing photoetching of poly(ethylene-terephthalate) films by far ultraviolet excimer laser-radiation. Appl Phys Lett 41:576–578

Tomos AD, Sharrock RA (2001) Cell sampling and analysis (SiCSA): metabolites measured at single cell resolution. J Exp Bot 52:623–630

Tsuji H, Aya K, Ueguchi-Tanaka M, Shimada Y, Nakazono M, Watanabe R, Nishizawa NK, Gomi K, Shimada A, Kitano H, Ashikari M, Matsuoka M (2006) GAMYB controls different sets of genes and its differentially regulated by microRNA in aleurone cells and anthers. Plant J 47:427–444

Ventzek PLG, Gilgenbach RM, Ching CH, Lindley RA (1992) Schlieren and dye-laser resonance-absorption photographic investigations of KrF excimer laser-ablated atoms and molecules from polyimide, polyethylenterephthalate, and aluminium. J Appl Phys 72:1696–1706

Vogel A, Venugopalan V (2003) Mechanisms of pulsed laser ablation of biological tissues. Chem Rev 103:577–644

Vogel A, Busch S, Parlitz U (1996) Shock wave emission and cavitation bubble generation by picosecond and anaosecond optical breakdown in water. J Acoust Soc Am 100:148–165

Vogel A, Noack J, Nahen K, Theisen D, Busch S, Parlitz U, Hammer DX, Noojin GD, Rockwell BA, Birngruber R (1999) Energy balance of optical breakdown in water at nanosecond to femtosecond time scales. Appl Phys B 68:271–280

Vogel A, Noack J, Huttman G, Paltauf G (2005) Mechanisms of femtosecond laser nanosurgery of cells and tissues. Appl Phys B 81:1015–1047

Walsh JT, Deutsch TF (1991) Measurement of Er-YAG laser ablation plume dynamics. Appl Phys B 52:217–224

Woll K, Borsuk LA, Stransky H, Nettleton D, Schnable PS, Hochholdinger F (2005) Isolation, characterization, and pericycle-specific transcriptome analyses of the novel maize lateral and seminal root initiation mutant *rum1*. Plant Physiol 139:1255–1267

Wu S-L, Hancock WS, Goodrich GG, Kunitake ST (2003) An approach to the proteomic analysis of a breast cancer cell line (SKBR-3). Proteomics 3:1037–1046

Wu YR, Machado AC, White RG, Llewellyn DJ, Dennis ES (2006) Expression profiling identifies genes expressed early during lint fibre initiation in cotton. Plant Cell Physiol 47:107–127

Xu BJ, Caprioli RM, Sanders ME, Jensen RA (2002) Direct analysis of laser capture microdissected cells by MALDI mass spectrometry. J Am Soc Mass Spectrom 13:1292–1297

Zang L, Toy DP, Hancock WS, Sgroi DC, Karger BL, J. (2004) Proteomic analysis of ductal carcinoma of the breast using laser capture microdissection, LC-MS, and O-16/O-18 isotopic labeling. J Proteome Res 3:604–612

Zhu XL, Shaw PN, Pritchard J, Newbury J, Hunt EJ, Barrett DA (2005) Amino acid analysis by micellar electrokinetic chromatography with laser-induced fluorescence detection: Application to nanolitre-volume biological samples from *Arabidopsis thaliana* and *Myzus persicae*. Electrophoresis 26:911–919

Zweig AD (1991) A thermomechanical model for laser ablation. J Appl Phys 70:1684–1691

Plasma Membrane Redox Systems: Lipid Rafts and Protein Assemblies

Sabine Lüthje

Contents

Abstract Electron transport processes are mainly associated with photosynthesis and respiratory chain. Nowadays the occurrence and possible roles of electron transport systems in plant plasma membranes appear well established. Besides quinones, NAD(P)H dehydrogenases, and *b*-type cytochromes have been purified and characterized in detail. Members of the flavocytochrome *b* family have been identified by nucleotide sequence analysis in several plant species and shown to be involved in iron uptake and oxidative stress including biotic interactions, abiotic stress factors and plant development. Recent work supports not only the existence of microdomains (so-called lipid rafts) in plant plasma membranes but also the occurrence of redox systems therein. Furthermore participation of plasma membrane-bound redox enzymes in protein–protein interactions and high molecular mass protein assemblies has been suggested. For this reason it is the aim of this overview to summarize the current knowledge about plasma membrane-bound redox systems with a special focus on their possible structures and functions, their occurrence in lipid rafts and their participation in protein assemblies.

Sabine Lüthje
Biozentrum Klein Flottbek und Botanischer Garten, Universität Hamburg, Ohnhorststrasse 18, D-22609 Hamburg, Germany
e-mail: s.luthje@botanik.uni-hamburg.de

U. Lüttge et al. (eds.), *Progress in Botany 69.*

1 Introduction

Nowadays the occurrence and possible roles of oxidoreductase (redox) systems in plant plasma membranes appear well established. Since the discovery of electron transport activities at the plasma membrane (PM) of plants 35 years ago (e.g. Chaney et al. 1972; Bienfait 1985; Böttger and Lüthen 1986; Marschner et al. 1986; Rubinstein and Stern 1986), investigations have focussed mainly on the purification and biochemical characterization of redox components (see refs in Döring and Lüthje 1996; Lüthje et al. 1997, 2005; Asard et al. 1998; Bérczi and Møller, 2000) and their functions in iron uptake (see refs in Connolly and Guerinot 1998; Schmidt 1999; Grotz and Guerinot 2006) and oxidative stress, including biotic interactions, abiotic stress factors and plant development (see refs in Döring et al. 1998a, b; Lüthje et al. 2000; Torres and Dangl 2005). The recent progress in the research on PM redox systems has also been documented by a series of interdisciplinary and international conferences on "*(Plasma) Membrane redox systems and their role in biological stress and disease*" (special issues of *Protoplasma* 184, 205, 217, 221; *Biofactors* 20; *Acta Biologica Szegediensis* 50) and symposia on "*Iron nutrition and interactions in plants*" (*Plant Physiology and Biochemistry* 45(5), 2007).

However, proteomic approaches suggest that most PM redox proteins belong to the group of hypothetical proteins with unknown functions (e.g. Santoni et al. 1998, 2000; Alexandersson et al. 2004; Marmagne et al. 2004; Lilley and Dupree 2006; Morel et al. 2006). Consequently, complete amino acid sequences have only been presented for a minor part of these proteins (Mika 2005). Whereas others could be identified by N-terminal sequencing (Bérczi and Møller 1998; Cordobá-Pedregosa et al. 1998) or by nucleotide sequence analysis in comparison with genes of animal and yeast systems (Keller et al. 1998; Torres et al. 1998; Amicucci et al. 1999; Robinson et al. 1999; Yoshioka et al. 2001; Simon-Plas et al. 2002; Waters et al. 2002; Li et al. 2004).

According to the fluid mosaic model, all membranes have the same basic molecular organization (Singer and Nicolson 1972). The properties of proteins and the composition of lipids confer them with their unique functional characteristics, but they are variable for each membrane and the respective physiological conditions. Thus, the existence of lipid rafts and their interactions with specific membrane proteins has become the currently favored model for the molecular organization of plant plasma membranes (see refs in Bhat and Panstrugah 2005; Martin et al. 2005). PM-bound redox systems and lipid rafts are both suggested to be involved in signal transduction and membrane transport and some other functions. Evidence for the occurrence of redox systems in lipid rafts and their participation in protein assemblies have been presented (e.g. Bagnaresi et al. 1997; Trost et al. 1997; Borner et al. 2005; Morel et al. 2006). For this reason it is the aim of this review to summarize the current knowledge about PM redox systems with a special focus on their possible structures and functions, their occurrence in lipid rafts and participation in protein assemblies.

2 Transmembrane Electron Transport

The first evidence for a transmembrane electron transport system in the plant PM was presented by in vivo experiments. The application of artificial electron acceptors like ferricyanide (hexacyanoferrate III, HCF III) caused a depolarization of the membrane potential, a reduction of the probe and an increase in proton secretion (see Döring and Lüthje 1996 and refs therein). Two models were presented which aimed to explain these observations. Rubinstein and Stern (1986) suggested that a transmembrane iron reductase could activate H^+-ATPase. Simultaneously, Böttger and Lüthen (1986) postulated that an electron transport chain transfers protons to the apoplast, while oxygen is reduced by a two-electron transfer mechanism at the cytosolic surface. Consequently, the application of HCF III presents a short-cut in this electrogenic system and causes the observed depolarization of the membrane potential. Two years later, Bienfait and Lüttge (1988) postulated the existence of an electron transport chain involved in redox regulation and the production of reactive oxygen species (ROS). According to this model, thiol groups of inactivated transport proteins are reduced by the system and thereby reactivate the transporters. In all these models NADH or NADPH is suggested as the natural electron donors of the constitutive and transmembrane redox activity.

In recent years it became obvious that the grasses evolved an iron uptake strategy distinct from all other plants (see refs in Connolly and Guerinot 1998; Schmidt 1999). While so-called "strategy I plants" (dicots, non-grass monocots) reduce the iron before uptake by a transmembrane redox activity (Robinson et al. 1997, 1999; Waters et al. 2002), grasses (strategy II plants) release phytosiderophores of the mugineic acid (MA) family, chelate the Fe^{3+}, and reduce the Fe^{3+}-chelate inside the cell (Chaney et al. 1972; Takagi 1976; Takagi et al. 1984; Bienfait 1985; Marschner et al. 1986).

In addition, evidence has been presented for an inducible iron reductase in the PM of a range of chlorophyte algae (Weger and Espie 2000; Weger et al. 2002, 2006), including *Chlamydomonas* (see refs in Merchant et al. 2006). This strategy also appears to be used by some marine diatoms (Jones et al. 1987; Maldonado and Price 2000, 2001), whereas *Thalassiosira* appears to have an iron acquisition strategy that is distinct from chlorophytes and strategy I plants (Davey et al. 2003).

A second, constitutive redox system – the so-called standard system – has been detected in all plants (including grasses). It appears to fulfil a house-keeping function (see refs in Döring and Lüthje 1996; Lüthje et al. 1997; Bérczi and Møller 2000). Due to their distinct iron uptake strategy, grasses are ideal model organisms for the investigation of the standard system (see refs in Döring and Lüthje 1996) and a few years ago a significant transmembrane electron transport activity was observed with NAD(P)H-loaded and sealed right-side-out (apoplastic side out) PM vesicles isolated from maize (*Zea mays* L.) roots (Menckhoff and Lüthje 2004). Although this observation fits nicely with in vivo experiments on the standard system (see refs in Lüthje et al. 1997; Döring et al. 1998a), the underlying structural components of this transmembrane activity still have to be elucidated. So far none of the NAD(P)H oxidoreductases characterized from plant plasma membranes can provide a molecular basis for the observed transmembrane HCF III reduction of maize roots (Menckhoff and Lüthje 2004).

Besides HCF III reduction (126 nmol min^{-1} mg^{-1} protein), low rates of superoxide anion radical production (765 pmol min^{-1} mg^{-1} protein) were observed with NADPH-loaded vesicles. ROS are known as key regulators of metabolic and defense pathways (see refs in Neill et al. 2002; Kawano 2003; Desikan et al. 2004a; Laloi et al. 2004; Gevhev and Hille 2005; Torres et al. 2006). Thus the observed transmembrane activity was suggested to be involved in signaling (Menckhoff and Lüthje 2004).

3 PM-Bound Redox Systems

In the 1990s, aqueous two-phase partitioning was established as the state of the art method for the purification of highly enriched PM vesicles (Widell et al. 1982; Yoshida et al. 1983; Larsson et al. 1994). This break-through allowed the characterization of transport processes with PM vesicles of defined orientation (e.g. Palmgren et al. 1990; Horemans et al. 2000), the characterization of lipids (e.g. Uemura and Steponkus 1994; Uemura et al. 1995, 2006; Cowan et al. 1993; Bohn et al. 2001; Sperling et al. 2005; Bohn et al. 2006), proteins and redox components located therein (see refs in Lüthje et al. 1997; Asard et al. 1998; Bérczi et al. 1998; Bérczi and Møller 2000).

Although PM preparations are generally assumed to be comparatively pure (>95%) one has to keep in mind that: (a) soluble proteins are an often overlooked contaminant in PM preparations (Bérczi and Asard 2003; Mika et al. 2004) and (b) even minor contaminants (<5%) of endomembrane systems in PM preparations may result in the erroneous detection of a major protein of these membranes (e.g. Bérczi et al. 2001; Griesen et al. 2004; Kjell et al. 2004; Preger et al. 2004). As a consequence, low activities and/or highly varying amounts of a compound found with independent membrane preparations should be discussed cautiously.

For example, an oscillating, touch and diphenylene iodonium (DPI)-sensitive but cyanide-insensitive NADH oxidase-like activity was purified from PM preparations by Morré and co-worker (Brightman et al. 1988; Morré 1998). This protein has properties comparable with a protein disulfide isomerase (Chueh et al. 1997). Due to its very low NADH oxidase-like activity (e.g. Brightman et al. 1988), indistinct properties and a lack of sequence information for the plant system, further proof is needed for an unequivocal identification and localization of this protein (Lüthje et al. 1997; Bérczi and Møller 2000). So far all protein disulfide isomerases identified in PM preparations by proteomic approaches appear to be the result of contamination from the endoplasmatic reticulum (e.g. Borner et al. 2005). More convincing examples of redox components which have been identified, purified and characterized from plant PM are discussed in more detail in the following paragraphs (Table 1).

Table 1 Redox proteins purified from plant plasma membranes. Several redox proteins have been characterized after partial purification from plant plasma membranes. MW_{nat} Native molecular mass detected by SDS-PAGE or flagged with *(m)* modified SDS-PAGE, MW_{calc} molecular mass calculated from identified amino acid sequence, *PTM* post-translational modifications, *pI* point isoelectric, *NQR* NAD(P)H:quinone reductase, *pmPOX* plasma membrane-bound peroxidase, *MDAR* monodehydroascorbate reductase, *FCR* ferric chelate reductase, *N-glyco sites* putative N-glycosylation sites, *MDH* malate dehydrogenase, *ASC* ascorbate, *Cyt* cytochrome. References: *1* Serrano et al. (1994a), *2* Serrano et al. (1994b), *3* Serrano et al. (1995), *4* Trost et al. (1997), *5* Luster and Buckhout (1989), *6* Lüthje et al. (1998), *7* Mika and Lüthje (2003), *8* Mika (2005), *9* Bérczi and Møller (1998), *10* Holden et al. (1991), *11* Holden et al. (1995), *12* Bagnaresi et al. (1997), *13* Córdoba-Pedegrosa et al. (1998), *14* L. Menckhoff and S. Lüthje, unpublished data, *15* Trost et al. (2000), *16* Bérczi et al. (2003), *17* S. Lüthje, unpublished data

Enzyme	Species	Tissue	MW_{nat} (kDa)	MW_{calc} (kDa)	PTM	Cofactor, prostetic group	pI	References
NQR	*Allium cepa*	Root	27			NAD(P)H,FMN	6.0	1, 2, 3
	Cucurbita pepo	Hypocotyl	24			NAD(P)H,FMN		4
	Zea mays	Root	27			NAD(P)H,FMN		5, 6
NQR	*Allium cepa*	Root	31			NADH, FAD	8.0	1, 2, 3
	Zea mays	Root	31			NADH, FAD		6
pmPOX1	*Zea mays*	root	70(m)	38.3	3 N-glyco sites	*Heme*	6.8	7, 8
pmPOX2a	*Zea mays*	root	118(m)		*glycosylation*	*Heme*		7, 8
pmPOX2b	*Zea mays*	root	55(m)	33.4	3 N-glyco sites	*Heme*	8.9	7, 8
MDAR	*Spinacea oleracea*	leaf	45			NADH, FMN		9
FCR	*Lycopersicon esculentum*	root	200, 28			NADH	5.5, 6.2	10, 11, 12
MDH	*Allium cepa*	Root	40			NADH		13
	Zea mays	Root, leaf	41			NADH		6, 14
	Spinacea oleracea	Leaf	41			NADH		14
	Brassica oleracea	Inflorescences	41			NADH		14
ASC-reducible Cyt *b*	*Paseolus vulgaris*	Hook	55		*glycosylation*	*Heme*		15
	Arabidopsis thaliana	Leaf	27, 114			*Heme*		16
	Zea mays	Root	86, 110			*Heme*		17

3.1 *Flavocytochrome b Family*

The flavocytochrome *b* family is conserved throughout the animal, plant and fungal kingdoms (see refs in Lambeth 2002, 2004; Babior et al. 2002; Kimball and Saier Jr 2002; Philpott 2006; Kaplan et al. 2006) and its existence in red algae was demonstrated recently (Herve et al. 2006). The α-subunit of the phagocyte NADPH oxidase (NADPH oxidase homolog 2, NOX2; also named glycosylated protein rough molecular mass 91 kDa phagocyte oxidase, gp^{91phox}) is the most investigated member of this protein family (Henderson 2001; Lambeth 2002, 2004; Segal 2005). Fe^{3+}-chelate reductases (*f*erric chelate *r*eductase/*o*xidase, FRO) and *r*espiratory *b*urst *o*xidase *h*omologs (Rboh) have been identified by nucleotide sequence analysis in several plant species. Up to ten transmembrane domains, two *heme*, FAD and NADPH binding sites have been predicted for the gene products; additional EF-hand motifs suggest a regulation of Rboh by calcium (Keller et al. 1998; Torres et al. 1998; Amicucci et al. 1999; Robinson et al. 1999; Yoshioka et al. 2001; Simon-Plas et al. 2002; Waters et al. 2002; Li et al. 2004).

3.1.1 Fe^{3+}-Chelate Reductase (E.C. 1.16.1.7)

Iron uptake strategies of plants are illustrated in Fig. 1 and discussed in detail below. PM-bound Fe^{3+}-chelate reductase activity was demonstrated by in vivo (see refs in Bienfait 1985; Schmidt 1999) and in vitro experiments with isolated plasma

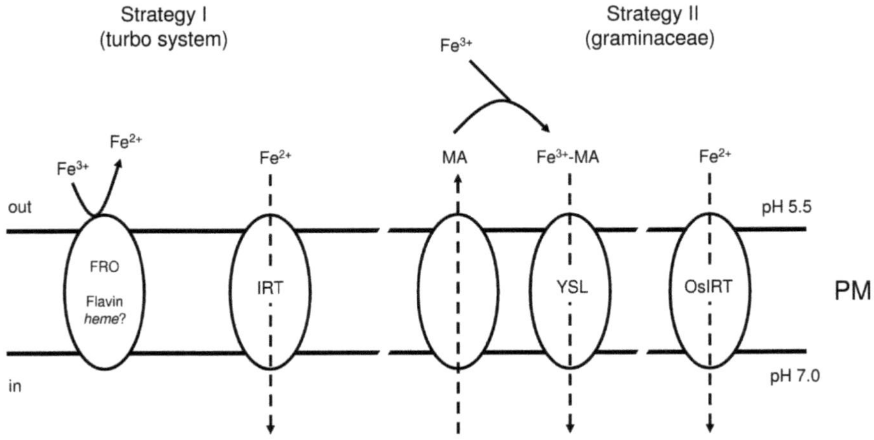

Fig. 1 Iron uptake strategies of plants. Strategy I plants (dicots and non-grass monocots) reduce the iron before uptake by a transmembrane redox activity (FRO, *f*erric chelate *r*eductase/*o*xidase). Once iron is reduced, it is transported across the PM by *i*ron *r*egulated *t*ransporters (IRT1), which is the major iron uptake system in *Arabidopsis*. Grasses (strategy II) release phytosiderophores of the MA family, chelate the Fe^{3+} and reduce the Fe^{3+}-chelate inside the cell. Fe^{3+}-chelates are taken up by MA transporters, i.e. yellow stripe 1-like (YSL) proteins. In addition to absorbing Fe^{3+}-MA, rice possesses an iron uptake system (OsIRT1) that directly absorbs Fe^{2+}

membranes (Holden et al. 1991, 1995). Iron-insufficient plants showed a significant increase in Fe^{3+}-chelate reductase activity (turbo system) and proton secretion. Both NADH-dependent and NADPH-dependent activities could be detected in plasma membranes of strategy I plants. However, only the NADH-dependent activity increased under iron deficiency. Initial attempts to purify and characterize the protein with this NADH-dependent Fe^{3+}-chelate reductase activity (Table 1) have been presented for tomato (*Lycopersicon esculentum* Mill.) roots (Holden et al. 1991, 1995; Bagnaresi et al. 1997).

Major progress in this field has been made by molecular biological approaches (Connolly and Guerinot 1998; Mori 1999; Connolly et al. 2003; Curie and Briat 2003; Bauer et al. 2004; Durrett et al. 2006; Mukherjee et al. 2006; Vasconcelos et al. 2006). As a result, the tomato gene coding for LeFRO1 (81 kDa) was isolated and characterized in detail (Li et al. 2004). The first Fe^{3+}-chelate reductase gene identified was *FRO2* from *Arabidopsis thaliana* L. Heyn. (Robinson et al. 1999). Overexpression of the *FRO2* transcript and reduced growth of a *FRO2* delition mutant were observed under conditions of iron deficiency. Meanwhile, eight genes for Fe^{3+}-chelate reductases (*FRO1* to *FRO8*) and two for *i*ron *r*egulated *t*ransporters (*IRT1*, *IRT2*) have been isolated from *Arabidopsis* (Eide et al. 1996; Robinson et al. 1997, 1999; Vert et al. 2001, 2002; Hall and Williams 2003; Wu et al. 2005; Colangelo and Guerinot 2006; Grotz and Guerinot 2006). Tissue-specific expression profiles of *AtFRO* genes suggest that AtFRO2 and AtFRO3 are Fe^{3+}-chelate reductases with roles in iron acquisition and metabolism in roots, whereas AtFRO5, AtFRO6, AtFRO7 and AtFRO8 are required for iron homeostasis in different tissues of shoots (Wu et al. 2005). After iron transport across the PM, members of several different transport families are implicated in the intracellular and intercellular transport of iron, including the *N*atural *r*esistance *a*ssociated *m*acrophage *p*rotein (Nramp) family and the YSL family (Eckhardt et al. 2001; Bereczky et al. 2003; Hall and Williams 2003; Grotz and Guerinot 2006; Puig et al. 2006).

Sequence analysis of a putative FRO predicts eight to ten transmembrane domains, *heme*, flavin and NADPH binding-sites (Robinson et al. 1999; Waters et al. 2002). However, such predictions have to be treated with caution because, in the case of the topology of FRO2, recently published experimental data revealed significant differences from those obtained by theoretical predictions (Schägerlof et al. 2006).

Nonetheless a highly conserved membrane domain, present in the whole flavocytochrome *b* family, seems to consist of a four-helix bundle. Helices V and VII contain the invariant histidine residues, which are responsible for *heme* binding in FRE1, the yeast homolog of FRO2 (Thrasher and Keep 1994; Finegold et al. 1996). An additional transmembrane helix, helix VIII, is present at the start of the water-soluble domain (Schägerlof et al. 2006). The water-soluble domain, which contains NADPH, FAD and oxidoreductase sequence motifs, is located on the inner (cytoplasmic) surface of the PM and between helices VIII and IX. Helices I, II, IX and X are not conserved within the whole protein family. Evidence for a transmembrane location for the oxidoreductase motif region was not found in this study, although it had been suggested by others (Waters et al. 2002; Schwacke et al. 2003; Li et al. 2004).

The increase in proton secretion measured under iron deficiency suggests activation of the PM-bound H^+-ATPase (see refs in Schmidt 1999). However, a possible function of FRO in proton channelling appears to be another explanation for this observation (Kimball and Saier Jr 2002; Lüthje et al. 2005). Both hypotheses and biochemical properties of FRO have to be further elucidated. *FRO* homologs have never been shown in strategy II plants. In contrast to strategy I plants, grasses increase the production and secretion of MA in response to iron deficiency, and Fe^{3+}-MA complexes are taken up by specific transporters (Fig. 1). The genes for MA synthesis have been isolated from several graminaeous plants (Herbik et al. 1999; Higuchi et al. 1999, 2001; Takahashi et al. 1999; Nakanishi et al. 2000; Kobayashi et al. 2001; Mizuno et al. 2003; Bashir et al. 2006) and high affinity Fe^{3+}-MA transporters (yellow stripe 1, *YS1*) have been isolated and characterized for maize, rice (*Oryza sativa* L.) and barley (*Hordeum vulgare* L.) roots (Curie et al. 2001; Koike et al. 2004; Roberts et al. 2004; Schaaf et al. 2004; Murata et al. 2006). In addition to absorbing Fe^{3+}-phytosiderophores, rice possesses an iron uptake system (OsIRT1) that directly absorbs Fe^{2+}, a strategy that is advantageous for growth in submerged conditions (Ishimaru et al. 2006).

3.1.2 Respiratory Burst Oxidase Homologs

So far excellent work has been published on the regulation and physiological functions of *Rboh*. It was demonstrated that a massive production of ROS (so-called oxidative burst) occurs during the first phase of pathogen defense; and expression profiles of *Rboh* indicated that its gene product may be involved in these reactions (Yoshioko et al. 2001, 2003; Simon-Plas et al. 2002; Torres et al. 2002; Kwak et al. 2003; Torres and Dangl 2005; Kobayashi et al. 2006). It was also suggested that production of ROS by Rboh may have a function in growth regulation and stress responses (Foreman et al. 2003; Kwak et al. 2003; Desikan et al. 2004a, b).

A PM localization of Rboh was shown by protein immuno-reaction with specific antibodies (e.g. Simon-Plas et al. 2002; Desikan et al. 2004b) and the detection of StRboh:green fluorescence protein (GFP) fusion constructs at the PM after transient expression in onion epidermal cells (Kobayashi et al. 2006). In contrast to the thoroughly investigated physiological functions of *Rboh*, a biochemical characterization of the gene product is still lacking. Rboh was predicted to be a transmembrane spanning protein with its NADPH binding-site at the cytoplasmic surface (Groom et al. 1996; Keller et al. 1998; Torres et al. 1998; Amicucci et al. 1999; Yoshioka et al. 2001; Simon-Plas et al. 2002).

NAD(P)H oxidase-like activity measured with sealed and right-side-out PM vesicles could be attributed to peroxidases (see refs in Vianello and Macri 1991; Mika et al. 2004). A calcium-sensitive and transmembrane ROS production was observed with NADPH-loaded and sealed right-side-out PM vesicles isolated from maize roots (Menckhoff and Lüthje 2004). However, the occurrence of superoxide dismutase (SOD) at the cell surface (Kukavica et al. 2005) could not be excluded in these experiments and its presence may significantly reduce the detectable amount of superoxide anion radicals. Nonetheless, the detected transmembrane activity may suggest the contribution of a constitutively Rboh in plasma membranes of maize roots, but the origin of this activity needs further investigation.

Initial attemps at purification of plant Rboh have been presented. A protein with NADPH oxidase-like activity was purified from PM of bean (*Phaseolus vulgaris* L.) hooks (Van Gestelen et al. 1997). Its enzymatic characteristics pointed to similarities with NOX2 with respect to inhibitor sensitivities, but the partially purified enzyme apparently contained FMN instead of FAD and no *b*-type cytochrome could be detected. Besides this enzyme, proteins with NAD(P)H-dependent quinone reductase (NQR) activity have been isolated from the PM of elicitor responsive been hooks (Van Gestelen et al. 1998). ROS production by these proteins was demonstrated and increased significantly in the presence of quinones. Unfortunately, specific antibodies against Rboh are not yet available for this plant species. Thus a possible contribution of a putative Rboh to any of these activities cannot be confirmed.

Arabidopsis RbohA was partially purified from microsomal fractions and detected by protein immuno-reaction (Keller et al. 1998). The protein could not be identified by N-terminal sequence analysis and no biochemical characterization of the partially purified AtRbohA was presented. In addition, it has to be mentioned that their documented cross-reactions with specific antibodies were stronger in the microsomal fractions than in PM fractions.

In contrast to this study, the accumulation of NtrbohD in the PM of tobacco leaves was clearly proved by Simon-Plas et al. (2002) whereas no band was immunodetected in any fraction isolated from elicited gp3T cells transformed with antisense constructs of NtrbohD cDNA. After elicitation with cryptogein, these tobacco cells showed the same extracellular alkalinization as the control, but no further production of ROS.

3.2 Standard System

ROS are produced by different enzymes and sources at the PM and in the apoplast (Lüthje et al. 2000; Mika et al. 2004). This production of ROS at the plant PM is thought to cause lipid peroxidation (Qiu and Liang 1995; Quartacci et al. 2001; Rawyler et al. 2002; Jalloul et al. 2002) which necessitates the existence of antioxidant and detoxifying strategies that regulate the production of ROS at the PM and protect the lipid bilayer against oxidative stress (Döring and Lüthje 1996; Döring et al. 1998b). One of the key players in this cell detoxification is SOD, which was recently shown to be associated with the PM (Alscher et al. 2002; Kukavica et al. 2005).

Judging from the currently available biochemical characterizations of redox components isolated from plasma membranes, it appears likely that the following electron transport chain (known as the "standard system") exists in the plant PM

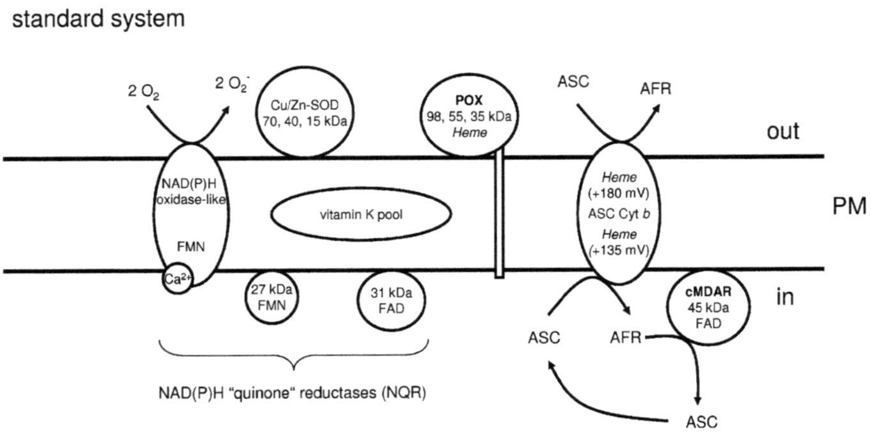

Fig. 2 Hypothetical structure of the standard system in plant plasma membranes. Although the structure of the ubiquitous and constitutive standard system is still unknown, increasing evidence exists for an electron transport chain with vitamin K_1 as mobile electron and proton carrier inside the lipid bilayer. Besides the function as electron carrier, vitamin K and ASC are antioxidants that protect the PM against oxidative stress. The other components were suggested to be involved in regeneration of these antioxidants, cell detoxification and redox regulation. SOD and PM-bound peroxidases detoxifies ROS at the cell surface. Further details are explained in the text

(Fig. 2). Electrons are transferred from cytosolic NAD(P)H via NQRs, vitamin K and a *b*-type cytochrome to the apoplast. In tandem with electrons, protons are transported to the apoplastic space by this transmembrane redox chain (Döring and Lüthje 1996; Lüthje et al. 1997).

The following subsection aims to summarize the most relevant findings which have been presented in the most recent reviews during recent years (Lüthje et al. 1997; Asard et al. 1998; Bérczi et al. 1998; Döring et al. 1998a, b; Stöhr 1998; Bérczi and Møller 2000) and to add progress on newly identified proteins in more detail.

3.2.1 NAD(P)H:Quinone Reductase Activities

At least two different proteins with NQR activities have been purified from plasma membranes (Table 1). They were isolated from plasma membranes of different species and tissues. An NAD(P)H dehydrogenase (24–27 kDa) capable of reducing HCF III and artificial quinones (Luster and Buckhout 1989; Serrano et al. 1994a, b, 1995; Trost et al. 1997; Lüthje et al. 1998). The protein showed no preference for NADH or NADPH in the presence of HCF III. However, when naphthoquinones (vitamin K) were the electron acceptor, the redox activity with NADPH was considerably higher than with NADH. Further investigations and comparison with soluble enzymes suggest the interaction of a soluble cytosolic NQR with the PM (Sparla et al. 1996, 1999; Trost et al. 1997).

An NADH-dependent NQR activity (31 kDa protein) was characterized in PM isolated from root tissues. This enzyme reduced HCF III > cytochrome *c* ≥ naphthoquinones > Fe^{3+}-chelates by a one-electron transfer mechanism (Serrano et al. 1994a, b, 1995; Sparla et al. 1997; Lüthje et al. 1998). In contrast to the protein with NQR activity purified from onion (*Allium cepa* L.) roots (Serrano et al. 1994a, b), the respective proteins isolated from plasma membranes of maize roots could be found in two separate peak fractions (Lüthje et al. 1998).

As shown in Fig. 2, the 27 kDa and 31 kDa NAD(P)H dehydrogenases with NQR activities appear to be peripheral proteins, located at the cytosolic surface of the PM (Lüthje et al. 1998). A further protein with NQR and NAD(P)H oxidase-like activity has been isolated from PM of elicitor responsive bean hooks (Van Gestelen et al. 1998). ROS production observed with sealed and NADPH-loaded PM vesicles isolated from maize roots (Menckhoff and Lüthje 2004) appears to be due to an integral NQR activity that showed properties comparable with that of the bean protein (N. Lepin, M. Menckhoff and S. Lüthje, unpublished data). Sequence analysis has not been presented for any of the characterized NQR isolated from plant plasma membranes so far.

3.2.2 Quinones

Quinones act as mobile electron and proton carriers in mitochondria, thylacoid membranes and animal PM (e.g. Kálen et al. 1987; Villalba et al. 1998). Together with ASC, ubiquinone (coenzyme Q, CoQ) acts as antioxidant and regenerates tocopherol (TQ) inside biomembranes. CoQ, however, has not been detected in plant PM (Askerlund 1990; Lüthje et al. 1998). Lipid analysis of plasma membranes isolated from carrot (*Daucus carrota* L.) cells and maize roots revealed a component that showed an absorbance spectrum comparable with that of vitamin K_1 (Barr et al. 1992; Lüthje et al. 1998) and a comparison of microsomal and PM fractions further suggested an accumulation of this component in the PM.

Due to its relatively high concentration, it appears possible that vitamin K in plant plasma membranes has an analogous function to CoQ (Döring and Lüthje 1996; Döring et al. 1998b; Lochner et al. 2003). Vitamin K has been found to be possess about 80% of the effectiveness of TQ as an antioxidant (Canfield et al. 1985). In addition a number of biological hydroquinones (including CoQ and vitamin K) have been found to regenerate TQ (Mukai et al. 1992, 1993; Fiorentini et al. 1997; Vervoort et al. 1997; Ortiz and Aranda 1999) which suggests that mixtures of TQ and these hydroquinones (as well as those of TQ and ASC) may function synergistically as antioxidants in various tissues and PM. Inhibition of lipid peroxidation was demonstrated in the order of CoQ-10 > K_2 > K_1 (Ohyashiki et al. 1991). Thus, vitamin K may act as a mobile electron and proton carrier and simultaneously as an antioxidant inside the plant PM (Döring and Lüthje 1996).

3.2.3 ASC-Reducible Cytochrome *b*

As shown in Fig. 2, *heme*-containing proteins (100–800 pmol *heme* mg^{-1} protein) have also been located in plant PM (see refs in Asard et al. 1998). The major component (60–80%) is an ascorbate (ASC)-reducible cytochrome *b* (α-band maximum 559–562 nm) with a standard redox potential around +150 mV (Askerlund et al. 1989; Asard et al. 1989). Several attempts to purify and characterize this protein have been made, but its amino acid sequence and function are still unknown (Trost et al. 2000; Berczi et al. 2001, 2003; Preger et al. 2004).

This compound does not appear to belong to the cytochrome *b*-561 family, which is known to exist in tonoplasts (Asard et al. 2001; Verelst et al. 2003, 2004; Griesen et al. 2004; Preger et al. 2004). This suggestion is further supported by genome studies of *Arabidopsis*. A cytochrome *b*-561 homolog has not been predicted for plasma membranes of this species.

Despite this fact, two cDNA species have been isolated for wild watermelon (*Citrullus lanatus* L.) cytochrome *b*-561, designated *CLb561A* and *CLb561B* (Nanasato et al. 2005). Levels of both *CLb561A* mRNA and protein increased significantly in the leaves during drought stress. A CLb561A-GFP fusion construct was shown at the PM. But the properties of the ASC-reducible cytochrome *b* isolated from plant plasma membranes (Table 1) were distinct from those of the cytochrome *b*-561 family – a finding which makes further investigations necessary (Trost et al. 2000; Bérczi et al. 2003). Besides ASC, vitamin K was able to reduce the PM-bound cytochrome *b* in the presence of NAD(P)H at the cytosolic surface, as were quinoles (Lüthje et al. 1998; Preger et al. 2004). This observation further supports the hypothesis of an electron transport chain in the PM (Fig. 2).

In addition to the ASC-reducible cytochrome *b*, two minor components have been identified in PM fractions with redox potentials of −20 mV to +20 mV and +220 mV (see refs in Asard et al. 1998; Lüthje et al. 2005). FRO and/or Rboh may contribute to the remaining 20–30% of the cytochrome content. But the presence of a *heme* group has never been directly demonstrated for members of the flavocytochrome *b* family in plant plasma membranes.

3.3 Peroxidases (E.C. 1.11.1.7)

PM-bound peroxidases (pmPOX) were shown for several plant materials (see refs in Vianello and Macri 1991; Mika et al. 2004) and characterized in detail for maize roots (Mika and Lüthje 2003). At least three isoperoxidases (pmPOX1, pmPOX2a, pmPOX2b) have been purified from plasma membranes of this plant species. The *heme*-containing enzymes belong to the classic secretory plant peroxidases (class III peroxidases), i.e. the superfamily of peroxidases from plants, bacteria and fungi (Welinder 1992). Complete amino acids sequences have been identified for pmPOX1 and pmPOX2b (Mika 2005). The deduced amino acid sequences showed three potential glycosylation sites, histidin residues for *heme* binding, eight structural cystein residues and a hydrophobic domain at the N-terminus. So far the data suggest a localization of the active center of both the enzymes pmPOX1 and pmPOX2b on the outer (apoplastic) surface of the membrane. Although the physiological functions of PM-bound peroxidases have to be further elucidated, a membrane protective function was suggested for these enzymes and appears most likely (Mika and Lüthje 2003; Mika et al. 2004).

3.4 Monodehydroascorbate Reductase (E.C. 1.6.5.4)

A cytochrome b_5 reductase (45 kDa) was isolated from PM of spinach (*Spinacea oleracea* L.) leaves (Askerlund et al. 1991). The protein was detected by cross-reaction with antibodies purified against cytochrome b_5 reductase from endoplasmatic reticulum of potato (*Solanum tuberosum* L.) tuber. However, N-terminal amino acid sequence analysis of the purified enzyme showed a strong similarity to cytosolic monodehydroascorbate reductase (MDAR; Berczi et al. 1995, 1998). Its specificity, kinetics, inhibitor sensitivity and cross-reactivity with anti-MDAR antibodies confirm this identification. FAD is a cofactor of MDAR, which appears to be strongly associated with the inner (cytoplasmic) surface of the PM under in vivo conditions. Together with the ASC-reducible cytochrome *b*, the PM-associated MDAR was suggested to have a function in the reduction of ASC in both the cytosol and the apoplast (Fig. 2). Thus, this antioxidant may be regenerated directly at the PM.

3.5 Malate Dehydrogenase (E.C. 1.1.1.37)

An oxaloacetate (OA) reductase activity was found in plasma membranes of different plant species and purified from onion and maize roots (Van Gestelen et al. 1997; Córdoba-Pedregosa et al. 1998, Lüthje et al. 1998; Hadzi-Taskovic Sukalovic et al. 1999). The relative molecular mass of PM-associated malate dehydrogenase (MDH; 40–41 kDa) was slightly higher than soluble cytosolic (c)MDH and other membrane-associated MDH (30–36 kDa) identified in peroxisomes (Struglics et al. 1993), inner mitochondrial membranes (Srere et al. 1997) and renal brush border membranes (Hanss et al. 2002). N-terminal sequence analysis of the PM-associated MDH showed a strong similarity to the amino acid sequence of cMDH (Córdoba-Pedregosa et al. 1998).

The structure, regulation and physiological functions of the PM-associated MDH are still unclear. Córdoba-Pedregrosa et al. (1998) suggested a regeneration of malate by the PM-associated MDH which could be transported by a malate shuttle across the membrane. According to this model, apoplastic OA crosses the PM and is reduced by the PM-associated MDH at the expense of cytosolic NADH. Together with cell wall MDH, this system could be involved in growth control (i.e.

cell wall cross-linking) or protection against environmental stresses (Martinoia and Rentsch 1994; Delhaize and Ryan 1995).

3.6 Nitrate Reductase Activity

A PM-bound protein with nitrate reductase activity (PM-NRA) was purified from algae and higher plants (see refs in Stöhr 1998; Lo Piero et al. 2003). The proteins isolated from *Chlorella* and sugar beet (*Beta vulgare* L.) were attached to the PM via a glycosylphosphatidylinositol (GPI)-anchor at the extracellular surface of the PM (Stöhr et al. 1995; Kunze et al. 1997). Some investigations suggest an additional association of a cytosolic isoenzyme at the inner surface of the PM (see refs in Stöhr 1998). Meanwhile, two isoenzymes (63 kDa, 93 kDa) that showed distinct substrate specificities were identified in roots and leaves (Stöhr and Ullrich 1997; Stöhr 1999). Peptide analysis of PM-NRA isolated from tobacco (*Nicotiana tabacum* L.) roots suggests the presence of an unknown protein; evidence for a true nitrate reductase protein, however, has not been found (Stöhr and Stremlau 2006). PM-NRA was postulated to act as a nitrate sensor and/or to contribute to nitric oxide (NO) production at the plant PM (Stöhr et al. 2001; Stöhr and Stremlau 2006).

3.7 Nitrite:NO Reductase Activity

Nitric oxide (NO) plays an important role in several physiological processes in plants and is produced by different enzymes and sources (see refs in Stöhr and Ullrich 2002; Shapiro 2005; Grün et al. 2006). NO formation has been shown for plasma membranes of tobacco roots (Stöhr et al. 2001). The protein (310 kDa) corresponding to this activity could be separated from PM-NRA. Nitrite:NO reductase (NI-NOR) activity appears to reduce apoplastic nitrite produced by PM-NRA in vivo and may have a function in nitrate signaling. Rates of NO production depend largely on the physiological conditions, mainly the availability of nitrate and oxygen (Stöhr and Stremlau 2006). In this context, NO is proposed to play a role during anoxia as an indicator of the external nitrate availability and in regulating symbiotic interactions at the root surface.

4 Molecular Organization of the PM

The physiological roles of biological membranes are defined by their protein and lipid profiles. According to the fluid mosaic model (Singer and Nicolson 1972), proteins have been proposed to be peripheral or integral components of the lipid bilayer (Vereb et al. 2003). Nowadays it appears that the PM is organized in microdomains, so-called lipid rafts (Simons and Ikonen 1997; Brown and London 1998; Pike 2004; Martin et al. 2005; Bhat and Panstrugah 2005).

Up to now approximately 550 polypeptides could be resolved from plant PM preparations by proteomic approaches (Masson and Rosignol 1995; Santoni et al. 1998, 2000; Alexandersson et al. 2004; Marmagne et al. 2004, 2006; Santoni 2007). A recent study identified 145 proteins in lipid rafts of tobacco leaves (Morel et al. 2006). Functional grouping of these proteins by homology searches revealed five groups, which were in agreement with the main physiological cellular processes involving the plant PM. It was further demonstrated that transport proteins seem to be under-represented in lipid rafts, whereas other groups of proteins are disproportionately more abundant. Among these are proteins involved in signaling and response to biotic and abiotic stress, cellular trafficking, cell wall synthesis and degradation, and metabolism. Nonetheless, a significant amount (~20%) of the PM-specific protein spots still corresponds to hypothetical proteins with unknown functions.

4.1 Lipid Rafts

Ten years ago, Simons and Ikonen (1997) and Brown and London (1998) put forward the hypothesis that PM lipids of eukaryotic cells do not form a homogeneous phase consisting of sterols and glycerophospholipids, but form a mosaic of domains with unique biochemical compositions . The controversy regarding the existence, size and molecular nature of lipid rafts is still a matter of debate (e.g. Pike 2004; Lagerholm et al. 2005). There is, however, a steadily increasing body of evidence for animal, plant and fungal cells that supports the raft hypothesis (Pike 2004; Bhat and Panstrugah 2005; Martin et al. 2005; Wachtler and Balasubramanian 2006).

Lipid rafts appear to have unique physicochemical properties that direct their organization into liquid-ordered phases floating in a liquid-crystalline ocean of glycerophospholipids (Brown and London 2000). These domains are resistant to solubilization by non-ionic detergents like Triton X-100 at 4 °C (*d*etergent *r*esistant *m*embranes, DRM; also named *d*etergent *i*nsoluble *m*embranes, DIM) and are destabilized by cholesterol- and sphingolipid-depleting agents (Brown and London 1998).

Lipid rafts have been morphologically characterized as small membrane patches (e.g. Brown and London 2000; Ishitsuka et al. 2005). Cellular and/or exogenous proteins that interact with lipid rafts can use them as transport shuttles on the cell surface. Thus, rafts act as molecular sorting machines capable of co-ordinating the spatiotemporal organization of signal transduction pathways, transport proteins and others within selected areas of the PM.

Lipid rafts are heterogeneous in terms of both their protein and their lipid content and can be localized to different regions of the cell (Degroote et al. 2004; Pike 2004). Sterols and (glycol-)sphingolipids were suggested for the outer half of rafts and saturated phospholipids for the cytosolic half. It was further postulated that rafts are embedded in unsaturated phospholipids. Apart from abundant integral proteins, the peripheral proteins in rafts are either GPI-anchored at the outer surface or associated by lipid–protein interactions at the cytosolic surface (see Fig. 2 in Pike 2004). Protein migration into lipid rafts seems to be mediated by lipid–lipid interactions, whereas protein–lipid and protein–protein interactions appear crucial for the specific localization of proteins in lipid rafts. Nowadays different models are under discussion for apical sorting of GPI-anchored proteins in polarized epithelial cells (e.g. Paladino et al. 2004; Tivodar et al. 2006). The PM of those cells is divided into two domains, apical and basolateral, which display different protein and lipid compositions, therefore resulting in specialized functions.

4.1.1 Lipid Rafts in Plants

To date, little is known about lipid rafts in plants (see refs in Bhat and Panstruga 2005; Martin et al. 2005). Phospholipids, sterols and (glycol-)sphingolipids in the form of glucocerebroside were thought to constitute the major lipid classes in plant plasma membranes (e.g. Yoshida and Uemura 1984; Grandmougin et al. 1989; Uemura and Steponkus 1994; Uemura et al. 1995, 2006; Bohn et al. 2001, 2006). These lipids and the proportion of unsaturated fatty acids define the fluidity of the membrane (e.g. Lynch and Phinney 1995; Warude et al. 2006). Very long chain fatty acids (VLCFA) may either increase the diameter or possibly regulate membrane curvature (Millar et al. 1998; Wallis et al. 2002). Sterols contribute to bilayer thickness and may influence the dynamic trafficking of proteins with polar distributions in the PM (Schroeder 1984; Hartmann 1998; Clouse 2002; Zuckermann et al. 2004; Betts and Moore 2003). In addition, the presence of non-bilayer lipids may also define the functionality of membrane proteins (De Kruijff 1997; Lee 2000) and the formation of hexagonal-II phases may hamper the most salient function of the PM the compartmentalization of cells (Webb et al. 1994).

Evidence for microdomains in plant plasma membranes has been provided by the following observations. In vitro studies demonstrated that plant sterols and sphingolipids promote the formation of ordered membrane domains (Xu et al. 2001). DIM were found to be greatly enriched in glycosylceramide and a mixture of different sterols, whereas phospho- and glycoglycerolipids of the PM were largely excluded (Mongrand et al. 2004; Borner et al. 2005; Laloi et al. 2006). However, a detailed analysis of long-chain sphingobases of different plant materials and species revealed that the resulting sphingobase profiles from cerebrosides and PM showed large qualitative and quantitative differences (Sperling et al. 2005). A major part of these lipids appears to contribute to hydrophobic anchors of PM proteins (i.e. diacylglycerol-based GPI anchors and inositol–phosphoryl–ceramide moieties with a different glycosyl headgroup, GIPC-anchors). Several hundred of those proteins were preliminarily identified by a genomic analysis of *Arabidopsis* (Borner et al. 2003) and GPI-anchored proteins have been identified in plasma membranes and DIM (Stöhr et al. 1995; Kunze et al. 1997; Peskan et al. 2000; Eisenhaber et al. 2003; Borner et al. 2003, 2005). Besides lipids, proteins of Triton X-100 insoluble membranes have been investigated by proteomic approaches. A recent study grouped DIM proteins by their physiological functions (Morel et al. 2006). Proteins with hydrophobic modifications (e.g. GPI-anchored, palmitoylation, myristoylation, etc.), integral proteins and signaling components have been identified (Peskan et al. 2000; Elortza et al. 2003, 2006; Shahollari et al. 2005; Morel et al. 2006; Sutter et al. 2006).

One major drawback of almost all studies dealing with lipid rafts in plant plasma membranes is the fact that most of them were limited to Triton X-100 insoluble membrane fractions from PM preparations with so-called "minor contaminations" (Peskan et al. 2000; Morel et al. 2006). However, those DIM fractions do not appear to be specific for PM (Phinney and Thelen 2005; Laloi et al. 2006) and the detection of major proteins of endomembrane systems in lipid rafts (e.g. Morel et al. 2006; Uemura et al. 2006) supports the existence of Triton X-100 insoluble membrane fractions containing membrane systems other than plasma membranes.

4.1.2 PM-Bound Redox Systems and Lipid Rafts

Triton X-100 and CHAPS are most effective in membrane solubilization and hence selectively extract glycerophospholipids (and some proteins), yielding a raft that is highly enriched in cholesterol and sphingolipids (Pike 2004). Most of the PM-bound redox proteins described so far appear to be poorly soluble in Trtion X-100 or CHAPS, with the exception of peripheral NQR, MDAR and PM-associated MDH. Although the ASC-reducible cytochrome *b* of bean hooks was easily solubilized by Triton X-100 from PM (Trost et al. 2000), the respective protein from plasma membranes of maize roots and *Arabidopsis* leaves was very resistant to solubilization with that detergent (Bérczi et al. 2001). Due to this observation, it was assumed that the composition of lipids in the PM of different tissues might influence the solubilization of proteins (Bérczi and Horvath 2003). Analysis of total lipids extracted from these tissues revealed that sterols and cerebrosides are present in much higher amounts in plasma membranes from *Arabidopsis* leaves than in those from etiolated bean hooks. However, the lipid profiles of plasma membranes isolated from *Arabidopsis* leaves were consistent with data obtained for maize roots (Bohn et al. 2001; Bérczi and Horvath 2003). Figure 3 shows redox components most likely associated with DIM. Besides the ASC-reducible cytochrome *b*, GPI-anchored proteins are predicted as components in lipid rafts. Thus raft localization appears sure for PM-NRA. Besides blue light signaling, a function of PM-NRA in the production of NO has been proposed (Stöhr 1998; Stöhr and Stremlau 2006). Both suggestions fit nicely with the proposed function of lipid rafts in signaling. As shown in Table 2, a GPI-anchored monocopper oxidase-like protein of the multicopper oxidase family was identified in PM and DIM preparations (Borner et al. 2005; Elorza et al. 2003, 2006; Morel et al. 2006). This observation is consistent with the detection of copper in plasma membranes of higher plants (Lüthje et al. 1995; Møller et al. 1991).

Detection of NtrbohD and its small cytosolic interaction partner NtRac5 in DIM further implicates a function of lipid rafts in ROS signaling (Morel et al. 2006). A possible spatiotemporal control of ROS production by Rboh for polarized growth was suggested for lipid rafts in root hair apices (Foreman et al. 2003; Mittler et al.

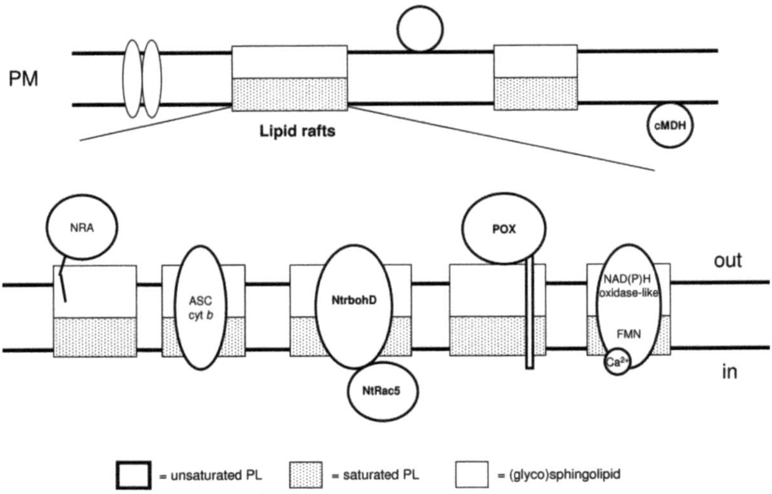

Fig. 3 Plasma membrane redox systems and lipid rafts. Evidence for the occurrence of PM-bound redox systems in lipid rafts has been presented. Since rafts may be heterogeneous in their composition, components may participate in different raft fractions. Nitrate reductase activity (NRA) is the protein most likely located in rafts, because of its GPI-anchor. NtrbohD and NtRac5 were identified by peptide sequence analysis of raft proteins, as were NQR (Table 2). Interaction of these peripheral NQR with lipid rafts has to be further investigated. PM-bound peroxidases, the ASC-reducible cytochrome *b,* and the integral NQR are poorly soluble by Triton X-100 or CHAPS, which suggest raft localization

Table 2 Redox proteins identified in lipid rafts. DIM were prepared from highly enriched PM fractions by Triton X-100. Proteins were separated by 2D-electrophorese techniques (isoelectric focussing/SDS-PAGE) and protein spots were analysed by mass spectroscopy. MW_{calc} Calculated molecular mass (kDa), *pI* point isoelectric, *PTM* post-translational modifications. *Ref. 1* Borner et al. (2005), *Ref. 2* Morel et al. (2006)

Species	Protein family	Accession number of gene	MWcalc	pI	PTM	Ref.
A. thaliana	Putative quinone reductase	At4g36750	28.7	5.17		1
	Flavin mononucleotide-binding quinone reductase (FQR1)	At5g54500	21.8	6.84		1
	FAD-binding domain	At4g20830	60.5			1
N. tabaccum	FAD-binding domain	At4g20830	60.5			1
	NtrbohD	Q8RVJ9	105.8	8.93		2
	Monocopper oxidase-like protein SKS1 (precursor)	Q8VXX5	65.9	6.90	GPI	2
	Putative monocopper oxidase SKU5 (precursor)	Q9SU40	65.6	7.74	GPI	2
	1,4-Benzoquinone-like reductase	Q9LSQ5	21.8	5.96		2

2004). In animal systems, NOX did not appear to be restricted to the PM and targeting of NADPH oxidase to discrete subcellular compartments was discussed as a mechanism of localizing ROS and activation of downstream redox signaling events that mediate various cell functions (Ushio-Fukai 2006).

Solubilization studies of the calcium-sensitive NQR – a putative integral protein of plant plasma membranes (Fig. 2) – point to an accumulation in Triton X-100 insoluble membranes. Although the precise nature and function of this protein still has to be investigated, superoxide anion radical production was observed in vitro and suggests a possible function in signaling (Menckhoff, Lepin and Lüthje, unpublished data). Due to the hydrophobic domain at the N-terminus, PM-bound peroxidases were found to be poorly soluble in Trition X-100 (Mika and Lüthje 2003). Localization in lipid rafts may allow PM-bound peroxidases to move next to ROS producing and detoxifying enzymes in the membrane (Fig. 3). Thus PM-bound peroxidases could probably not only detoxify H_2O_2 directly at the site of origin to guarantee the optimal protection of the membrane, but also protect specific functional regions of the PM, in a similar manner as discussed for the thylacoid ascorbate peroxidase (Asada 1992; Asada et al. 1996; Yoshimura et al. 1998; Kieselbach et al. 2000).

The identification of redox proteins in DIM (Table 2) fits nicely with postulated functions of these microdomains (Brown and London 2000; Bhat and Panstrugah 2005; Morel et al. 2006). Studies on DIM isolated from *Arabidopsis* and tobacco leaves identified at least three NQR: a putative quinone reductase, a benzoquinone-like reductase and a flavin mononucleotide-binding flavodoxin-like quinone reductase (FQR1; Borner et al. 2005; Morel et al. 2006). The theoretical molecular masses of these proteins were comparable with those of the peripheral NQRs purified and characterized from plant plasma membranes (Table 1). However, the latter proteins were solubilized with Triton X-100. FQR1 was characterized after heterologous expression in *Escherichia coli* and was suggested to be a soluble component (Laskowski et al. 2002). Interactions of this protein with membrane proteins or the PM have not been described so far and need further proof.

4.2 Protein Assemblies

Most physiological processes are not carried out by single proteins, but rather by protein assemblies (Alberts 1998). The importance of membrane-bound protein complexes is clearly illustrated by their involvement in different physiological processes, like photosynthesis, respiration, etc. (Kügler et al. 1997; Poetsch et al. 2000; Jänsch et al. 1996). Such complex formations and their respective activities are strongly regulated by parameters such as the participating proteins, cofactors, posttranslational modifications and interaction with inhibitory or activating compounds (Rajan et al. 2002; Helms 2002).

Nowadays blue native polyacrylamide gel electrophoresis (BN-PAGE) is one of the state of the art methods to study integral membrane protein complexes at the proteome scale (Schägger and von Jagow 1991; Schägger 2003). For example, several studies demonstrated not only the existence of complexes but also that of supercomplexes in plant mitochondria and thylacoide membranes by BN-PAGE (e.g. Aseeva et al. 2004; Komenda et al. 2004; Aro et al. 2005; Dudkina et al. 2006).

On the one hand, the existing lack of sequence information of PM-bound redox systems will hamper the investigation and identification of those enzymes in protein assemblies. On the other hand, evidence was presented for protein–protein interactions of PM-bound redox proteins. BN-PAGE applied to plant PM confirmed the existence of several high molecular mass protein complexes in spinach (*Spinacea oleracea* L.) leaves (Kjell et al. 2004). Preliminary data from our group suggest that these complexes vary with respect to the plant species and tissue investigated. So far redox proteins have not been identified, with the exception of a GPI-anchored monocopper oxidase-like protein, which was found in a high molecular mass protein complex in plasma membranes of *Arabidopsis* leaves (A. Schmitt, F. Buck and S. Lüthje, unpublished data). Although the aggregation of proteins or detergent–protein micelles has to be carefully excluded from those investigations, participation of PM-bound redox proteins in protein–protein interactions and protein complexes appears likely and has to be further elucidated.

Purification of Fe^{3+}-chelate reductase activity from plasma membranes of tomato roots revealed a 200 kDa complex (Bagnaresi et al. 1997). Subunits of soluble NQR are linked together by disulfide bridges and build a tetramer of 85–100 kDa (Sparla et al. 1999). The authors claimed that a similar quinone reductase participated in an intrinsic protein complex (300 kDa) in the PM of zucchini (*Cucurbita pepo* L.) hypocotyls (Trost et al. 1997). The complex was active with lipophilic quinones and markedly sensitive to the inhibitor DPI. Following purification the complex dissociated, which caused significant changes in the biochemical properties of the NQR.

Malate dehydrogenases are multimeric enzymes that consist of identical subunits mainly organized as dimers or tetramers (see refs in Minarik et al. 2002; Scheibe 2004). The PM-associated MDH co-purifies with an unidentified protein of 28 kDa (Córdoba-Pedregrosa et al. 1998; Lüthje et al. 1998). It was suggested that the smaller protein could be responsible for the association of soluble cMDH with the PM. This hypothesis will be worth further investigation, because a membrane-associated cMDH was purified from brush border plasma membranes (Hanss et al. 2002). This protein co-purifies with an anion channel and reconstitution experiments indicate that the membrane-associated cMDH confers the nucleic acid selectivity of the ion channel.

A tight association between NI-NOR and PM-NRA was suggested by solubilization studies (Meyer and Stöhr 2002); and finally soluble NtRac5 was identified by peptide mass analysis, together with NtRbohD in DIM of tobacco (Mongrand et al. 2004; Morel et al. 2006). The detection of NtRac5 in these preparations was suggested to be caused by protein–protein interaction with NtRbohD.

5 Conclusions

Despite the intensity of current research on PM-bound redox systems and their physiological functions, it has to be admitted that we still lack fundamental knowledge in this field. Several of these redox proteins have been isolated and characterized in detail, but their functions are still unknown. Complete amino acid sequences have been presented only for a minor part of these enzymes, which hampers further functional analysis. So far, investigations have focussed on the physiological functions of the flavocytochrome *b* family. Some basic biochemical properties of FRO have been documented, whereas the properties of Rboh are still unknown.

In spite of the fact that the presence of redox systems in plant plasma membranes seems to be well established, the molecular structure and function of the standard system still needs further clarification. The same applies to the plentiful evidence for redox systems in lipid rafts. The participation of those enzymes in protein–protein interactions of the plant PM is clearly evident but an unequivocal structural and functional characterization will necessitate a lot of further research work.

Acknowledgement This work is dedicated to Michael Böttger (Universität Hamburg) on the occasion of his 65th birthday.

References

Alberts B (1998) The cell as a collection of protein machines: preparing the next generation of molecular biologists. Cell 92:291–294

Alexandersson E, Saalbach G, Larsson C, Kjellbom P (2004) Arabidopsis plasma membrane proteomics identifies components of transport, signal transduction and membrane trafficking. Plant Cell Physiol 45:1543–1556

Alscher RG, Erturk N, Heath LS (2002) Role of superoxide dismutases (SODs) in controlling oxidative stress in plants. J Exp Bot 53:1331–1341

Amicucci E, Gaschler K, Ward JM (1999) NADPH oxidase genes from tomato (*Lycopersicum esculentum*) and Curlyleaf pondweed. Plant Biol 1:524–528

Aro EM, Suorsa A, Rokka A, Allahverdiyeva Y, Paakkarinen V, Saleem A, Battchikova N, Rintamäki E (2005) Dynamics of photosystem II: a proteomic approach to thylakoid protein complexes. J Exp Bot 56:347–356

Asada K (1992) Ascorbate peroxidase: a hydrogen peroxide – scavenging enzyme in plants. Physiol Plant 85:235–241

Asada K, Miyake C, Ogawa K, Hossain MA (1996) Microcompartmentation of ascorbate peroxidase and regeneration of ascorbate from ascorbate radical: its dual role in chloroplasts. In: Obinger C, Burner U, Ebermann R, Penel C, Greppin H (eds) Proceedings of the IV international symposium on plant peroxidases: biochemistry and physiology. University of Vienna/University of Geneva, pp 163–167

Asard H, Venken M Caubergs R, De Greef JA (1989) *b*-Type cytochromes in higher plant plasma membranes. Plant Physiol 90:1077–1083

Asard H, Horemans N, Preger V, Trost P (1998) Plasma membrane *b*-type cytochromes. In: Asard H, Bérczi A, Caubergs RJ (eds) Plasma membrane redox systems and their role in biological stress and disease. Kluwer, Dordrecht, pp 1–31

Asard H. Kapila Y, Verelst W, Bérczi A (2001) Higher-plant plasma membrane cytochrome b-561 a protein in search of a function. Protoplasma 217:77–93

Aseeva E, Ossenbühl F, Eichacker LA, Wanner G, Soll J, Vothknecht UC (2004) Complex formation of Vipp1 depends on its a-helical PspA-like domain. J Biol Chem 279:35535–35541

Askerlund P (1990) Redox processes of plant plasma membranes. PhD thesis, University of Lund, Lund.

Askerlund P, Larsson C, Widell S (1989) Cytochromes of plant plasma membranes. characterization by absorbance difference spectrophotometry and redox titration. Physiol Plant 76:123–134

Askerlund P, Laurent P, Nakagawa H, Kadar JC (1991) NADH-ferricyanide reductase of leaf plasma membranes. Partial purification and immunological relation to potato tuber microsomal NADH-ferricyanide reductase and spinach leaf NADH-nitrate reductase. Plant Physiol 95:6–13

Babior BM, Lambeth JD, Nauseef W (2002) The neutrophil NADPH oxidase. Arch Biochem Biophys 397:342–344

Bagnaresi P, Basso B, Pupillo P (1997) The NADH-dependent Fe^{3+}-chelate reductases of tomato roots. Planta 202:427–434

Barr R, Pun RS, Crane FL, Brightman AO, Morré DJ (1992) Destruction of vitamin K_1 of cultured carrot cells by ultraviolett radiation and its effect on plasma membrane electron transport reactions. Biochem Int 27:449–456

Bashir K, Inoue H, Nagasaka S, Takahashi M, Nakanishi H, Mori S, Nishizawa NK (2006) Cloning and characterization of deoxymugineic acid synthase genes from Graminaceous plants. J Biol Chem 281:32395–32402

Bauer P, Thiel T, Klatte M, Bereczky Z, Brumbarova T, Hell R, Grosse I (2004) Analysis of sequence, map position, and gene expression reveals conserved essential genes for iron uptake in *Arabidopsis* and tomato. Plant Physiol 136:4169–4183

Bérczi A, Asard A (2003) Soluble proteins, an often overlooked contaminant in plasma membrane preparations. Trends Plant Sci 8:250–251

Bérczi A, Horvat, G (2003) Lipid rafts in the plant plasma membranes? Acta Biol Szegediensis 47: 7–10

Bérczi A, Møller IM (1998) NADH-monodehydroascorbate oxidoreductase is one of the redox enzymes in spinach leaf plasma membranes. Plant Physiol 116:1029–1036

Bérczi A, Møller IM (2000) Redox enzymes in the plant plasma membrane and their possible roles. Plant Cell Environ 23:1287–1302

Bérczi A, Fredlund KM, Møller IM (1995) Purification and characterization of an NADH-hexacyanoferrate (III) reductase from spinach leaf plasma membrane. Arch Biochem Biophys 320:65–72

Bérczi A, Getselen P van, Pupillo P (1998) NAD(P)H-utilizing flavo-enzymes in the plant plasma membrane. In: Asard H, Bérczi A, Caubergs RJ (eds) Plasma membrane redox systems and their role in biological stress and disease. Kluwer, Dordrecht, pp 33–67

Bérczi A, Lüthje S, Asard H (2001) *b*-Type cytochromes in plasma membranes of *Phaseolus vulgaris* hypocotyls, *Arabidopsis thaliana* leaves, and *Zea mays* roots. Protoplasma 217:50–55

Bérczi A, Caubergs RJ, Asard H (2003) Partial purification and characterization ofan ascorbat-reducible b-type cytochrom from the plasma membrane of Arabidopsis thaliana leaves. Protoplasma 221:47–56

Bereczky Z, Wang HY, Schubert V, GAnal M, Bauer P (2003) Differential regulation of nramp and irt metal transporter genes in wild type and iron uptake mutants of tomato. J Biol Chem 278:24697–24704

Betts H, Morre I (2003) Plant cell polarity: the ins and outs of sterol transport. Curr Biol 13: R781–R783

Bhat RA, Panstruga, R (2005) Lipid rafts in plants. Planta 223:5–19

Bienfait F (1985) Regulated redox processes at the plasmalemma of plant root cells and their function in iron uptake. J Bioenerg Biomembr 17:73–83

Bienfait F, Lüttge U (1988) On the function of two systems that can transfer electrons across the plasma membrane. Plant Physiol Biochem 26:665–671

Bohn M, Heinz E, Lüthje S (2001) Lipid composition and fluidity of plasma membranes isolated from corn (*Zea mays* L.) roots. Arch Biochem Biophys 387:35–40

Bohn M, Lüthje S, Sperling P, Heinz E, Dörffling K (2006) Plasma membrane lipid alterations induced by cold acclimation and isothermal abscisic acid treatment of winter wheat seedlings differing in frost resistance. J Plant Physiol 164:146–156

Borner GHH, Lilley KS, Stevens TJ, Dupree P (2003) Idenification of glycosylphosphatidylinosit ol-anchored proteins in *Arabidopsis*. A proteomic and genomic analysis. Plant Physiol 132:568–577

Borner GHH, Sherrier DJ, Weimar T, Michaelson LV, Hawkins ND, MacAskill A, Napier JA, Beale MH, Lilley KS, Dupree P (2005) Analysis of detergent-resistant membranes in *Arabidopsis*. Evidence for plasma membrane lipid rafts. Plant Physiol 137:104–116
Böttger M, Lüthen H (1986) Possible linkage between NADH-oxidation and proton secretion in *Zea mays* L. roots. J Exp Bot 37:666–675
Brightman AO, Barr R, Crane FL, Morre DJ (1988) Auxin-stimulated NADH oxidase purified from plasma membrane of soybean. Plant Physiol 86:1264–1269
Brown DA, London E (1998) Functions of lipid rafts in biological membranes. Annu Rev Cell Dev 14:111–136
Brown DA, London E (2000) Structure and function of sphingolipid- and cholesterol-rich membrane rafts. J Biol Chem 275:17221–17224
Canfield LM, Davy LA, Thomas GL (1985) Anti-oxidant/pro-oxidant reactions of vitamin K. Biochemi Biophys Res Commun 128:211–219
Chaney RL, Brown JC, Tiffin LO (1972) Obligatory reduction of ferric chelates in iron uptake by soybeans. Plant Physiol 50:208–213
Chueh PJ, Morré DM, Penel C, DeHahn T, Morré DJ (1997) The hormone-responsive NADH oxidase of the plant plasma membrane has properties of a NADH:protein disulfide reductase. J Biol Chem 272:11221–11227
Clouse SD (2002) *Arabidopsis* mutants reveal multiple roles for sterols in plant development. Plant Cell 14:1995–2000
Colangelo E, Guerinot ML (2006) Put the metal to the petal: metal uptake and transport throughout plants. Curr Opin Plant Biol 9:322–330
Connolly EL, Guerinot ML (1998) Reduction and uptake of iron in plants. In: Asard H, Bérczi A, Caubergs R (eds) Plasma membrane redox systems and their role in biological stress and disease. Kluwer, Dordrecht, pp 179–192
Connolly EL, Campbell NH, Grotz N, Prichard CL, Guerinot ML (2003): Overexpression of the FRO2 ferric chelate reductase confers tolerance to growth on low iron and uncovers posttranscriptional control. Plant Physiol 133:1102–1110
Córdoba-Pedregosa MC, González-Reyes JA, Serrano A, Villalba JM, Navas P, Córdoba F (1998) Plasmalemma-associated malate dehydrogenase activity in onion root cells. Protoplasma 205:29–36
Cowan DSC, Cooke DT, Clarkson DT, Hall JL (1993) Lipid and sterol composition of plasma membranes isolated from stele and cortex of maize roots. J Exp Bot 44:991–994
Curie C, Briat JF (2003) Iron transport and signaling in plants. Annu Rev Plant Biol 54:183–206
Curie C, Panaviene Z, Loulergue C, Dellaporta SL, Briat JF, Walker EL (2001) Maize yellow stripe1 encodes a membrane protein directly involved in Fe (III) uptake. Nature 409:346–349
Davey MS, Suggett DJ, Geider RJ, Taylor AR (2003) Phytoplankton plasma membrane redox activity: effect of iron limitation and interaction with photosynthesis. J Phycol 39:1132–1144
De Kruijff B (1997) Lipids beyond the bilayer. Nature 386:129–130
Degroote S, Wolthoorn J, Meer G van (2004) The cell biology of glycosphingolipids. Semin Cell Dev Biol 15:375–387
Delhaize E, Ryan PR (1995) Aluminum toxicity and tolerance in plants. Plant Physiol 107:315–321
Desikan R, Cheung MK, Bright J, Henson D, Hancock JT, Neill SJ (2004a) ABA, hydrogen peroxide and nitric oxide signaling in stomatal guard cells. J Exp Bot 55:205–212
Desikan R, Cheung MK, Clarke A, Golding S, Sagi M, Fluhr R, Rock C, Hancock J, Neill S (2004b) Funct Plant Biol 31:913–920
Döring O, Lüthje S (1996) Molecular components and biochemistry of electron transport in plant plasma membranes. Mol Membr Biol 13:127–142
Döring O, Lüthje S, Böttger M (1998a) To be or not to be – a question of plasma membrane redox? Prog Bot 59:328–354
Döring O, Böttger M, Lüthje S (1998b) The constitutive electron transport system in plant plasma membranes: a role in defense against oxidative stress in the apoplast and within the plasma membrane? – A hypothesis. In: Noga G, Schmitz M (eds) Antioxidants in higher plants: biosynthesis, characteristics, actions and specific functions in stress defense.Shaker, Aachen, pp 19–30

Dudkina NV, Heinemeyer J, Sunderhaus S, Boekema EJ, Braun HP (2006) Respiratory chain super complexes in the plant mitochondrial membrane. Trends Plant Sci 11:232–240

Durrett TP, Connolly EL, Rogers EE (2006) *Arabidopsis* cpFtsY mutants exhibit pleiotropic defects including an inability to increase iron deficiency-inducible root Fe(III) chelate reductase activity. Plant J 47:467–479

Eckhardt U, Mas Marques A, Buckhout TJ (2001) Two iron-regulated cation transporters from tomato complement metal uptake-deficient yeast mutants. Plant Mol Biol 45:437–448

Eide D, Broderius M, Fett J, Guerinot ML (1996)A novel iron regulated metal transporter from plants identified by functional expression in yeast. Proc Natl Acad Sci USA 93:5624–5628

Eisenhaber B, Wildpaner M, Schultz CJ, Borner GHH, Dupree P, Eisenhaber F (2003) Glycosylp hosphatidylinositol lipid anchoring of plant proteins. Sensitive prediction from sequence- and genome-wide studies for *Arabidopsis* and rice. Plant Physiol 133:1691–1701

Elortza F, Nühse TS, Foster LJ, Stensballe A, Peck SC, Jensen ON (2003) Proteomic analysis of glycosylphosphatidylinositol-anchored membrane proteins. Mol Cell Proteomics 2:1261–1270

Elortza F, Mohammed S, Bunkenborg J, Foster LJ, Nühse TS, Brodbeck U, Peck SC, Jensen ON (2006) Modification-specific proteomics of plasma membrane proteins: Identification and characterization of glycosylphosphatidylinositol-anchored proteins released upon phospholipse D treatment. J Proteomics Res 5:935–943

Finegold AA, Shatwell KP, et al (1996) Intramembrane bis-heme motif for transmembrane electron transport conserved in a yeast iron reductase and the human NADPH oxidase. J Biol Chem 271:31021–31024

Fiorentini D, Cipilline M, Galli MC, Landi L (1997). Antioxidant activity of reduced menadione in solvent solution and in model membranes. Free Radical Res 26:419–429

Foreman J, Demidchik V, Bothwell JHF, Mylona P, Medema H, Torres MA, Linstead P, Costa S, Brownlee C, Jones JDG, Davies JM, Dolan L (2003) Reactive oxygen species produced by NADPH oxidase regulate plant cell growth. Nature 422:442–446

Gevhev TS, Hille J (2005) Hydrogen peroxide as a signal controlling plant programmed cell death. J Cell Biol 168:17–20

Grandmougin A, Bouvier-Nave P, Ullmann P, Benveniste P, Hartmann MA (1989) Plant Physiol 90:591–597

Griesen D, Su D, Bérczi A, Asard H (2004) Localization of an ascorbate-reducible cytochrome *b*561 in the plant tonoplast. Plant Physiol 134: 726–734

Groom QJ, Torres MA, Fordham Skelton AP, Hammond-Kosack KE, Robinson NJ, Jones JDG (1996) RbohA, a rice homoloug of the mammalian gp^{91}phox respiratory burst oxidase gene. Plant J 10:515–522

Grotz N, Guerinot ML (2006): Molecular aspects of Cu, Fe and Zn homeostasis in plants. Biochim Biophys Acta 1763:595–608

Grün S, Lindermayr C, Sell S, Durner J (2006) Nitric oxide and gene regulation in plants. J Exp Bot 57:507–516

Hadži-Taškovič Šukalović V, Vuletić M, Ignjatović-Micić D, Vučinić Z (1999) Plasma membrane-bound malate dehydrogenase activity in maize roots. Protoplasma 207:203–212

Hall JL, Williams LE (2003) Transition metal transporters in plants. J Exp Bot 54:2601–2613

Hanss B, Leal-Pinto E, Teixeira A, Christian RE, Shabanowitz J, Hunt DF, Klotman PE (2002) Cytosolic malate dehydrogenase confers selectivity of the nucleic acid-conducting channel. Proc Natl Acad Sci USA 99:1707–1712

Hartmann, MA (1998). Plant sterols and the membrane environment. Trends Plant Sci 3:170–175

Helms V (2002) Attraction within the membrane. Forces behind transmembrane protein folding and supramolecular complex assembly. EMBO Rep 3:1133–1138

Henderson LM (2001) NADPH oxidase subunit gp^{91phox}: a proton pathway. Protoplasma 217:37–42

Herbik A, Koch G, Mock HP, Dashkow D, Czihal A, Thielmann J, Stephan UW, Baumlein H (1999) Isolation, characterization and cDNA cloning of nicotianamine synthase from barley. A key enzyme for iron homeostasis in plants. Eur J Biochem 265:231–239

Herve C, Tonon T, Collen J, Corre E, Boyen C (2006) NADPH oxidase in eukaryotes: red algae provide new hints! Curr Genet 49:190–204

Higuchi K, Suzuki K, Nakanishi H, Yamaguchi H, Nishizawa NK, Mori S (1999) Cloning of nicotianamine synthase, novel genes involved in the biosynthesis of phytosiderophores. Plant Physiol 119:471–480

Higuchi K, Watanabe S, Takahashi M, Kawasaki S, Nakanishi H, Nishizawa NK, Mori S (2001) Nicotianamine synthase gene expression differs in barley and rice under Fe-deficient conditions. Plant J 25:159–167

Holden MJ, Luster DG, Chaney RL, Buckhout TJ, Robinson C (1991) Fe^{3+}-chelate reductase activity of plasma membranes isolated from tomato (*Lycopersicon esculentum* Mill.) roots. Plant Physiol 97:537–544

Holden MJ, Crimmins TJ, Chaney RL (1995) Cu^{2+} reduction by tomato root plasma membrane vesicles. Plant Physiol 108:1093–1098

Horemans N, Foyer CH, Asard H (2000) Transport and action of ascorbate at the plant plasma membrane. Trends Plant Sci 5:263–267

Ishimaru Y, Suzuki M, Tsukamoto T, Suzuki K, Nakazono M, Kobayashi T, Wada Y, Watanabe S, Matsuhashi S, Takahashi M, Nakanishi H, Mori S, Nishizawa NK (2006) Rice plants take up iron as an Fe^{3+}-phytosiderophore and as Fe^{2+}. Plant J 45:335–346

Ishitsuka R, Sato SB, Kobayashi T (2005) Imaging lipid rafts. J Biochem 137:249–254

Jalloul A, Montillet JL, Assigbetse K, Agnel JP, Delannoy E, Triantaphylides C, Daniel JF, Marmey P, Geiger JP, Nicole M (2002) Lipid peroxidation in cotton: *Xanthomonas* interactions and the role of lipoxygenases during the hypersensitive reaction. Plant J 32:1–12

Jänsch L, Kruft V, Schmitz UK, Braun H-P (1996) New insights into the composition, molecular mass and stoichiometry of the protein complexes of plant mitochondria. Plant J 9:357–368

Jones G J, Palenik BP, Morel FMM (1987) Trace-metal reduction by phytoplankton – the role of plasmalemma redox enzymes. J Phycol 23:237–244

Kalén A, Norling B, Appelkvist EL, Dallner G (1987) Ubiquinone biosynthesis by the microsomal fraction from rat liver. Biochim Biophys Acta 926:70–78

Kaplan J, McVey Ward D, Crisp RJ, Philpott CC (2006) Iron-dependent metabolic remodeling in *S. cerevisiae*. Biochim Biophys Acta 1763:646–651

Kawano T (2003) Roles of the reactive oxygen species-generating peroxidase reactions in plant defense and growth induction. Plant Cell Rep 21:829–837

Keller T, Damude HG, Werner D, Doerner P, Dixon RA, Lamb C (1998) A plant homolog of the neutrophil NADPH oxidase gp^{91}phox subunit gene encodes a plasma membrane protein with Ca^{2+} binding motifs. Plant Cell 10:255–266

Kieselbach T, Bystedt M, Hynds P, Robinson C, Schroder WP (2000) A peroxidase homologue and novel plastocyanin located by proteomics to the *Arabidopsis* chloroplast thylakoid lumen. FEBS Lett 480:271–276

Kimball RA, Saier MH Jr (2002) Voltage-gated H+ channels associated with human phagocyte superoxide-generating NADPH oxidases: sequence comparisons, structural predictions, and phylogenetic analyses. Mol Membr Biol 19:137–147

Kjell J, Rasmusson AG, Larsson H, Widell S (2004) Protein complexes of the plant plasma membrane resolved by blue native PAGE. Physiol Plant 121:546–555

Kobayashi M, Kawakita K, Maeshima M, Doke N, Yoshioka H (2006) Subcellular localization of Strboh proteins and NADPH-dependent O_2^--generating activity in potato tuber tissues. J Exp Bot 57:1373–1379

Kobayashi T, Nakanishi H, Takahashi M, Kawasaki S, Nishizawa NK, Mori S (2001) In vivo evidence that Ids3 from *Hordeum vulgare* encodes a dioxygenase that converts 2′-deoxymugineic acid to mugineic acid in transgenic rice. Planta 212:864–871

Koike S, Inoue H, Mizuno D, Takahashi M, Nakanishi H, Mori S, Nishizawa NK (2004) OsYSL2 is a rice metal-nicotianamine transporter that is regulated by iron and expressed in the phloem. Plant J 39:415–424

Komenda J, Reisinger V, Müller BC, Dobáková M, Granvofel B, Eichacker LA (2004) Accumulation of the D2 protein is a key regulator step for assembly of the photosystem II reaction center complex in *Synechocystis* PCC 6803. J Biol Chem 279:48620–48629

Kügler M, Jänsch L, Kruft V, Schmitz UK, Braun H-P (1997) Analysis of the chloroplast protein complexes by blue-native polyacrylamide gel electrophoresis (BN-PAGE). Photosynth Res 53:35–44

Kukavica B, Vučinić Ž, Vuletić M (2005) Superoxide dismutase, peroxidase, and germin-like protein activity in plasma membranes and apoplast of maize roots. Protoplasma 226:191–197

Kunze M, Riedel J. Lange U, Hurwitz R, Tischner R (1997) Evidence for the presence of GPI-anchored PM-NR in leaves of *Beta vulgaris* and for PM-NR in barley leaves. Plant Physiol Biochem 35:507–512

Kwak JM, Mori IC, Pei ZM, Leonhard N, Torres AM, Dangl JL, Bloom RE, Bodde S, Jones JDG, Schroeder JL (2003) NADPH oxidase AtrbohD and AtrbohF genes function in ROS-dependent ABA signalling in *Arabidopsis*. EMBO J 22:2623–2633

Lagerholm BC, Weinreb GE, Jacobson K, Thompson NL (2005) Detetcting microdomains in intact cell membranes. Annu Rev Phys Chem 56:309–336

Laloi C, Apel K, Danon A (2004) Reactive oxygen signalling: the latest news. Curr Opin Plant Biol 7:323–328

Laloi M, Perret AM, Chatre L, Melser S, Cantrel , Vaultier MN, Zachowski A, Bathany K, Schmitter JM, Vallet M, Lessire R, Hartmann MA, Moreau P (2006) Insights into the role of specific lipids in the formation and delivery of lipid microdomains to the plasma membrane of plant cells. Plant Physiol 143:461–472

Lambeth JD (2002) Nox/Duox family of nicotinamide adenine dinucleotide (phosphate oxidases. Curr Opin Hematol 9:11–17

Lambeth JD (2004) NOX enzymes and the biology of reactive oxygen. Nat Rev Immunol 4:181–189

Larsson C, Sommarin M, Widell S (1994) Isolation of highly purified plant plasma membranes and separation of inside-out and right-side-out vesicles. Methods Enzymol 228:451–469

Laskowski M, Dreher KA, Gehring MA, Abel S, Gensler AL, Sussex IM (2002) FQR1, a novel primary auxin-response gene, encodes a flavin mononucleotide-binding quinone reductase. Plant Physiol 128:578–590

Lee AG (2000) Membrane lipids: it's only a phase. Curr Biol 10:R377–R380

Li L, Cheng X, Ling HQ (2004) Isolation and characterization of Fe(III)-chelate reductase gene LeFRO1 in tomato. Plant Mol Biol 54:125–136

Lilley KS, Dupree P (2006) Methods of quantitative proteomics and their application to plant organelle characterization. J Exp Bot 57:1493–1500

Lo Piero AR, Cultrone A, Monachello D, Petrone G (2003) Different kinetic and regulatory properties of soluble and membrane-bound nitrate reductases in tomato leaves. Plant Sci 165:139–145

Lochner K, Döring O, Böttger M (2003) Phylloquinone, what can we learn from plants? BioFactors 18:73–78

Luster DG, Buckhout TJ (1989) Purification and identification of plasma membrane associated electron transport protein from maize (*Zea mays* L.) roots. Plant Physiol 91:1014–1019

Lüthje S, Niecke M, Böttger M (1995) Iron and copper in plasma membranes of maize (*Zea mays* L.) roots investigated by proton induced X-ray emission (PIXE). Protoplasma 184:145–150

Lüthje S, Döring O, Heuer S, Lüthen H, Böttger M (1997) Oxidoreductases in plant plasma membranes (review). Biochim Biophys Acta 1331:81–102

Lüthje S, Van Gestelen P, Córdoba-Pedregosa MC, González-Reyes JA, Asard H, Villalba JM, Böttger M (1998) Quinones in plant plasma membranes – a missing link? Protoplasma 205:43–51

Lüthje S, Böttger M, Döring O (2000) Are plants stacked neutrophiles? Pathogen-induced oxidative burst in plants and mammals in comparison. Prog Bot 61:187–222

Lüthje S, Böttger M, Döring O (2005) Proton channelling *b*-Type cytochromes in plant plasma membranes? Prog Bot 66:187–207

Lynch DV, Phinney AJ (1995) The transbilayer distribution of glucosylceramide in plant plasma membrane. In: Kader JC, Mazliak P (eds) Plant lipid metabolism. Kluwer, Dordrecht, pp 239–241

Maldonado MT, Price NM (2000) Nitrate regulation of Fe reduction and transport by Fe-limited *Thalassiosira oceanica*. Limnol Oceanogr 45:814–826

Maldonado MT, Price NM (2001) Reduction and transport of organically bound iron by *Thalassiosira oceanica* (Bacillariophyceae). J Phycol 37:298–309

Marmagne A, Rouet MA, Ferro M, Rolland N, Alcon C, Joyard J, Garin J, Barbier-Brygoo H, Ephritikhine G (2004) Identification of new intrinsic proteins in Arabidopsis plasma membrane proteome. Mol Cell Proteomics 3:675–691

Marmagne A, Salvi D, Rolland N, Ephritikhine G, Joyard J, Barbier-Brygoo H (2006) Purification and fractionation of membranes for proteomic analyses. Methods Mol Biol 323:403–420

Marschner H, Römheld V, Kissel M (1986) Different strategies in higher plants in mobilization and uptake of iron. J Plant Nutr 9:695–713

Martin SW, Glover BJ, Davies JM (2005) Lipid microdomains – plant membranes get organized. Trends Plant Sci 10:263–265

Martinoia E, Rentsch D (1994) Malate compartmentation: responses to a complex metabolism. Annu Rev Plant Physiol Plant Mol Biol 45:447–467

Masson F, Rossignol M (1995) Basic plasticity of protein expression in tobacco leaf plasma membrane. Plant J 8:77–85

Menckhoff M, Lüthje S (2004) Transmembrane electron transport in sealed and NAD(P)H-loaded right-side-out plasma membrane vesicles isolated from maize (*Zea mays* L.) roots. J Exp Bot 55:1343–1349

Merchant SS, Allen MD, Kropat J, Moseley JL, Lpng JC, Tottey S, Terauchi AM (2006) Between rock and a hard place: trace element nutrition in *Chlamydomonas*. Biochim Biophys Acta 1763:578–594

Meyer C, Stöhr C (2002) Nitrate reductase and nitrite reductase. In: Foyer C, Noctor G (eds) Photosynthetic nitrogen assimilation and associated carbon metabolism. Advances in photosynthesis. Kluwer, Dordrecht, pp 49–62

Mika A (2005) Biochemische Charakterisierung und bioinformatische Sequenzanalyse plasmamembrangebundener Peroxidasen aus Maiswurzeln (*Zea mays* L.). PhD thesis, University of Hamburg, Hamburg

Mika A, Lüthje S (2003) Properties of guaiacol peroxidase activities isolated from corn root plasma membranes. Plant Physiol 132:1489–1498

Mika A, Minibayava F, Beckett R, Lüthje S (2004) Possible functions of extracellular peroxidases in stress-induced generation and detoxification of active oxygen species. Phytochem Rev 3:173–193

Millar AA, Wrischer M, Kunst L (1998) Accumulation of very-long-chain fatty acids in membrane glycerolipids is associated with dramatic alterations in plant morphology. Plant Cell 11:1889–1902

Minarik P, Tomaskova N, Kollarova M, Antalik M (2002) Malate dehydrogenase – structure and function. Gen Physiol Biophys 21:257–265

Mittler R, Vanderauwera S, Gollery M, Van Breusegem F (2004) Reactive oxygen gene network of plants. Trends Plant Sci 9:490–498

Mizuno D, Higuchi K, Sakamoto T, Nakanishi H, Mori S, Nishizawa NK (2003) Three nicotianamine synthase genes isolated from maize are differentially regulated by iron nutritional status. Plant Physiol 132:1989–1997

Mongrand S, Morel J, Laroche J, Claverol S, Carde JP, Hartmann MA, Bonneu M, Simon-Plas F (2004) Lipid rafts in higher plant cells. J Biol Chem 279:36277–36286

Morel J, Claverol S, Mongrand S, Furt F, Fromentin J, Bessoule JJ, Blein JP, Simon-Plas F (2006) Proteomics of plant detergent-resistant membranes. Mol Cell Proteomics 5:1396–1411

Mori S (1999) Iron acquisition by plants. Curr Opin Plant Biol 2:250–253

Morré DJ (1998) NADH oxidase: a multifunctional ectoprotein of the eukaryotic cell surface. In: Asard H, Bérczi A, Caubergs RJ (eds) Plasma membrane redox systems and their role in biological stress and disease. Kluwer, Dordrecht, pp 121–156

Mukai K, Itho S, Morimoto H (1992) Stopped-flow kinetic study of vitamin E regeneration reaction with biological hydroquinones (reduced forms of ubiquinone, vitamin K and tocopherol quinone) in solution. J Biol Chem 267:22277–22281

Mukai K, Morimoto H, Kikuchi S, Nagaoka S (1993) Kinetic study of free radical-scavenging action of biological hydroquinones (reduced forms of ubiquinone, vitamin K and tocopherol quinone) in solution. Biochim Biophys Acta 1157:313–317

Mukherjee I, Campbell NH, Ash JS, Connolly EL (2006) Expression profiling of the *Arabidopsis* ferric chelate reductase (FRO) gene family reveals differential regulation by iron and copper. Planta 223:1178–1190
Murata Y, Ma JF, Yamaji N, Ueno D, Nomoto K, Iwashita T (2006) A specific transporter for iron (III)-phytosiderophore in barley roots. Plant J 46:563–572
Møller IM, Askerlund P, Widell S (1991) Electron transport constituents in the plasma membrane. In: Crane FL, Morre DJ, Löw HE (eds) Oxidoreduction at the plasma membrane: relation to growth and transport. (Plants, vol 2). CRC, Boca Raton, pp 35–59
Nakanishi H, Yamaguchi H, Sasakuma T, Nishizawa NK, Mori S (2000) Two dioxygenase genes, Ids3 and Ids2, from *Hordeum vulgare* are involved in the biosynthesis of mugineic acid family phytosiderophores. Plant Mol Biol 44:199–207
Nanasato Y, Akashi K, Yokota A (2005) Co-expression of cytochrome b561 and ascorbate oxidase in leaves of wild watermelon under drought and high light conditions. Plant Cell Physiol 46:1515–1524
Neill SJ, Desikan R, Hanckokc JT (2002) Hydrogen peroxide signalling. Curr Opin Plant Biol 5:388–395
Ohyashiki T, Yabunaka Y, Matsui K (1991) Antioxidant effect of vitamin K homologues on ascorbic acid/Fe^{2+}-induced lipid peroxidation of lecithin liposomes. Chem Pharm Bull 39:976–979
Ortiz A, Aranda FJ (1999) The influence of vitamin K_1 on the structure and phase behaviour of model membrane systems. Biochim Biophys Acta 1418:206–220
Paladino S, Sarnataro D, Pillich R, Tivodar S, Nitsch L, Zurzolo C (2004) Protein oligomerization modulated raft partitioning and apical sorting of GPI-anchored proteins. J Cell Biol 167:699–709
Palmgren MG, Askerlund P, Fredrikson K, Widell S, Sommarin M, Larsson C (1990) Sealed inside-out and right-side-out plasma membrane vesicles. Plant Physiol 92:871–880
Peskan T, Westermann M, Oelmüller R (2000) Identification of low-density Triton X-100-insoluble plasma membrane microdomains in higher plants. Eur J Biochem 267:6989–6995
Philpott CC (2006) Iron uptake in fungi: a system for every source. Biochim Biophys Acta 1763:636–645
Phinney BS, Thelen JJ (2005) Proteomic characterization of a triton-insoluble fraction from chloroplasts defines a novel group of proteins associated with macromolecular structures. J Proteome Res 4:497–506
Pike LJ (2004) Lipid rafts: heterogeneity on the high seas. Biochem J 378:281–292
Poetsch A, Neff D, Seelert H, Schägger H, Dencher NA (2000) Dye removal, catalytic activity and 2D crystallization of chloroplast H^+-ATP synthase purified by blue native electrophoresis. Biochim Biophys Acta 1466:339–349
Preger V, Scagliarini S, Pupillo P, Trost P (2004) Identification of an ascorbate-dependent cytochrome bof the tonoplast membrane sharing biochemical features with members of the cytochrome b561 family. Planta 220:365–375
Puig S, Andrés-Colás N, García-Molina A, Peñarubia L (2006) Copper and iron homeostasis in *Arabidopsis*: responses to metal deficiencies, interactions and biotechnological applications. Plant Cell Environ 30:271–290
Qiu QS, Liang HG (1995) Lipid peroxidation caused by the redox system of plasma membranes from wheat roots. J Plant Physiol 145:261–265
Quartacci MF, Cosi E, Navari-Izzo F (2001) Lipids and NADPH-dependent superoxide production in plasma membrane vesicles from roots of wheat grown under copper deficiency or excess. J Exp Bot 52:77–84
Rajan S, Preisig-Muller R, Wischmeyer E, Nehring R, Hanley PJ, Renigunta V, Musset B, Schlichthorl G, Derst C, Karschin A, Daut J (2002) Interaction with 14-3-3 proteins promotes functional expression of the potassium channels TASK-1 and TASK-3. J Physiol 545:13–26
Rawyler A, Arpagaus S, Braendle R (2002) Impact of oxygen stress and energy availability on membrane stability of plant cells. Ann Bot 90:499–507
Roberts LA, Pierson AJ, Panaviene Z, Walker EL (2004) Yellow stripe1. Expanded roles for the maize iron-phytosiderophore transporter. Plant Physiol 135:112–120

Robinson NJ, Sadjuga MR, Groom QJ (1997) The froh gene family from *Arabidopsis thaliana*: putative iron-chelate reductases. Plant Soil 196:245–248

Robinson NJ, Procter CM, Connolly EL, Guerinot ML (1999) A ferric-chelate reductase for iron uptake from soils. Nature 397:694–697

Rubinstein B, Stern AI (1986) Relationship of transplasmalemma redox activity to proton and solute transport by roots of *Zea mays*. Plant Physiol 80:805–811

Santoni V (2007) Plant plasma membrane protein extraction and solubilization for proteomic analysis. Methods Mol Biol 355:93–109

Santoni V, Rouquie D, Doumas P, Mansion M, Boutry M, Degand H, Dupree P, Packman L, Sherrier J, Prime T, Bauw G, Posada E, Rouze P, Dehais P, Sahnoun I, Barlier I, Rossignol M (1998) Use of a proteome strategy for taggin proteins present at the plasma membrane. Plant J 16:633–641

Santoni V, Kieffer S, Desclaux D, Masson F, Rabilloud T (2000) Membrane proteomics: use of additive main effects with multiplicative interaction model to classify plasma membrane proteins according to their solubility and electrophorectic properties. Electrophoresis 21:3329–3344

Schaaf G, Ludewig U, Erenoglu BE, Mori S, Kitahara T, Wiren N von (2004) ZmYS1 functions as a proton-coupled symporter for phytosiderophore- and nicotianamine-chelated metals. J Biol Chem 279:9091–9096

Schägerlof U, Wilson G, Herbert H, Al-Karadaghi S, Hägerhäll C (2006) Transmembrane topology of FRO2, a ferric chelate reductase from *Arabidopsis thaliana*. Plant Mol Biol 62:215–221

Schägger H (2003) Blue native electrophoresis. In: Hunte C, Michel H (eds) Membrane protein purification and crystallization. Elsevier Science, London, pp 105–130

Schägger H, Jagow G von (1991) Blue native electrophoresis for isolation of membrane protein complexes in enzymatically active form. Anal Biochem 199:223–231

Scheibe R (2004) Malate valves to balance cellular energy supply. Physiol Plant 120:21–26

Schmidt W (1999) Mechanisms and regulation of reduction-based iron uptake in plants. New Phytol 141:1–26

Schroeder F (1984) Fluorescent sterols: probe molecules of membrane structure and function. Prog Lipid Res 23:97–113

Schwacke R, Flügge UI, Kunze R (2003) Plant membrane proteom databases. Plant Physiol Biochem 42:1023–1034

Segal AW (2005) How neutrophils kill microbes. Annu Rev Immunol 23:197–223

Serrano A, Córdoba F, Gonzáles-Reyes JA, Navas P, Villalba JM (1994a) Purification and characterization of two distinct NAD(P)H dehydrogenases from onion (*Allium cepa* L.) root plasma membrane. Plant Physiol 106:87–96

Serrano A, Villalba JM, González-Reyes JA, Navas P, Córdoba F (1994b) Two distinct NAD(P)H-dependent redox enzymes isolated from onion root plasma membranes. Biochem Mol Biol Int 32:841–849

Serrano A, Cordoba F, González-Reyes JA, Santos C, Navas P, Villalba JM (1995) NADH-specific dehydrogenase from onion root plasma membrane: purification and characterization. Protoplasma 184:133–139

Shahollari B, Varma A, Oelmüller R (2005) Expression of a receptor kinase in *Arabidopsis* roots is stimulated by the basidomycete *Piriformospora indica* and the protein accumulates in Triton X-100 insoluble plasma membrane microdomains. J Plant Physiol 162:945–958

Shapiro AD (2005) Nitric oxide signaling in plants. Vitam Horm 72:339–398

Simon-Plas F, Elmayan T, Blein JP (2002) The plasma membrane oxidase NtrbohD is responsible for AOS production in elicited tobacco cells. Plant J 31:137–147

Simons K, Ikonen E (1997) Functional rafts in cell membranes. Nature 387:569–572

Singer SJ, Nicolson GL (1972) The fluid mosaic model of the structure of cell membranes. Sci 175:720–731

Sparla F, Tedeschi G, Trost P (1996) NAD(P)H:(quinone-acceptor) oxidoreductase of tobacco leaves is a flavin mononucleotide-containing flavoenzyme. Plant Physiol 112:249–258

Sparla F, Bagnaresi P, Scagliarini S, Trost P (1997) NADH:Fe(III)-chelate reductase of maize roots is an active cytochrome b_5 reductase. FEBS Lett 414:571–575
Sparla F, Tedeschi G, Pupillo P, Trost P (1999) Cloning and heterologous expression of NAD(P)H:quinone reductase of *Arabidopsis thaliana*, a functional homologue of animal DT-diaphorase. FEBS Lett 463:382–386
Sperling P, Franke S, Lüthje S, Heinz E (2005) Are glucocerebrosides the predominant shingolipids in plant plasma membranes? Plant Physiol Biochem 43:1031–1038
Srere PA, Sherry AD, Malloy CR, Sumegi B (1997) Channelling in the Krebs tricarboxylic acid cycle. In: Agius L, Sherratt HSA (eds) Channeling in intermediary metabolism, Portland. London, pp 201–217
Stöhr C (1998) Plasma membrane-bound nitrate reductase in algae and higher plants. In: Asard H, Bérczi A, Caubergs RJ (eds) Plasma membrane redox systems and their role in biological stress and disease, Kluwer, Dordrecht, pp 103–119
Stöhr C (1999) Relationship of nitrate supply with growth rate plasma membrane-bound and cytosolic nitrate reductase, and tissue nitrate content in tobacco plants. Plant Cell Environ 22:169–177
Stöhr C, Stremlau S (2006) Formation and possible roles of nitric oxide in plant roots. J Exp Bot 57:463–470
Stöhr C, Ullrich WR (1997) A succinate-oxidizing nitrate reductase is located at the plasma membrane of plant roots. Planta 203:129–132
Stöhr C, Ullrich WR (2002) Generation and possible roles of NO in plant roots and their apoplastic space. J Exp Bot 53:2293–2303
Stöhr C,Schuler F, Tischner R (1995) Glycosyl-phosphatidylinositol-anchored proteins exist in the plasma membrane of *Chlorella sorokiniana*. Planta 196:284–287
Stöhr C, Strube F, Marx G, Ullrich WR, Rockel P (2001) A plasma membrane-bound enzyme of tobacco roots catalyses the formation of nitric oxide from nitrite. Planta 212:835–841
Struglics A, Fredlund KM, Rasmussin AG, Møller IM (1993) The presence of a short redox chain in the membrane of intact potato tuber peroxisomes and the association of malate dehydrogenase with the peroxisomal membrane. Physiol Plant 88:19–28
Sutter JU, Campanoni P, Tyrrell M, Blatt MR (2006) Selective mobility and sensitivity to SNAREs is exhibited by the *Arabidopsis* KAT1 K^+ channel at the plasma membrane. Plant Cell 18:935–954
Takagi S (1976) Naturally occurring iron-chelating compounds in oat- and rice-root washing. I. Activity measurement and preliminary characterization. Soil Sci Plant Nutr 22:423–433
Takagi S, Nomoto K, Takemoto T (1984) Physiological aspect of mugineic acid, a possible phytosiderophore of graminaceous plants. J Plant Nutr 7:1–5
Takahashi M, Yamaguchi H, Nakanishi H, Nishizawa NK, Mori S (1999) Cloning two genes for nicotianamine aminotransferase, a critical enzyme in iron acquisition (strategy II) in graminaceous plants. Plant Physiol 121:947–956
Thrasher AJ, Keep NH, et al (1994) Chronic granulomatousdisease. Biochim Biophys Acta Mol Basis Dis 1227:1–24
Tivodar S, Paladino S, Pillich R, Prinetti A, Chigomo V, Merr G van, Sonnino S, Zurzolo C (2006) Analysis of detergent-resistant membranes associated with apical and basolateral GPI-anchored proteins in polarized epithelial cells. FEBS Lett 580:5702–5712
Torres MA, Dangl JL (2005) Functions of the respiratory burst oxidase in biotic interactions, abiotic stress and development. Curr Opin Plant Biol 8:397–403
Torres MA, Onouchi H, Hamada S, Machida C, Hammond-Kossack KE, Jones DJG (1998) Six *Arabidopsis thaliana* homologues of the human respiratory burst oxidase (gp^{91phox}). Plant J 14:365–370
Torres MA, Dangl J, Jones JDG (2002) *Arabidopsis* gp^{91}phox homologues AtrbohD and AtrbohF are required for accumulation of reactive oxygen intermediates in the plant defense response. Proc Natl Acad Sci USA 99:517–522
Torres MA, Jones JDG, Dangl JL (2006) Reactive oxygen species signaling in response to pathogens. Plant Physiol 141:373–378

Trost P, Foscarini S, Preger S, Bonora P, Vitale L, Pupillo P (1997) Dissecting the diphenylene iodonium-sensitive NAD(P)H:quinone oxidoreductase of zucchini plasma membrane. Plant Physiol 114:737–746
Trost P, Bérczi A, Sparla F, Sponza G, Marzadori B, Asard H, Pupillo P (2000) Purification of cytochrome *b*-561 from bean hypocotyls plasma membrane. Evidence for the presence of two heme centers. Biochim Biophys Acta 1468:1–5
Uemura M, Steponkus PL (1994) A contrast of the plasma membrane lipid composition of oat and rye leaves in relation to freezing tolerance. Plant Physiol 104:479–496
Uemura M, Joseph RA, Steponkus PL (1995) Cold acclimation of *Arabidopsis thaliana*: effect on plasma membrane lipid composition and freeze-induced lesions. Plant Physiol 109:15–30
Uemura M, Tominaga Y, Nakagawara C, Shigematsu S, Minami A, Kawamura Y (2006) Responses of the plasma membrane to low temperatures. Physiol Plant 126:81–89
Ushio-Fukai M (2006) Localizing NADPH oxidase-derived ROS. Sci Signal Transduct Knowl Environ 349:re8
Van Gestelen P, Asard H, Caubergs RJ (1997) Solubilization and separation of a plant plasma membrane NADPH-O_2^- synthase from other NAD(P)H oxidoreductases. Plant Physiol 115:543–550
Van Gestelen P, Asard H, Horemans N, Caubergs RJ (1998) Superoxide-producing NAD(P)H-oxidases in plasma membrane vesicles from elicitor responsive bean plants. Physiol Plant 104:653–660
Vasconcelos M, Eckert H, Arahana V, Graef G, Grusak MA, Clemente T (2006) Molecular and phenotypic characterization of transgenic soybean expressing the *Arabidopsis* ferric chelate reductase gene, FRO2. Planta 224:1116–1128
Vereb G, Szöllösi J, Matko J, Nagy P, Farkas T, Vigh L, Mátyus L, Waldmann TA, Damjanovich S (2003) Dynamic, yet structured: the cell membrane three decades after Singer–Nicholson model. Proc Natl Acad Sci USA 100:8053–8058
Verelst W, Asard H (2003) A phylogenetic study of cytochrome b561 proteins. Genet Biol 4:38
Verelst W, Asard H (2004) Analysis of an *Arabidopsis thaliana* protein family, structurally related to cytochromes b561 and potentially involved in catecholamine biochemistry in plants. J Plant Physiol 161:175–181
Vert G, Briat J,Curie C (2001) *Arabidopsis* IRT2 gene encodes a root periphery iron transporter. Plant J 26:181–189
Vert G, Grotz N, Dedaldechamp F, Gaymard F, Guerinot ML, Briat,JF, Curie C (2002) IRT1, an *Arabidopsis* transporter essential for iron uptake from the soil and for plant growth. Plant Cell 14:1223–1233
Vervoort LM, Ronden JE, Thijssen HH (1997) The potent antioxidant activity of vitamin K cycle in microsomal lipid peroxidation. Biochem Pharm 54:871–876
Vianello A, Macri F (1991) Generation of superoxide anion hydrogen peroxide at the surface of plant cells. J Bioenerg Biomembr 23:409–423
Villalba JM, Crane FL, Navas P (1998) Antioxidative role of ubiquinone in animal plasma membranes. In: Asard H, Bérczi A, Caubergs R (eds) Plasma membrane redox systems and their role in biological stress and disease. Kluwer, Dodrecht, pp 247–265
Wachtler V, Balasubramanian MK (2006) Yeast lipid rafts? – an emerging view. Trends Cell Biol 16:1–4
Wallis JG, Watts J, Browse J (2002) Polyunsaturated fatty acid synthesis: what will they think of next? Trends Biol Sci 27:467–473
Warude D, Joshi K, Harsulkar A (2006) Polyunsaturared fatty acids: biotechnology. Crit Rev Biotechnol 26:83–93
Waters BM, Blevins DG, Eide DJ (2002) Characterization of FRO1, a pea ferric-chelate reductase involved in root iron acquisition. Plant Physiol 129:85–94
Webb MS, Uemura M, Steponkus PL (1994) A comparison of freezing injury in oat and rye: two cereals at the extremes of freezing tolerance. Plant Physiol 104:467–478
Weger HG, Espie GS (2000) Ferric reduction by iron-limited *Chlamydomonas* cells interacts with both photosynthesis and respiration. Planta 210:775–781

Weger HG, Middlemiss JK, Petterson CD (2002) Ferric chelate reductase activity as affected by the iron-limited growth rate in four species of unicellular green algae (Chlorophyta). J Phycol 38:513–519

Weger HG, Matz CJ, Magnus RS, Walker CN, Fink MB, Treble RG (2006) Differences between two green algae in biological availability of iron bound to strong chelators. Can J Bot 84:400–411

Welinder KG (1992) Superfamily of plant, fungal and bacterial peroxidases. Curr Opin Struct Biol 2:388–393

Widell S, Lundborg T, Larsson C (1982) Plasma membranes from oats prepared by partition in an aqueous polymer two-phase system: on the use of light-induced cytochrome *b* reduction as a marker for the plasma membrane. Plant Physiol 70:1429–1435

Wu H, Li L, Du J, Yuan Y, Cheng X, Ling H-Q (2005) Molecular and biochemical characterization of the Fe(III) chelate reductase gene family in *Arabidopsis thaliana*. Plant Cell Physiol 46:1505–1514

Xu X, Bittman R, Duportail G, Heissler D, Vilcheze C, London E (2001) Effect of the structure of natural sterols and sphingolipids on the formation of ordered sphingolipid/sterol domains (rafts). J Biol Chem 276:33540–33546

Yoshida S, Uemura M (1984) Protein and lipid composition of isolated plasma membranes from orchard grass (*Dactylis glomerata* L.) and changes during cold acclimation. Plant Physiol 75:31–37

Yoshida S, Uemura M, Niki T, Sakai A, Gusta LV (1983) Partition of membrane particles in aqueous two-polymer phase systems and its practical use for purification of plasma membranes from plants. Plant Physiol 72:105–114

Yoshimura K, Ishikawa T, Nakamura Y, Tamoi M, Takeda T, Tada T, Nishimura K, Shigeoka S (1998) Comparative study on recombinant chloroplastic and cytosolic ascorbate peroxidase isozymes of spinach. Arch Biochem Biophys 353:55–63

Yoshioka H, Sugie K, Park HJ, Maeda H, Tsuda N, Kawakita K, Doke N (2001) Induction of plant gp91 *phox* homolog by fungal cell wall, arachidonic acid, and salicylic acid in Potato. Mol Plant Microb Interact 14:725–736

Yoshioka H, Numata N, Nakajima K, Katou S, Kawakita K, Rowland O, Jones JDG, Doke N (2003) *Nicotiana benthamiana* gp^{91phox} homologs *Nb*rbohA and *Nb*rbohB participate in H_2O_2 accumulation and resistance to *Phytophthora infestans*. Plant Cell 15:706–718

Zuckermann MJ, Ipsen JH, Miao L, Mouritsen OG, Nielsen M, Polson J, Thewalt J, Vattulainen I, Zhu H (2004) Modeling lipid–sterol bilayers: applications to structural evolution, lateral diffusion, and rafts. Methods Enzymol 383:198–229

Subcellular Sites of Environmental Sensing

Margarete Baier, Andrea Kandlbinder, Karl-Josef Dietz, and Dortje Golldack(✉)

Contents

Abstract Plants sense environmental signals by various sensory mechanisms distributed over all subcellular compartments. Plasma membrane integrated receptors

Dortje Golldack
Department of Biochemistry and Physiology of Plants, W5/117 Faculty of Biology, University of Bielefeld, Universitaetstrasse 25, D-33615 Bielefeld, Germany
e-mail: dortje.golldack@uni-bielefeld.de

U. Lüttge et al. (eds.), *Progress in Botany 69.*

sense pathogens, excess salinity, hormonal signals and light. Through membrane transporters, soluble environmental signals are imported into the cytosol, where the ion, redox and pH homeostasis is controlled. Endomembranes and microbodies form small cellular sub-compartments involved in environmental sensing and signal transduction. Light sensing, which strongly controls plant metabolism, is separated into qualitative light sensing by photoreceptor proteins at the plasma membrane and in the cytosol and quantitative light sensing through photosynthesis. Embedded into the cytosol, besides light sensing, the organelles mainly integrate environmental signals, respond to cytosolic changes and trigger metabolite fluxes, the energy supply and developmental programmes such as cell death. Covering a selection of environmental sensing mechanisms, recent insights into the diversity of subcellular sites and mechanisms of environmental sensing will be reviewed.

1 Introduction

Plants continuously adapt to their changing abiotic and biotic environment. Key parameters are specifically monitored by sensors and receptors in different cellular compartments and activate adaptive responses via signal transduction networks in a co-ordinated manner (Fig. 1). Many physical and chemical constraints as well as

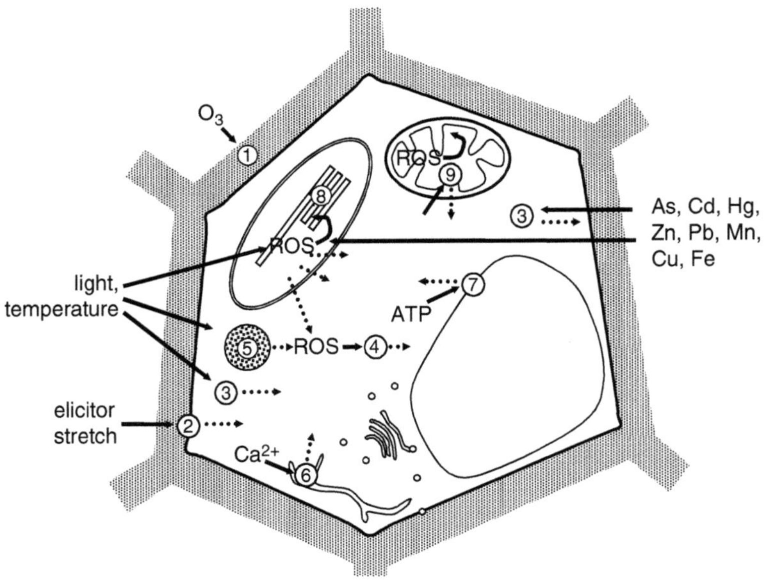

Fig. 1 Subcellular sites of signal perception. Exemplarily, nine sites are indicated by circled numbers. They are localized to the apoplast ①, plasma membrane ②, cytosol ③, ④, peroxisome ⑤, endomembrane system ⑥, tonoplast ⑦, chloroplast ⑧ and mitochondrion ⑨. Some input parameters are named and characterized by a *solid line* with an *arrow head*; output signals are denoted by *dotted lines*. *ROS* Reactive oxygen species

biotic attack first interact with the extracellular matrix (apoplast) and the plasma membrane. The extracellular matrix of plant cells consists of carbohydrate constituents, such as celluloses, hemicelluloses and (up to 10% by dry weight) proteins, for example structural proteins, hydrolases, peroxidases and pathogenesis-related proteins, as well as receptor proteins that are involved in diverse developmental and environmental sensing (Dietz 1996).

The plasma membrane separates the cell from its environment. It is the receptor site for many signals and it controls the uptake of chemical signals sensed inside the cell. In the cytosol extracellular signals are integrated with metabolic signals controlling the redox, pH and ion homeostasis. Cell organelles also function in signal perception and represent intracellular integration points of signal transduction. Chloroplasts and mitochondria are part of the cellular metabolic network, producing cellular ATP, reducing power and C-intermediates and having a specific role in the biosynthesis of nucleotides, amino acids, fatty acids, vitamin cofactors and iron–sulfur clusters (Foyer and Quick 1997). Co-ordination is achieved by reciprocal communication between the different cellular compartments (Foyer and Noctor 2003). For example, mitochondrial respiration and photosynthesis in the chloroplast are intimately linked (Carrari et al. 2003; Raghavendra and Padmasree 2003; Van Lis and Atteia 2004; Nunes-Nesi et al. 2005). Besides, mitochondria and chloroplasts are implicated in cell signalling (Vandecasteele et al. 2001; Logan and Knight 2003; Baier and Dietz 2005) and the induction of programmed cell death (Jones 2000; Youle and Karbowski 2005; Yao and Greenberg 2005). In addition chloroplasts are the main sites for quantitative light sensing and integrate many environmental signals, such as temperature, heavy metals and CO_2 availability through imbalances between the biophysical processes of photosynthesis and chloroplast biochemical reactions. The review attempts a broad approach to the topic of environmental signal perception. Its focus is taken from a cell-centred perspective, i.e. the involvement of diverse subcellular sites. This rather ambitious claim implies necessarily that we only describe examples of site-specific environmental sensing. For this reason, we apologize to anyone who misses a specific stressor or signal perception mechanism.

2 Apoplast and Plasma Membrane

2.1 ROS and Pathogen Sensing in the Apoplast

Although the apoplast has long been considered to have no function in cellular signalling, the picture now emerges that this cellular compartment might play an essential role in perceiving and triggering environmental signals to the cell (Table 1). The apoplast is, for example, a major site of H_2O_2 synthesis by apoplastic NADPH oxidases and nitric oxide (NO) is generated non-enzymatically at acidic pH in the apoplast next to its enzymatic synthesis at the plasma membrane, in the cytosol, in mitochondria and in chloroplasts (Bethke et al. 2004; Wendehenne

Table 1 Sensing of environmental factors in the apoplast and at the plasma membrane

Environmental factor	Subcellular site of sensing	Sensor	Primary response	Signalling pathway components	Literature
Wounding	Apoplast	Receptor-like kinases?	ROS synthesis	Changes in cytosolic calcium	Neill et al. (2002)
Pathogens			Generation of NO	Salicylic acid	Wendehenne et al. (2004)
				Jasmonic acid	Lamotte et al. (2006)
				ABA	Bright et al. (2006)
				Ethylene	Desikan et al. (2006)
					Takabatake et al. (2006)
Ozone	Apoplast		ROS synthesis	Changes in cytosolic calcium	Rao and Davis (2001)
	Plasma membrane			Salicylic acid	Torres and Dangl (2005)
				Jasmonic acid	
				Ethylene	
Cold	Plasma membrane	Receptor-like kinases?	Changes in membrane fluidity	Changes in cytosolic calcium	Zhang et al. (2005),
Heat					Chinnusamy et al. (2004)
Osmotic Stress				ABA	Navarro-Avino and Bennett (2005)
Salt				Ethylene	Cerana et al. (2006)
					He et al. (2004)
					Osakabe et al. (2005)
Light	Plasma membrane	Phototropins			Im et al. (2006)

et al. 2004). Nitric oxide functions as a component of signal transduction pathways in response to wounding and in defence; in addition, nitric oxide activates intracellular Ca^{2+} permeable channels and has a role in salicylic- and jasmonic acid-mediated signalling and ABA-induced closure of guard cells (Neill et al. 2002; Wendehenne et al. 2004; Lamotte et al. 2006). Interestingly, ABA-induced generation of nitric oxide is dependent on the synthesis of H_2O_2, as was shown in an *Arabidopsis thaliana* double mutant of the NADPH oxidases *AtRbohD* and *AtRbohF* in which both the synthesis of nitric oxide and the ABA-induced stomatal closure were reduced (Bright et al. 2006). In the *A. thaliana* ethylene receptor mutants *etr1-1* and *etr1–3* the closure of stomata was insensitive to H_2O_2 and ethylene, indicating a cross-talk of both signalling components during stomatal movements (Desikan et al. 2006). In response to pathogen attacks, NADPH oxidase-derived reactive oxygen species contribute as second messengers to oxidative bursts and hypersensitive response, and can suppress salicylic acid-dependent cell death in *A. thaliana* (Torres and Dangl 2005; Torres et al. 2005). Ozone damages plant cells by oxidation of biomolecules in the apoplast and by lipid peroxidation of the plasma membrane. As an adaptational response, the application of ozone activates an oxidative burst and a hypersensitive response, induces changes in Ca^{2+} fluxes and triggers intracellular signal transduction that involves ethylene, jasmonic and salicylic acid (Samuel et al. 2000; Rao et al. 2000; Rao and Davis 2001; Torres and Dangl 2005).

2.2 Sensing of Salt and Temperature Changes by Plasma Membrane Fluidity

The fatty acid composition of membranes is of crucial importance for membrane integrity and function under various environmental conditions; and plants respond to many abiotic and biotic stresses by a modulation of the biosynthesis of fatty acids, indicating a key role of these biomolecules for environmental sensing and adaptation. By time-resolved X-ray scattering, Amenitsch et al. (2004) showed in situ that oriented lipid multilayers under salinity display phase changes due to water diffusion followed by relaxation, suggesting that membranes may act as primary sensors for osmotic stress. After decreasing the membrane fluidity in *A. thaliana* by mutation of oleate and linoleate desaturase, the cold-induced activation of the diacylglycerol signalling pathway was affected, suggesting that membrane rigidification is a first step of cold sensing in plant cells (Vaultier et al. 2006). Over-expression of the ω-3 desaturases FAD3 and FAD8 in tobacco increased the ratio of linolenic to linoleic acids in membranes that enhanced heat sensitivity and improved tolerance to drought and osmotic stress (Zhang et al. 2005). Changes in the fluidity of the plasma membrane may induce Ca^{2+} influx by modulating the activity of Ca^{2+} channels that subsequently induces intracellular signalling cascades and leads to metabolic and physiological adaptation of cells.

2.3 The Signalling Function of Calcium in Plasma Membrane Localized Sensing

Transient cytosolic Ca^{2+} bursts as second messengers mediate responses to cold, heat, drought, salt, UV light and wounding, and activate intracellular signalling cascades that involve calcium-dependent protein kinases (CDPKs), CDPK-related kinases (CRKs), calmodulin-dependent protein kinases (CaMKs), and SnRKs that are related to SNF1 from yeast (Knight 2000; Hepler 2005; Kaplan et al. 2006). For example, *A. thaliana* SnRK3 kinases are implicated in salt stress responses. Protein kinases of the SnRK2 and SnRK3 type are involved in ABA signalling (Halford et al. 2000; Guo et al. 2001, 2002; Shen et al. 2001; Gong et al. 2002; Mustilli et al. 2002; Yoshida et al. 2002). Ca^{2+} sensing is a molecular mechanism in the early salt stress response. In *A. thaliana* the Ca^{2+} binding protein SOS3 activates the serine/threonine protein kinase SOS2 and the plasma membrane Na^+/H^+ antiporter SOS1 that are involved in regulation of the intracellular Na^+-homeostasis (Chinnusamy et al. 2004). Expression of the plasma membrane-localized Ca^{2+}-ATPase *LCA1* of tomato is induced by ABA and might function in modulating Ca^{2+} signals by restoring steady-state Ca^{2+} levels and thus affecting the frequency and amplitude of Ca^{2+} bursts (Navarro-Avino and Bennett 2005). Interestingly, Beffagna et al. (2005) observed co-ordinated changes in Ca^{2+} influx, extracellular generation of reactive oxygen species (ROS) and cytosolic pH. In response to hypo-osmotic treatment, in *A. thaliana* Ca^{2+} influx was accompanied by H_2O_2 accumulation in the apoplast, changed K^+ and H^+ net fluxes across the plasma membrane and cytoplasmic pH changes, indicating a role of the plasma membrane H^+-ATPase and Ca^{2+}-ATPase in signal transduction triggering ROS production under environmental stresses (Beffagna et al. 2005). Expression of two isoforms of plasma membrane Ca^{2+}-ATPases in *A. thaliana* (*AT-ACA8*, *AT-ACA9*) was stimulated by ABA, suggesting a role for plasma membrane Ca^{2+}-ATPases also in ABA signalling (Cerana et al. 2006). Based on transcriptome analyses, Kaplan et al. (2006) identified an abscisic acid-responsive element (ABRE) as a potential Ca^{2+}-responsive *cis* element that mediates transcriptional activation in response to transient changes in cytosolic Ca^{2+}.

2.4 Light Sensing at the Plasma Membrane

An important signal for plant development and environmental adaptation is light that is sensed by photoreceptors in the classes of red/far-red light absorbing phytochromes and the blue light absorbing cryptochromes and phototropins (Briggs et al. 2001; Frankhauser 2001). The plasma membrane-associated phototropin proteins (Phot) contain two flavin monocleotide (FMN) binding domains that are sensitive to light, oxygen and voltage (LOV domains), as well as a serine/threonine kinase domain; and autophosphorylation of Phot receptors in response to blue light has been shown (Fedorov et al. 2003). The Phot photoreceptors are involved in phototropism, chloroplast movement and stomatal movement, as well as light-regulated transcription of enzymes of chlorophyll and carotenoid biosynthesis, as described for *Clamydomonas reinhardtii* (Im et al. 2006; Cho et al. 2007). Interestingly, the *A. thaliana* cryptochrome double mutant *cry1 cry2* also showed a reduced stomatal reponse to blue light, indicating additive effects of CRY to Phot in stomatal opening (Mao et al. 2005).

2.5 Sensing of Wounding, Abscisic Acid and Osmotic Changes by Plasma Membrane-Integrated Receptors

Another class of proteins that has a central function in sensing environmental signals and in activating adaptational downstream signalling pathways are the plasma membrane-integrated receptor-like kinases (RLK). RLKs belong to a large protein family that consists of more than 600 members in *A. thaliana*, whereas in rice more than 1100 proteins of this type have been identified (Shiu et al. 2004; Morillo and Tax 2006). RLKs contain an intracellular cytoplasmic kinase domain that is linked by a transmembrane domain to an extracellular region that probably functions in perceiving intercellular and environmental signals. RLKs play an essential role in plant growth and development, such as for example the RLK *CLAVATA1* that regulates the shoot meristem maintenance (Kessler and Sinha 2004). In addition, RLKs function in plant defence and in pathogen and abiotic stress responses (Morillo and Tax 2006). Suppression of the plasma membrane-localized leucine-rich repeat RLK WRK (wound-induced RLK) in transgenic tobacco decreased the wound-induced expression of pathogenesis-related genes in a jasmonic acid-dependent manner, indicating involvement of the kinase in sensing and mediating wound signals (Takabatake et al. 2006). Interestingly, perception of phytohormones that have a key function in plant development and environmental adaptation by plasma membrane localized receptors was reported recently. Yamazaki et al. (2003) visualized abscisic acid (ABA) binding sites on the surface of *Vicia faba* guard cell protoplasts by administering biotinylated ABA and fluorescence-labelled avidin. ABA binding was blocked by pre-treatment with proteinase K. The results of this experimental approach elegantly demonstrate ABA perception by plasma membrane-localized receptor proteins that possibly mediate ABA-induced signalling to intracellular pathways. The *A. thaliana* RLK AtLecRK2 is localized at the plasma membrane and is characterized by an extracellular lectin-like domain (He et al. 2004). Salt stress induced the expression of *AtLecRK2* but it was inhibited in transgenic *A. thaliana* over-expressing the salt-inducible ethylene receptor *NTHK1*, suggesting that *AtLecRK2* is involved in a salt-responsive signalling pathway that is influenced by ethylene (He et al. 2004). Antisense repression of the *A. thaliana* leucine-rich repeat receptor-like kinase *RPK1* decreased the sensitivity of the transgenic plants to ABA; and down-regulation of ABA-inducible genes was observed, indicating involvement of the kinase in ABA perception and signalling (Osakabe et al. 2005). As an example for two-component signal transduction in response to environmental stress, the hybrid histidine kinase *ATHK1* that functions as an osmo-sensor has been identified; and in a *Populus* species a HK1 homologous protein was inducible by osmotic stress and interacted with histidine-containing phosphotransfer proteins that might be another component of the plant osmo-sensing pathway (Chefdor et al. 2005).

3 Cytosol

The cytosol integrates many diverse types of stimuli released from different compartments or entering from the extracellular space via the plasma membrane, particularly chemical stimuli such as protons, transient Ca^{2+} spikes or metal ion accumulation. Chemical sensors need to have a very high specificity and sufficiently high sensitivity for binding the signalling molecule. Following binding, the specific response must be triggered.

3.1 Sensing Metal Ions

Along these lines activation of phytochelatin synthase is an exemplary straightforward sensing mechanism of a defined abiotic stressor with subsequent induction of specific defence activity upon exposure to some metal pollutants. Detoxification of Cu, Cd, Hg and, to a lesser extent, Zn but also arsenate is linked to binding the metals to phytochelatin (PC) which is an oligopeptide synthesized ribosome-independently from glutathione by phytochelatin synthase PCS (Grill et al. 1989) and consisting of repeating units of γ-glutamylcysteine and a C-terminal Gly, Ser or beta-Ala residue [$(\gamma\text{-Glu-Cys})_n$-Xaa; $n = 3$–11]. Immunopurified AtPCS1 binds Cd^{2+} at high affinity ($K_d = 0.54 \pm 0.20\,\mu M$) and high capacity (stoichiometric ratio $=7.09 \pm 0.94$ mol mol^{-1}). The binding sites are localized within the protein structure (Maier et al. 2003). In the absence of bound metal ions, the PCS protein is catalytically inactive (Vatamaniuk et al. 1999). Upon ligand binding, the enzyme is switched on and catalyses phytochelatin synthesis. Relative to Cd-dependent activation which was set to 100%, Ha et al. (1999) reported activitation to 253% with Cu, 230% with Ag, 97% with Hg and 10% with Zn. Accumulating PCs sequester metals, thereby lowering free metal ion activity in the cytosol and enabling transport, e.g. export of metal–PC complexes to the vacuolar lumen in yeast (Ortiz et al. 1992) and probably also from the mitochondrion to the cytosol in plants (Kim et al. 2006). Activation of PCS by metal ions may be considered as a minimized signalling pathway, with the cytosol being the site of sensing and response to adapt to an environmental constraint.

3.2 pH Sensing

Cytosolic pH alterations are caused by metabolic changes, for example during the diurnal cycle of crassulacean acid metabolism (Hafke et al. 2001), under anoxia (Sanders and Slayman 1982), after signal-induced changes in transport processes (Felle 1988), after exposure to acidic gases such as SO_2 (Yin et al. 1990) or as a consequence of pathogen or rhizobia infection, in the first case leading to alkalization, in the latter to acidification (Felle et al. 1996; Mathieu et al. 1996). The cellular pH stat counteracts any pH disturbance. It consists of two functionally distinct components: the short-term physical pH stat linked to membrane transport and the biochemical pH stat which depends on metabolic reactions and is involved in long-term pH adjustment (Davis 1986). The activity of enzymes that are involved in cytoplasmic pH homeostasis responds to alterations in pH. The phospho*enol*pyruvate carboxylase (PEPC) is considered to be a main player in readjustment of proper cytoplasmic pH after alkalization. PEPC activity in barley mesophyll protoplasts increases with increasing pH in the range pH 7.2–8.8 (Kromer et al. 1996). The presence of malate increases the sensitivity of PEPC towards changes in cytosolic pH. However, the physiological significance of PEPC pH dependency and thus as a pH sensory mechanism may be low due to three characteristics:

1. The cytosolic pH changes are transient.
2. In the presence of physiological glucose-6-phosphate concentrations the pH dependency is greatly reduced.
3. The phosphorylation state of PEPC protein correlates with kinetic properties of PEPC, i.e. the affinity to its substrate PEP increases with increasing phosphorylation state of the protein.

At least in the context of nitrogen metabolism, where PEPC activity was suggested to balance NO_3^--dependent acidification of the cytosol, its role in pH homeostasis was questioned and reduced to the anaplerotic role in refilling the citric acid cycle to generate carbon skeletons for amino acid synthesis (Britto and Kronzucker 2005). In any case, protonatable amino acids in proteins function as pH sensors with implications in homeostasis and signalling.

3.3 Redox Sensing

Cells are redox systems that function in a narrow range of redox potentials. The redox environment of the cell is mostly defined by a few important redox couples. S-containing organic compounds that undergo thiol-disulfide transitions constitute a ubiquitous cell redox system at millimolar concentrations. Deviations from metabolic equilibrium often alter the redox state of specific subcellular protein targets or cause rather general oxidative stress (Jacob et al. 2006; Ströher and Dietz 2006). Enzymatic antioxidants counteract the redox imbalance. Peroxiredoxins and glutathione peroxidases are thiol-based peroxidases that detoxify hydrogenperoxide, alkylhydroperoxide and peroxinitrite (Dietz 2003b). During the catalytic cycle of 2-Cys peroxiredoxins, the peroxidatic thiol group is oxidized and a disulfide is formed with a second thiol group. The affinity to the substrates is high. Thus, peroxiredoxins trap toxic peroxides. Thiol reductases such as thioredoxins, glutaredoxins and cyclophilins with conserved Cys residues regenerate the active form of reduced peroxiredoxins. Recently this mechanism was suggested to enable peroxiredoxins to function as redox sensors transmitting information on peroxide concentration to interacting targets (Dietz et al. 2006). Drainage of electrons into peroxide reduction through peroxiredoxins shifts the redox state of the electron donor to a more oxidized state and indirectly also alters the redox states and functions of other targets of the respective donor. Specificity versus non-specific thiol oxidation and redundancy of pathways is a central question in redox regulation. The *Arabidopsis thaliana* genome encodes large-size gene families involved in thiol-disulfide homeostasis, e.g. disulfide dithiol reductases with 30 glutaredoxin-like genes (Lemaire 2004) and 42 thioredoxin genes (Meyer et al. 2005), as well as thiol-dependent antioxidants with potential redox sensor function, i.e. ten genes encoding peroxiredoxins (Horling et al. 2003) and five glutathione peroxidases (Rouhier and Jacquot 2005). These high figures imply a significant degree of specificity. In yeast the Yap1 transcription factor regulates hydroperoxide homeostasis. Yap1 is not directly oxidized by hydroperoxide but through a glutathione peroxidase (GPX)-like enzyme Gpx3. Apparently Gpx3 serves as sensor and transducer of the hydroperoxide signal to Yap1 by inducing formation of a disulfide bond. Thioredoxin turns off the pathway by reducing both sensor and regulator (Delaunay et al. 2002). The systems allows for specific sensing and transmission of redox information to decisive regulators. Similar systems are also likely to exist in plants. At least one type II Prx (in *Arabidopsis thaliana* three, i.e. PrxII B, C, D, with a yet uncharacterized chimeric PrxII A that possibly is not expressed; Dietz 2003a) and one 1-Cys Prx reside in the cytosol of higher plants and may be involved in redox-sensing mechanisms.

Apart from these three selected examples, the cytosol is the site of sensing many other chemical and physical cues. In fact it is likely that the significance of the cytoplasm as a site of environmental signal perception and signal integration is far from being understood comprehensively. Light sensors such as phytochrome (Franklin and Whitelam 2004), their integration with the circadian clock (Schoning and Staiger 2005), cytoplasmic Ca sensing (White and Broadley 2003) or the signal-controlled nucleocytoplasmic partitioning of gene regulatory proteins (Kircher et al. 1999; Meier 2005) are other prominent examples for cytoplasmic signal perception.

4 Endomembranes

The endomembrane system comprises the endoplasmic reticulum, Golgi apparatus and vacuoles which are connected by dynamic intracellular vesicle trafficking. Its role in intracellular signalling is established, e.g. as site of Ca release from internal stores through phosphoinositol-sensitive Ca^{2+} release channels, and at the tonoplast also through cADPR-activated Ca^{2+} channels. These processes are involved in signal transduction. Endomembranes also emerge as site of immediate signal perception.

1. Recent evidence suggests that the guard cell tonoplast senses osmotic gradients and responds to water flow into the vacuole by increased vacuolar ion efflux to minimize cytoplasmic dilution. An aquaporin is suggested to function as sensor involved in the signal transduction chain (MacRobbie 2006).
2. The endocytic network participates in plant gravisensing and graviresponses. Endocytosis proteins and endomembrane functions are modified by the gravitational input signal that affect transcellular auxin transport (Samaj et al. 2005). Mutations in candidate proteins such as the VTI11-SNARE, a protein involved in vesicular fusion, alters gravisensing of *Arabidopsis* shoots (Yano et al. 2003).
3. The vacuolar V-ATPase functions as temperature sensor in chilling sensitive plant species. Acidification of the cytosol upon lowering the temperature is linked to the inhibition of V-ATPase (Yoshida et al. 1999). Cytoplasmic acidification might involve temperature-dependent dissociation of the V1 and V0 domains of V-type ATPase.

5 Peroxisomes

Peroxisomes are the site of oxygen-dependent metabolism. Significant rates of ROS are released, particularly in oxidase-linked metabolism such as photorespiration. Foyer and Noctor (2003) estimated H_2O_2 generation in the light to amount to about 216 nmol m^{-2} s^{-1} in mitochondria, 4030 nmol m^{-2} s^{-1} in chloroplasts and about 10 000 nmol m^{-2} s^{-1} in peroxisomes. High-efficiency ROS detoxification is mediated by a set of catalase isoforms and ascorbate peroxidase (del Rio et al. 2003). Catalase is sensitive to light inhibition. Inactivation by blue light occurs in achlorophyllous plant tissues. In addition it was shown for winter rye leaves (*Secale cereale* L.) that catalase is inactivated by both blue and red light. While the former is caused by direct light absorption through the catalase prosthetic heme group, the latter is mediated by photosynthesis-dependent release of ROS and possibly of protonated superoxide radical, HO_2, from the chloroplast and is correlated with the reduction state of the plastoquinone pool (Shang and Feierabend 1999). As a consequence of catalase inhibition, H_2O_2 may be released from the peroxisome at increased rates, with the risk of photo-oxidative damage. The photo-inactivation of catalase usually remains latent due to continuous de novo synthesis of catalase. In a type of compensatory reaction post-transcriptional mechanisms affect the rate of de novo catalase synthesis. This includes a light-regulated reversible modification of the catalase mRNA (cat1) by methylation. The thereby increased translation efficiency of the rye cat1 mRNA by cap N-7 methylation is stronger than that of a *lhcb* transcript (Schmidt et al. 2006). Photo-inhibition of catalase may be considered as a sensor mechanism linking environmental stimuli to ROS signalling. Transcriptome analysis of transgenic plants expressing catalase on a low level demonstrated the importance of peroxisomal ROS as signals modifying nuclear gene expression (Vandenabeele et al. 2003).

6 Chloroplast

Endosymbiosis of photosynthetic prokaryotes combines the advantages of eukaryotic cell biology with photo-autotrophism. Embedded in the cytosol and surrounded by two additional membranes, chloroplasts are less affected by chemical signals than the apoplasm, the plasma membrane and the cytosol, but are the main light and CO_2 sensors and integrate quantitative light signals with signals triggered by biotic and abiotic stressors (Fig. 2, Table 2).

6.1 *Thylakoid Membrane*

In the light-harvesting complexes, carotenoids and chlorophylls perceive light energy and "sense" the light intensity and quality. X-ray and electron diffraction

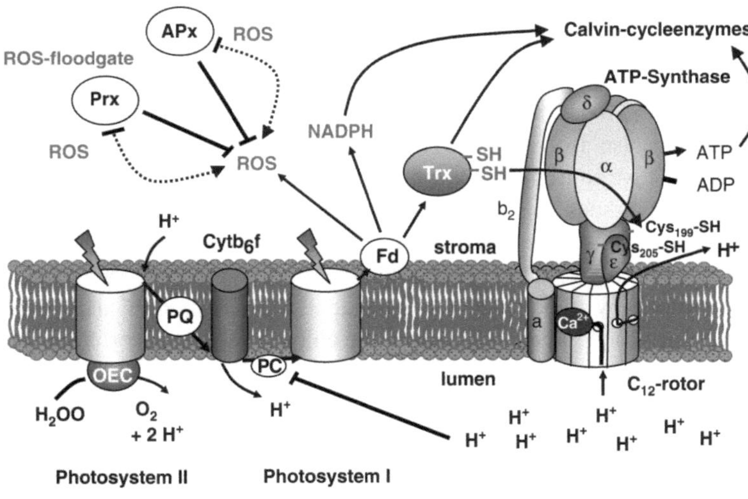

Fig. 2 pH, redox and Ca^{2+}-sensing in chloroplasts. The luminal pH regulates chloroplast ATP-synthase and inhibits plastocyanin (*PC*). In addition ATP-synthase is sensitive to Ca^{2+} and to the stromal redox poise, which is triggered through NADPH and thioredoxin by photosynthetic electron transport relative to the metabolic activity and the antioxidant capacity. Peroxiredoxins (*Prx*) and ascorbate peroxidase (*Apx*) control floodgates regulating chloroplast ROS accumulation

Table 2 Sensing environmental signals in chloroplasts

Sensor	Signal	Response	Literature
Thylakoid membrane			
ATP-synthase	Luminal pH	ATP synthesis	Schwarz and Strotman (1998)
	Ca^{2+}-sensing	Increased luminal pH, modulation of grana stacking	
PsbS	Luminal pH	Activation of non-photochemical quenching	Haripal et al. (2006), Niyogi et al. (2005)
VDE	Luminal pH	Activation of non-photochemical quenching	Emanuelsson et al. (2003)
Plastocyanin	Luminal pH	De-coupling of photosynthetic electron transport	Sas et al. (2006)
Plastoquinone	Electron transport efficiency	State transition via LHCII-phosphorylation, transcription of psaAB	Pfannschmidt et al. (1999), Bellafiore et al. (2005), Bonadi et al. (2005)
TCP34	?	Regulation of PEP-activity	Weber et al. (2005)
Oxygen evolving complex	Heavy metal sensing	Regulation of electron transport	Shutilova (2006)
Chloroplast stroma			
PS-I acceptor site	Electron pressure on ferredoxin	Reduction of small redox proteins (thioredoxins, glutaredoxins, cyclophilins)	Summarized by Buchanan and Luan (2005), Dietz (2003a, b)
		Reduction of $NADP^+$	
		Regulation of metabolism and gene expression	Baier et al. (2004), Piippo et al. (2006)
	generation of ROS	Regulation of ascorbate peroxidase and peroxiredoxin floodgates	Miyake et al. (2006), König et al. (2003), Dietz et al. (2006)
		Triggering hypersensitive cell death	Kodama and Sano (2006)
	Temperature sensing	Regulation of gene expression	Huner et al. (1996), Ensminger et al. (2006)
Rubisco	CO_2-sensing	Regulation of carbohydrate metabolism and photorespiration	
PII-homologues	Carbon:nitrogen sensing	Regulation of amino acid biosynthesis	Sugiyama et al. (2004)
	ABA-sensing	Regulation of drought and cold responses	Shen et al. (2006)
Membranes			
Membrane	Cd^{2+}-sensing	Membrane stability	Patra et al. (2004)

crystallography provided atomic resolution of the reaction centres involved and pigment localization (Kühlbrandt et al. 1994), from which the energy transfer pathway between the pigments has been concluded (Broess et al. 2006; Frahmcke and Walla 2006). Excitation of chlorophyll molecules in the reaction centres triggers photosynthetic electron transport, which generates a proton motif force over the thylakoid membrane and redox signals in the chloroplast stroma. Passage of protons along the proteolipid ring of chloroplast ATPase induces a rotatory moment to the CF_1 complex ($\alpha_3\beta_3\gamma_1\delta_1\varepsilon_1$; Storck et al. 1999), which endows the β-subunits with low, medium and high nucleotide affinities and regulates ATP release, binding of ADP and inorganic phosphate and ATP synthesis (Abrahams et al. 1994). ATP synthase is activated in a two-stage process with pK values of 6.8 and 4.9. Low thylakoid acidification activates ATP synthesis, while strong acidification destabilizes the complex (Schwarz and Strotmann 1998). The pH sensitivity depends on reducing conditions, which couple the ε-subunit to γ rotation (Richter 2004) by cleavage of the disulfide bond between Cys199 and Cys205 of the γ-subunit (Samra et al. 2006).

In H^+ domains sequestered within the proteo-lipid core of the thylakoid membrane (Ewy and Dilley 2000) protonation of E122 and E226 of PsbS supports zeaxanthin binding (Haripal et al. 2006) and facilitates energy transfer between pigment protein complexes (Niyogi et al. 2005). Lumen acidification also regulates protonation of histidine residues in violaxanthin de-epoxidase (VDE) (Emanuelsson et al. 2003) and regulates binding of VDE to the thylakoids (Gisselsson et al. 2004), where it reduces epoxidized end-group moieties in xanthophylls (Eskling et al. 1997). In addition, below pH 5.3, reduced plastocyanin performs structural changes at the metal binding site, leading to reversible inhibitory aggregation of the electron transmitter protein (Sas et al. 2006).

Because plastoquinone oxidation is the slowest redox reaction within the photosynthetic electron transport chain (Allen 1993), the redox state of the plastoquinone pool sensitively shifts to more reduced values whenever there is an imbalance between photoreaction centre activation and electron consumption. As recently confirmed by isolation of mutants in *Arabidopsis thaliana* and *Chlamydomonas reinhardtii* (Bellafiore et al. 2005; Bonardi et al. 2005), plastoquinone reduction activates a protein kinase which triggers state transition by phosphorylation of the light-harvesting complexes. In parallel, transcription of the psaAB operon is induced (Pfannschmidt et al. 1999), which encodes the two reaction centre proteins of PS-I. Expressional activation of PS-I alleviates the electron pressure on the plastoquinone pool and, therefore, is part of the long-term acclimation mechanism of plants to increased light intensities or changes in the light quality. Co-regulation has been suggested for several nuclear encoded genes, but may be dominated either by ROS-ABA-signalling, the carbohydrate status or by redox signals depending on the acceptor availability at PS-I (Piippo et al. 2006; Rossel et al. 2006).

Two-component redox regulatory systems have been postulated to sense photosynthetic signals and to adjust chloroplast function (Allen 1993). Forsberg et al. (2001) reported the identification of nucleus-encoded chloroplast-targeted histidine sensor kinases and aspartate response regulators for higher plants. Recently, TCP34 response regulators were shown to be part of a thylakoid His/Asp-phospho-relay system. Supercomplex formation of phosphorylated TCP34 with plastid-encoded RNA polymerase (PEP; Weber et al. 2006) suggests that TCP34 translates information from the photosynthetic membrane into transcriptional activity.

In parallel with the loss of all prokaryotic genes for two-component response regulators from the plastome, nucleus-encoded redox-sensitive electron transmitters have diversified. Chloroplast targeting of thioredoxins, glutaredoxins and cyclophilins (summarized by Dietz 2003a) suggests that the prokaryotic sensors are modularized and translocated into the chloroplast stroma.

6.2 Stromal Site of the Thylakoid Membrane and Chloroplast Stroma

6.2.1 The Acceptor Site of Photosystem I

At the stromal site of the thylakoid membrane, the reduction of ferredoxin provides the reducing power used for reduction of $NADP^+$ and thioredoxin. NADPH is involved in many metabolic reactions, such as the Calvin cycle and nitrite reduction, and thioredoxin is the major electron coupling protein regulating the reduction state of protein thiols (Buchanan and Luan 2005). Depending on the photosynthetic electron pressure, NADPH and thioredoxin distribute the reduction signal all over the chloroplast and the extra-plastid compartments. Combined transcript and metabolite profiling of *Arabidopsis* leaves (Kolbe et al. 2006) demonstrated that the total biosynthesis of organic acids, amino acids, starch, proteins and cell wall increases upon reduction, while the levels of metabolites consumed in these biosynthetic pathways, such as pyruvate, 2-ketoglutarate and hexose-phosphates, decrease. On the transcript level, the strongest changes were observed for thioredoxins and glutaredoxins. In addition, experimental data accumulate that the expression of various nuclear genes encoding, e.g. thylakoid proteins, Calvin cycle enzymes and antioxidant enzymes, correlates with the redox state at the acceptor site of PS-I (Baier et al. 2004; Piippo et al. 2006). In the promoter of the 2-Cys peroxiredoxin-A gene, a 216-bp redox-sensitive promoter domain was identified (Baier et al. 2004) which is presently under investigation. Most likely metabolite exchange, such as the antiport of malate and oxaloacetate (Scheibe and Stitt 1988), couples the redox poise in chloroplasts with that in the cytosol (Baier and Dietz 2005) and may trigger cytosolic signalling cascades such as phospho-relays (Baier and Dietz 2005).

Temperature is mainly sensed through imbalances between the less temperature-sensitive biophysical processes of photosynthesis and the highly temperature-sensitive chloroplast biochemical reactions (Huner et al. 1996), which decrease the redox state of photosynthetic electron transport components (Salvucci and Crafts-Brandner 2004; Ensminger et al. 2006). Consistently, for example, 2-cys peroxiredoxin-A transcription is induced by the combination of light and cold, while cold alone does not activate expression (Baier et al. 2004).

6.2.2 ROS Sensors in the Chloroplast Stroma

If PET intensity exceeds the acceptor capacities of the $NADP^+$ and thioredoxin-related systems, e.g. in response to excess light, drought or nutrient deficiency, increasing amounts of electrons are transferred to alternative acceptor molecules, such as molecular oxygen (Asada 2000). Increased superoxide formation leads to D1 inactivation (Aro et al. 1993), decreased Calvin cycle activity (Salvucci and Crafts-Brandner 2004) and re-organization of nuclear gene expression (Vanderauwera et al. 2005). In response to reactive oxygen species (ROS), the transcriptional activity of genes encoding proteins of the photosynthetic membrane and Calvin cycle enzymes decreases (Rizhsky et al. 2003), while e.g. expression of lipoxygenase is stimulated (Kiddle et al. 2003).

Low molecular weight antioxidants and antioxidant enzymes, such as superoxide dismutases and peroxidases, counteract the generation of ROS (Asada 2000). However, despite their antioxidant function, especially the ascorbate peroxidases and peroxiredoxins are sensitive to ROS. Excess H_2O_2 oxidizes the catalytic iron ion in the reaction centre of ascorbate peroxidase, which inhibits the enzyme (Miyake et al. 2006). If the redox poise of the chloroplast stroma shifts, peroxiredoxin activity is also lost. Over-oxidation of the catalytic cysteinyl residues inactivates the enzyme and increases thylakoid attachment (König et al. 2003). Decreased ascorbate peroxidase and peroxiredoxin activities accelerate H_2O_2 accumulation. Therefore, these enzymes may serve as floodgates in the induction of oxidative bursts.

In addition, the redox state of low molecular weight antioxidants is influenced by ROS. A signalling function has been postulated, especially for glutathione and ascorbate (Irihimovitch and Shapira 2000; Kiddle et al. 2003), whose redox information can be transmitted by antioxidant-dependent enzymatic reactions, such as violaxanthin de-epoxidation (Eskling et al. 1997) or act via modification of the protein thiol/disulfide status. Glutathione redox signals can also specifically be transmitted via protein glutathionylation. For example, the MAP-kinase-kinase MEKK1, which transmits wounding responses (Hadiarto et al. 2006), is inactivated by glutathiolation of the ATP-binding site (Cross and Templeton 2004).

A recent study suggests that hypersensitive cell death is regulated by chloroplast wound-induced WIN4 (Kodama and Sano 2006). The 17-kDa putative basic helix-loop-helix protein is hypothesized to be a chloroplast transcriptional or translational regulator which evolved from a nuclear transcription factor. Constitutive over-expressors of WIN4 died, transient over-expressors developed chlorosis, while RNAi-lines exhibited reduced hypersensitive cell death, demonstrating that WIN4 is involved in cell death regulation.

6.2.3 Sensing CO_2 Availability

The CO_2 availability can be directly sensed in the chloroplast stroma by Rubisco, whose activity feeds back on the NADPH/NADP$^+$ ratio and is regulated by the chloroplast thiol system. Depending on the relative concentrations of CO_2 and O_2, either the Calvin cycle or photorespiration is initiated, which gives different carbohydrate signatures.

Under most environmental conditions, CO_2 availability is limiting and plants adapted by the evolution of carbon concentration mechanisms (CCM). In many photosynthetically active organisms, e.g. in the unicellular green alga *Chlamydomonas reinhardtii*, the CCM is regulated by CO_2 availability. Screens for mutants with altered CCM induction allowed the isolation of mutants defective in the post-translationally regulated transcription factors Ccm1 (Fukuzawa et al. 2001) and Cia5 (Wang et al. 2005) and demonstrated that CCM induction is regulated extra-plastidially.

6.2.4 Metabolite Sensing

Availability of excess reductive energy facilitates a high biosynthetic potential of plants. Homeostatic regulation upon changes of enzyme activities provides flux control. NADP-glyceraldehyde 3-P dehydrogenase (GAPDH) is one of the key enzymes sensing the metabolic status of chloroplasts. It combines energy and thioredoxin- and NADPH-dependent redox information and translates it into a specific carbohydrate signature. Triose-phosphate export couples the Calvin cycle regulation to extraplastidic metabolism depending on the availability of inorganic phosphate (Flügge and Heldt 1984). In the chloroplast, GAPDH is assembled in two complexes with molecular masses of 550 kDa and 600 kDa (Scheibe et al. 2002). The 600-kDa A_8B_8-complex is redox-regulated and sensitive to 1,3-bisphosphoglycerate. The 550-kDa assembly of GAPDH, phosphoribulokinase and CP12 binds 60–70% of GAPDH, but is insensitive to 1,3-bisphosphoglycerate (Scheibe et al. 2002), which facilitates differential relaying of the regulatory metabolite 1,3-bisphosphoglycerate into GAP-DH activity. In addition to energy and redox sensing, chloroplasts sense the carbon:nitrogen status. Following reduction of nitrite to ammonium, 2-oxoglutarate is aminated to glutamate. Nitrogen storage metabolites, such as glutamine and asparagine, are formed in subsequent trans-amination reactions. In bacteria, PII monitors the carbon:nitrogen balance by sensing 2-oxoglutarate availability (Ninfa and Atkinson 2000). Recently, a PII-homologous protein was identified in rice (Sugiyama et al. 2004). It interacts with N-acetylglutamate kinase-1, which regulates arginine biosynthesis (Caldovic and Tuchman 2003), suggesting that conserved sensory and regulatory mechanisms adjust N-metabolism to the carbon availability in plants and bacteria.

6.3 Transition Metal and Heavy Metal Sensing at the Water-Splitting Complex and at Membranes

In chloroplasts, transition metals (such as Cu, Fe, Mn) are essential components of the active sites of various antioxidant enzymes, such as superoxide dismutases and ascorbate peroxidase, and of the water-splitting complex at PS-II. For uptake, various ion channels have been recorded electrophysiologically (Goetze et al. 2006; Hemmler et al. 2006). Chloroplast compartmentalization of various heavy metal ions is so effective that Ruiz et al. (2003) even suggested the use of chloroplast-transformed plants expressing mercuric ion reductase (merA) and organomercurial lyase (merA) for the phytoremediation of Hg^{2+}.

Lead and mercury are most strongly sensed in chloroplasts due to changes in membrane permeability, inactivation of sulfhydryl groups and replacement of essential ions (Patra et al. 2004). Hg^{2+}, like Zn^{2+}, Cd^{2+} and Pb^{2+}, also efficiently binds to the oxygen-evolving complex, presumably to hydroxyl residues of the ligands in the electron-donor part (Shutilova 2006). These ions are sensed by inhibition of PET (Shutilova 2006), which leads to increased photo-oxidative damage and finally to chlorosis and necrosis.

7 Mitochondria

7.1 Heat Shock Transcription Factors as Sensors of Reactive Oxygen Species/Oxidative Stress

Oxidative stress is a key underlying component of most abiotic stresses (Apel and Hirt 2004; Kuzniak and Sklodowska 2004; Mittova et al. 2004). Production of reactive oxygen species (ROS) increases dramatically during abiotic stress conditions. Oxidative stress occurs when the rate of ROS production outstrips the capacity of antioxidant systems to detoxify them. The result is oxidative damage to proteins, lipids and DNA, leading to cellular dysfunction and ultimately cell death (Halliwell 2006). In order to avoid these cell death scenarios, plants invoke a molecular response that allows them to cope with the oxidative stress situation. The precise nature of this molecular response remains poorly characterized. Many stress-related genes, antioxidant enzymes and metabolites present in plants have been identified (Mittler 2002). It is also apparent that there are specific transcriptomic responses to oxidative stress (Gadjev et al. 2006) and specific regulatory elements of this transcriptomic response are beginning to be uncovered (Davletova et al. 2005). However, knowledge about the different sensor types responsible for the induction of the pathways is still far from comprehensive (Kacperska 2004).

Kuzmin et al. (2004) reported that maize respiration-deficient mutants showed a reduced mitochondrial transmembrane potential, leading to constitutive expression of heat-shock proteins. Miller and Mittler (2006) hypothesized recently that certain heat-shock transcription factors function as molecular sensors that directly sense reactive oxygen species and control the expression of oxidative stress response genes during oxidative stress.

7.2 *Metabolite Sensing – Involvement of Metabolic Changes Within the TCA Cycle*

In yeast, it is well established that metabolic change plays a central role in its response to oxidative stress. For example, upon H_2O_2 treatment, carbon is diverted away from respiratory pathways and into the oxidative pentose phosphate pathway (OPPP) to provide reductant for antioxidant metabolism (Godon et al. 1998). According to Liu et al. (2005) this re-routing event is accompanied by a distinct gene expression pattern regulated by specific redox-sensitive transcription factors (McCammon et al. 2003; Butow and Avadhani 2004). However, the role of metabolism during oxidative stress in plants remained rudimentary. The sensitivity of enzymes in the TCA cycle, such as aconitase, pyruvate- and 2-oxoglutarate dehydrogenase to oxidative inhibition (Verniquet et al. 1991; Sweetlove et al. 2002), as well as the connection between the oxidative modification of proteins (Heazlewood et al. 2003; Millar et al. 2004, 2005; Møller and Kristensen 2004, 2006) and the loss of protein function have been demonstrated (Hancock et al. 2005). Feeding the redox-active quinone, menadione, to heterotrophic *Arabidopsis* cells and thereby inducing oxidative stress (Thor et al. 1982) caused a pronounced metabolic phenotype originating in the mitochondrion (Baxter et al. 2006). The TCA cycle and large sectors of amino acid metabolism were inhibited. The sequential accumulation of metabolites in specific pathways led to a subsequent strengthening of glycolysis and a diversion of carbon into the oxidative pentose phosphate pathway. Also, transcripts encoding enzymes of inhibited pathways such as the TCA cycle and amino acid biosynthesis were down-regulated (avoiding wastage of energy in producing proteins that would only be immediately oxidatively damaged or have no function in anabolic pathways starved of precursors). Transcripts involved in catabolic pathways (such as lipid and amino acid breakdown) which lead to a mobilization of internal carbon reserves or reconfiguring metabolic fluxes to bypass inhibited pathways were induced. Finally, an increased expression of key antioxidant enzymes and a re-routing of glycolytic carbon flow into the oxidative pentose phosphate pathway were monitored (Baxter et al. 2006).

7.3 Two-Component System-Like Sensors?

In bacteria, adaptation to changing environmental conditions is mediated through conserved two-component regulatory systems (Chang and Stewart 1998; Mann 2000). These systems are commonly composed of a histidine sensor kinase that perceives the environmental signal and acts as a transmitter and a response regulator. The perception of a signal leads to activation of the transmitter domain and autophosphorylation of a conserved histidine residue of the histidine sensor kinase. The phosphate group is then transferred to the aspartate residue of response regulators resulting in the activation of an "output" domain (e.g. transcription factor) that may regulate different target genes. Like in bacteria, histidine phosphorylation is also present in higher plants. Forsberg et al. (2001) identified nuclear encoded histidine sensor kinases and aspartate response regulators similar to bacterial response regulators that seem to be targeted to chloroplast or mitochondrial membranes. For example, a 37-kDa histidine phosphoprotein of plant mitochondria represents the α-subunit of the succinyl-CoA synthase that is phosphorylated on a histidine residue (Håkansson et al. 1995). Additionally, the pyruvate dehydrogenase kinase located in the mitochondrial matrix contains a bacterial histidine kinase domain.

7.4 AOX and UCP: Role of Respiratory Bypass Proteins

Mitochondrial alternative oxidases (AOXs) and plant uncoupling mitochondrial proteins (UCPs) are two mitochondrial energy-dissipating systems, encoded by multigenic families in plants (Nogueira et al. 2005). These systems allow a fine tuning of the mitochondrial membrane potential and can thus decrease ROS production caused by ETC over-reduction (Camacho et al. 2004; Borecky et al. 2006). AOXs and UCPs do not pump protons; and proton flow through uncoupling proteins is not coupled to ATP synthesis. Therefore, flux through the alternative pathways does not contribute to respiratory ATP production. Respiratory bypass proteins are implicated in several physiological processes, including thermogenesis (Siedow and Day 2000), the prevention of reactive oxygen species formation (Fernie et al. 2004; Møller 2001) and the dissipation of excess redox equivalents (Raghavendra and Padmasree 2003). Indirect evidence also suggests that rapid adjustments in the respiratory chain occur primarily through regulation of the alternative pathways: the multiprotein basal respiratory complexes appear to be relatively insensitive to external factors, while the respiratory bypasses display rapid quantitative changes in response to various biotic and abiotic stimuli (Finnegan et al. 1997; Simons et al. 1999; Svensson and Rasmusson 2001; Yu et al. 2001; Escobar et al. 2004). Clifton et al. (2006) hypothesized that AOX may play a central role as a mediator of a cellular response to changing conditions. Its rapid induction and subsequent changes that result from its activity may act as the signals to induce the changes in expression of a variety of other components to achieve an overall cellular response to changing conditions.

7.5 *Programmed Cell Death*

Biotic and abiotic stressors often activate programmed cell death (PCD), which is intricately connected with signalling from and ROS production in the mitochondrion (Jones 2000; Vianello et al. 2007). In mammalian cells mitochondrial ROS acts as an important intracellular signal in PCD and disease (Hoebrichts and Woltering 2003) involving mitochondrial thioredoxin (Tanaka et al. 2002) and possibly AOX (Robson and Vanlerberghe 2002). This is probably also the case in plant cells. Preceding PCD is a decrease in the mitochondrial transmembrane potential, representing a permeability transition that can be attenuated by cyclosporin A. Plant mitochondria are known to contain the key components that constitute the animal permeability transition pore (PTP), cyclophilin D-homologue, adenine nucleotide translocator and a voltage-dependent anion channel (Godbole et al. 2003; Romano et al. 2004). Under PTP-inducing conditions, swelling of purified potato mitochondria proceeded with kinetics similar to that in animals and this resulted in selective rupture of the outer mitochondrial membrane and release of intermembrane space proteins, including cytochrome c (depending on mitochondrial Ca^{2+} concentrations; Arpagaus et al. 2002). Vacca et al. (2006) showed recently that cytochrome c release can be blocked by antioxidants, an observation that can be taken as evidence that cytochrome c release depends on ROS. Metabolic responses activated in cells differ greatly not only depending on the different kinds of ROS generated under environmental stimuli, but also depending on both the intensity of the oxidative stress and the different timing of ROS production (de Pinto et al. 2006). Under anoxia (leading to accelerated ATP depletion and increased levels of Pi), the onset of the Ca^{2+}-induced swelling as well as the rate of the process are accelerated, indicating that, in vitro, plant mitochondria undergo a faster mitochondrial permeability transition (Arpagaus et al. 2002; Virolainen et al. 2002). Additionally, it has been shown that, in response to oxidative stress, increases in electron transport in mitochondria trigger H_2O_2 production, depletion of ATP, opening of PTP and cell death (Tiwari et al. 2002).

8 Summary and Outlook

Environmental sensing mechanisms are of diverse natures and are distributed among all subcellular compartments. The cell is a multi-responsive sensing system, in which a wide range of environmental stimuli are perceived, inducing homeostasis control mechanisms and acclimation responses. The small selection of sensing mechanisms presented here demonstrates that the same signal, e.g. heavy metal concentrations, alters activities and functions in various compartments by diverse sensory proteins, while e.g. redox poise and pH homeostasis are sensed and controlled by conserved mechanisms in different subcellular compartments. Recent developments in bioinformation-based systems biology will help to combine signalling processes from different compartments in a cellular concept of environmental sensing and cellular responsiveness and to generate models on cellular signalling.

References

Abrahams JP, Leslie AGW, Lutter R, Walker JE (1994) Structure at 2.8 A resolution of F1-ATPase from bovine heart mitochondria. Nature 370:594–595

Allen JF (1993) Redox control of gene expression and the function of chloroplast genomes – an hypothesis. Photosynth Res 36:95–102

Amenitsch H, Rappolt M, Teixeira CV, Majerowicz M, Laggner P (2004) In situ sensing of salinity in oriented lipid multilayers by surface X-ray scattering. Langmuir 20:4621–4628

Apel K, Hirt H (2004) Reactive oxygen species: metabolism, oxidative stress, and signal transduction. Annu Rev Plant Biol 55:373–399

Aro EM, Virgin I, Andersson B (1993) Photoinhibition of photosystem-2: inactivation, protein damage and turnover. Biochim Biophys Acta 1143:113–134

Arpagaus S, Rawyler A, Braendle R (2002) Occurrence and characteristics of the mitochondrial permeability transition in plants. J Biol Chem 277:1780–1787

Asada K (2000) The water–water cycle as alternativ photon and electron sink. Philos Trans R Soc Lond 355:1419–1431

Baier M, Dietz K-J (2005) Chloroplasts as source and target of cellular redox regulation: a discussion on chloroplast redox signals in context of plant physiology. J Exp Bot 56:1449–1462

Baier M, Ströher E, Dietz K-J (2004) The acceptor availability at photosystem I and ABA control nuclear expression of 2-Cys peroxiredoxin-A in *Arabidopsis thaliana*. Plant Cell Environ 45:997–1006

Baxter CJ, Redestig H, Schauer N, Repsilber D, Patil KR, Nielsen J, Selbig J, Liu J, Fernie AR, Sweetlove LJ (2006) The metabolic response of heterotrophic *Arabidopsis* cells to oxidative stress. Plant Physiol (in press)

Beffagna N, Buffoli B, Busi C (2005) Modulation of reactive oxygen species production during osmotic stress in *Arabidopsis thaliana* cultured cells: involvement of the plasma membrane Ca^{2+}-ATPase and H^+-ATPase. Plant Cell Physiol 46:1326–1339

Bellafiore S, Bameche F, Peltier G, Rochaix JD (2005) State transition and light adaptation require chloroplast thylakoid protein kinase STN7. Nature 433:892–895

Bethke PC, Badger MR, Jones RL (2004) Apoplastic synthesis of nitric oxide by plant tissues. Plant Cell 16:332–341

Bonardi V, Pesaresi P, Becker T, Schleiff E, Wagner R, Pfannschmidt T, Jahns P, Leister D (2005) Photosystem II core phosphorylation and photosynthetic acclimation require two different protein kinases. Nature 437:1179–1182

Borecky J, Nogueira FT, de Oliveira KA, Maia IG, Vercesi AE, Arruda P (2006) The plant energy-dissipating mitochondrial systems: depicting the genomic structure and the expression profiles of the gene families of uncoupling protein and alternative oxidase in monocots and dicots. J Exp Bot 57:849–864

Briggs WR, Christie JM, Salomon M (2001) Phototropins: a new family of flavin-binding blue light receptors in plants. Antioxid Redox Signal 3:775–788

Bright J, Desikan R, Hancock JT, Weir IS, Neill SJ (2006) ABA-induced NO generation and stomatal closure in *Arabidopsis* are dependent on H_2O_2 synthesis. Plant J 45:113–122

Britto DT, Kronzucker HJ (2005) Nitrogen acquisition, PEP carboxylase, and cellular pH homeostasis: new views on old paradigms. Plant Cell Environ 28:1396–1409

Broess K, Trinkunas G, Van der Weij–de Wit CD, Dekker JP, Hoek A van, Amerongern H van (2006) Excitation energy transfer and charge separation in photosystem II membranes revised. Biophys J 91:3776–3786

Buchanan BB, Luan S (2005) Redox regulation in the chloroplast thylakoid lumen: a new frontier in photosynthesis research. J Exp Bot 56:1439–1447

Butow RA, Avadhani NNG (2004) Mitochondrial signaling: the retrograde response. Mol Cell 14:1–15

Caldovic L, Tuchman M (2003) N-Acetylglutamate and its changing role through evolution. Biochem J 372:279–290

Camacho A, Moreno-Sanchez R, Bernal-Lugo I (2004) Control of superoxide production in mitochondria from maize mesocotyls. FEBS Lett 570:52–56

Carrari F, Nunes-Nesi A, Gibon Y, Lytovchenko A, Ehlers Loureiro M, Fernie AF (2003) Reduced expression of aconitase results in an enhanced rate of photosynthesis and marked shifts in carbon partitioning in illuminated leaves of wild species tomato. Plant Physiol 133:1322–1335

Cerana M, Bonza MC, Harris R, Sanders D, De Michelis MI (2006) Abscisic acid stimulates the expression of two isoforms of plasma membrane Ca^{2+}-ATPase in *Arabidopsis thaliana* seedlings. Plant Biol 8:572–578

Chang C, Stewart RC (1998) The two-component system. Regulation of diverse signaling pathways in prokaryotes and eukaryotes. Plant Physiol 117:723–731

Chefdor F, Benedetti H, Depierreux C, Delmotte F, Morabito D, Carpin S (2005) Osmotic stress sensing in *Populus*: components identification of a phosphorelay system. FEBS Lett 580:77–81

Chinnusamy V, Schumaker K, Zhu JK (2004) Molecular genetic perspectives on cross-talk and specificity in abiotic stress signalling in plants. J Exp Bot 55:225–236

Cho HY, Tseng TS, Kaiserli E, Sullivan S, Christie JM, Briggs WR (2007) Physiological Roles of the LOV Domains of Phototropin 1 and Phototropin 2 in *Arabidopsis*. Plant Physiol 143:517–529

Clifton R, Millar AH, Whelan J (2006) Alternative oxidases in *Arabidopsis*: a comparative analysis of differential expression in the gene family provides new insights into function of non-phosphorylating bypasses. Biochim Biophys Acta 1757:730–741

Cross JV, Templeton DJ (2004) Oxidative stress inhibits MEKK1 by site-specific glutathionylation in the ATP-binding domain. Biochem J 381:675–683

Davis DD (1986) The fine control of cytosolic pH. Physiol Plant 67:702–706

Davletova S, Schlauch K, Coutu J, Mittler R (2005) The zinc-finger protein zat12 plays a central role in reactive oxygen and abiotic stress signaling in *Arabidopsis*. Plant Physiol 139:847–856

De Pinto MC, Paradiso A, Leonetti P, De Gara L (2006) Hydrogen peroxide, nitric oxide and cytosolic ascorbate peroxidase at the crossroad between defence and cell death. Plant J 48:784–795

Delaunay A, Pflieger D, Barrault MB, Vinh J, Toledano MB (2002) A thiol peroxidase is an H_2O_2 receptor and redox-transducer in gene activation. Cell 111:471–481

Desikan R, Last K, Harrett-Williams R, Tagliavia C, Harter K, Hooley R, Hancock JT, Neill SJ (2006) Ethylene-induced stomatal closure in *Arabidopsis* occurs via AtrbohF-mediated hydrogen peroxide synthesis. Plant J 47:907–916

Dietz KJ (1996) Functions and responses of the leaf apoplast under stress. Prog Bot 58:221–254

Dietz KJ (2003a) Plant peroxiredoxins. Annu Rev Plant Biol 54:93–107

Dietz KJ (2003b) Redox control, redox signalling, and redox homeostasis in plant cells. Int Rev Cytol 228:141–193

Dietz KJ, Jacob S, Oelze ML, Laxa M, Tognetti V, Miranda SMN de, Baier M, Finkemeier I (2006) The function of peroxiredoxins in plant organelle redox metabolism. J Exp Bot 57:1697–1709

Emanuelsson A, Eskling M, Åkerlund HE (2003) Chemical and mutational modification of histidines in violaxanthin de-epoxidase from *Spinacia oleracea*. Physiol Plant 119: 97–104

Ensminger I, Busch F, Huner NPA (2006) Photostasis and cold acclimation: sensing low temperature through photosynthesis. Physiol Plant 126:28–44

Escobar MA, Franklin KA, Svensson AS, Salter MG, Whitelam GC, Rasmusson AG (2004) Light regulation of the *Arabidopsis* respiratory chain. Multiple discrete photoreceptor responses contribute to induction of type II NAD(P)H dehydrogenase genes. Plant Physiol 136:2710–2721

Eskling M, Arvidsson P-O, Åkerlund H-E (1997) The xanthophyll cycle, its regulation and components. Plant Physiol 100:806–816

Ewy RG, Dilley RA (2000) Distinguishing between luminal and localized proton buffering pools in thylakoid membranes. Plant Physiol 122:583–595

Fedorov R, Schlichting I, Hartmann E, Domratcheva T, Fuhrmann M, Hegemann P (2003) Crystal structures and molecular mechanism of a light-induced signaling switch: the Phot-LOV1 domain from *Chlamydomonas reinhardtii*. Biophys J 84:2474–2482

Feierabend J, Schaan C, Hertwig B (1992) Photoinactivation of catalase occurs under both high- and low-temperature stress conditions and accompanies photoinhibition of photosystem II. Plant Physiol 100:1554–1561
Felle HH (1988) Auxin causes oscillations of cytosolic free calcium and pH in *Zea mays* coleoptiles. Planta 174:495–499
Felle HH, Kondorosi É, Kondorosi Á, Schultze M (1996) Rapid alkalinization in alfalfa root hairs in response to rhizobial lipochitooligosaccharide signals. Plant J 10:295–301
Fernie AR, Carrari F, Sweetlove LJ (2004) Respiratory metabolism: glycolysis, the TCA cycle and mitochondrial electron transport chain. Curr Opin Plant Biol 7:254–261
Finnegan PM, Whelan J, Millar AH, Zhang QS, Smith MK, Wiskich JT, Day DA (1997) Differential expression of the multigene family encoding the soybean mitochondrial alternative oxidase. Plant Physiol 114:455–466
Flügge UI, Heldt HW (1984) The phosphate-triose-phosphoglycerate translocator of the chloroplast. Trends Biochem Sci 9:530–533
Forsberg J, Rosenquist M, Fraysse L, Allen JF (2001) Redox signalling in chloroplasts and mitochondria: genomic and biochemical evidence for two-component regulatory stystems in bioenergetic organelles. Biochem Soc Trans 29:403–407
Foyer CH, Noctor G (2003) Redox sensing and signalling associated with reactive oxygen in chloroplasts, peroxisomes and mitochondria. Physiol Plant 119:355–364
Foyer CH, Quick WP (1997) A molecular approach to primar metabolism in higher plants. Taylor and Francis, London
Frahmcke JS, Walla PJ (2006) Coulombic couplings between pigments in the major light-harvesting complex LHCII calculated by the transition density cube method. Chem Phys Lett 430:397–403
Frankhauser C (2001) The phytochromes, a family of red/far-red absorbing photoreceptors. J Biol Chem 276:11453–11456
Franklin KA, Whitelam GC (2004) Light signals, phytochromes and cross-talk with other environmental cues. J Exp Bot 55:271–276
Fukuzawa H, Miura K, Ishizaki K, Kucho K, Saito T, Kohinata T, Ohyama K (2001) Ccm1, a regulatory gene controlling the induction of a carbon-concentrating mechanism in *Chlamydomonas reihardtii* by sensing CO_2 availability. Proc Natl Acad Sci USA 98:5347–5352
Gadjev I, Vanderauwera S, Gechev TS, Laloi C, Minkov IN, Shulaev V, Apel K, Inze D, Mittler R, Van Breusegem F (2006) Transcriptomic footprints disclose specificity of reactive oxygen species signaling in *Arabidopsis*. Plant Physiol 141:436–445
Gisselsson A, Szilagyi A, Akerlund HE (2004) Role of histidine in the binding of violaxanthin de-epoxidase to the thylakoid membrane as studied by site-directed mutagenesis. Physiol Plant 122:337–343
Godbole A, Varghese J, Sarin A, Mathew MK (2003) VDAC is a conserved element of death pathways in plant and animal systems. Biochim Biophys Acta 1642:87–96
Godon C, Lagniel G, Lee J, Buhler JM, Kieffer S, Perrot M, Boucherie H, Toledano MB, Labarre J (1998) The H_2O_2 stimulon in *Saccharomyces cerevisiae*. J Biol Chem 273:22480–22489
Goetze TA, Phillipar K, Ilkavets I, Soll J, Wagner R (2006) OEP37 is a new member of the chloroplast outer membrane ion channels. J Biol Chem 281:17989–17998
Gong D, Guo Y, Jagendorf AT, Zhu JK (2002) Biochemical characterization of the *Arabidopsis* protein kinase SOS2 that functions in salt tolerance. Plant Physiol 130:256–264
Grill E, Loffler S, Winnacker EL, Zenk MH (1989) Phytochelatins, the heavy-metal-binding peptides of plants, are synthesized from glutathione by a specific gamma-glutamylcysteine dipeptidyl transpeptidase (phytochelatin synthase). Proc Natl Acad Sci USA 86: 6838–6842
Guo Y, Halfter U, Ishitani M, Zhu JK (2001) Molecular characterization of functional domains in the protein kinase SOS2 that is required for plant salt tolerance. Plant Cell 13:1383–1400
Guo Y, Xiong L, Song CP, Gong D, Halfter U, Zhu JK (2002) A calcium sensor and its interacting protein kinase are global regulators of abscisic acid signaling in *Arabidopsis*. Dev Cell 3:233–244

Ha SB, Smith AP, Howden R, Dietrich WM, Bugg S, O'Connell MJ, Goldsbrough PB, Cobbett CS (1999) Phytochelatin synthase genes from *Arabidopsis* and the yeast *Schizosaccharomyces pombe*. Plant Cell 11:1153–1164

Hadiarto T, Nanmori T, Matsuoka D, Iwasaki T, Sato K, Fukami Y, Azuma T, Yasuda T (2006) Activation of *Arabidopsis* MAPK kinase kinase (AtMEKK1) and induction of AtMEKK1–ATMEK1 pathway by wounding. Planta 223:708–713

Hafke JB, Neff R, Hutt MT, Luttge U, Thiel G (2001) Day-to-night variations of cytoplasmic pH in a crassulacean acid metabolism plant. Protoplasma 216:164–170

Håkansson G, Allen JF (1995) Histidine and tyrosine phosphorylation in pea mitochondria: evidence for protein phosphorylation in respiratory redox signalling. FEBS Lett 372:238–242

Halford NG, Bouly JP, Thomas M (2000) SNF1-related protein kinases (SnRKs): regulators at the heart of the control of carbon metabolism and partitioning. Adv Bot Res 32:405–434

Halliwell B (2006) Reactive species and antioxidants. Redox biology is a fundamental theme of aerobic life. Plant Physiol 141:312–322

Hancock JT, Henson D, Nyirenda M, Desikan R, Harrison J, Lewis M, Hughes J, Neill SJ (2005) Proteomic identification of glyceraldehyde 3-phosphate dehydrogenase as an inhibitory target of hydrogen peroxide in *Arabidopsis*. Plant Physiol Biochem 43:828–835

Haripal PK, Raval HK, Raval MK, Rawal RM, Biswal B, Biswal UC (2006) Three-dimensional model of zeaxanthin binding PsbS protein associated with nonphotochemical quenching of excess quanta of light energy absorbed by the photosynthetic apparatus. J Mol Model 12:847–853

He XJ, Zhang ZG, Yan DQ, Zhang JS, Chen SY (2004) A salt-responsive receptor-like kinase gene regulated by the ethylene signaling pathway encodes a plasma membrane serine/threonine kinase. Theor Appl Genet 109:377–383

Heazlewood JL, Howell KA, Whelan J, Millar AH (2003) Towards an analysis of the rice mitochondrial proteome. Plant Physiol 132:230–242

Hemmler R, Becker T, Schleiff E, Bolter B, Strahl T, Soll J, Gotze TA, Braams S, Wagner R (2006) Molecular properties of Oep21, an ATP-regulated anion-selective solute channel from the outer chloroplast membrane. J Biol Chem 281:12020–12029

Hepler PK (2005) Calcium: a central regulator of plant growth and development. Plant Cell 17:2142–2155

Hoebrichts FA, Woltering EJ (2003) Multiple mediators of plant programmed cell death: interplay of conserved cell death mechanisms and plant-specific regulators. Bioessays 25:47–57

Horling F, Lamkemeyer P, Konig J, Finkemeier I, Kandlbinder A, Baier M, Dietz KJ (2003) Divergent light-, ascorbate-, and oxidative stress-dependent regulation of expression of the peroxiredoxin gene family in *Arabidopsis*. Plant Physiol 131:317–325

Huner NPA, Maxwell DP, Gray GR, Savitch LV, Krol M, Ivanov AG, Falk S (1996) Sensing environmenatl temperature changes through imbalances between energy supply and energy consumption: redox state of photosystem II. Physiol Plant 98:358–364

Im CS, Eberhard S, Huang K, Beck CF, Grossman AR (2006) Phototropin involvement in the expression of genes encoding chlorophyll and carotenoid biosynthesis enzymes and LHC apoproteins in *Chlamydomonas reinhardtii*. Plant J 48:1–16

Irihimovitch V, Sapira M (2000) Glutathione redox potential modulated by reactive oxygen species regulates translation of Rubisco large subunit in the chloroplast. J Biol Chem 275:16389–16295

Jacob C, Knight I, Winyard PG (2006) Aspects of the biological redox chemistry of cysteine: from simple redox responses to sophisticated signalling pathways. Biol Chem 387:1385–1397

Jones A (2000) Does the plant mitochondrion integrate cellular stress and regulate programmed cell death? Trends Plant Sci 5:225–230

Kacperska A (2004) Sensor types in signal transduction pathways in plant cells responding to abiotic stressors: do they depend on stress intensity? Physiol Plant 122:159–168

Kaplan B, Davydov O, Knight H, Galon Y, Knight MR, Fluhr R, Fromm H (2006) Rapid transcriptome changes induced by cytosolic Ca^{2+} transients reveal ABRE-related sequences as Ca^{2+}-responsive cis elements in *Arabidopsis*. Plant Cell 18:2733–2748

Kessler S, Sinha N (2004) Shaping up: the genetic control of leaf shape. Curr Opin Plant Biol 7:65–72
Kiddle G, Pastori GM, Bernard S, Pignocchi C, Antoniw J, Verrier PJ, Foyer CH (2003) Effects of leaf ascorbate content on defense and photosynthesis gene expression in *Arabidopsis thaliana*. Antioxid Redox Signal 5:23–32
Kim DY, Bovet L, Kushnir S, Noh EW, Martinoia E, Lee Y (2006) AtATM3 is involved in heavy metal resistance in *Arabidopsis*. Plant Physiol 140:922–932
Kircher S, Wellmer F, Nick P, Rugner A, Schafer E, Harter K (1999) Nuclear import of the parsley bZIP transcription factor CPRF2 is regulated by phytochrome photoreceptors. J Cell Biol 144:201–211
Knight H (2000) Calcium signaling during abiotic stress in plants. Int Rev Cytol 195:269–324
Kodama Y, Sano H (2006) Evolution of a basic helix-loop-helix protein from the transcriptional repressor to a plastid-resident regulatory factor – involvement in hypersensitive cell death in tobacco plants. J Biol Chem 281:35369–35380
Kolbe A, Olivier SN, Fernie AR, Stitt M, Dongen JT van, Geigenberger P (2006) combined transcript and metabolite profiling of Arabidopsis leaves reveals fundamental effects of the thiol-disulfide status on plant metabolism. Plant Physiol 141:412–422
König J, Lotte K, Plessow R, Brockhinke A, Baier M, Dietz K-J (2003) Reaction mechanism of plant 2-Cys peroxiredoxin: role of the c-Terminus and the quaternary structure. J Biol Chem 278:24409–24420
Kromer S, Gardestrom P, Samuelsson G (1996) Regulation of the supply of cytosolic oxaloacetate for mitochondrial metabolism via phospho enolpyruvate carboxylase in barley leaf protoplasts. 1. The effect of covalent modification on PEPC activity, pH response, and kinetic properties. Biochim Biophys Acta 1289:343–350
Kühlbrandt W, Wang DN, Fujiyoshi Y (1994) Atomic structure of plant light-harvesting complex determined by electron crystallography. Nature 367:614–621
Kuzmin EV, Karpova OV, Elthon TE, Newton KJ (2004) Mitochondrial respiratory deficiencies signal up-regulation of genes for heat shock proteins. J Biol Chem 279:20672–20677
Kuzniak E, Sklodowska M (2004) The effect of *Botrytis cinerea* infection on the antioxidant profile of mitochondria from tomato leaves. J Exp Bot 55:605–612
Lamotte O, Courtois C, Dobrowolska G, Besson A, Pugin A, Wendehenne D (2006) Mechanisms of nitric-oxide-induced increase of free cytosolic Ca^{2+} concentration in *Nicotiana plumbaginifolia* cells. Free Radic Biol Med 40:1369–1376
Lemaire SD (2004) The glutaredoxin family in oxygenic photosynthetic organisms. Photosynth Res 79:305–318
Liu H, Colavitti R, Rovira, II, Finkel T (2005) Redox-dependent transcriptional regulation. Circ Res 97:967–974
Logan DC, Knight MR (2003) Mitochondrial and cytosolic calcium dynamics are differentially regulated in plants. Plant Physiol 133:21–24
MacRobbie EAC (2006) Osmotic effects on vacuolar ion release in guard cells. Proc Natl Acad Sci USA 103:1135–1140
Maier T, Yu C, Kullertz G, Clemens S (2003) Localization and functional characterization of metal-binding sites in phytochelatin synthases. Planta 218:300–308
Mann NH (2000) Detecting the environment. In: Whitton BA, Potts M (ed) The ecology of cyano-bacteria. Kluwer Academic, Dordrecht, pp 367–395
Mao J, Zhang YC, Sang Y, Li QH, Yang HQ (2005). A role for *Arabidopsis* cryptochromes and COP1 in the regulation of stomatal opening. Proc Natl Acad Sci USA 102:12270–12275
Mathieu Y, Lapous D, Thomine SC, Guern J (1996) Cytoplasmic acidication as an early phosphorylation-dependent response to tobacco cells to elicitors. Planta 199:416–424
McCammon MT, Epstein CB, Przybyla-Zawislak B, McAlister Henn L, Butow RA (2003) Global transcription analysis of krebs tricarboxylic acid cycle mutants reveals an alternating pattern of gene expression and effects on hypoxic and oxidative genes. Mol Biol Cell 14:958–972
Meier I (2005) Nucleocytoplasmic trafficking in plant cells. Int Rev Cytol 244:95–135

Meyer Y, Reichheld JP, Vignols F (2005) Thioredoxins in *Arabidopsis* and other plants. Photosynth Res 86:419–433
Millar AH, Trend AE, Heazlewood JL (2004) Changes in the mitochondrial proteome during the anoxia to air transition in rice focus around cytochrome-containing respiratory complexes. J Biol Chem 279:39471–39478
Millar AH, Heazlewood JL, Kristensen BK, Braun HP, Møller IM (2005) The plant mitochondrial proteome. Trends Plant Sci 10:36–43
Miller G, Mittler R (2006) Could heat shock transcription factors function as hydrogen peroxide sensors in plants? Ann Bot 98:279–288
Mittler R (2002) Oxidative stress, antioxidants and stress tolerance. Trends Plant Sci 7:405
Mittova V, Guy M, Tal M, Volokita M (2004) Salinity up-regulates the antioxidative system in root mitochondria and peroxisomes of the wild salt-tolerant tomato species *Lycopersicon pennellii*. J Exp Bot 55:1105–1113
Miyake C, Shinzaki Y, Nishioka M, Horiguchi S, Tomizawa KI (2006) Photoinactivation of ascorbate peroxidase in isolated tobacco chloroplasts. *Galderia partia* APX maintains the electron flux through the water–water cycle in transplastomic tobacco plants. Plant Cell Physiol 47:200–210
Møller IM (2001) Plant mitochondria and oxidative stress: electron transport, NADPH turnover, and metabolism of reactive oxygen species. Annu Rev Plant Physiol Plant Mol Biol 52:561–591
Møller IM, Kristensen BK (2006) Protein oxidation in plant mitochondria detected as oxidized tryptophan. Free Radic Biol Med 40:430–435
Morillo SA, Tax FE (2006) Functional analysis of receptor-like kinases in monocots and dicots. Curr Opin Plant Biol 9:460–469
Mustilli AC, Merlot S, Vavasseur A, Fenzi F, Giraudat J (2002). *Arabidopsis* OST1 protein kinase mediates the regulation of stomatal aperture by abscisic acid and acts upstream of reactive oxygen species production. Plant Cell 14:3089–3099
Navarro-Avino JP, Bennett AB (2005) Role of a Ca^{2+}-ATPase induced by ABA and IAA in the generation of specific Ca^{2+} signals. Biochem Biophys Res Comm 329:406–415
Neill SJ, Desikan R, Clarke A, Hancock JT (2002) Nitric oxide is a novel component of abscisic acid signaling in stomatal guard cells. Plant Physiol 128:13–16
Ninfa AJ, Atkinson MR (2000) PII signal transduction proteins. Trends Microbiol 8:172–179
Niyogi KK, Li X-P, Rosenberg V, Jung H-S (2005) Is PsbS the site of non-photochemical quenching in photosynthesis? J Exp Bot 56:375–382
Nogueira FT, Borecky J, Vercesi AE, Arruda P (2005) Genomic structure and regulation of mitochondrial uncoupling protein genes in mammals and plants. Biosci Rep 25:209–226
Nunes-Nesi A, Carrari F, Lytovchenko A, Smith AMO, Ehlers Loureiro M, Ratcliffe RG, Sweetlove LJ, Fernie AF (2005) Enhanced photosynthetic performance and growth as a consequence of decreasing mitochondrial malate dehydrogenase activity in transgenic tomato plants. Plant Physiol 137:611–622
Ortiz DF, Kreppel L, Speiser DM, Scheel G, McDonald G, Ow DW (1992) Heavy metal tolerance in the fission yeast requires an ATP-binding cassette-type vacuolar membrane transporter. EMBO J 11:3491–3499
Osakabe Y, Maruyama K, Seki M, Satou M, Shinozaki K, Yamaguchi-Shinozaki (2005) Leucine-rich repeat receptor-like kinase1 is a key membrane-bound regulator of abscisic acid early signaling in *Arabidopsis*. Plant Cell 17:1105–1119
Patra M, Bhowmik N, Bandopadhyay B, Sharma A (2004) Comparison of mercury, lead and arsenic with respect to genotoxic effects on plant systems and the development of genetic tolerance. Environ Exp Bot 52:199–223
Pfannschmidt T, Nilsson A, Tullberg A, Link G, Allen JF (1999) Direct transcriptional control of the chloroplast genes psbA and psaAB adjust photosynthesis to light energy distribution in plants. IUBMB Life 48:271–276
Piippo M, Allahverdiyeva Y, Paakkarinen V, Suoranta UM, Battchikova N, Aro EM (2006) Chloroplast-mediated regulation of nuclear genes in *Arabidopsis thaliana* in the absence of light stress. Physiol Genomics 25:142–152

Raghavendra AS, Padmasree K (2003) Beneficial interactions of mitochondrial metabolism with photosynthetic carbon assimilation. Trends Plant Sci 8:546–553
Rao MV, Davis KR (2001) The physiology of ozone induced cell death. Planta 213:682–690
Rao MV, Koch JR, Davis KR (2000) Ozone: a tool for probing programmed cell death in plants. Plant Mol Biol 44:345–358
Richter ML (2004) Gamma-epsilon interactions regulate the chloroplast ATP synthase. Photosynth Res 79:319–329
Rio LA del, Corpas FJ, Sandalio LM, Palma JM (2003) Plant peroxisomes, reactive oxygen metabolism and nitric oxide. IUBMB Life 55:71–81
Rizhsky L, Liang H, Mittler R (2003) The water–water cycle is essential for chloroplast protection in the absence of stress. J Biol Chem 278:38921–38925
Robson CA, Vanlerberghe GC (2002) Transgenic plant cells lacking mitochondrial alternative oxidase have increased susceptibility to mitochondria-dependent and -independent pathways of programmed cell death. Plant Physiol 129:1908–1920
Romano PGN, Horton P, Gray JE (2004) The *Arabidopsis* cyclophilin gene family. Plant Physiol 134:1268–1282
Rossel JB, Walter PB, Hendrickson L, Chow WS, Poole A, Mullineaux PM, Pogson BJ (2006) A mutation affecting ascorbate peroxidase 2 gene expression reveals a link between responses to high light and drought tolerance. Plant Cell Environ 29:269–281
Rouhier N, Jacquot JP (2005) The plant multigenic family of thiol peroxidases. Free Radic Biol Med 38:1413–1421
Ruiz ON, Hussein HS, Terry N, Daniell H (2003) Phytoremidiation of organomercurial compounds via chloroplast genetic engeneering. Plant Physiol. 132:1344–1352
Salvucci ME, Crafts-Brandner SJ (2004) Inhibition of photosynthesis by heat stress: the activation state of Rubisco as a limiting factor in photosynthesis. Physiol Plant 120:179–186
Samaj J, Read ND, Volkmann D, Menzel D, Baluska (2005) The endocytic network in plants. Trends Cell Biol 15:425–433
Samra HS, Gao F, He F, Hoang E, Chen ZG, Gegenheimer PA, Berrie CL, Richter ML (2006) Structural analysis of the regulatory dithiol-containing domain of the chloroplast ATP synthase gamma subunit. J Biol Chem 281:31041–31049
Samuel MA, Miles GP, Ellis BE (2000) Ozone treatment rapidly activates MAP kinase signalling in plants Plant J 22:367–376
Sanders D, Slayman CL (1982) Control of intracellular pH. Predominant role of oxidative metabolism, not proton transport in the eukaryotic micro-organism *Neurospora*. J Gen Physiol 80:377–380
Sas KN, Halrup A, Hemmingsen L, Danielsen E, Ogendal LH (2006) pH-dependent structural change of reduced spinach plastocyanin studied by perturbed angular correlation of gamma-rays and dynamic light scattering. J Biol Inorg Chem 11:409–418
Scheibe R, Stitt M (1988) Comparison of NADP-malate dehydrogenase activation, Q_A reduction and O_2-evolution in spinach leaves. Plant Physiol Biochem 26:473–481
Scheibe R, Wedel N, Vetter S, Emmerlich V, Sauermann S-M (2002) Co-existance of two regulatory NADP-glyderaldehyde 3-P dehydrogenase complexes in higher plant chloroplasts. Eur J Biochem 269:5617–5624
Schmidt M, Grief J, Feierabend J (2006) Mode of translational activation of the catalase (cat1) mRNA of rye leaves (*Secale cereale* L.) and its control through blue light and reactive oxygen. Planta 223:835–846
Schoning JC, Staiger D (2005) At the pulse of time: protein interactions determine the pace of circadian clocks. FEBS Lett 579:3246–3252
Schwarz O, Strotmann H (1998) Control of chloroplast ATP synthase (CF_0CF_1) activity by ΔpH. Photosynth Res 57:287–295
Shang W, Feierabend J (1999) Dependence of catalase photoinactivation in rye leaves on light intensity and quality and characterization of a chloroplast-mediated inactivation in red light. Photosynth Res 59:201–213

Shen Q, Gomez-Cadenas A, Zhang P, Walker-Simmons MK, Sheen J, Ho TH (2001) Dissection of abscisic acid signal transduction pathways in barley aleurone layers. Plant Mol Biol 47:437–448

Shiu SH, Karlowski WM, Pan R, Tzeng YH, Mayer KF, Li WH (2004) Comparative analysis of the receptor-like kinase family in *Arabidopsis* and rice. Plant Cell 16:1220–1234

Shutilova NI (2006) Molecular mechanisms of the inhibitory action exerted by heavy metals on the oxygen-evolving pigment–lipoprotein complex of chloroplast membranes. Biol Membr 23:355–363

Siedow JN, Day DA (2000) Respiration and photorespiration. In: Buchanan BB, GruissemW, Jones RL (eds) Biochemistry and molecular biology of plants. American Society of Plant Physiologists, Rockville, pp 676–728

Simons BH, Millenaar FF, Mulder L, Van Loon LC, Lambers H (1999) Enhanced expression and activation of the alternative oxidase during infection of *Arabidopsis* with *Pseudomonas syringae* pv tomato. Plant Physiol 120:529–538

Storck D, Leslie AGW, Walker JE (1999) Molecular architecture of the rotary motor in ATP synthase. Science 286:1700–1705

Ströher E, Dietz KJ (2006) Concepts and approaches towards understanding the cellular redox proteome. Plant Biology 8:407–418

Sugiyama K, Hayakawa T, Kudo T, Ito T, Yamaya T (2004) Interaction of N-acetylglutamate kinase with a PII-like protein in rice. Plant Cell Physiol 45:1768–1778

Svensson AS, Rasmusson AG (2001) Light-dependent gene expression for proteins in the respiratory chain of potato leaves. Plant J 28:73–82

Sweetlove LJ, Heazlewood JL, Herald V, Holtzapffel R, Day DA, Leaver CJ, Millar AH (2002) The impact of oxidative stress on Arabidopsis mitochondria. Plant J 32:891–904

Takabatake R, Seo S, Ito N, Gotoh Y, Mitsuhara I, Ohashi Y (2006) Involvement of wound-induced receptor-like protein kinase in wound signal transduction in tobacco plants. Plant J 47:249–257

Tanaka T, Hosoi F, Yamaguchi-Iwai Y, Nakamura H, Masutani H, Ueda S, Nishiyama A, Takeda S, Wada H, Spyrou G, Yodoi J (2002) Thioredoxin-2 (TRX-2) is an essential gene regulating mitochondria-dependent apoptosis. EMBO J 21:1695–1703

Thor H, Smith MT, Hartzell P, Bellomo G, Jewell SA, Orrenius S (1982) The metabolism of menadione (2-methyl-1,4-naphthoquinone) by isolated hepatocytes. A study of the implications of oxidative stress in intact cells. J Biol Chem 257:12419–12425

Tiwari BS, Belenghi B, Levine A (2002) Oxidative stress increased respiration and generation of reactive oxygen species, resulting in ATP depletion, opening of mitochondrial permeability transition, and programmed cell death. Plant Physiol 128:1271–1281

Torres MA, Dangl JL (2005) Functions of the respiratory burst oxidase in biotic interactions, abiotic stress and development. Curr Opin Plant Biol 8:397–403

Torres MA, Jones JD, Dangl JL (2005) Pathogen-induced, NADPH oxidase-derived reactive oxygen intermediates suppress spread of cell death in *Arabidopsis thaliana*. Nat Genet 37:1130–1134

Vacca RA, Valenti D, Bobba A, Merafina RS, Passarella S, Marra E (2006) Cytochrome c is released in a reactive oxygen species-dependent manner and is degraded via caspase-like proteases in tobacco BY-2 cells en route to heat-shock induced cell death. Plant Physiol 141:208–219

Van Lis R, Atteia A (2004) Control of mitochondrial function via hotosynthetic redox signals. Photosynth Res 79:133–148

Vandecasteele G, Szabadkai G, Rizzuto R (2001) Mitochondrial calcium homeostasis: mechanisms and molecules. IUBMB Life 52:213–219

Vandenabeele S, Van Der Kelen K, Dat J, Gadjev I, Boonefaes T, Morsa S, Rottiers P, Slooten L, Van Montagu M, Zabeau M, Inze D, Van Breusegem F (2003) A comprehensive analysis of hydrogen peroxide-induced gene expression in tobacco. Proc Natl Acad Sci USA 100:16113–16118

Vanderauwera S, Zimmermann P, Rombauts S, Vandenabeele S, Langebartels C, Gruissem W, Inze D, Van Breusegem F (2005) Genome-wide analysis of hydrogen peroxide-regulated gene expression in *Arabidopsis* reveals a high light-induced transcriptional cluster involved in anthocyanin biosynthesis. Plant Physiol 139:806–821

Vatamaniuk OK, Mari S, Lu YP, Rea PA (1999) AtPCS1, a phytochelatin synthase from *Arabidopsis*: isolation and in vitro reconstitution. Proc Natl Acad Sci USA 96:7110–7115

Vaultier MN, Cantrel C, Vergnolle C, Justin AM, Demandre C, Benhassaine-Kesri G, Cicek D, Zachowski A, Ruelland E (2006) Desaturase mutants reveal that membrane rigidification acts as a cold perception mechanism upstream of the diacylglycerol kinase pathway in *Arabidopsis* cells. FEBS Lett 580:4218–4223

Verniquet F, Gaillard J, Neuburger M, Douce R (1991) Rapid inactivation of plant aconitase by hydrogen peroxide. Biochem J 276:643–648

Vianello A, Zancani M, Peresson C, Petrussa E, Casolo V, Krajnakova J, Patui S, Braidot E, Macri F (2007) Plant mitochondrial pathways leading to programmed cell death. Physiol Plant 129:242–252

Virolainen E, Blokhina O, Fagersedt K (2002) Ca^{2+}-induced high amplitude swelling and cytochrome c release from wheat (*Triticum aestivum* L.) mitochondria under anoxic stress. Ann Bot 90:509–516

Wang Y, Sun ZH, Horken KM, Im CS, Xiang YB, Grossman AR, Weeks DP (2005) Analysis of CIA5, the master regulator of the carbon-concentrating mechanism in *Chlamydomonas reihardtii*, and its control of gene expression. Can J Bot 83:765–779

Weber P, Fulgosi H, Piven I, Muller L, Krupinska K, Duong VH, Herrmann RG, Sokolenko A (2006) TCP34, a nuclear-encoded response regulator-like TPR protein of higher plant chloroplasts. J Mol Biol 357:535–549

Wendehenne D, Durner J, Klessig DF (2004) Nitric oxide: a new player in plant signalling and defense responses. Curr Opin Plant Biol 7:449–455

White PJ, Broadley MR (2003) Calcium in plants. Ann Bot 92:487–511

Yamazaki D, Yoshida S, Asami T, Kuchitsu K (2003) Visualization of abscisic acid-perception sites on the plasma membrane of stomatal guard cells. Plant J 35:129–139

Yano D, Sato M, Saito C, Sato MH, Morita MT, Tasaka M (2003) A SNARE complex containing SGR3/AtVAM3 and ZIG/VTI11 in gravity-sensing cells is important for *Arabidopsis* shoot gravitropism. Proc Natl Acad Sci USA 100:8589–8594

Yao N, Greenberg JT (2005) Arabidopsis ACCELERATED CELL DEATH2 modulates programmed cell death. Plant Cell 18:397–411

Yin ZH, Neimanis S, Wagner U, Heber U (1990) Light-dependent pH changes in leaves of C-3 plants. (1) Recording of pH changes in various cellular compartments. Planta 182:244–252

Yoshida R, Hobo T, Ichimura K, Mizoguchi T, Takahashi F, Alonso J, Ecker JR, Shinozaki K (2002) ABA-activated SnRK2 protein kinase is required for dehydration stress signaling in *Arabidopsis*. Plant Cell Physiol 43:1473–1483

Yoshida S, Hotsubo K, Kawamura Y, Murai M, Arakawa K, Takezawa D (1999) Alterations of intracellular pH in response to low temperature stresses: J Plant Res 112:225–236

Youle RJ, Karbowski M (2005) Mitochondrial fission in apoptosis. Nat Rev Mol Cell Biol 6:657–663

Yu JP, Nickels R, McIntosh L (2001) A genome approach to mitochondrial–nuclear communication in Arabidopsis. Plant Physiol Biochem 39:345–353

Zhang M, Barg R, Yin M, Gueta-Dahan Y, Leikin-Frenkel A, Salts Y, Shabtai S, Ben-Hayyim G (2005) Modulated fatty acid desaturation via overexpression of two distinct omega-3 desaturases differentially alters tolerance to various abiotic stresses in transgenic tobacco cells and plants. Plant J 44:361–371

Oxidative Stress and Salt Tolerance in Plants

Cai-Hong Pang and Bao-Shan Wang(✉)

Contents

Abstract Salt stress can induce ionic stress and osmotic stress in plant cells. A direct result of these primary effects is the enhanced accumulation of reactive oxygen species (ROS) that are harmful to plant cells at high concentrations. To cope with the oxidative stress resulting from the ROS, higher plants have developed a complex scavenging system including enzymatic and non-enzymatic (antioxidants) system. In plant cells, specific ROS producing and scavenging systems are found in different organelles such as chloroplasts, mitochondria, and peroxisomes; and the ROS-scavenging pathways from different cellular compartments are coordinated. Relatively low levels of ROS can be used for signaling molecules to control abiotic stress responses. Coordinated work of ROS-scavenging pathways from different

Bao-Shan Wang
College of Life Sciences, Shandong Normal University, Jinan 250014, Shandong Province, PR China
e-mail: bswang@sdnu.edu.cn

U. Lüttge et al. (eds.), *Progress in Botany 69.*

cellular compartments in modulating the level of ROS in cells preventing cellular damage and controlling ROS signaling may play a key role in plant salt tolerance. Here we attempt to summarize the recent researches on ROS and the mechanism of salt tolerance of plants under salt stress, and we also propose some perspectives involved in ROS and plant salt tolerances in the future.

1 Introduction

Salinity is a worldwide problem that limits distribution and production of major crops. One-fifth of irrigated agriculture is adversely affected by soil salinity (Flowers and Yeo 1995). So increasing crop salt tolerance is essential for sustaining food production.

Salinity can induce ionic stress and osmotic stress. As a consequence of these primary effects, secondary stresses such as oxidative stress often occur. Salt stress induces the accumulation of reactive oxygen species (ROS) that are harmful to plant cells at high concentrations. They cause oxidative damage to membrane lipids, proteins, and nucleic acids (Gómez et al. 1999; Hernández et al. 2001). Therefore, it is important to understand how plants respond and adapt to oxidative stress. Antioxidant resistance mechanisms may provide a strategy to enhance plant salt tolerance. In this review, recent progress in research on oxidative stress induced by salinity and salt tolerance is discussed.

Molecular oxygen (O_2) has a relatively low reactivity toward most biological substances. Partially reduced forms of O_2 are extremely reactive and may oxidize biological molecules. Reduction of O_2 to an active oxygen molecule results from the addition of one, two, or three electrons to form a superoxide radical (O_2^-), hydrogen peroxide (H_2O_2), or hydroxyl radical ($OH^•$), respectively (Asada and Takahashi 1987; Apel and Hirt 2004).

The first step in O_2 reduction produces O_2^-. The half-life for O_2^- is approximately 2–4 ms. O_2^- can oxidize specific amino acids, such as histidine, methionine, and tryptophane. In the cellular environment, O_2^- also causes lipid peroxidation, thereby weakening cell membranes. Via the Mehler reaction, more harmful ROS such as $OH^•$ are produced if O_2^- accumulates in the cells.

The second O_2 reduction generates H_2O_2, a relatively long-lived molecule (1 ms) that can diffuse some distance from its site of production (Levine et. al. 1994; Willekens et. al. 1997). H_2O_2 can oxidize SH groups. The biological toxicity of H_2O_2 can be enhanced in the presence of metal catalysts through Haber–Weiss or Fenton-type reactions.

The third production of reduced oxygen is that of the hydroxyl radical ($OH^•$). It is the most reactive ROS causing oxidative damage and has a half-life of <1 ms. As result, it has a very high affinity for biological molecules at its site of production, reacting at almost diffusion-controlled rates. There is no specific antioxidant to scavenging the $OH^•$. So the cell should be equipped with a perfect scavenging system for O_2^- and H_2O_2 to protect itself from oxidative damage.

Under high salinity conditions, the production of ROS is increased dramatically and the physiological homeostasis of the cell is disrupted. To cope with the oxidative stress resulting from the ROS, higher plants have developed a complex antioxidant system consisting of low molecular weight antioxidants, including carotenoids, ascorbate, glutathione (GSH), and tocopherol (V_E), as well as antioxidant-enzymes such as superoxide dismutases (SOD), catalase (CAT), and the ascorbate–glutathione cycle (Asada 1999).

2 Production of ROS in Plant Cells Under Salinity

Under normal conditions, the production and removal of ROS are at an equilibrium. Many stresses such as salt stress, drought, chilling, heat shock, and high-light stress disturb the balance and enhance the production of ROS. In plants, ROS are continuously produced predominantly in chloroplasts, mitochondria, and peroxisomes.

2.1 Chloroplasts

The chloroplast is considered to be a focal point of ROS metabolism because, under light, the oxygen pressure in chloroplasts is much higher than in other organelles. Therefore, the chloroplast is considered as a major producer of O_2^- and H_2O_2 (Davletova et al. 2005). In chloroplast thylakoids, the reaction centers of PSI and PSII are the major generation sites of ROS.

Superoxide (O_2^-) is generated by the one-electron reduction of molecular oxygen in the plastoquinone (PQ) pool by plastosemiquinone, by ferredoxin (Fd) or by iron–sulfur redox centers in the electron transport chain within PSI (Dat et al. 2000). The O_2^- formed is rapidly converted to hydrogen peroxide (H_2O_2) either spontaneously or by SOD. Production of ROS by this source is enhanced in plants by conditions limiting CO_2 fixation, such as salt stress, as well as by the combination of these conditions with high-light stress (Mittler et al. 2004). H_2O_2 can also lead to the production of the hydroxyl radical ($OH^{\bullet}$). Singlet oxygen ($^1O_2$*) is mainly produced by the reaction centre of PSII (P680). Oxygen in the ground (triplet) state ($^3O_2$) is excited to the singlet state ($^1O_2$) by excited triplet chlorophyll molecules in the reaction center (Fryer et al. 2002; Hideg et al. 2002). The yield of $^1O_2$* is increased by high light or UV and this is accompanied by the photoinhibition of PSII (Hideg et al. 2002).

2.2 Mitochondria

The mitochondrial electron transport chain (mtETC) contains four electron transporting complexes (complexes I–IV) and one H^+-translocating ATP synthetic complex (complex V). Two of these complexes were shown to be responsible for much of the O_2^- generated: complex I (the NADH ubiquinone oxidoreductase) and complex III (the ubiquinol–cytochrome *c* oxidoreductase; Boveris and Chance 1977; Takeshiga and Minakami 1979; Beyer 1991; Mittler 2002). In contrast, O_2^- generation via the mitochondrial respiratory chain in most organisms is probably also produced by a non-enzymatic mechanism. In the course of electron transport reactions in the respiratory chain, ubisemiquinone (UQ10) species donate electrons to oxygen and provide a constant source of O_2^- (Raha and Robinson 2000).

O_2^- generated from the respiratory chain is reduced by dismutation to H_2O_2 and O_2. H_2O_2 is a relatively low-toxic compound. It can react with reduced Fe^{2+} and Cu^+ to produce highly toxic hydroxyl radicals and diffuse from the mitochondrion into other cellular parts (Grene 2002; Sweetlove and Foyer 2004).

In green tissue ROS production in the mitochondrion is very low, compared with mitochondria in mammalian cells. One reason is that plant mitochondria have an alternative oxidase (AOX). AOX competes with the cytochrome bc1 complex for electrons and catalyzes the tetravalent reduction of O_2 by ubiquinone producing H_2O, which reduces ROS production (Apel and Hirt 2004). However, in the dark or in non-green tissues, mitochondria are a major source of ROS (Puntarulo et al. 1988).

Through perturbing mtETC and increasing mtROS, salt stress is at least partly responsible for oxidative stress and plant responses (Hernández et al. 1993; Mittova et al. 2003).

2.3 *Peroxisomes*

In plant cells, peroxisomes are probably the major intracellular H_2O_2-producing organelle. In recent years, it was demonstrated that O_2^- is also produced in peroxisomes. There are at least two sites of O_2^- generation (del Río et al. 2006). One site occurs in the peroxisomal matrix. Xanthine oxidase catalyzes the oxidation of xanthine and hypoxanthine to uric acid and produces O_2^- radicals (Halliwell and Gutteridge 2000). The other site is the peroxisomal membrane, where a small electron transport chain appears to be involved (del Río et al. 2006). Three integral peroxisomal membrane polypeptides (PMPs) of this electron transport chain with molecular masses of 18, 29, and 32 kDa have been characterized and demonstrated to be responsible for O_2^- generation (López-Huertas et al. 1999).

H_2O_2 production in peroxisomes has at least two pathways: one is the disproportionation of O_2^- generated in this organelle. The other is a direct pathway. During photorespiration glycolate is catalyzed by glycolate oxidase, yielding H_2O_2. Fatty acid β-oxidation, the enzymatic reaction of flavin oxidases, can also produce H_2O_2 (Baker and Graham 2002; del Río et al. 2002).

Salt stress can increase lipid peroxidation of peroxisomes. However, peroxisomal H_2O_2 content is unaffected by salinity (Mittova et al. 2003). This may be correlated with porins in the peroxisomal membranes (Reumann et al. 1997). Further studies are needed in order to find direct evidence for the relationship between peroxisomal ROS production and salt stress.

2.4 *Other Sources*

Plasma membrane NADPH oxidases contain a multimeric flavocytochrome that forms an electron transport chain capable of reducing O_2 to O_2^-. This enzyme could be the enzymatic source participating in ROS accumulation during NaCl stress (Hernández et al. 2001). In addition to NADPH oxidases, pH-dependent cell wall peroxidases, germin-like oxalate oxidases, and amine oxidases are proposed to generate ROS in the apoplast (Bolwell and Wojtaszek 1997; Hu et al. 2003; Walters 2003). Cell wall peroxidases produce O_2^- at the expense of NADH in a Mn^{2+}-dependent reaction (Elstner and Oßwald 1994).

3 ROS Scavenging Systems and Salt Tolerance in Plants

Salt stress enhances ROS production in plant cells. In order to keep the balance between ROS production and scavenging, plants developed scavenging systems against ROS, involving both enzymatic and non-enzymatic (antioxidants) systems. Major ROS-scavenging enzymes include SOD, ascorbate peroxidase (APX), CAT, glutathione peroxidase (GPX), glutathione reductase (GR), and peroxiredoxins (Prxs). Antioxidants include ascorbic acid (AsA), GSH, carotenoids, and V_E. Interestingly, higher plants also developed specific ROS-scavenging systems in different organelles to efficiently remove the ROS produced in these cellular parts; and, in particular under environmental stress such as salt stress, they coordinately work to provide plant cells with a highly efficient machinery for detoxifying ROS.

3.1 *Chloroplasts*

Thylakoid SOD (tSOD), thylakoid APX (tAPX), and Fd-dependent reduction of mondehydroascorbate (MDA) form the thylakoidal scavenging system, which functions as the first defense against ROS (Mittler et al. 2004). The O_2^- photogenerated by the PSI complex is in situ disproportionated to H_2O_2 and O_2, catalyzed by tSOD. Then the H_2O_2 is reduced to water by AsA, catalyzed with tAPX, and AsA is oxidized to the MDA radical. All the above reactions make up the water–water cycle. Subsequently, MDA is directly reduced to AsA by either reduced ferredoxin (redFd; Miyake and Asada 1994) or spontaneously disproportionated to dehydroascorbate (DHA). Then MDA or DHA is reduced to AsA by the ascorbate–glutathione cycle (AsA-GSH cycle). Reduction of H_2O_2 by Prx is another mechanism to inactive ROS at PSI (Dietz 2003).

ROS that escape from the thylakoid or are produced in the stroma undergo detoxification by stromal SOD, stromal APX, and the stromal AsA-GSH cycle. PrxR and GPX cycle are also involved in H_2O_2 removal in the stroma (Mittler et al. 2004).

H_2O_2 can convert to $OH^{\bullet}$ by the Haber–Weiss reaction (Imlay and Linn 1988). $OH^{\bullet}$ is the most reactive ROS, causing oxidative damage to chloroplast components. No specific scavenging enzyme is available for $OH^{\bullet}$. In fact, damage by H_2O_2 and O_2^- results primarily from their role in hydroxyl radical production rather than from direct action (Imlay and Linn 1988). So the cell should be equipped with a perfect scavenging system for O_2^- and H_2O_2 to protect itself from oxidative damage.

In the chloroplasts $^1O_2$ is a ROS without unpaired electron. Once formed, $^1O_2$ rapidly reacts with nearby molecules, causing oxidative damage to proteins, lipids, and DNA. In the PSII reaction center two molecules of β-carotene participate in the quenching of $^1O_2$ generated via $^3P680^*$. Tocopherols (V_E) can also quench $^1O_2$, but the rate is two orders of magnitude lower than that with β-carotene (Krasnovsky Jr 1998).

Non-enzymatic antioxidants such as AsA and GSH are the major cellular redox buffers. AsA concentration in chloroplasts is high. It can react directly with hydroxyl radicals, superoxide, and singlet oxygen (Buettner and Jurkiewicz 1996).

Chloroplastic SOD, APX, and GR play a central role in the enzymatic scavenging system. They all have two types: thylakoid-bound and stromal type. CuZn-SOD is the major isoform of SOD in the chloroplast. Several plants such as tobacco also contain Fe-SOD (Kurepa et al. 1997). It is exclusively localized in the chloroplast stroma. Our results showed that thylakoid-bound SOD, APX, GR, and stromal SOD, APX, GR in the chloroplasts of the halophyte *Suaeda salsa* L. are markedly enhanced under high salinity (Pang et al. 2005; Zhang et al. 2005). This is possibly an important mechanism of salt tolerance in halophytes. Overexpression of tAPX (Yabuta et al. 2002; Murgia et al. 2004) can increase tolerance to methyl viologen (MV)-induced oxidative stress. Transgenic plants overexpressing *Escherichia coli* catalase in chloroplasts show enhanced resistance to photooxidative stress by MV (Miyagawa et al. 2000). Plants with cytosolic APX overexpressed in chloroplasts show enhanced tolerance to salt and drought stresses (Badawi et al. 2004). Transgenic tobacco plants expressing both SOD and APX in chloroplasts have enhanced tolerance against MV-mediated oxidative stress (Kwon 2002). Mutants of tAPX are thought to be lethal (Yabuta et al. 2002). Plants with reduced tAPX activity are sensitive to MV stress (Tarantino et al. 2005) or paraquat-induced photooxidative stress (Tarantino et al. 2005). All these results suggest that the ROS enzyme scavenging system in chloroplasts plays a central role to protect the chloroplast from oxidative damage induced by environmental stresses.

3.2 Mitochondria

ROS produced in mitochondria can damage mitochondrial lipids, proteins, and DNA. So the overproduced ROS must be scavenged in time. Mitochondrial Mn-SOD catalyzing the production of H_2O_2 from O_2^- is the first step in scavenging the O_2^- generated from the mitochondrial electron transport chain (Kliebenstein et al. 1998; Møller 2001; Mittler et al. 2004). O_2^- can also be converted to H_2O_2 by spontaneous dismutation. Consequently, H_2O_2 is removed by mitochondrial APX through the AsA-GSH cycle (Jiménez et al. 1997; Chew et al. 2003) or CAT, which was reported in maize

(*Zea mays*) mitochondria (Sweetlove and Foyer 2004). Prxs also could reduce mitochondrial H_2O_2. These enzymes use reduced thioredoxins as reductant sources, which in turn are reduced by thioredoxin reductase (Sweetlove and Foyer 2004).

In addition to directly detoxifying ROS, plant mitochondria also can modulate superoxide production from mtETC. There are two modulation mechanisms. In the first, AOX acts to maintain a basal ubiquinone pool reduction state as initially proposed by Purvis and Shewfelt (1993) and diminishes mtROS production (Popov et al. 1997; Purvis 1997; Maxwell et al. 1999). Second, uncoupling protein (UCP) uncouples by facilitating a proton leak across the membrane, consequently removes inhibition of the mtETC (Hourton-Cabassa et al. 2004; Sluse and Jarmuszkiewicz 2004), and then decreases ROS formation.

When oxidative damage is produced, some enzymes such as GST and the type II Prx /thioredoxin system may be involved in repairing lipid peroxidation and some forms of protein oxidation (Rhoads et al. 2006).

Salinity up-regulates the levels of ASA and GSH and the activities of SOD, APX MDHAR, DHAR, and GPX in root mitochondria of the wild salt-tolerant tomato species *Lycopersicon pennellii* (Mittova et al. 2004). Transcript levels of mitochondrial MnSOD were strongly induced by salt treatment in the salt-tolerant variety but not in the NaCl-sensitive variety (Hernández et al. 2000). Overexpression of mitochondrial Mn-SOD from *Nicotiana plumbagnifolia* in *Nicotiana tabacum* mitochondria protected the latter from oxidative damage (Bowler et al. 1991). A yeast strain deficient in mitochondrial MnSOD regained its resistance to oxidative stress when plant mitochondrial MnSOD was expressed in it (Scandalios 1993; Zhu and Scandalios 1992). Among several potential mechanisms for ROS detoxification in plant mitochondria, only MnSOD is firmly established. Further experiments are needed to testify other ROS detoxification mechanisms in plant mitochondria and their roles in salt tolerance.

3.3 *Peroxisomes*

Plant peroxisomal SOD has been reported in at least nine different plant species (del Río et al. 2002). Three of them have been purified and characterized (del Río et al. 2002): a CuZn-SOD and a Mn-SOD from watermelon and a Mn-SOD from pea leaves. They convert the O_2^- generated in peroxisomes to H_2O_2 and O_2. Peroxisomal H_2O_2 can be converted to H_2O through CAT and the AsA-GSH cycle in peroxisomes (Mittler et al. 2004). APX, monodehydroascorbate reductase (MDAR), dehydroascorbate reductase (DHAR), and GR of the cycle in peroxisomes were purified from pea leaves and tomato leaves and roots (del Río et al. 1998; Mittova et al. 2004; Kużniak and Skłodowska 2005). The intraperoxisomal distribution and function of the AsA-GSH cycle have also been identified (del Río et al. 2006). DHAR and GR are found in the matrix of peroxisomes, whereas APX and MDAR are bound to the cytosolic side of the peroxisomal membrane. The presence of APX and MDAR in the leaf peroxisomal membrane suggests a dual complementary function in the peroxisomal metabolism of these membrane-bound antioxidant enzymes. First, MDAR reoxidizes NADH to maintain a constant supply of NAD^+ for peroxisomal metabolism. Second, the membrane-bound antioxidant enzymes prevent H_2O_2 leaking from the peroxisomes (del Río et al. 2002). GPX has been reported in leaf peroxisomes of tomato plants (Kużniak and Skłodowska 2005) and a putative Prx with a molecular mass of 60 kDa has been localized in the matrix of pea leaf peroxisomes (Corpas et al. 2003). They can also decrease the level of H_2O_2.

Salt-induced peroxisomal oxidative stress probably results from salt-induced stomatal closure. Under these conditions, the photorespiratory peroxisomal glycolate oxidase increases the rate of ROS generation and leads to oxidative stress (Smirnoff 1993). Salinity up-regulates the levels of ASA and GSH and the activities of SOD, APX, CAT, and MDAR in root peroxisomes of the wild salt-tolerant tomato species *Lycopersicon pennellii* (Mittova et al. 2004). Overexpression of an *Arabidopsis* peroxisomal APX gene in tobacco increases protection against oxidative stress caused by aminotriazole which inhibits catalase activity. However, the overexpression cannot protect plants from oxidative damage caused by paraquat, which leads to the production of reactive oxygen species in chloroplasts (Wang et al. 1999). These results indicated that the protection provided by the targeted expression of antioxidant genes may be selective, depending on the nature of oxidative stress.

3.4 Apoplast

The enzymatic components responsible for ROS detoxification in the apoplast and cell wall are only partially known. Previous studies have failed to detect the antioxidant enzymes of the AsA-GSH cycle in the apoplast (Castillo and Greppin 1988; Polle et al. 1990; Luwe 1996; Vanacker et al. 1998a, b). However, recently the presence not only of SOD activity, but also of the AsA-GSH cycle and CAT in the apoplast of both barley (*Hordeum vulgare*) and oat (*Avena sativa*) leaves were identified. Little or no GSH has been found in the apoplast of plant cells (Luwe 1996; Vanacker et al. 1998a, b). Salt stress can change the apoplastic ROS-scavenging components of two pea cultivars (Hernández et al. 2001).

4 Salt Stress, Oxidative Stress, ROS Signal Perception/Sensing and Salt Tolerance in Plants

A high concentration of ROS can lead to phytotoxicity whereas relatively low levels can be used for signaling. Under salt stress, ROS production in plant cells, including extracellular and intracellular spaces, can be sensed by ROS sensors. Salt-induced ROS are predominantly formed as H_2O_2 and this process occurs inside the cell. A possible signal transduction pathway under salt stress is proposed (Fig. 1) which is maybe different from the pathway induced by extracellular H_2O_2

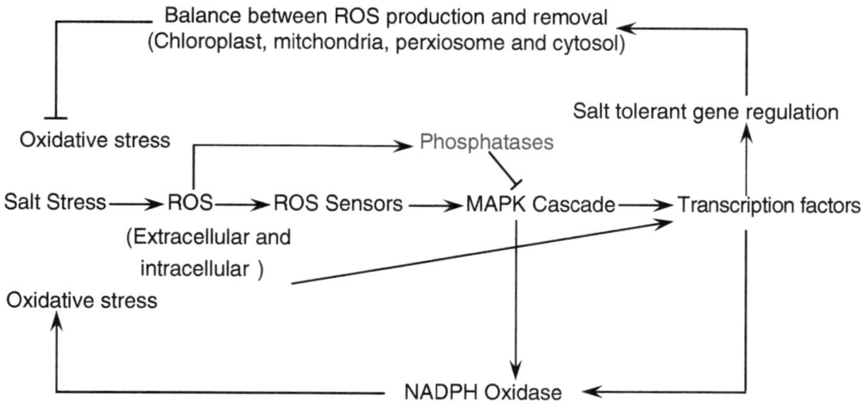

Fig. 1 A possible ROS signal transduction pathway in plant cells under salt stress

(Avsian-Kretchmer et al. 2004). Salt stress rapidly activates the mitogen-actived protein kinase (MAPK) casade. Intracellular ROS can also influence the ROS-induced MAPK signal pathway through inhibition of phosphatases or downstream transcription factors. Then gene expression in response to salt stress can be regulated. In salt-tolerant plants, ROS production and removal was found to be kept in balance and the level of oxidative stress to be low. NADPH oxidases possibly amplify ROS signaling (Mittler et al. 2004).

5 Perspectives

Different plant species have different responses to salt stress and their capacity for resistance to salt stress is also not identical. Salt stress can secondarily induce oxidative stress and lead to the accumulation of ROS. The mechanism of ROS production and scavenging in different organelles has been described in many published papers. There are still many uncertainties in understanding signal transduction under salt stress conditions.

5.1 Sensors of ROS Perception

To date, no ROS receptor has been unambiguously identified in plants. Plant cells sense ROS possibility via at least three different mechanisms:

1. ROS sensors: redox-sensitive transcription factors, such as NPR1(non-expressor of pathogenesis-related genes 1), control the onset of systemic acquired resistance, plant immunity, to a broad spectrum of pathogens that is normally established after a primary exposure to avirulent pathogens.

2. Heat-shock transcription factors (HSFs) are the terminal components of the signal transudation pathway for gene activation in response to heat stress.
3. Direct inhibition of phosphatases by ROS (Mittler et al. 2004).

How the plant cells sense ROS under salt stress is still an open question.

5.2 Key Components in ROS Signal Transduction Cascade

Salt stress rapidly (within 5–10 min) activates *Arabidopsis* mitogen-activated protein kinase kinase kinase (AtMEKK1; Ichimura et al. 1998), mitogen activated protein kinase kinase (AtMKK2; Teige et al. 2004), and MAPKs (ATMPK3, ATMPK4, ATMPK6; Mizoguchi et al. 1996; Ichimura et al. 2000). A role for the MAPK module consisting of MEKK1–MKK2–MPK4/MPK6 has now been confirmed in cold and salt stress (Nakagami et al. 2005). Mutant plants *mkk2*-null are hypersensitive to salt stress (Nakagami et al. 2005). Plants with overexpression of *ANP1* (a MAPKKK; Kovtun et al. 2000) and *MKK2* (Nakagami et al. 2005) are more tolerant to salt stress. Different transcript factors (WRKY, Zat, RAV, GRAS, Myb families) induced or activated by the MAPK casade regulate the ROS-scavenging and ROS-producing pathways (Mittler et. al. 2004).

H_2O_2 is sensed via the modification of thiol groups in certain proteins. The inactivation of *Arabidopsis* protein tyrosine phosphatases (AtPTP1) by H_2O_2 may be mediated by the oxidation state of the active-site cysteine. AtPTP1 can inactivate *Arabidopsis* MPK6 (Gupta and Luan 2003). When oxidative stress occurs, AtPTP1 is inactivated by H_2O_2. Then the MAPK casade is activated and transcription factors are regulated.

Thus, it can be seen that transcription factors and the MAPK casade are key components in ROS signal transduction. In *Arabidopsis*, the role of the MAPK casade has been investigated under salt stress. Nevertheless its role in other plants is still unclear. The expression of different transcription factors is enhanced by ROS and this includes members of the WRKY, Zat, RAV, GRAS, and Myb families (Mittler et al. 2004). To find out which transcription factors could be regulated by salinity is a challenge. Large-scale transcriptome analyses coupled with proteomic and metabolomic analyses of plants perturbed at the levels of individual or multiple components of the salt induced ROS network will be essential for future studies.

5.3 The Mechanism of Salt Tolerance Involved in Antioxidants in Plants

Salt stress does not induce oxidative damage in salt-tolerant plants (Hernández et al. 2000; Mittova et al. 2004; Wang et al. 2004; Pang et al. 2005; Zhang et al. 2005). This is due to an enhancement of the ROS scavenging system. Conversely, salt-sensitive plants can be damaged by salt stress-induced ROS (Hernández et al. 2000; Mittova et al. 2004). There are perhaps different signal transduction or transcript factors in the salt-tolerant and salt-sensitive plants. It is also possible that there is considerable difference in non-coding regions of antioxidant enzyme genes, such as SOD, between halophytes and nonhalophytes. This needs more direct proof.

5.4 Coordinated Work of the Different Cellular Parts Under Salinity

Different organelles have their special ROS scavenging systems. The function of the components of the different organelles' ROS scavenging systems has been clarified. It is considered that the different compartments are protected by their own ROS-scavenging systems at first, and they then also somehow work coordinately to protect plant cells from oxidative damage. It is reported that the cytosolic H_2O_2–scavenging enzyme ascorbate peroxidase 1 (APX1) can protect chloroplasts during light stress (Davletova et al. 2005). This provides evidence for cross-compartment protection of thylakoid and stromal/mitochondrial APXs by cytosolic APX1. At present there is a new view of the ROS network: the coordinated function of ROS-scavenging pathways from different cellular compartments in modulating the level of ROS in cells prevents cellular damage and controls ROS signaling. In the future, research on the relationship between the different subcellular ROS-scavenging pathways will be a powerful approach for understanding plant salt tolerance at the cellular level.

Acknowledgements We are grateful for financial support from the National Natural Science Research Foundation of China (NSFC, projects No. 30670177, No. 30070069) and the Specialized Research Fund for the Doctoral Program of Higher Education (SRFDP, 20050445003).

References

Apel K, Hirt H (2004) Reactive oxygen species: metabolism, oxidative stress, and signal transduction. Annu Rev Plant Biol 55:373–99

Asada K (1999) The water–water cycle in chloroplasts: scavenging of active oxygen and dissipation of excess photons. Annu Rev Plant Physiol Plant Mol Biol 50:601–639

Asada K, Takahashi M (1987) Production and scavenging of active oxygen in photosynthesis. In: Kyle DJ, Osmond CB, Arntzen CJ (eds) Photoinhibition. Elsevier, Amsterdam, pp 227–287

Avsian-Kretchmer O, Gueta-Dahan Y, Lev-Yadun S, Gollop R, Ben-Hayyim G (2004) The salt-stress signal transduction pathway that activates the *gpx1* promoter is mediated by intra-cellular H_2O_2, different from the pathway induced by extracellular H_2O_2. Plant Physiol 135:1685–1696

Badawi GH, Kawano N, Yamauchi Y, Shimada E, Sasaki R, Kubo A, Tanaka K (2004) Overexpression of ascorbate peroxidase in tobacco chloroplasts enhances the tolerance to salt stress and water deficit. Physiol Plant 121:231–238

Baker A, Graham I (2002) Plant peroxisomes. Biochemistry, cell biology and biotechnological applications. Kluwer, Dordrecht

Beyer R (1991) An analysis of the role of coenzyme Q in free radical generation and as an antioxidant. Biochem Cell Biol 70:390–343

Bolwell GP, Wojtaszek P (1997) Mechanisms for the generation of reactive oxygen species in plant defence – broad perspective. Physiol Mol Plant Pathol 51:347–366

Boveris A, Chance B (1977) Mitochondrial production of superoxide radical and hydrogen peroxide. In: Reivich M, Coburn R, Lahiri S, Chance B (eds) Tissue hypoxia and ischemia. Plenum, New York, pp 67–82

Bowler C, Slooten L, Vandenbranden S, De Rycke R, Botterman J (1991) Manganese superoxide dismutase can reduce cellular damage mediated by oxygen radicals in transgenic plants. EMBO J 10:1723–1732

Buettner GR, Jurkiewicz BA (1996) Chemistry and biochemistry of ascorbic acid. In: Cadenas E, Packer L (eds) Handbook of antioxidants. Dekker, New York, pp 91–115

Castillo FJ, Greppin H (1988) Extracellular ascorbic acid and enzyme activities related to ascorbic acid metabolism in *Sedum album* L. after ozone exposure. Environ Exp Bot 28:231–238

Chew O, Whelan J, Millar AH (2003) Molecular definition of the ascorbate–glutathione cycle in *Arabidopsis* mitochondria reveals dual targeting of antioxidant defenses in plants. J Biol Chem 278:46869–46877

Corpas FJ, Pedrajas JR, Sandalio LM, León AM, Carreras A, Palma JM, Valderrama R, Río LA del, Barroso JB (2003) Localization of peroxiredoxin in peroxisomes from pea leaves. Free Radical Res 37[Suppl 2]:19

Dat J, Vandenabeele S, Vranová E, Montagu MV, Inzé D, Breusegem FV (2000) Dual action of the active oxygen species during plant stress responses. Cell Mol Life Sci 57:779–795

Davletova S, Rizhsky L, Liang HJ, Zhong SQ, Oliver DJ, Coutu J, Shulaev V, Schlauch K, Mittler R (2005) Cytosolic ascorbate peroxidase 1 is a central component of the reactive oxygen gene network of *Arabidopsis*. Plant Cell 17:268–281

Dietz KJ (2003) Plant peroxiredoxins. Annu Rev Plant Biol 54:93–107

Elstner EF, Oßwald W (1994) Mechanisms of oxygen activation during plant stress. Proc R Soc Edin B 102:131–154

Flowers TJ, Yeo AR (1995) Breeding for salinity resistance in crop plants. Where next? Aust J Plant Physiol 22:875–884

Fryer MJ, Oxborough K, Mullineaux PM, Baker NR (2002) Imaging of photo-oxidative stress responses in leaves. J Exp Bot 53:1249–1254

Gómez JM, Hernández JA, Jiménez A, Río LA del, Sevilla F (1999) Differential response of antioxidative enzymes of chloroplast and mitochondria to long term NaCl stress of pea plants. Free Radical Res 31:S11–S18

Grene R (2002) Oxidative stress and acclimation mechanisms in plants. In: Somerville CR, Myerowitz EM (eds) The *Arabidopsis* book. American Society of Plant Biologists, Rockville. Available at: www.aspb.org/publications/Arabidopsis/

Gupta R, Luan S (2003) Redox control of protein tyrosine phosphatases and mitogen-activated protein kinases in plants. Plant Physiol 132:1149–1152

Halliwell B, Gutteridge JMC (2000) Free radicals in biology and medicine. Oxford University Press, Oxford

Hernández JA, Corpas FJ, Gómez M, Río LA del, Sevilla F (1993) Salt induced oxidative stress mediated by activated oxygen species in pea leaf mitochondria. Physiol Plant 89:103–110

Hernández JA, Jiménez A, Mullineaux P, Sevilla F (2000) Tolerance of pea (*Pisum sativum* L.) to long-term salt stress is associated with induction of antioxidant defenses. Plant Cell Environ 23:853–862

Hernández JA, Ferrer MA, Jiménez A, Barcelo AR, Sevilla F (2001) Antioxidant systems and O_2^-/H_2O_2 production in the apoplast of pea leaves. Its relation with salt-induced necrotic lesions in minor veins. Plant Physiol 127:817–831

Hideg E, Barta C, Kalai T, Vass I, Hideg K, Asada K (2002) Detection of singlet oxygen and superoxide with fluorescent sensors in leaves under stress by photoinhibition or UV radiation. Plant Cell Physiol 43:1154–1164

Hourton-Cabassa C, Matos AR, Zachowski A, Moreau F (2004) The plant uncoupling protein homologues: a new family of energy-dissipating proteins in plant mitochondria. Plant Physiol Biochem 42:283–290

Hu X, Bidney DL, Yalpani N, Duvick JP, Crasta O, Folkerts O, Lu GH (2003) Overexpression of a gene encoding hydrogen peroxide-generating oxalate oxidase evokes defense responses in sunflower. Plant Physiol 133:170–181

Ichimura K, Mizoguchi T, Irie K, Morris P, Giraudat J, Matsumoto K, Shinozaki K (1998) Isolation of ATMEKK1 (a MAP kinase kinase kinase)-interacting proteins and analysis of a MAP kinase cascade in *Arabidopsis*. Biochem Biophys Res Commun 253:532–543

Ichimura K, Mizoguchi T, Yoshida R, Yuasa T, Shinozaki K (2000) Various abiotic stresses rapidly activate *Arabidopsis* MAP kinases ATMPK4 and ATMPK6. Plant J 24:655–665

Imlay JA, Linn S (1988) DNA damage and oxygen radical toxicity. Science 240:1302–1309

Jiménez A, Hernández JA, Río LA del, Sevilla F (1997) Evidence for the presence of the ascorbate-glutathione cycle in mitochondria and peroxisomes of pea leaves. Plant Physiol 114:275–284

Kliebenstein DJ, Monde RA, Last RL (1998) Superoxide dismutase in *Arabidopsis*: an eclectic enzyme family with disparate regulation and protein localization. Plant Physiol 118:637–650

Kovtun Y, Chiu WL, Tena G, Sheen J (2000) Functional analysis of oxidative stress-activated mitogen-activated protein kinase cascade in plants. Proc Natl Acad Sci USA 97:2940–2945

Krasnovsky AA Jr (1998) Singlet molecular oxygen in photobiochemical systems: IR phosphorescence studies. Membr Cell Biol 12:665–690

Kurepa J, Hérouart D, Van Montagu M, Inzé D (1997) Differential expression of CuZn- and Fe-superoxide dismutase genes of tobacco during development, oxidative stress and hormonal treatments. Plant Cell Physiol 38:463–470

Kuźniak E, Skłodowska M (2005) Fungal pathogen-induced changes in the antioxidant systems of leaf peroxisomes from infected tomato plants. Planta 222:192–200

Kwon SY, Jeong YJ, Lee HS, Kim JS, Cho KY, Allen RD, Kwak SS (2002) Enhanced tolerances of transgenic tobacco plants expressing both superoxide dismutase and ascorbate peroxidase in chloroplasts against methyl viologen-mediated oxidative stress. Plant Cell Environ 25:873–882

Levine A, Tenhaken R, Dixon R, Lamb C (1994) H_2O_2 from the oxidative burst orchestrates the plant hypersensitive disease resistance response. Cell 79:583–593

López-Huertas E, Corpas FJ, Sandalio LM, Río LA del (1999) Characterization of membrane polypeptides from pea leaf peroxisomes involved in superoxide radical generation. Biochem J 337:531–536

Luwe M (1996) Antioxidants in the apoplast and symplast of beech (*Fagus sylvatica* L.) leaves: seasonal variations and responses to changing ozone concentration in air. Plant Cell Environ 19:321–328

Maxwell DP, Wang Y, McIntosh L (1999) The alternative oxidase lowers mitochondrial reactive oxygen production in plant cells. Proc Natl Acad Sci USA 96:8271–8276

Mittler R (2002) Oxidative stress, antioxidants and stress tolerance. Trends Plant Sci 7:405–410

Mittler R, Vanderauwera S, Gollery M, Breusegem FV (2004) Reactive oxygen gene network of plants. Trends Plant Sci 9:490–498

Mittova V, Tal M, Volokita M, Guy M (2003) Up-regulation of the leaf mitochondrial and peroxisomal antioxidative systems in response to salt-induced oxidative stress in the wild salt-tolerant tomato species. Plant Cell Environ 26:845–856

Mittova V, Guy M, Tal M, Volokita M (2004) Salinity up-regulates the antioxidative system in root mitochondria and peroxisomes of the wild salt-tolerant tomato species *Lycopersicon pennellii*. J Exp Bot 55:1105–1113

Miyagawa Y, Tamoi M, Shigeoka S (2000) Evaluation of the defense system in chloroplasts to photooxidative stress caused by paraquat using transgenic tobacco plants expressing catalase from *Escherichia coli*. Plant Cell Physiol 41:311–320

Miyake C, Asada K (1994) Ferredoxin-dependent photoreduction of monodehydroascorbate radicals in spinach thylakoids. Plant Cell Physiol 35:539–549

Mizoguchi T, Irie K, Hirayama T, Hayashida N, Yamaguchi-Shinozaki K, Matsumoto K, Shinozaki K (1996) A gene encoding a mitogen-activated protein kinase kinase kinase is induced simultaneously

with genes for a mitogen-activated protein kinase and an S6 ribosomal protein kinase by touch, cold, and water stress in *Arabidopsis thaliana*. Proc Natl Acad Sci USA 93:765–769
Murgia I, Tarantino D, Vannini C, Bracale M, Carrabvieri S, Soave C (2004) *Arabidopsis thaliana* plants overexpressing thylakoidal ascorbate peroxidase show resistance to paraquat-induced photooxidative stress and to nitric oxide-induced cell death. Plant J 38:940–995
Møller IM (2001) Plant mitochondria and oxidative stress: electron transport, NADPH turnover, and metabolism of reactive oxygen species. Annu Rev Plant Physiol Plant Mol Biol 52:561–591
Nakagami H, Pitzschke A, Hirt H (2005) Emerging MAP kinase pathways in plant stress signalling. Trends Plant Sci 10:339–346
Pang CH, Zhang SJ, Gong ZZ, Wang BS (2005) NaCl treatment markedly enhances H_2O_2-scavenging system in leaves of halophyte *Suaeda salsa*. Physiol Plant 125:490–499
Polle A, Chakrabarti K, Schümann W, Rennenberg H (1990) Composition and properties of hydrogen peroxide decomposing systems in extracellular and total extracts from needles of Norway Spruce (*Picea abies* L., Karst.). Plant Physiol 94:312–319
Popov VN, Simonina RA, Skulachev VP, Starkov AA (1997) Inhibition of the alternative oxidase stimulates H_2O_2 production in plant mitochondria. FEBS Lett 415:87–90
Puntarulo S, Sánchez RA, Boveris A (1988) Hydrogen peroxide metabolism in soybean embryonic axes at the onset of germination. Plant Physiol 86:626–630
Purvis AC (1997) Role of the alternative oxidase in limiting superoxide production in plant mitochondria. Physiol Plant 100:165–170
Purvis AC, Shewfelt RL (1993) Does the alternative pathway ameliorate chilling injury in sensitive plant tissues? Physiol Plant 88:712–718
Raha S, Robinson BH (2000) Mitochondria, oxygen free radicals, disease and ageing. Trends Biochem Sci 25:502–508
Reumann S, Bettermann M, Benz R, Heldt HW (1997) Evidence for the presence of a porin in the membrane of glyoxysomes of castor bean. Plant Physiol 115:891–899
Rhoads DM, Umbach AL, Subbaiah CC, Siedow JN (2006) Mitochondrial reactive oxygen species. Contribution to oxidative stress and interorganellar signaling. Plant Physiol 141:357–366
Río LA del, Pastori GM, Palma JM, Sandalio LM, Sevilla F, Corpas FJ, Jiménez A, López-Huertas E, Hernández JA (1998) The activated oxygen role of peroxisomes in senescence. Plant Physiol 116:1195–1200
Río LA del, Corpas FJ, Sandalio LM, Palma JM, Gómez M, Barroso JB (2002) Reactive oxygen species, antioxidant systems and nitric oxide in peroxisomes. J Exp Bot 53:1255–1272
Río LA del, Sandalio LM, Corpas FJ, Palma JM, Barroso JB (2006) Reactive oxygen species and reactive nitrogen species in peroxisomes. Production, scavenging, and role in cell signaling. Plant Physiol 141:330–335
Scandalios JC (1993) Oxygen stress and superoxide dismutases. Plant Physiol 101:7–12
Sluse FE, Jarmuszkiewicz W (2004) Regulation of electron transport in the respiratory chain of plant mitochondria. In: Day DA, Millar AH, Whelan J (eds) Plant mitochondria: from genome to function, vol 1. (Advances in photosynthesis and respiration) Kluwer, Dordrecht, pp 229–245
Smirnoff N (1993) The role of active oxygen in response of plants to water deficit and desiccation. New Phytol 125:27–58
Sweetlove LJ, Foyer CH (2004) Roles for reactive oxygen species and antioxidants in plant-mitochondria. In: Day DA, Millar AH, Whelan J (eds) Plant mitochondria: from genome to function, vol 1. (Advances in photosynthesis and respiration) Kluwer, Dordrecht, pp 307–320
Takeshiga K, Minakami S (1979) NADH and NADPH dependent formation of superoxide anions by bovine heart submitochondrial particles and NADH-ubiquinone reductase preparation. Biochem J 180:129–135
Tarantino D, Vannini C, Bracale M, Camp M, Soave C, Murgia I (2005) Antisense reduction of thylakoidal ascorbate peroxidase in *Arabidopsis* enhances paraquat-induced photooxidative stress and nitric oxide induced cell death. Planta 221:757–765
Teige M, Scheikl E, Eulgem T, Doczi R, Ichimura K, Shinozaki K, Dangl JL, Hirt H (2004) The MKK2 pathway mediates cold and salt stress signaling in *Arabidopsis*. Mol Cell 15:141–152

Vanacker H, Carver TLW, Foyer CH (1998a) Pathogen induced changes in the antioxidant status of the apoplast in barley leaves. Plant Physiol 117:1103–1114
Vanacker H, Harbinson J, Carver TLW, Foyer CH (1998b) Antioxidant defenses of the apoplast. Protoplasma 205:129–140
Wang BS, Lüttge U, Ratajczak R (2004) Specific regulation of SOD isoforms by NaCl and osmotic stress in leaves of the C3 halophyte *Suaeda salsa* L. J Plant Physiol 161:285–293
Wang J, Zhang H, Allen RD (1999) Overexpression of an *Arabidopsis* peroxisomal ascorbate peroxidase gene in tobacco increases protection against oxidative stress. Plant Cell Physiol 40:725–732
Walters DR (2003) Polyamines and plant disease. Phytochemistry 64:97–107
Willekens H, Chamnongpol S, Davey M, Schraudner M, Langebartels C, Van Montagu M (1997) Catalase is a sink for H_2O_2 and is indispensable for stress defence in C_3 plants. EMBO J 16:4806–4816
Yabuta Y, Motoki T, Yoshimura K, Takada T, Ishikawa T, Shigeoka S (2002) Thylakoid membrane-bound ascorbate peroxidase is a limiting factor of antioxidative systems under photo-oxidative stress. Plant J 32:912–925
Zhang QF, Li YY, Pang CH, Lu CM, Wang BS (2005) NaCl enhances thylakoid-bound SOD activity in the leaves of C3 halophyte *Suaeda salsa L.* Plant Sci 168:423–430
Zhu D, Scandalios JG (1992) Expression of the maize *MnSod* (*Sod3*) gene in MnSOD-deficient yeast rescues the mutant yeast under oxidative stress. Genetics 131:803–809

Crassulacean Acid Metabolism: a Cause or Consequence of Oxidative Stress in Planta?

Ewa Niewiadomska and Anne M. Borland(✉)

Contents

Abstract The photosynthetic specialization of crassulacean acid metabolism (CAM) is typically found in plants growing in environments where water and/or CO_2 is limiting and, by analogy, where irradiance and daytime temperatures may be high. Such abiotic factors are known to lead to the generation of reactive oxygen species (ROS) in planta which can elicit potentially damaging oxidative stress and/or act as signals for engaging mechanisms that ameliorate oxidative stress. It has been proposed that CAM prevents the production of ROS, since the daytime CO_2 concentrating effect prevents over-energization of the photosynthetic machinery under water-limited conditions. However, CAM per se has the potential to elevate the oxidative burden in planta as a consequence of sustained electron transport behind closed stomata, which can elevate internal O_2 concentrations to around 42%. This review considers and discusses evidence for the photoprotective function of CAM, alongside considerations of the extent of photorespiration and other potential sinks for O_2 consumption. Anti-oxidant metabolism in CAM species is also reviewed and considered along with the potential role of ROS in triggering the induction or up-regulation of this photosynthetic specialization in limiting environments. The overall aim of the review is to assess whether or not CAM alleviates the oxidative burden in plants exposed to potentially limiting environmental conditions.

Anne M. Borland
School of Biology, Institute for Research on Environment and Sustainability,
Devonshire Building, Newcastle University, Newcastle Upon Tyne, NE1 7RU, UK
e-mail: a.m.borland@ncl.ac.uk

U. Lüttge et al. (eds.), *Progress in Botany 69.*

1 Introduction

Crassulacean acid metabolism (CAM) is a specialized mode of photosynthetic carbon acquisition that has evolved many times in response to exceptional environmental conditions. In CAM plants, CO_2 is taken up at night (phase I) via the enzyme phospho*enol*pyruvate carboxylase (PEPC) and the products of dark carboxylation (i.e. organic acids) are subsequently decarboxylated during the day (phase III), thereby elevating [CO_2] around Rubisco and behind closed stomata. Depending on environmental conditions, stomata may open at the start (phase II) and end (phase IV) of the day, permitting direct uptake of atmospheric CO_2 and thereby supplementing the overall carbon balance of CAM plants. At an ecological level, CAM is typically found in plants growing in environments where water and/or CO_2 is limiting and, by analogy, where irradiance and daytime temperatures may be high (Winter and Smith 1996a; Cushman and Bohnert 1997; Cushman and Borland 2002). Such abiotic factors are known to lead to the generation of reactive oxygen species (ROS) in planta which can elicit potentially damaging oxidative stress and/or act as signals for engaging mechanisms that ameliorate oxidative stress (Foyer and Noctor 2005). It has been proposed that CAM prevents the production of ROS, since the daytime CO_2 concentrating effect prevents over-energization of the photosynthetic machinery under water-limited conditions (Griffiths 1989). Indeed, the anti-oxidant and photoprotective role of CAM has been proposed as a major driver for the evolution of the pathway in water-limited environments (Gil 1986). Such considerations have also prompted more recent suggestions that ROS are implicated in signaling the induction of CAM in facultative species, including the model halophyte *Mesembryanthemum crystallinum* (Ślesak et al. 2003; Borland et al. 2006). Further, CAM per se has the potential to elevate the oxidative burden in planta as a consequence of sustained electron transport behind closed stomata, which can elevate internal O_2 concentrations to around 42% (Spalding et al. 1979). It seems reasonable to speculate that the rate of oxygen release, as well as its consumption, influences the probability of ROS formation and the extent of oxidative stress in CAM plants. Here we review evidence for the photoprotective function of CAM, alongside considerations of the extent of photorespiration and other potential sinks for O_2 consumption. We also discuss anti-oxidant metabolism in CAM species and consider the role of ROS in triggering this photosynthetic specialization with an overall aim of assessing whether or not CAM alleviates oxidative burden in planta.

2 Photoinhibition in CAM Plants

According to the general view, CAM plants are typically expected to inhabit exposed arid and semi-arid environments. This might imply that CAM species should, in general, be well adapted to high irradiance. However, a review by Lüttge (2000) revealed a rather limited capacity of CAM plants to acclimate to high irradiances. Moreover, a comparison of the diverse habitats favoured by CAM species implied that high irradiance is not a persistent stress factor for many species (Lüttge 2004). Whilst some C_3–CAM intermediate species like *Clusia minor* and *Guzmania monostachia* can switch rapidly into CAM in response to high irradiance, both species prefer semi-shaded sites (Grams et al. 1997; Lüttge 2000; Maxwell 2002; Miszalski et al. 2006). Moreover, the above-ground morphology of many CAM species implies that self-shading and leaf/cladode orientation reduce PFD interception in the most exposed habitats. Such observations question the dogma that CAM species are truly tolerant of exposed, high-light environments.

Measurements of chlorophyll a fluorescence have been widely used to evaluate the photoprotection afforded by CAM. Maximum quantum yield of PSII (F_v/F_m) after dark acclimation is the fluorescence parameter most commonly used to assess light-use efficiency, where values below 0.83 are indicative of PSII photoinhibition (Björkman and Demmig 1987). Rapidly reversible (acute) photoinhibition is considered a photoprotective down-regulation of the amount of absorbed light that is proportional to the generation of heat in the xanthophyll cycle (Horton et al. 1996; Niyogi 2000; Müller et al. 2001; Ruban et al. 2001). In most physiological conditions, this parameter of light acclimation is described by non-photochemical quenching (NPQ). Slowly reversible photoinhibition (monitored by pre-dawn measurements) is attributed to restitution of the D1 protein of PSII reaction centres. Under conditions of extreme photodamage when degradation of D1 protein is promoted, chronic photoinhibition develops (Osmond 1994).

2.1 Does CAM Confer Photoprotection?

The carbon-concentrating consequences of organic acid decarboxylation imply a beneficial role of CAM in alleviating midday photoinhibition (for a review, see Lüttge 2000). This effect may be attributed to the increased 'sink' capacity of the photosynthetic carbon reduction (PCR) cycle at the elevated concentrations of CO_2 which predominate during phase III of the CAM cycle. The photoprotective impacts of the carbon-concentrating components of CAM are indicated by comparative studies of sympatric C_3 and CAM species of the same (as in the case of *Clusia* or *Guzmania*) or different genera (Skillman and Winter 1997; Lüttge 2002; Maxwell 2002). A study on a C_3–CAM intermediate, *Talianum triangulare*, also illustrated that induction of CAM played a crucial role in the protection of photosynthetic machinery against photoinhibition (Pieters et al. 2003). Adams and Osmond (1988) and Griffiths et al. (2002) provided direct support for the photoprotective function of the carbon-concentrating phase of CAM in *Kalanchoe pinnata* and *Clusia fluminensis* by treating plants with CO_2-free air overnight in order to manipulate the extent and duration of phase III decarboxylation. Thus, when nocturnal accumulation of malate was curtailed, photosynthetic light-use efficiency was reduced.

The photoprotective benefits of CAM however only appear to be evident during phase III, since a decrease in photochemical quenching (qP) and an increase in NPQ have been observed during phases II and IV when stomata are open at the start and end of the day respectively (Heber et al. 1996; Maxwell et al. 1998, 1999b; Griffiths et al. 2002). In CAM-induced *M. crystallinum* at moderate irradiance (350 μE m^{-2} s^{-1}), the photoinhibition which was noted pre-dawn was at least partly reversible during the remainder of the photoperiod (Schöttler et al. 2002). This was accompanied by a slow relaxation of P_{700}, indicating nocturnal reduction of the PQ pool. It seems that this pre-dawn phenomenon of photoinhibition might be promoted by conditions of extended night-time (Schöttler et al. 2002) or low irradiance (E. Niewiadomska, unpublished data) and is associated with a decreased rate of photosynthetic electron transport (PET) in the proximity of PSII.

A decreased rate of PET at the onset of the day has been repeatedly shown in various CAM plants (Skillman and Winter 1997; de Mattos et al. 1999; Maxwell 2002). This may in part be attributed to a continued activation of PEPC for several hours after dawn (Borland et al. 1993; Borland and Griffiths 1997) and a delay in Rubisco activation (Maxwell et al. 1999b). The maintenance of PEPC activation over phase II ensures continued accumulation of malate up to a threshold level, at which point PEPC is inactivated/de-phosphorylated (Borland et al. 1999). In cases where PEPC inactivation is delayed, the commencement of net malate decarboxylation is deferred until several hours after dawn when irradiance has increased and the risk of photodamage is amplified. Thus, it would appear that the dynamic shifts in photon-use efficiency over the daytime phases of CAM indicate an increased risk of photoinhibition during the non-CO_2 concentrating phases at the start and end of the day. However, it is important to recognize that, under natural field conditions, the lower levels of irradiance that prevail at the start and end of the day are less likely to invoke a requirement for photoprotective strategies than the maximal levels of irradiance experienced at midday.

Whilst the carbon concentrating action of CAM over phase III clearly increases the sink capacity of the PCR cycle, in many field situations where plants experience moderate to high light, other strategies are required to optimize light-use efficiency and curtail oxidative stress. In many epiphytic and understorey CAM plants, extraordinarily high rates of electron transport have been reported that are in excess of that predicted from the re-fixation of CO_2 during phase III decarboxylation (Skillman and Winter 1997). This phenomenon is associated with a high capacity for NPQ (Skillman et al. 2005) and appears to be beneficial in the case of photosynthetic acclimation to short and intensive sun flecks, but not for prolonged exposure to high light. When high light is experienced in conjunction with severe water deficits, CAM does not curtail photoinhibition, as demonstrated by the inferior performance of the C_3–CAM intermediate *C. minor* compared with the constitutive C_3 species *C. multiflora* (Grams et al. 1997; Miszalski et al. 2006). Moreover, despite the induction of CAM by high light in the epiphytic bromeliad *Guzmania monostachia*, this was accompanied by a range of other photoprotective strategies that included a strong reduction in chlorophyll and thylakoid content, as well as granal stacking, in concert with the protective aggregation of LHCII

(Horton and Ruban 1994; Maxwell et al. 1999b). In the obligate CAM plant *Kalanchoë daigremontiana* exposed to drought under high irradiance (i.e. 1500 μmol m^{-2} s^{-1}), acute photoinhibition occurred at midday (Lu et al. 2003). This was not attributed to photodamage but rather to photoprotection, whereby the xanthophyll cycle played a key role in dissipating excess electron transport energy. At this point it is also important to note that elevated irradiance and temperature can lead to a shortening of the carbon-concentrating phase of CAM by more rapid consumption of malate (Skillman and Winter 1997; Lüttge 2000; Barker et al. 2004). Hence, it would appear that the carbon-concentrating function of CAM is just one component of an integrated suite of photoprotective mechanisms that are required to facilitate photosynthetic acclimation to changes in the availability of light and water. These include mechanisms that adjust PSII performance and the composition of photosynthetic membranes in order to down-regulate and balance energy absorption with the shifting requirements for utilization of ATP and reductants over the phases of CAM (Barker et al. 2004).

3 Photorespiration in CAM Plants

As a major sink for electrons in C_3 plants, photorespiration is a key strategy for protection from photoinhibition (Kozaki and Takeba 1996; Wingler et al. 1999; Badger et al. 2000). The question arises as to whether or not the carbon-concentrating consequences of CAM dispense with photorespiration as a potential mechanism of photoprotection. During phase III, the decarboxylation of organic acids generates an elevated partial pressure of CO_2 at Rubisco active sites (P_c) and thereby increases carboxylation efficiency through a reduction in oxygenase activity. However, high rates of oxygen evolution and consumption have been documented during phase III of CAM (Robinson et al. 1992; Maxwell et al. 1998). According to the estimates of Edwards et al. (1996), the amount of CO_2 generated internally from malate decarboxylation during phase III could only account for 5–10% of O_2 evolved in CAM-performing *M. crystallinum*. Other studies on *K. pinnata*, using stable isotopes of O_2 to distinguish between $^{18}O_2$ uptake and $^{16}O_2$ evolution have shown that O_2 uptake during phase III represents 21–25% of gross O_2 evolution and is light-dependant (Maxwell et al. 1998; Osmond et al. 1999). The possibility of photorespiration occurring during the carbon-concentrating phase III of CAM has to be considered, particularly in light of the very low transfer conductance of CO_2 that is presented by the densely packed, highly vacuolated leaves of CAM species (Maxwell et al. 1997). This low CO_2 conductance, together with the high (>40%) O_2 concentrations that are generated during phase III (Spalding et al. 1979), indicate that P_c may not be high enough to preclude O_2 uptake by Rubisco. Indeed, gross O_2 uptake was found to be CO_2 sensitive in the CAM species *Hoya carnosa* during phases III and IV, suggesting Rubisco oxygenase activity (Maxwell et al. 1997). From the time of maximal decarboxylation in phase III to the time when direct uptake of atmospheric CO_2 occurs during phase IV, a dramatic drop (up to

450-fold) in P_c can occur in a matter of a few hours. In leaves of *K. daigremontiana* for example a P_c of only 109 μbar (1 μbar = 100 Pa) was calculated during phase IV (Maxwell et al. 1997). Thus, the rate of photorespiration during phase IV is likely to be considerable for many CAM species. Comparisons between plants of *C. minor* operating in either C_3 or CAM modes of photosynthesis indicated that, at the beginning of phase IV, photorespiratory O_2 uptake was higher by ~39% in CAM-performing plants (Duarte and Lüttge 2007). Such results appear to confirm that the high internal [O_2] that remains during the transition from phases III to IV can exacerbate Rubisco oxygenase activity (Lüttge 2007). Other metabolic sinks for O_2 consumption during phase III are considered below (see Section 5).

Other lines of evidence indicate that CAM does not entirely curtail photorespiration. The activities of several enzymes linked to photorespiratory metabolism, namely glycolate oxidase, NAD-dependent hydroxypyruvate reductase, glutamine synthetase, and glutamate dehydrogenase, remained unchanged in *M. crystallinum* after CAM was induced (Whiteside et al. 1991). Thus, in principal, the enzymatic capacity for photorespiration appears largely unchanged with the induction of CAM. Catalase is necessary for the decomposition of photorespiratory H_2O_2 (Chamnongpol et al. 1996; Willekens et al. 1997; Dat et al. 2003) and may also serve as an indicator of photorespiration. Although a decrease in extractable CAT activity was noted in the facultative species *Sedum album* and *M. crystallinum* after CAM was induced by drought/salinity (Castillo 1996; Niewiadomska et al. 1999), under high irradiance the CAM-associated decline in CAT activity was less pronounced (Broetto et al. 2002). This appears to indicate that, in CAM species, photorespiratory flux can be rapidly adjusted in line with prevailing environmental conditions.

Although a measure of enzyme activity in vitro can provide an indication of the inherent capacity for photorespiration, the activities of photosynthetic and photorespiratory enzymes in vivo are likely to be modulated in line with the dynamic changes in [CO_2] and [O_2] that occur over the diurnal phases of CAM. It is now apparent that Rubisco exhibits dynamic changes in activity and activation state over the daytime phases of CAM in a number of species (Maxwell et al. 1999b; Griffiths et al. 2002). The carbamylation of Rubisco, i.e. formation of a carbamate-Mg^{2+} complex at the active site of the enzyme, has been found to be extremely protracted in some CAM species; and it peaks much later in the day than C_3 species grown under the same environmental conditions (Maxwell et al. 1999b). CAM species maintain a high carbamylation state even under high [CO_2] (phase III), whilst in C_3 plants carbamylation generally decreases in response to elevated CO_2, as a result of limitation in RuBP and/or Pi (Maxwell et al. 1999b). The delayed activation of Rubisco noted in CAM plants might suggest the slow removal of an inhibitor, which in turn could depend on the activity of Rubisco activase. It is believed that activase acts as a chloroplast-localized ATPase which is activated by ATP and is, presumably, dependent on the redox state and ATP energy charge in the chloroplasts (Jensen 2000; Portis and Salvucci 2002; Portis 2003). Significant nocturnal increases in the ATP/ADP ratio have been reported in the starch-forming CAM plants (Chen and Nose 2003; Niewiadomska et al. 2004), suggesting that activase may not be limited by the availability of ATP in phase II. Rather, the slow activation of Rubisco during phase II appears to be regulated (at least in part) via transcriptional control of activase since the abundance of activase transcripts and protein are low at this time and increase during phases III and IV (Griffiths et al. 2002). Such observations are in line with suggestions that circadian and metabolite control of gene transcription plays a central role in controlling many of the metabolic, transport, and physiological components of CAM (Borland and Taybi 2004; Boxall et al. 2005).

In conclusion, CAM species appear to possess an enzymatic capacity for photorespiration that is at least equivalent to that found in C_3 plants. However, the magnitude of photorespiration in vivo is determined by diffusional limitation, the prevailing $[CO_2]:[O_2]$, and by transcriptional regulation of photosynthetic gene expression. In the morning phase II of CAM, photorespiration could be limited by the slow activation of Rubisco. The magnitude of photorespiration appears to be maximal during phase IV, when malate stores are exhausted and when Rubisco has reached a maximum activation state.

4 ROS Formation and Scavenging

The light reactions of photosynthesis are the most important sources of ROS in illuminated mesophyll cells (Mittler et al. 2004; Foyer and Noctor 2005) and environmental factors that reduce CO_2 availability, i.e. drought, salinity, and elevated temperatures, can enhance the generation of these potentially damaging species (Zhu 2002; Parida and Das 2005). This could imply an enhanced risk of ROS generation in CAM plants which typically inhabit environments where CO_2 is limiting and where the succulent nature of photosynthetic organs presents diffusional constraints to CO_2 uptake during the daytime phases II and IV, when stomata are open. In addition, during phase III, enhanced ROS generation might be expected due to the oxygen production by photosynthetic water splitting in the absence of photorespiration (Osmond et al. 1999; Lüttge 2002). As concluded in Section 2 above, the functioning of the CCM of CAM precludes photooxidative stress only under moderate irradiance. However, when CAM plants are exposed to conditions of extreme water and/or osmotic stress plus high irradiance, susceptibility to over-excitation of PSII and to photooxidative stress seems to be enhanced (Miszalskiet al. 2001). All of these considerations point to a clear requirement for efficient protection against ROS formation and photooxidative stress in CAM plants.

4.1 CAM Induction and Antioxidant Metabolism

A number of studies have compared the extractable activities of a range of antioxidant enzymes in facultative CAM species operating in either the C_3 or CAM modes of photosynthesis. Castillo (1996) was the first to document stimulation in the activity of several antioxidative enzymes, including superoxide dismutase

(SOD), ascorbate peroxidase (APX), monodehydroascorbate radical reductase (MDHAR), and glutathione reductase (GR), as well as a decline in the activity of CAT during the drought-induced transition from C_3 to CAM in *Sedum album*. More recently, investigations have tended to focus on the salt-induced CAM exhibited by the halophytic species *M. crystallinum*. Treatment with high salinity revealed that induction of CAM in *M. crystallinum* was accompanied by the stimulation of total SOD activity (Miszalski et al. 1998). However, the effect on SOD was light-dependent and, when salinity treatment was performed at low light (100 μmol m^{-2} s^{-1} and CAM Δ malate = 0.1 M), SOD activity actually declined (Miszalski et al. 2001). Three isoforms of SOD are found in *M. crystallinum*: chloroplastic FeSOD, mitochondrial MnSOD, and putatively cytosolic CuZnSOD (Miszalski et al. 1998). The activity of FeSOD increased rapidly under high salinity; however, this enhanced activity was not prolonged (Miszalski et al. 1998; Broetto et al. 2002). This may suggest that an enhanced oxidative load in chloroplasts is an initial response to salinity and declines as CAM is induced. Surprisingly, only a weak stimulation of the transcript abundance of FeSOD occurred after treatment with excessive light (2000 μmol m^{-2} s^{-1} for 1 h), whilst FeSOD transcripts declined in response to treatment with dibromothymoquinone (DBMiB), an inhibitor of PQ oxidation which mimics conditions of high light (Ślesak et al. 2003). In contrast, FeSOD transcript abundance was stimulated by treatment with DCMU, a compound that mimics low irradiance by oxidation of the plastoquinone pool. Such observations might indicate the up-regulation of other photoprotective strategies in salted *M. crystallinum* under extreme conditions (Broetto et al. 2002; Ślesak et al. 2003). The activity, amount, and transcript abundance of a putatively cytosolic CuZnSOD was also stimulated by salinity in a light-independent manner (Miszalski et al. 1998; Broetto et al. 2002; Hurst et al. 2004). Similarly to FeSOD, accumulation of CuZnSOD mRNA did not occur in plants subjected to treatment with DBMIB or in plants exposed to high light (2000 μmol m^{-2} s^{-1} for 1 h), whilst transcript abundance of CuZnSOD increased in response to feeding leaves with DCMU. This response is opposite to that documented in the C_3 species *Arabidopsis thaliana* and *Nicotiana tabacum* (Pfannschmidt 2003; Ślesak et al. 2003), and consequently it was proposed that C_3 and CAM species may differ in terms of the signals and metabolic pathways that underpin acclimation to photooxidative stress (for a review, see Ślesak et al. 2002).

Interestingly, the activity of mitochondrial MnSOD was found to be substantially higher that that of FeSOD and CuZnSOD in leaves of *M. crystallinum* plants with MnSOD activity responding positively to salinity and high irradiance (Borland et al. 2006). Moreover, induction of CAM was associated with the appearance of diurnal fluctuations in MnSOD activity, characterized by a nocturnal decline and an increase over the photoperiod (Niewiadomska et al. 2004). This suggests enhanced mitochondrial oxidative metabolism during the light period in CAM plants. Mitochondrial cyanide-insensitive respiration that is catalysed by alternative oxidase (AOX) is also considered to play a role in antioxidant metabolism by curtailing the generation of ROS (in comparison with that potentially generated by cyanide-sensitive respiration; Maxwell et al. 1999a). Mitochondria isolated from

CAM plants were reported to possess a high in vitro capacity of AOX (Rustin and Queiroz-Claret 1985). Moreover, measurements using an O_2 electrode linked to a mass spectrometer demonstrated that AOX activity in vivo peaked in phase III of CAM (Robinson et al. 1992). More recently unpublished observations of our own indicated an increase in transcript and protein abundance of AOX and MnSOD in CAM-induced *M. crystallinum* (E. Niewiadomska and A. Borland, unpublished data). Considering that ROS are implicated in signaling for the enhanced transcription of *AOX* and *MnSOD* genes (Maxwell et al. 1999a; Pastori and Foyer 2002), these observations could suggest increased generation of ROS in the mitochondria as CAM is induced.

The imposition of drought and/or salinity that is commonly used to induce CAM presents the difficulty of separating the influence of CAM on antioxidant metabolism from the impacts of the oxidative load imposed by drought or salinity. Inducing CAM in *M. crystallinum* via exposure to high irradiance (800–1000 µmol m^{-2} s^{-1}) has been adopted by some workers in order to avoid salinity- or drought-induced impacts on oxidative load (Broetto et al. 2002; Ślesak et al. 2003). However, this tactic instead brings about the effects of light per se, such as light-activation or photoinhibition of enzymes, elevated rates of photorespiration, increased risk of chronic photoinhibition, and many others. An alternative approach for dissecting the anti-oxidant requirements imposed by salinity (or other abiotic stressors) from those that are a consequence of the elevated [O_2] generated during phase III of CAM emerged following the isolation of a mutant of *M. crystallinum* that fails to induce CAM in response to salinity (Branco et al. 2003). Exposure to salinity resulted in elevated activities of several anti-oxidant markers in this mutant compared with the wild type (MnSOD, APX, CAT, GR, glutathione content; Borland et al. 2006; E. Niewiadomska, B. Pater, A. Borland, unpublished data). The data suggest that the oxidative burden in salted plants was higher in the absence of CAM. However, there was no evidence of oxidative stress or photodamage in the mutant, indicating that CAM induction is just one component in a suite of interactive strategies that underpin acclimation to salinity in *M. crystallinum*.

4.2 Do ROS Induce CAM?

Early stimulation of antioxidant enzymes during the C_3–CAM shift points to increased generation of ROS. This is supported by the analysis of expressed sequence tags (ESTs) induced during 30 h and 48 h of salinity treatment in *M. crystallinum* leaves, showing that genes encoding stress-related and antioxidant proteins are among the fastest to be induced (Kore-eda et al. 2004). It is also well documented that ROS play a central role in stress acclimation by signaling the activation/up-regulation of genes and proteins involved in ameliorating oxidative stress (Mittler et al. 2004, Halliwell 2006). Thus, could ROS induce CAM as a component of the defence strategy for ameliorating oxidative stress? This hypothesis has been tested using exogenously applied treatments that elevate oxidative load in planta. Treatment of

M. crystallinum C_3 plants with 80, 150, and 200 ppb gaseous ozone-enhanced oxidative burden, as evidenced by stimulation in the activity of APX, MnSOD, and CuZnSOD (Niewiadomska et al. 2002). However, despite a marked stimulation in the activity of CAM-related enzymes, such as PEPC, NADP, and NAD-malic enzymes, several starch-degrading enzymes together with an increased transcript abundance of *Ppc1*, the CAM-specific isoform of PEPC, and its dedicated activating kinase *Ppck*, exposure to ozone failed to induce functional CAM in *M. crystallinum* (Niewiadomska et al. 2002; Hurst et al. 2004; Borland et al. 2006). Such observations prompted the suggestion that key metabolic components required for CAM may have evolved as part of a general strategy for meeting increased respiratory costs and alleviating potential damage generated by increased oxidative load under water/CO_2-limited conditions (Borland et al. 2006).

Treatments with 200 ppb ethylene, as well as 1 ppm or 6 ppm SO_2, also proved ineffective in triggering day/night fluctuations of malate (Hurst et al. 2004; Surówka et al. 2006). Production of ethylene can be considered a marker of oxidative damage (Overmyer et al. 2003; Fujita et al. 2006). By analysing transcript abundance for 1-aminocyclopropane-1-carboxylate oxidase (ACC oxidase), a rate-limiting enzyme for ethylene production, Hurst et al. (2004) revealed that imposition of salinity and high irradiance did not elicit oxidative damage in *M. crystallinum*. Thus, it appears clear that a range of mechanisms (that likely include CAM) curtail the generation of oxidative stress in leaves of *M. crystallinum* plants during acclimation to salinity and high irradiance.

On the basis of the above results it would appear that enhanced oxidative load per se does not result in the induction of CAM. However, Ślesak et al. (2003) presented data indicating that the addition of H_2O_2 into the root medium resulted in the appearance of nocturnal acidity in *M. crystallinum* leaves. It remains to be verified whether this putative signal for CAM is a direct signal from the elevated oxidative load imposed by H_2O_2 or whether H_2O_2 fed to roots creates an osmotic stress which signals the induction of CAM. Moreover, given the failure of gaseous ozone to induce CAM, it may be that the cellular site at which elevated oxidative load is perceived is critical in determining whether or not CAM is induced.

5 Redox Adjustment and Other Protective Strategies

Since CAM species do not appear to suffer prolonged or damaging oxidative stress even under extremely limiting environmental conditions, it may be speculated that a re-adjustment of whole cell redox homeostasis may occur with CAM induction to prevent the increased generation of ROS. In this respect, it is worth noting that both CAM and C_4 photosynthesis necessitates a changed stoichiometry of the ATP and NADPH requirement for CO_2 fixation in comparison with C_3 plants (Noctor and Foyer 1998). CAM requires more ATP and reductants for the gluconeogenic recovery of pyruvate in the light, for the costs of regeneration of PEP from pyruvate in the dark, for the nocturnal reduction of OAA to malate, and to drive the active transport of malate into the vacuole (Nobel 1991; Winter and Smith 1996b; Lüttge 2004). In CAM plants, the precise stoichiometry of ATP/NAD(P)H: net CO_2 fixation depends on the type of carbohydrates produced in the light (chloroplastic starch or extra-chloroplastic sucrose/hexose), the nature of organic acids accumulated overnight (malate, citrate, isocitrate), the pathway of decarboxylation (mitochondrial NAD-ME, cytosolic NADP-ME, or PEP carboxykinase, PEPCK), and mode of vacuolar transport (tonoplast ATPase and/or inorganic pyrophosphatase PPase; Winter and Smith 1996b; Holtum et al. 2005). The ATP and NADPH budget of CAM plants is also influenced by the magnitude and duration of phases II and IV when direct uptake of CO_2 (and O_2) is mediated via Rubisco (Borland and Griffiths 1996).

The inherent flexibility in ATP and NADPH stoichiometry presented by the CAM cycle suggests that, among the various strategies that curtail photooxidative stress, mechanisms that modify the ATP/NAD(P)H ratio (for reviews, see Badger et al. 2000; Cruz et al. 2005) are of particular importance in CAM plants. Three chloroplastic processes, namely the water–water cycle (WWC), cyclic electron flow (CEF), and chlororespiration compete for electrons from linear electron transport, and ultimately can determine the overall ATP/NADPH ratio (Asada 2000; Badger et al. 2000; Nixon and Mullineaux 2001; Heber 2002; Cruz et al. 2005). Of these, CEF and chlororespiration may also use electrons from NAD(P)H due to the activity of NAD(P)H-PQ oxidoreductase (NDH). Whilst the WWC and chlororespiration direct electrons to O_2, activation of these processes is dependent on the redox state of the donor side of PSI.

The WWC derives its name due to the extraction of electrons from water by PSII, the use of these electrons to reduce O_2, and the re-oxidation of electrons to water via the ascorbate peroxidase cycle. According to the estimates of Osmond and Grace (1995), the WWC was proposed to account for up to 50% of whole chain ETR during the phase III CO_2-concentrating phase of CAM. However, further studies showed that the role of the WWC in CAM plants could be largely diminished when the oxygenation of Rubisco is suppressed (Maxwell et al. 1998). A study on C_3 plants (Ruuska et al. 2000) estimated that oxygen uptake assigned to WWC was no more than 10% of gross O_2 uptake when measured at high CO_2. Thus, in CAM plants, low engagement of the WWC may be hypothesized during the carbon-concentrating phase III, although it remains to be established whether engagement of the WWC varies dynamically over the diurnal phases.

A stimulation of cyclic or pseudocyclic electron flow is hypothesized to cover the extra ATP costs incurred by CAM (Köster and Winter 1985; Winter and Smith 1996b). As known from studies on C_3 plants, a shift in the activity and/or stoichiometry of PSII and PSI, as well as redox poising occurring at the donor side of PSI, is necessary to activate CEF (Bendall and Manasse 1995; Heber 2002). In *M. crystallinum* no distinct changes in the stoichiometry and activities of PSII and PSI were found when C_3 and CAM modes were compared (Schöttler et al. 2002). However, an increase in the proportion of the slow-relaxing component of chlorophyll fluorescence may suggest a stimulation of state transition after CAM is induced (Keiller et al. 1994). In contrast when NDH is present in the thylakoid membrane, this could distribute electrons between CEF and plastidic terminal oxidase (PTOX). Chlororespiration is considered an important electron sink under low and variable CO_2 availability in algae and cyanobacteria (Bendall and Manasse 1995; Badger et al. 2000; Nixon and Mullineaux 2001). The contribution made by chlororespiration to photoprotection and optimization of photosynthesis was recently shown in C_3 plants (Peltier and Cournac 2002; Streb et al. 2005). Elevated abundance of NDH was reported in the chloroplasts of C_4 bundle sheath cells (Kubicki et al. 1996), but no data are currently available for CAM plants in this regard.

Other chloroplastic strategies for redox regulation involve the consumption of excess reductants by reductive assimilation of nitrogen (Noctor and Foyer 1998). However, the salt-induced switch to CAM in *M. crystallinum* is accompanied by a substantial (i.e. 3-fold) reduction in nitrate reductase activity measured in vivo (S. Haider and A. Borland, unpublished data). Thus, in this species at least, it seems unlikely that reductive assimilation of nitrogen makes a major contribution to redox balance in CAM leaves. Another potentially important sink for excessive reductants generated in the light is mediated via a shuttling to the cytoplasm in the form of malate, the so called 'malate valve' (Fridyland et al. 1998; Scheibe 2004). In this process a chloroplastic light-activated enzyme NADP-MDH plays a key role. Salt-induced CAM in *M. crystallinum* is accompanied by increased transcript abundance of NADP-MDH (J. Cushman, personal communication), suggesting a role for this enzyme in salt acclimation.

Mitochondrial respiration may also participate in oxygen consumption during the photoperiod and help to optimize photosynthesis under changing CO_2 supply, in a manner that is well documented for C_3 plants (Drake et al. 1999; Igamberdiev et al. 2001; Padmasree et al. 2002; Noctor et al. 2004; van Lis and Atteia 2004). Environmental conditions are known to influence the relative engagement of the cytochrome and alternative oxidase branches of mitochondrial respiration in C_3 plants (van Lis and Atteia 2004). A predominant stimulation of the alternative respiratory branch was documented during the photoperiod in conjunction with photorespiration in several C_3 species (Igamberdiev et al. 2001; Noguchi et al. 2001; Millenaar and Lambers 2003). In CAM plants the involvement of mitochondrial metabolism during the light period seems to be dependant on the pathway of malate decarboxylation, as hypothesized by Maxwell et al. (1998). A comparison of proposed redox fluxes and the involvement of mitochondrial metabolism during the decarboxylation phase of CAM in NADP-ME and NAD-ME CAM plants is illustrated in Fig. 1. When oxidative decarboxylation of malate takes place in the mitochondria, as in C_4 and CAM NAD-ME plants, a stimulation of the alternative respiratory pathway has been shown (Robinson et al. 1992; Agostino et al. 1996). This might be attributed to the activation of AOX caused by the elevated generation of reducing power in mitochondria and/or increased release of pyruvate in this compartment. In contrast, when oxidative decarboxylation of malate is carried out in the cytosol, as in *M. crystallinum*, an NADP-ME plant, a stimulation of the cytochrome respiratory branch was documented (Niewiadomska et al. 2004). Considering that import of reducing power into the mitochondria is not likely due to the unfavourable redox gradient (Hanning and Heldt 1993;

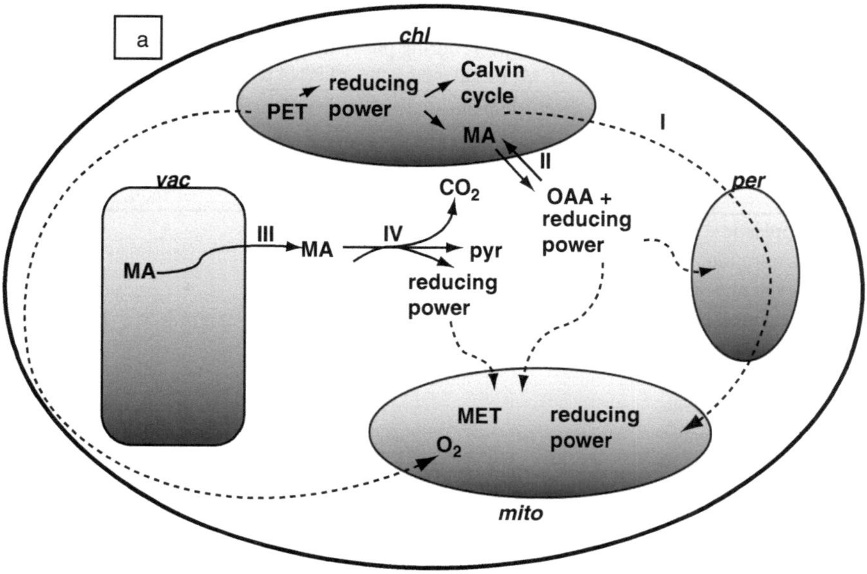

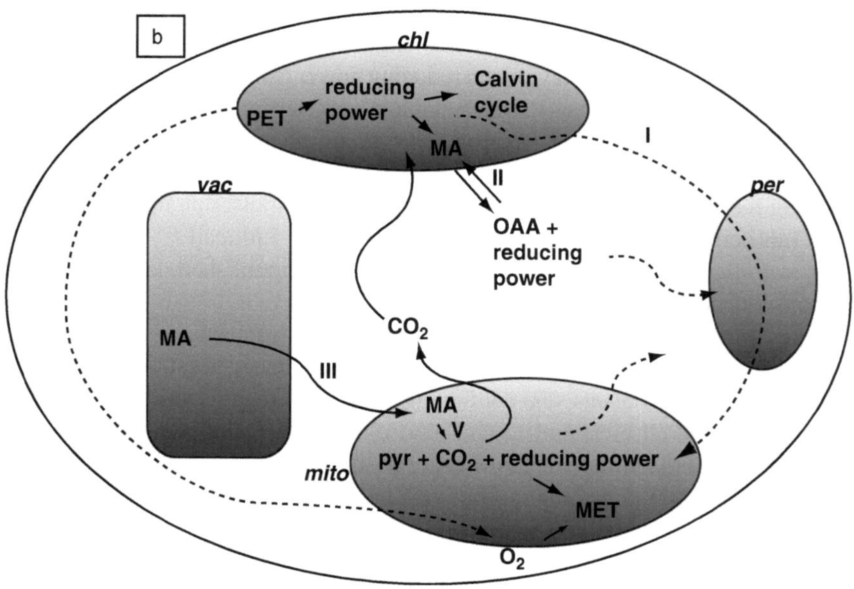

Fig. 1 Metabolic pathways, redox fluxes and involvement of mitochondrial metabolism during the decarboxylation phase III of CAM in NADP-ME plants (**a**) and in NAD-ME plants (**b**). Numbering of reactions/processes: *I* photorespiration, *II* malate valve, *III* passive efflux of malate from vacuole, *IV* NADP-dependent oxidative decarboxylation of malate, *V* NAD-dependent oxidative decarboxylation of malate. *Dashed lines* illustrate hypothetical routes. Abbreviations: *chl* chloroplasts, *mito* mitochondria, *vac* vacuole, *per* peroxisomes, *MA* malate, *OAA* oxaloacetate, *pyr* pyruvate, *PET* photosynthetic electron transport, *MET* mitochondrial electron transport

Table 1 A comparison of several processes, operating in the mesophyll cell, which may play a role in the optimization of photosynthesis during phase III of CAM, via supporting the carbon concentrating mechanism (CCM) or balancing the ATP/NADPH ratio

Process	Key elements underpinning photosynthetic efficiency during phase III of CAM		
	Net O_2 consumption	ATP production	Net consumption of reducing power
ChlToroplastic			
Photorespiration	+	–	–
Water–water cycle	–	+	+
Cyclic electron flow	–	+	–
Chlororespiration	+	–	+
Mitochondrial			
Cytochrome pathway	+	+	+
Alternative pathway	+	–	+

Raghavendra et al. 1998), this restriction could be by-passed by the activity of external dehydrogenase, located on the outer edge of the inner mitochondrial membrane (van Lis and Atteia 2004). In agreement with this supposition, mitochondria from CAM plants possess high activities of a rotenone-resistant NADH dehydrogenase that may accept reducing power from the *outer* mitochondrial space (Arron et al. 1979; Day 1980; Peckmann and Herppich (1998). The possible involvement of mitochondrial electron transport in photosynthesis of NADP-ME type CAM plants is illustrated in Fig. 1. However, an additional complication arises following recent observations of a stimulation in AOX transcript and protein abundance (an active monomeric form) after the salinity-induced C_3–CAM shift in *M. crystallinum* (A. Borland and E. Niewiadomska, unpublished data). This could be a consequence of increased oxidative load and/or serve as a 'safety valve' which facilitates the rapid engagement of AOX when CCM is exhausted and photorespiration is stimulated. In general, there is no conclusive evidence to indicate a definitive role for AOX in CAM plants, and conflicting data in this regard illustrate an absolute necessity to relate all biochemical and physiological measurements of respiratory parameters to the magnitude of CAM expression over the diel cycle. The influence that various oxygen-consuming processes have on the generation of reducing power is summarized in Table 1.

6 Conclusions

CAM appears to confer a degree of photoprotection under moderate levels of irradiance. However, at higher light intensities that may shorten the duration of the carbon concentrating phase III, other photoprotective strategies that include phase IV photorespiration assume greater importance. Since there is little evidence for oxidative stress/damage in CAM and/or during CAM induction, a protective re-adjustment of whole-cell redox homeostasis is likely to occur. We propose that there is a finely tuned interplay between: (a) the various processes that consume oxygen and reductant and (b) the enzymes and metabolites responsible for the scavenging of reactive oxygen species; and this interplay optimizes photosynthetic performance in line with the dynamic shifts in [CO_2] and [O_2] that occur over the diurnal phases of CAM. Whilst certain metabolic components required for CAM appear to be up-regulated in response to an increase in oxidative load, there is no conclusive evidence to indicate that CAM creates an increase in oxidative burden over and above that imposed by the potentially limiting conditions under which CAM species prevail. Given the inherent plasticity for modulating the magnitude and duration of the four phases of CO_2 exchange over the day/night cycle, CAM represents an under-exploited model system for dissecting the mechanisms that underpin redox homeostasis in a changing environment. Indeed, to reiterate Osmond (2006) in the previous volume in this series, 'why would one look beyond the CAM system to discover what really matters in oxidative stress'?

References

Adams WW III, Osmond CB (1988) Internal CO_2 supply during photosynthesis of sun and shade grown CAM plants in relation to photoinhibition. Plant Physiol 86:117–123

Agostino A, Heldt HW, Hatch MD (1996) Mitochondrial respiration in relation to photosynthetic C_4 acid decarboxylation in C_4 species. Aust J Plant Physiol 23:1–7

Arron GP, Spalding MH, Edwards GE (1979) Isolation and oxidative properties of intact mitochondria from the leaves of *Sedum praeltum*, a crassulacean acid metabolism plant. Plant Physiol 64:182–186

Asada K (2000) The water–water cycle as alternative photon and electron sinks. Phil Trans R Soc Lond B Biol Sci 355:1419–1431

Badger MR, Caemmerer S von, Ruuska S, Nakano H (2000) Electron flow to oxygen in higher plants and algae: rates and control of direct photoreduction (Mehler reaction) and rubisco oxygenase. Phil Trans R Soc Lond B Biol Sci 355:1433–1446

Barker DH, Marszalek J, Zimpfer JF, Adams WW III (2004) Changes in photosynthetic pigment composition and absorbed energy allocation during salt stress and CAM induction in *Mesembryanthemum crystallinum*. Funct Plant Biol 31:781–787

Bendall DS, Manasse RS (1995) Cyclic photophosphorylation and electron transport. Biochim Biophys Acta 1410:248–261

Björkman O, Demmig B (1987) Photon yield of O_2 evolution and chlorophyll fluorescence at 77K among plants of diverse origin. Planta 170:489–504

Borland AM, Griffiths H (1996) Variations in the phases of CAM and regulation of carboxylation patterns determined by carbon-isotope-discrimination techniques. In: Winter K, Smith JAC (eds) Crassulacean acid metabolism. Biochemistry, ecophysiology and evolution. Springer, Berlin Heidelberg NewYork, pp 230–249

Borland AM, Griffiths H (1997) A comparative study on the regulation of C_3 and C_4 carboxylation processes in the constitutive CAM plant *Kalanchoë daigremontiana* and the C_3–CAM intermediate *Clusia minor*. Planta 201:368–378

Borland AM, Taybi T (2004) Synchronization of metabolic processes in plants with crassulacean acid metabolism. J Exp Bot 55:1255–1265

Borland AM, Griffiths H, Broadmeadow MSJ, Fordham MC, Maxwell C (1993) Short-term changes in carbon isotope discrimination in the C_3–CAM intermediate *Clusia minor* L. growing in Trinidad. Oecologia 95:444–453

Borland AM, Hartwell J, Jenkins GI, Wilkins MB, Nimmo HG (1999) Metabolite control overrides circadian regulation of PEPc kinase and CO_2 fixation in crassulacean acid metabolism (CAM). Plant Physiol 121:889–896

Borland AM, Elliott S, Patterson S, Taybi T, Cushman J, Pater B, Barnes J (2006) Are the metabolic components of crassulacean acid metabolism up-regulated in response to an increase in oxidative burden? J Exp Bot 57:319–328

Boxall SE, Foster JM, Bohnert HJ, Cushman JC, Nimmo HG, Hartwell J (2005) Conservation and divergence of circadian clock operation in a stress-inducible crassulacean acid metabolism species reveals clock compensation against stress. Plant Physiol 137:969–982

Branco JB, Agarie S, Elliott S, Borland AM, Cushman JC (2003) Isolation and analysis of CAM-defective mutants in the common ice plant, *Mesembryanthemum crystallinum*. In: American Society of Plant Biologists (ed) ASPB Plant Biology 2003, ASPB, Honolulu, poster 403

Broetto F, Lüttge U, Ratajczak R (2002) Influence of light intensity and salt-treatment on mode of photosynthesis and enzymes of the antioxidative response system of *Mesembryanthemum crystallinum*. Funct Plant Biol 29:13–23

Castillo FJ (1996) Antioxidative protection in the inducible CAM plant *Sedum album* L. following the imposition of severe water stress and recovery. Oecologia 107:469–477

Chamnongpol S, Willekens H, Langebartels C, Van MontagueM, Inzé D, Van Camp W (1996) Transgenic tobacco with a reduced catalase activity develops necrotics lesions and induces pathogenesis-related expression under high light. Plant J 10:491–503

Chen L-S, Nose A (2003) Day–night changes of energy-rich compounds in crassulacean acid metabolism (CAM) species utilizing hexose and starch. Ann Bot 94:449–455

Cruz JA, Avenson TJ, Kanazawa A, Takizawa K, Edwards GE, Kramer DM (2005) Plasticity in light reactions of photosynthesis for energy production and photoprotection. J Exp Bot 56:395–406

Cushman JC, Bohnert HJ (1997) Molecular genetics of crassulacean acid metabolism. Plant Physiol 113:667–676

Cushman JC, Borland AM (2002) Induction of crassulacean acid metabolism by water limitation. Plant Cell Physiol 25:295–310

Dat JF, Pellinen R, Beckman T, Van de Cotte B, Langebartels C, Kangasjärvi J, Inzé D, Van Breusegem F (2003) Changes in hydrogen peroxide homeostasis trigger an active cell death process in tobacco. Plant J 33:621–632

Day DA (1980) Malate decarboxylation by *Kalanchoë daigremontiana* mitochondria and its role in crassulacean acid metabolism. Plant Physiol 65:675–679

Drake BG, Azcon-Bieto J, Berry J, Bunce J, Dijkstra P, Farrar J, Gifford RM, Gonzales-Meler MA, Koch G, Lambers H, Siedow J, Wullschleger S (1999) Does elevated atmospheric CO_2 concentration inhibit mitochondrial respiration in green plants? Plant Cell Environ 22:649–657

Duarte HM, Lüttge U (2007) Gas-exchange, photorespiration and spatio-temporal dynamics of relative quantum use efficiency in leaves of the C_3-photosynthesis/crassulacean acid metabolism-intermediate species *Clusia minor* L. (Clusiaceae) in both modes of photosynthesis. Trees (in press)

Edwards GE, Dai Z, Cheng SH, Ku MSB (1996) Factors affecting the induction of crassulacean acid metabolism in *Mesembryanthemum crystallinum*. In: Winter K, Smith JAC (eds) Crassulacean acid metabolism. Biochemistry, ecophysiology and evolution. Springer, Berlin Heidelberg NewYork, pp 119–134

Foyer CH, Noctor G (2005) Oxidant and antioxidant signalling in plants: a re-evaluation of the concept of oxidative stress in a physiological context. Plant Cell and Environ 28:1056–1071

Fridyland LE, Backhausen JE, Scheibe R (1998) Flux control of the malate valve in leaf cells. Arch Biochem Biophys 349:290–298

Fujita M, Fujita Y, Noutoshi Y, Takahashi F, Narusaka Y, Yamaguchi-Shinozaki K, Shinozaki K (2006) Crosstalk between abiotic and biotic stress responses: a current view from the points of convergence in the stress signaling networks. Curr Opin Plant Sci 4:436–442
Gil F (1986) Origin of CAM as an alternative photosynthetic carbon fixation pathway. Photosynthetica 29:494–507
Grams TEE, Haag-Kerwer A, Olivares E, Ball E, Arndt S, Popp M, Medina E, Lüttge U (1997) Comparative measurements of chlorophyll a fluorescence, acid accumulation and gas exchange in exposed and shaded plants of *Clusia minor* L and *Clusia multiflora* HBK in the field. Trees Struct Funct 11:240–247
Griffiths H (1989) Carbon dioxide concentrating mechanisms and the evolution of CAM in vascular epiphytes. In: Lüttge U (ed) Vascular plants as epiphytes. Springer, Berlin Heidelberg New York, pp 42–86
Griffiths H, Helliker B, Roberts A, Haslam RP, Girnus J, Robe WE, Borland AM, Maxwell K (2002) Regulation of Rubisco activity in crassulacean acid metabolism plants: better late than never. Funct Plant Biol 29:689–696
Halliwell B (2006) Reactive species and antioxidants, redox biology is a fundamental theme of aerobic life. Plant Physiol 141:312–322
Hanning I, Heldt HW (1993) On the function of mitochondrial metabolism during photosynthesis in spinach (*Spinacia oleracea* L.) leaves (partitioning between respiration and export of redox equivalents and precursors for nitrate assimilation products). Plant Physiol 103:1147–1154
Heber U (2002) Irrungen, Wirrungen? The Mehler reaction in relation to cyclic electron transport in C_3 plants. Photosynth Res 73:223–231
Heber U, Neimanis S, Kaiser WM (1996) Regulation of crassulacean acid metabolism in *Kalanchoë pinnata* as studied by gas exchange and measurements of chlorophyll fluorescence. In: Winter K, Smith JAC (eds) Crassulacean acid metabolism. Biochemistry, ecophysiology and evolution. Springer, Berlin Heidelberg NewYork, pp 78–96
Holtum JAM, Smith JAC, Neuhaus HE (2005) Intracellular transport and pathway of carbon flow in plants with crassulacean acid metabolism. Funct Plant Biol 32:429–449
Horton P, Ruban A (1994) The role of light-harvesting complex II in energy quenching. In: Baker NR, Bowyer JR (eds) Photoinhibition of photosynthesis. From molecular mechanisms to the field. Bios Scientific, London, pp 111–128
Horton P, Ruban A, Walters R (1996) Regulation of light harvesting in green plants. Annu Rev Plant Physiol Plant Mol Biol 47:655–684
Hurst AC, Grams TEE, Ratajczak R (2004) Effects of salinity, high irradiance, ozone, and ethylene on mode of photosynthesis, oxidative stress and oxidative damage in the C-3/CAM intermediate plant *Mesembryanthemum crystallinum* L. Plant Cell Environ 27:187–197
Igamberdiev AU, Bykova NV, Lea PJ, Gardeström P (2001) The role of photorespiration in redox and energy balance of photosynthetic plant cells: a study with a barley mutant deficient in glycine decarboxylase. Physiol Plant 111:427–438
Jensen RG (2000) Activation of Rubisco regulates photosynthesis at high temperature and CO_2. Proc Natl Acad Sci USA 97:12937–12938
Keiller DR, Slocombe SP, Cockburn W (1994) Analysis of chlorophyll-a fluoresence in C3 and CAM forms of *Mesembryanthemum crystallinum*. J Exp Bot 45:325–334
Kore-eda S, Cushman MA, Akselrod I, Bufford D, Fredrickson M, Clark E, Cushman JC (2004) Transcript profiling of salinity stress responses by large-scale expressed sequence tag analysis in *Mesembryanthemum crystallinum*. Gene 341:83–92
Köster S, Winter K (1985) Light scattering as an indicator of the energy state in leaves of the crassulacean acid metabolism plant *Kalanchoë pinnata*. Plant Physiol 79:520–524
Kozaki A, Takeba G (1996) Photorespiration protects C_3 plants from photooxidation. Nature 384:557–560
Krause GH, Weis E (1991) Chlorophyll fluorescence and photosynthesis. Annu Rev Plant Physiol Plant Mol Biol 42:313–349
Kubicki A, Funk E, Westhoff P, Steinmüller K (1996) Differential expression of plastome-encoded *ndh* genes in mesophyll and bundle-sheath chloroplasts of the C_4 plant *Sorghum bicolor*

indicates that the complex I-homologous NAD(P)H-plastoquinone oxidoreductase is involved in cyclic electron transport. Planta 199:276–281
Lis R van, Atteia A (2004) Control of mitochondrial function via photosynthetic redox signals. Photosynth Res 79:133–148
Lu C, Qiu N, Lu Q, Wang B, Kuang T (2003) PSII photochemistry, thermal energy dissipation, and the xanthophyll cycle in *Kalanchoë daigremontiana* exposed to a combination of water stress and high light. Physiol Plant 118:173–182
Lüttge U (2000) Light-stress and crassulacean acid metabolism. Phyton Annu Rev Bot 40:65–82
Lüttge U (2002) CO_2-concentrating: consequences in crassulacean acid metabolism. J Exp Bot 53:2131–2142
Lüttge U (2004) Ecophysiology of crassulacean acid metabolism (CAM). Ann Bot 93:629–652
Lüttge U (2007) Photosynthesis. In: Lüttge U (ed) Clusia. A woody neotropical genus of remarkable plasticity and diversity. (Ecological studies, vol 194) Springer, Berlin Heidelberg New York, pp 135–186
Mattos EA de, Herzog B, Lüttge U (1999) Chlorophyll fluorescence during CAM-phases in *Clusia minor* L. under stress. J Exp Bot 50:253–262
Maxwell K (2002) Resistance is useful: diurnal patterns of photosynthesis in C_3 and crassulacean acid metabolism epiphytic bromeliads. Funct Plant Biol 29:679–697
Maxwell K, Caemmerer S von, Evans JR (1997) Is a low internal conductance to CO_2 diffusion a consequence of succulence in plants with crassulacean acid metabolism? Aust J Plant Physiol 24:777–786
Maxwell K, Badger MR, Osmond CB (1998) A comparison of CO_2 and O_2 exchange patterns and the relationship with chlorophyll fluorescence during photosynthesis in C_3 and CAM plants. Aust J Plant Physiol 25:45–52
Maxwell DP, Wang Y, McIntosh L (1999a) The alternative oxidase lowers mitochondrial reactive oxygen production in plant cells. Proc Natl Acad Sci USA 96:8271–8276
Maxwell K, Borland AM, Haslam RP, Helliker BR, Roberts A, Griffiths H (1999b) Modulation of Rubisco activity during the diurnal phases of the crassulacean acid metabolism plant *Kalanchoe daigremontiana*. Plant Physiol 121:849–856
Millenaar FF, Lambers H (2003) The alternative oxidase: in vivo regulation and function. Plant Biol 5:2–15
Miszalski Z, Ślesak I, Niewiadomska E, Baczek-Kwinta R, Lüttge U, Ratajczak R (1998) Subcellular localization and stress response of superoxide dismutase isoform from leaves in the C_3–CAM intermediate halophyte *Mesembryanthemum crystallinum* L. Plant Cell Environ 21:169–179
Miszalski Z, Niewiadomska E, Ślesak I, Lüttge U, Kluge M, Ratajczak R (2001) The effect of irradiance on carboxylating/decarboxylating enzymes and fumarase activities in *Mesembryanthemum crystallinum* L. leaves exposed to salinity stress. Plant Biol 3:17–23
Miszalski Z, Kornaś A, Gawrośska K, Ślesak I, Niewiadomska E, Kruk J, Christian AL, Fisher-Schliebs E, Krisch R, Lüttge U (2006) Superoxide dismutase activity in C_3 and C_3/CAM intermediate species of *Clusia*. Biol Plant 51:86–92
Mittler R, Vanderauwera S, Gollery M, Breusegem F van (2004) Reactive oxygen gene network of plants. Trends Plant Sci 9:490–498
Müller P, Li X-P, Niyogi KK (2001) Non-photochemical quenching. A response to excess light energy. Plant Physiol 125:1558–1566
Niewiadomska E, Miszalski Z, Ślesak I, Ratajczak R (1999) Catalase activity during C_3–CAM transition in *Mesembryanthemum crystallinum* L. leaves. Free Radic Res 31:S251–S256
Niewiadomska E, Pater B, Miszalski Z (2002) Does ozone induce a C_3–CAM transition in *Mesembryanthemum crystallinum* leaves? Phyton 42:69–78
Niewiadomska E, Karpinska B, Romanowska E, Ślesak I, Karpinski S (2004) A salinity-induced C_3–CAM transition increases energy conservation in the halophyte *Mesembryanthemum crystallinum* L. Plant Cell Physiol 45:789–794
Nixon PJ, Mullineaux CW (2001) Regulation of photosynthetic electron transport. In: Aro E-M, Andersson B (eds) Regulation of photosynthesis. Kluwer Academic, Dordrecht, pp 533–555
Niyogi KK (2000) Safety valves for photosynthesis. Curr Opin Plant Biol 3:455–460

Nobel PS (1991) Achievable productivities of certain CAM plants: basis for high values compared with C_3 and C_4 plants. New Phytol 119:183–205
Noctor G, Foyer CH (1998) A reevaluation of the ATP:NADPH budget during C_3 photosynthesis: a contribution from nitrate assimilation and its associated respiratory activity. J Exp Bot 49:1895–1908
Noctor G, Dutilleul C, De Paepe R, Foyer CH (2004) Use of mitochondrial electron transport mutants to evaluate the effects of redox state on photosynthesis, stress tolerance and the integration of carbon/nitrogen metabolism. J Exp Bot 55:49–57
Noguchi K, Go C-S, Terashima I, Ueda S, Yoshinari T (2001) Activities of the cyanide-resistant respiratory pathway in leaves of sun and shade species. Aust J Plant Physiol 28:27–35
Osmond CB (1994) What is photoinhibition? Some insights from comparisons of shade and sun plants. In: Baker NR, Bowyer JR (eds) Photoinhibition of photosynthesis. From molecular mechanisms to the field. Bios Scientific, London, pp 1–24
Osmond CB (2006) Crassulacean acid metabolism: now and then. In: Esser K, Lüttge UE, Beyschlag W, Murata J (eds) Progress in botany, vol 68. Springer, Berlin Heidelberg New York, pp 3–34
Osmond CB, Grace SC (1995) Perspectives on photoinhibition and photorespiration in the field: quintessential inefficiencies of the light and dark reactions of photosynthesis. J Exp Bot 46:1351–1362
Osmond B, Maxwell K, Popp M, Robinson S (1999) On being thick: fathoming apparently futile pathways of photosynthesis and carbohydrate metabolism in succulent CAM plants In: Bryant JA, Burrell MM, Kruger NJ (eds) Plant carbohydrate biochemistry. Bios, Oxford, pp 183–200
Overmyer K, Brosché M, Kangasjärvi J (2003) Reactive oxygen species and hormonal control of cell death. Trends Plant Sci 8:337–342
Padmasree K, Padmavathi L, Raghavendra AS (2002) Essentiality of mitochondrial oxidative metabolism for photosynthesis: optimization of carbon assimilation and protection against photoinhibition. Crit Rev Biochem Mol Biol 37:71–119
Parida AK, Das AB (2005) Salt tolerance and salinity effects on plants: a review. Ecotoxicol Environ Saf 60:324–349
Pastori GM, Foyer CH (2002) Common components, networks, and pathways of cross-tolerance to stress. The central role of "redox" and abscisic acid-mediated controls. Plant Physiol 129:460–468
Peckmann K, Herppich WB (1998) Effects of short-term drought and rewatering on the activity of mitochondrial enzymes and the oxidative capacity of leaf mitochondria from a CAM plant, *Aptenia cordifolia*. J Plant Physiol 152:518–524
Peltier G, Cournac L (2002) Chlororespiration. Annu Rev Plant Biol 53:523–550
Pfannschmidt T (2003) Chloroplast redox signals: how photosynthesis controls its own genes. Trends Plant Sci 8:33–421
Pieters J, Tezara W, Herrera A (2003) Operation of the xanthophyll cycle and degradation of D1 protein in the inducible CAM plant, *Talinum triangulare*, under water deficit. Ann Bot 92:393–399
Portis AR (2003) Rubisco activase – Rubisco's catalytic chaperone. Photosynth Res 75:11–27
Portis AR, Salvucci ME (2002) The discovery of Rubisco activase – yet another story of serendipity. Photosynth Res 73:257–264
Raghavendra AS, Reumann S, Heldt HW (1998) Participation of mitochondrial metabolism in photorespiration. Plant Physiol 116:1333–1337
Robinson SA, Yakir D, Ribas-Carbo M, Giles L, Osmond CB, Siedow JN, Berry JA (1992) Measurements of engagement of cyanide-resistant respiration in the crassulacean acid metabolism plant *Kalanchoe daigremontiana* with the use of on-line oxygen isotope discrimination. Plant Physiol 100:1087–1091
Ruban AV, Wentworth M, Horton P (2001) Kinetic analysis of nonphotochemical quenching of chlorophyll fluorescence. 1. Isolated chloroplasts. Biochemistry 40:9896–9901
Rustin P, Queiroz-Claret C (1980) Changes in the oxidative properties of *Kalanchoë blossfeldiana* leaf mitochondria during development of crassulacean acid metabolism. Planta 164:415–422

Ruuska S, Badger M, Andrews T, Caemmerer S von (2000) Photosynthetic electron sinks in transgenic tobacco with reduced amounts of Rubisco: little evidence for significant Mehler reaction. J Exp Bot 51:357–368

Scheibe R (2004) Malate valves to balance cellular energy supply. Physiol Plant 120:21–26

Schöttler MA, Kirchhoff H, Siebke K, Weis E (2002) Metabolic control of photosynthetic electron transport in crassulacean acid metabolism-induced *Mesembryanthemum crystallinum*. Funct Plant Biol 29:697–705

Skillman JB, Winter K (1997) High photosynthetic capacity in a shade-tolerant crassulacean acid metabolism plant – implications for sunfleck use, nonphotochemical energy dissipation, and susceptibility to photoinhibition. Plant Physiol 113:441–450

Skillman JB, Garcia M, Virgo A, Winter K (2005) Growth irradiance effects on photosynthesis and growth in two co-occurring shade-tolerant neotropical perennials of contrasting photosynthetic pathways. Am J Bot 92:1811–1819

Ślesak I, Miszalski Z, Karpinska B, Niewiadomska E, Ratajczak R, Karpinski S (2002) Redox control of oxidative stress responses in the C_3–CAM intermediate plant *Mesembryanthemum crystallinum*. Plant Physiol Biochem 40:669–677

Ślesak I, Karpinska B, Surówka E, Miszalski Z, Karpinski S (2003) Redox changes in the chloroplast and hydrogen peroxide are essential for regulation of C_3–CAM transition and photooxidative stress responses in the facultative CAM plant *Mesembryanthemum crystallinum* L. Plant Cell Physiol 44:573–581

Spalding MH, Stumpf DK, Ku MSB, Burris RH, Edwards GE (1979) Crassulacean acid metabolism and diel variations of internal CO_2 and O_2 concentrations in *Sedum praeltum* DC. Aust J Plant Physiol 6:557–567

Streb P, Josse E-M, Gallouët E, Baptist F, Kuntz M, Cornic G (2005) Evidence for alternative electron sinks to photosynthetic carbon assimilation in the high mountain plant species *Ranunculus glacialis*. Plant Cell Environ 28:1123–1135

Surówka E, Karolewski P, Niewiadomska E, Libik M, Miszalski Z (2006) The antioxidative response of *Mesembryanthemum crystallinum* plants to exogenous SO_2 application. Plant Sci 172:76–82

Whiteside DG, Rogers WJ, Tobin AK (1991) Photorespiratory enzyme activities in C_3 and CAM forms of the facultative CAM plant, *Mesembryanthemum crystallinum* L. J Exp Bot 42:485–492

Willekens H, Chamnongpol S, Davey M, Schraudner M, Langebartels C, Van Montagu M, Inzé D, Van Camp W (1997) Catalase is a sink for H_2O_2 and is indispensable for stress defence in C_3 plants. EMBO J 16:4806–4816

Wingler A, Quick WP, Bungard RA, Bailey KJ, Lea PJ, Leegood RC (1999) The role of photorespiration during drought stress: an analysis utilizing barley mutants with reduced activities of photorespiratory enzymes. Plant Cell Environ 22:361–373

Winter K, Smith JAC (1996a) An introduction to crassulacean acid metabolism. Biochemical principles and ecophysiological diversity. In: Winter K, Smith JAC (eds) Crassulacean acid metabolism. Biochemistry, ecophysiology and evolution. Springer, Berlin Heidelberg NewYork, pp 1–13

Winter K, Smith JAC (1996b) Crassulacean acid metabolism: current status and perspectives. In: Winter K, Smith JAC (eds) Crassulacean acid metabolism. Biochemistry, ecophysiology and evolution. Springer, Berlin Heidelberg NewYork, pp 389–426

Zhu JK (2002) Salt and drought stress signal transduction in plants. Annu Rev Plant Biol 53:247–273

Cuscuta spp: "Parasitic Plants in the Spotlight of Plant Physiology, Economy and Ecology"

Markus Albert (✉), Xana Belastegui-Macadam, Marc Bleischwitz, and Ralf Kaldenhoff

Contents

Abstract *Cuscuta* spp represent a unique group of holoparasitic dicotyledonous plants which can infect nearly all dicotyledonous species. Lacking roots and leaves, these parasitic plants are completely dependent on nutrients, carbohydrates and water from host plants. The physical connection between parasite and host is mediated by specific organs, called haustoria, which connect the vascular tissues from both plants. Here we review the fascinating life cycle of *Cuscuta* spp and also the ecological aspects and problems related to *Cuscuta* spp infestations. For prevention purposes, different biotechnological approaches including hosts which show a resistance to *Cuscuta* spp are suggested.

1 Introduction

In general, plants are autotrophic, obtaining all their necessary resources from the environment. However, about 1% of angiosperms are parasitic and depend on other plants resources (Kuijt 1969; Atsatt 1983). Parasites are common in most plant communities throughout the world. Plant parasitism has arisen independently several times in the evolution of the angiosperms (Wrobel and Yoder 2001) and parasites are present in 16 families (Musselman and Press 1995). All parasitic plants have a common feature, which is the haustoria that serve to attach, invade and parasitise the host plant.

Markus Albert
ZMBP–AG Felix,University of Tübingen, Auf der Morgenstelle 5, D-72076 Tübingen, Germany
e-mail: markus.albert@zmbp.uni-tuebingen.de

U. Lüttge et al. (eds.), *Progress in Botany 69.*

The genus *Cuscuta*, also known as dodder, includes 170 parasitic species with a worldwide distribution (Dawson 1987). Members of the *Cuscuta* genus are holoparasitic, dicotyledonous plants that depend on nutrients, water and carbohydrates from their host plants (Dawson et al. 1994). *Cuscuta* spp lack roots or leaves and possess specific penetrating organs, the so-called haustoria, which are fully developed when a successful interaction between parasite and host is established. Then *Cuscuta* can propagate and grow rapidly, causing massive damage in the case of crop fields. In general the control of infected crops is difficult. Although several methods can be applied to control the infestation (see below), these are not always successful and lead to loss of the harvest. Besides the problems dodder may cause to crops, it represents an important and fascinating part of natural communities where it contributes to the ecological equilibrium like any other species.

Cuscuta spp are a point of interest for several different reasons. Ancient Indian and Chinese medicine uses this plant as an herbal remedy. Further, the parasite receives some attention because of its negative action on crops. Although some techniques to control growth of the parasite are available, the infection process itself is still not completely understood.

2 *Cuscuta* spp Life Cycle

2.1 Cuscuta spp Early Developmental Stages

Cuscuta is an annual plant producing seeds that germinate on or near the soil surface. The seeds have a hard coat which maintains their viability in the soil for many years. The seed dormancy is broken through soil microbial activity, weathering or grazing. Germination is influenced by temperature, with an optimum at 30 °C, but not by light. This influence is reported for *Cuscuta campestris* in a recent study which also points to seed dormancy as a main survival strategy which allows persistence in agro-ecosystems (Benvenuti et al. 2005). In contrast to other parasitic plants, the germination process of *Cuscuta* is independent of the presence of chemical compounds released by host plants (Dawson 1987). *Cuscuta* seed size is around 2 mm which makes it difficult to separate these from other crop seeds, like legumes. Dodder seeds have been spread worldwide by humans through contaminated equipment, clothes or crop seed. As *Cuscuta* seeds remain viable in soil for long periods once a seed bank is established, the control of the parasite becomes difficult. The seeds then germinate during the warm season and may get easily mixed with crop plants.

Right after *Cuscuta* seed germination, a root- and leafless stem of up to 7 cm emerges and grows. *Cuscuta* seedlings perform a rotational movement (circumnutation) in a counter-clockwise direction in order to find a host stem within a 6–8 cm distance. *Orobanche* and *Striga* are root parasite species which are able to somehow select among potential hosts (Kelly 1990, 1992; Pennings and Callaway 2002;

Koch et al. 2004) but the involved mechanisms are still not totally understood, although host-derived metabolites are known to trigger parasite growth and attachment to the host (Bouwmeester et al. 2003; Cook et al. 1966; Yoder 2001). Although *Cuscuta* seedlings appear to coil indiscriminately around any vertically elongated object (Dawson et al. 1994), the nastic movements that the parasite performs in order to reach or avoid a potential host before any stem penetration takes place (Kelly 1990, 1992) might be related to yet unknown chemical cues released by the hosts. Recent studies by Runyon et al. (2006) showed directed growth of *Cuscuta pentagona* to tomato plants and also identified several tomato volatile compounds which were able to attract the parasite's growth. Due to its almost complete lack of photosynthesis, the seedling will die if a suitable host is not found within a few days. Once *Cuscuta* is attached to the host plant, temperature and light determine its rate of growth (Allred and Tingey 1964; Dawson 1966). Any part of *Cuscuta* is in principle able to attach to a suitable host and generate a new infestation. *Cuscuta* spp flower from late spring through fall and each plant is able to produce several thousand seeds. From the produced seeds a small percentage will germinate the next year and the rest can remain dormant, yet viable, in the soil for more than 10 years.

The group of suitable hosts for *Cuscuta* spp include several cultivated and ornamental plants, as well as weeds and some woody species. Among crops, some of *Cuscuta*'s common hosts are different legumes, like alfalfa and clover, tomato, tobacco, potato, carrot and sugarbeet (Dawson et al. 1990; Lanini and Kogan 2005). Cranberry, blueberry, raspberry, coffee and citrus are some of the potential woody dicotyledonous hosts. A very small group of monocotyledonous species can be parasitised by *Cuscuta* spp, like for example onion. In general, grasses are not suitable hosts for *Cuscuta* spp plants. In relation to host selection, Runyon et al. (2006) showed the preferential growth towards tomato (dicotyledonous suitable host) as compared with wheat (monocotyledonous non-suitable host), related to the volatile compounds released by each species. When the connection between the parasite and the host is established, the seedling loses its connection to the soil and lives entirely on the host. The processes taking part in the development of the parasite–host connection were reported in detail by Vaughn (2002, 2003) and are described in the following pages. Under favourable conditions *Cuscuta* grows continuously and attaches to new host stems, spreading rapidly from plant to plant.

As stated before, the parasite is a leaf- and rootless plant and has very low or no photosynthesis. Some *Cuscuta* spp lack chlorophyll and therefore do not show photosynthesis. Others, like *Cuscuta reflexa* are capable of photosynthesis, but its plastids appear to have fewer thylakoids than normal photosynthetic plastids (Machado and Zetsche 1990). Data by Hibberd et al. (1998) show that, even though plastids of *C. reflexa* have fewer thylakoids and less chlorophyll, they are functional at the level of light harvesting and energy dissipation by both photochemical and non-photochemical quenching. This evidence is supported by the increased de-epoxidation state of the xanthophyll-cycle pigments on exposure to high light. The authors suggest that the ability of *C. reflexa* to photosynthesise allows the fixation of CO_2 lost from respiration, which contributes to the plant's carbon budget. Despite these data, the general contribution of photosynthesis to the overall *Cuscuta* metabolism is not well understood and, in any case, *Cuscuta* plants are reliant on carbohydrates withdrawn from the host plant (Hibberd et al. 1998; Ma chado and Zetsche 1990). Consequently *Cuscuta* seedlings are considered holoparasites: they coil around stems and leaves of hosts and develop an efficient connection with the host in order to survive.

2.2 Interaction of Cuscuta spp with Compatible Host Plants

The most important step in the *Cuscuta* spp life cycle is building up a connection to host vascular tissue. This is initiated right after attachment to a host stem. The development of prehaustoria starts at contact sites with the differentiation of a secondary meristem from parasitic epidermal and parenchymal cells (Heidejorgensen 1991). The attachment of the parasite is forced by adhesive substances such as pectins and related polysaccharides which are secreted by prehaustoria (Vaughn 2002). But the parasite does not only produce its own sticky substances, it also coerces the host plant to produce attachment-enhancing components. Albert et al. (2006) showed that, after *Cuscuta* attachment to tomato plants, the host plant synthesises an arabinogalactan protein, attAGP, which plays an important role for successful interaction. Expression of this attAGP is restricted to the penetration sites and it was demonstrated that the protein promotes the adherence of the parasite to the host plant.

After the attachment phase, the infection process continues with the penetration phase. The host tissue is invaded by the haustoria through a fissure in the host stem. The developing haustoria overcome epidermal and hypodermal tissue and grow inside the host plant to contact the xylem and phloem tubes. For penetration and host cell wall degradation, *Cuscuta* uses hydrolytic enzymes, such as methylesterases (Srivastava et al. 1994), pectinases (Albert et al., unpublished data) or loosening particles, which are complexes of lytic enzymes (Vaughn 2003). At the tip of the in-growing haustoria, searching hyphae trying to reach the vascular bundles can be observed. These parasitic cells form chimerical cell walls with the host cells and are conducted by plasmodesmata (Vaughn 2003). Thus, the fascinating case of a cytoplasmic connection between two higher organisms occurs. When searching hyphae find a sieve cell of the host, they start to grow around the cell like the fingers of a hand. Consequently, the searching hyphae are structurally differentiated while their cell surfaces are enlarged 20 times and more. In this way, the modified cells have an ambivalent character, functioning as sieve element and/or transfer cell (Dorr 1972). Birschwilks et al. (2006) demonstrated that there is a cytoplasmic connection between host phloem cells and *Cuscuta* spp haustorial cells in different *Cuscuta* spp–host plant relations by using phloem mobile GFP-fusion proteins and phloem mobile fluorescent dyes (Birschwilks et al. 2006; Haupt et al. 2001). Moreover it was proven that *Cuscuta* spp could function as a "bridge" between two plants, where substances and even viruses could be transported from one to another host plant (Bennett 1940; Heintz 1989; Kaminska and Korbin 1999). For xylem connection, the contact starts as a synchronised development of a parasitic and host cell of the xylem parenchyma. These two cells accrete and build up a continuous xylem tube from host to parasite.

Through the established connection, water, nitrogen compounds, assimilates and even proteins or plant viruses are transferred into the parasitic plant (Haupt et al. 2001; Hibberd et al. 1998; Jeschke et al. 1994a; Jeschke et al. 1994b; Machado et al. 1990). Data by Roney et al. (2007) showed that also phloem-mobile mRNAs were transferred from the tomato host plant to *Cuscuta pentagona* Engelm. The *Cuscuta* absorption system for host substances appears to be very efficient; and it was shown that, during host plant fruit development, *Cuscuta* competes for assimilates and acts as a much stronger sink than the fruit itself (Wolswinkel 1984). Under favourable conditions, *Cuscuta* grows continuously and attaches to new host stems, spreading rapidly from plant to plant.

3 Plant Defences Against *Cuscuta* spp

Although *Cuscuta* spp have a very broad host spectrum – they could infect nearly every herbaceous dicotyledonous plant – some plants remain which show an active defence mechanism and prevent successful penetration or development of the haustoria.

During the interaction of *Cuscuta* spp with the Malvaceae *Gossypium hirsutum* or *Hibiscus rosa sinensis*, parasitic haustoria were able to penetrate epidermal tissue but the infection process was interrupted by the formation of wound tissue by the host plant (Capderon et al. 1985). In this case, the parasitic haustorial cells and specially the searching hyphae die, the junction to host vascular tissue cannot be established and the parasite dies eventually.

The interaction between *C. reflexa* and some resistant tomato varieties is one of the best studied examples of incompatible plant parasite–host relation. Interestingly, the process is interrupted early within the attachment phase due to the active defence mechanism displayed by the tomato. The defence reaction is visible on tomato stems as necrotic plaques directly at the sites where parasitic haustoria try to penetrate the tomato epidermal tissue (Ihl et al. 1988; Sahm et al. 1995). The elongation of hypodermal cells is a very characteristic feature of the process, although it is unclear whether this is a necessary part of the resistance reaction. During the incompatible interaction, a microscopic approach showed that the haustoria cannot overcome the host's epidermal tissue due to the presence of a barrier which appears about 2 days after the parasite's onset. This barrier can be seen as a fluorescing secondary tissue under UV light or as a red-stained band after phloroglucine/HCl staining (Ihl et al. 1988). This protective tissue consists of phenolic and aliphatic compounds, as described for wound suberin (Bernards 2002). The aliphatic constituents are mainly di-fatty acids, longchain ω-hydroxy fatty acids or long chain alkan di-alcohols. In addition, components of the phenylpropanoid pathway and elements of the plant cell wall, such as pectins or wall proteins, are present and strongly linked with the above-described constitutes, which is also indicated by an increased peroxidase activity at the infection site (Sahm et al. 1995). These linked constituents build up an impenetrable network for *Cuscuta* haustoria and form an indigestible barrier for the parasite's secreted hydrolytic enzymes. The whole resistance reaction was described as a hypersensitive reaction (Ihl et al. 1988) which prevents the parasite from connecting to the host xylem and phloem. Consequently, *C. reflexa* dies of starvation.

First results from molecular biology studies gave interesting insights into tomato's defence strategies, although the basis of the signal perception and the key resistance proteins of tomato are not yet characterised. A subtracted tomato cDNA library made of mRNAs from infection sites was built up and differentially screened for genes which are induced during the resistance reaction (Albert et al. 2004; Werner et al. 2001). Two aquaporin genes, LeAQP2 and LeTRAMP (tomato ripening associated membrane protein), are induced at the infection site in tomato upon *C. reflexa* onset and also systemically (Werner et al. 2001). The authors suggested that the aquaporins could be involved in cell elongation at the infection site, but it is unclear whether they play a key role for the successful resistance reaction. The expression of several genes encoding cell wall proteins or cell wall-modifying proteins, such as expansins, prolinerich proteins, endo-β-1,4-glucanase or endotransglycosylases, was also observed and indicates that cell wall modifications play an important role for the tomato resistance reaction. The xyloglucan endotransglycosylase/hydrolase LeXTH1 was expressed and active in tomato at *C. reflexa* infection sites (Albert et al. 2004). Taking into account the observed increase in auxin levels at infection sites in tomato (Loffler et al. 1999), the effect of this hormone on LeXTH1 gene expression was analysed and an induction was verified. However, a parallel study on the expression of LeXTH1 in the auxin signalling mutant tomato *Diageotropica* also showed expression of the gene and the resistance reaction of tomato to *C. reflexa* was not influenced. The auxin level which increases at the infection site in tomato within 24 h seems to be responsible for cell elongation (Loffler et al. 1999) but not for LeXTH1 expression or for the resistance of tomato (Albert et al. 2004). To determine other processes of the resistance reaction, different tomato mutants or transgenic tomato plants were tested for their reaction after *Cuscuta* spp onset. These plants were mutants of the jasmonic acid pathway, the brassinosteroid pathway and the ethylene pathway. Moreover allenoxide-cyclase antisense and salicylate hydroxylase overexpressing transgenic tomato plants were tested in order to elucidate the resistance reaction. All tested tomatoes showed the same resistance reaction as wild-type plants, indicating that not all common signalling pathways related to wound response or systemic acquired resistance are important for the successful defence of tomato against *C. reflexa*. It is likely that a very specific resistance reaction occurs, according to the model of a gene-for-gene interaction or basal resistance. In this situation, a receptor protein can recognise a protein or another elicitor and then lead to the hypersensitive reaction at the tomato infection site. However, such a model would argue that this resistance reaction is specific for the *C. reflexa–Lycoperpersicon esculentum* interaction and not for the interaction of other *Cuscuta* species with tomato. This would explain why other closely related Solanaceae did not show a defence response, with the exception of *Solanum nigrum* and *Atropa bella donna*.

4 Ecology, Control and Uses of *Cuscuta* spp

Parasitic plants like *Cuscuta* spp are important components of natural ecosystems. *C. rostrata* for example was found in disturbed habitats where the forest canopy had opened. In this case, water and soil movement are probably important in seed spreading. Parasitic plants have a great influence on the natural communities they inhabit, mainly due to the effect they exert on the host. Because of reducing host performance, parasitic plants affect also the community structure, diversity and vegetation cycling (Pennings and Callaway 2002). Further, water and nutrient uptake from the host can have consequences for organisms like herbivores or pollinators. The impact of parasitic plants in natural communities is extensively reviewed by Press and Phoenix (2005).

As stated earlier, in spite of its fascinating physiology and scientific interest *Cuscuta* often causes problems in agriculture. *Cuscuta* spp are amongst the most damaging pests worldwide due to their broad distribution and host range (Parker 1991; Parker and Riches 1993) and they are included on the US Department of Agriculture's Top Ten Weeds List. *Cuscuta* spp parasitism is reported to reduce more than 50% tomato and carrot crop production (Bewick et al. 1988; Lanini 2004).

Nowadays, the available tools to control *Cuscuta* spp infestations are not fully effective and include the application of prevention methods, mechanical and/or hand removal and the use of resistant varieties or herbicides.

Preventive methods include certified crop seed use and the deep cleaning and revising of the agricultural machinery in contact with the field. A non-expensive and effective method is so-called crop rotation. Because *Cuscuta* spp do not parasitise members of the Gramineae, forage grasses and cereal grains, these species can be sown out alternatively to the main crop. All the *Cuscuta* spp seeds which germinate then die in that season due to the lack of a suitable host.

Once a field is invaded, in most cases radical methods like mowing or complete burning must be applied. Mowing was reported to increase yield by 32% in dodder-infected alfalfa patches in comparison with untreated alfalfa (Cudney et al. 1992). In contrast, burning reduces or totally destroys the cultivated crops and some *Cuscuta* seeds can still survive in the top layers of the soil.

The chemical control of *Cuscuta* spp has also been approached but until now, a specific herbicide for *Cuscuta* spp is unavailable. Nadler-Hassar et al. (2004) reported that the application of the herbicide glyphosate reduced *C. campestris* growth without severely affecting the host plant development. But this effect on *Cuscuta* is due to reduced nutrients uptake from the host and not to EPSPS (5-enolpyruvylshikimate-3-phosphate synthase) inhibition (Nadler-Hassar et al. 2004). However the possibility that glyphosate influences and damages the crop plant must be taken into account. The use of contact herbicides, like paraquat which is a non-selective product, is useful for erasing scattered patches of *Cuscuta* spp in crop fields but the application also reduces yield. Other herbicides, like chlorpropham, can be useful when *Cuscuta* spp seedlings emerge, resulting in a lower effect on the crop itself. Despite all the available herbicides, little is known about their actual effect on both *Cuscuta* and the host plant and therefore more studies on this subject would be desirable.

Cuscuta plants have also some positive features with regard to practical application. For example, some species of *Cuscuta* are used to produce dyes (Dawson et al. 1994) and both the plant and the seeds are used in Chinese and Indian medicine. In addition, *Cuscuta* is used in research studies as a tool to transfer pathogens from plant to plant (Bennet 1944; Johnson 1941).

Recently, the use of *C. campestris* as a biological control agent was analysed (Lian et al. 2006). Lian and coworkers (2006) demonstrated a positive result of the use of *C. campestris* for control of the parasitic vine *Mikania micrantha*, which is a serious problem in Southern China landscapes. *C. campestris* parasitism leads to a decrease in the *M. micrantha* biomass and to a significant increase in the biodiversity of the studied area.

5 Future Prospects and Molecular Biological Solutions for Crop Protection

Many aspects of the physiology, infection process and control of *Cuscuta* remain unclear and poorly studied. A closer look on the enzymatic and molecular biological processes taking place during the development and infection process of *Cuscuta* is essential in order to improve the available techniques to control the parasite.

The biotechnological or genetic control of *Cuscuta* spp needs a better understanding of the infection process on a molecular level. Research on compatible and incompatible interactions of *Cuscuta* spp with different hosts is therefore of key importance. As said before, Albert et al. (2006) showed that a decrease in the expression of the attAGP gene via RNAi or virus-induced gene silencing reduces the attachment of *C. reflexa* to tomato stem. Attachment of *Cuscuta* represents the initial groundbreaking step and is essential for the success of the infection process. Thus, any successful attempt to stop this stage of the infection process will avoid crop infestation. The described gene technological approach shows a chance to prevent *Cuscuta* spp infestation in cultivated crops.

Studying the expression pattern of *C. reflexa* genes during the infection process and haustoria formation, Ralf Kaldenhoff (Darmstadt) and co-workers identified several genes whose expression was up-regulated. Further analysis of corresponding proteins would help to discover the function and role for an infection process and might lead to approaches for successful crop protection from *Cuscuta* infestation. Genes which are up-regulated during haustoria formation include hydrolytic enzymes, such as pectinase, cellulase and proteases. For several parasitic organisms, cysteine proteases (CPs) are described as an important virulence factor. The amoeba *Entamoeba histolytica* expresses a cysteine protease when it attacks a compatible host. Inhibition of the protease results in a loss of successful parasitism (Moncada et al. 2006). Preliminary studies on *C. reflexa* obtained by Ralf Kaldenhoff (Darmstadt) and co-workers showed the involvement of a cysteine protease expressed during the infection process. The cysteine protease is translated as an inactive precursor protein, which consists of a pre- and a propeptide and the mature enzyme. The propeptide inhibits the enzyme when it is translocated to the site of action. These "proinhibitors" seem to be very powerful and specific, as shown by Fox et al. (1992). Further studies should verify the putative use of the propeptide as an inhibitor of the infection process.

Another efficient way to protect plants from parasitism is to obtain transgenic plants which express receptor genes which could trigger a plant resistance response. However, until now none of these receptors has been identified. Several lines of research can be pursued in order to obtain useful information on *Cuscuta* physiology, development and strategy to infect. Resolving these fascinating secrets will finally lead to solutions for successful crop protection.

References

Albert M, Werner M, Fry SC, Proksch P, Kaldenhoff R (2004) The cell wall-modifying xyloglucan endotransglycosylase/hydrolase LeXTH1 is expressed during the defence reaction of tomato against the plant parasite *Cuscuta reflexa*. Plant Biology 6;402–407

Albert M, Belastegui-Macadam X, Kaldenhoff R (2006) An attack of the plant parasite *Cuscuta reflexa* induces the expression of attAGP, an attachment protein of the host tomato. Plant J 48:548–556

Allred KR, Tingey DC (1964) Germination and spring emergence of dodder as influenced by temperature. Weeds 12:45–48

Atsatt PR (1983) Host–parasite interactions in higher plants. In: Lange OL, Nobel PD, Osmond CB, Ziegler H (eds) Response to the chemical and biological environment. (Encyclopedia of plant physiology. Physiological plant ecology III) Springer, Berlin Heidelberg New York, pp 519–535

Bennett C (1940) Acquisition and transmission of viruses by dodder (*Cuscuta subinclusa*). Phytopathology 30:2–11

Bennet C (1944) Studies of dodder transmission of plant viruses. Phytophatology 34:905–932

Benvenuti S, Dinelli G, Bonetti A, Catizone P (2005) Germination ecology, emergence and host detection in *Cuscuta campestris*. Weed Res 45:270–278

Bernards MA (2002) Demystifying suberin. Can J Bot 80:227–240

Bewick TA, Binning LK, Dana MN (1988) Postattachment control of swamp dodder (*Cuscuta gronovii*) in cranberry (*Vaccinium macrocarpon*) and carrot (*Daucus carota*). Weed Technol 2:166–169

Birschwilks M, Haupt S, Hofius D, Neumann S (2006) Transfer of phloem-mobile substances from the host plants to the holoparasite *Cuscuta* sp. J Exp Bot 57:911–921

Bouwmeester HJ, Matusova R, Sun ZK, Beale MH (2003) Secondary metabolite signalling in host–parasitic plant interactions. Curr Opin Plant Biol 6:358–364

Capderon M, Fer A, Ozenda P (1985) About an unreported system leading to the expulsion of a parasite – *Cuscuta* on cotton-plant (*Cuscuta lupuliformis* Krock on *Gossypium hirsutum* L). C R Acad Sci Ser III 300:227–232

Cook CE, Whichard LP, Turner B, Wall ME (1966) Germination of witchweed (*Striga lutea* Lour) – isolation and properties of a potent stimulant. Science 154:1189–1190

Cudney DW, Orloff SB, Reints JS (1992) An integrated weed management procedure for the control of dodder (*Cuscuta indecora*) in alfalfa (*Medicago sativa*). Weed Technol 6:603–606

Dawson JH (1966) Response of field dodder to shade. Weeds 14:4–5

Dawson JH (1987) *Cuscuta* (Convolvulaceae) and its control. in: Weber HC, Forstreuter W (eds) Parasitic flowering plants. Philipps-Universität, Marburg, pp 137–149

Dawson JH (1994) Biology and control of *Cuscuta*. Rev Weed Sci 6:265–317

Dawson JH, Musselman, LJ, Wolswinkel P, Dorr I (1990) Biology and control of *Cuscuta*. Weed Sci 6:265–317
Dorr I (1972) Contact of *Cuscuta* hyphae with sieve tubes of its host plants. Protoplasma 75:167–187
Fox T, De Miguel E, Mort JS, Storer AC (1992) Potent slow-binding inhibition of cathepsin B by its propeptide. Biochemistry 31:12571–12576
Haupt S, Oparka KJ, Sauer N, Neumann S (2001) Macromolecular trafficking between *Nicotiana tabacum* and the holoparasite *Cuscuta reflexa*. J Exp Bot 52:173–177
Heidejorgensen HS (1991) Cuticle development and ultrastructure – evidence for a procuticle of high osmium affinity. Planta 183:511–519
Heintz W (1989) Transmission of a new mycoplasma-like organism (mlo) from *Cuscuta odorata* (Ruiz et Pav) to herbaceous plants and attempts to its elimination in the vector. J Phytopathol 125:171–186
Hibberd JM, BungardRA, Press MC, Jeschke WD, Scholes JD, Quick WP (1998) Localization of photosynthetic metabolism in the parasitic angiosperm *Cuscuta reflexa*. Planta 205:506–513
Ihl B, Tutakhil N, Hagen A, Jacob F (1988) Studies on *Cuscuta reflexa* Roxb. 7. Defense mechanisms of *Lycopersicon esculentum* Mill. Flora 181:383–393
Jeschke WD, Rath N, Baumel P, Czygan FC, Proksch P (1994a) Modeling of the flows and partitioning of carbon and nitrogen in the holoparasite *Cuscuta reflexa* Roxb. and its host *Lupinus albus* L. 2. Flows between host and parasite and within the parasitized host. J Exp Bot 45:801–812
Jeschke WD, Rath N, Baumel P, Czygan FC, Proksch P (1994b) Modeling the flow and partitioning of carbon and nitrogen in the holoparasite *Cuscuta reflexa* Roxb. and its host *Lupinus albus* L. 1. Methods for estimating net flows. J Exp Bot 45:791–800
Johnson F (1941) Transmission of plant viruses by dodder. Phytophatology 31:649–656
Kaminska M, Korbin M (1999) Graft and dodder transmission of phytoplasma affecting lily to experimental hosts. Acta Physiol Plant 21:21–26
Kelly CK (1990) Plant foraging – a marginal value model and coiling response in *Cuscuta subinclusa*. Ecology 71:1916–1925
Kelly CK (1992) Resource choice in *Cuscuta europaea*. Proc Natl Acad Sci USA 89:12194–12197
Koch AM, Binder C, Sanders IR (2004) Does the generalist parasitic plant *Cuscuta campestris* selectively forage in heterogeneous plant communities? New Phytol 162:147–155
Kuijt J (1969) Observations on *Wallrothiella arceuthobii*, a fungus parasite of dwarf mistletoes. Can J Bot 47:1359–1360
Lanini WT (2004) Economical methods of controlling dodder in tomatoes. Proc Calif Weed Sci Soc 56:57–59
Lanini WT, Kogan M (2005) Biology and management of *Cuscuta* in crops. Cienc Invest Agr 32:127–141
Lian J, Ye WH, Cao HL, Lai ZM, Wang ZM, Cai CX (2006) Influence of obligate parasite *Cuscuta campestris* on the community of its host *Mikania micrantha*. Weed Res 46:441–443
Loffler C, Czygan FC, Proksch P (1999) Role of indole-3-acetic acid in the interaction of the phanerogamic parasite *Cuscuta* and host plants. Plant Biol 1:613–617
Machado MA, Zetsche K (1990) A structural, functional and molecular analysis of plastids of the holoparasites *Cuscuta reflexa* and *Cuscuta europaea*. Planta 181:91–96
Moncada D, Keller K, Ankri S, Mirelman D, Chadee K (2006) Antisense inhibition of *Entamoeba histolytica* cysteine proteases inhibits colonic mucus degradation. Gastroenterology 130:721–730
Musselman LJ, Press MC (1995) Introduction to parasitic plants. In: M.G.J. Press (ed) Parasitic plants. Chapman and Hall, London, pp 1–13
Nadler-Hassar T, Goldshmidt A, Rubin B, Wolf S (2004) Glyphosate inhibits the translocation of green fluorescent protein and sucrose from a transgenic tobacco host to *Cuscuta campestris* Yunk. Planta 219:790–796
Parker C (1991) Protection of crops against parasitic weeds. Crop Prot 10:6–22

Parker C, Riches C (1993) Parasitic weeds of the world: biology and control. CABI, Wallingford
Pennings SC, Callaway RM (2002) Parasitic plants: parallels and contrasts with herbivores. Oecologia 131:479–489
Press MC, Phoenix GK (2005) Impacts of parasitic plants on natural communities. New Phytol 166:737–751
Roney JK, Khatibi PA, Westwood JH (2007) Cross-species translocation of mRNA from host plants into the parasitic plant dodder. Plant Physiol 143:1037–1043
Runyon JB, Mescher MC, De Moraes CM (2006) Volatile chemical cues guide host location and host selection by parasitic plants. Science 313:1964–1967
Sahm A, Pfanz H, Grunsfelder M, Czygan FC, Proksch P (1995) Anatomy and phenylpropanoid metabolism in the incompatible interaction of *Lycopersicon esculentum* and *Cuscuta reflexa*. Bot Acta 108:358–364
Srivastava S, Nighojkar A, Kumar A (1994) Multiple forms of pectin methylesterase from *Cuscuta reflexa* filaments. Phytochemistry 37:1233–1236
Vaughn KC (2002) Attachment of the parasitic weed dodder to the host. Protoplasma 219:227–237
Vaughn KC (2003) Dodder hyphae invade the host: a structural and immunocytochemical characterization. Protoplasma 220:189–200
Werner M, Uehlein N, Proksch P, Kaldenhoff R (2001) Characterization of two tomato aquaporins and expression during the incompatible interaction of tomato with the plant parasite *Cuscuta reflexa*. Planta 213:550–555
Wolswinkel P (1984) Phloem unloading and sink strength – the parallel between the site of attachment of *Cuscuta* and developing legume seeds. Plant Growth Regul 2:309–317
Wrobel RL, Yoder JI (2001) Differential RNA expression of alpha-expansin gene family members in the parasitic angiosperm *Triphysaria versicolor* (Scrophulariaceae). Gene 266:85–93
Yoder JI (2001) Host-plant recognition by parasitic Scrophulariaceae. Curr Opin Plant Biol 4:359–365

Ecology

Bayesian Data–Model Integration in Plant Physiological and Ecosystem Ecology

Kiona Ogle(✉) and Jarrett J. Barber

Contents

Abstract This paper reviews and illustrates the use of modern methods for integrating diverse data sources with process-based models for learning about plant physiological and ecosystem processes. The particular focus is on how such data sources and models can be coupled within a hierarchical Bayesian modeling framework. This framework, however, has been underutilized in physiological and ecosystem ecology, despite its great potential for data–model integration in these areas. This paper provides a summary of the use of Bayesian methods in ecological research and gives detailed examples highlighting existing and potential uses of Bayesian and hierarchical Bayesian methods in plant physiological and ecosystem ecology. This paper also provides an overview of the statistical theory underlying the development of hierarchical Bayesian methods for analyzing complex ecological problems.

Kiona Ogle
Department of Botany and Department of Statistics, Department of Botany, Box 3165, University of Wyoming 1000E. University Ave. Laramie, WY 82071, USA
e-mail: kogle@uwyo.edu

U. Lüttge et al. (eds.), *Progress in Botany 69.*

The methods are applied to specific examples that include a detailed illustration of a hierarchical Bayesian analysis of leaf-level gas exchange data that are integrated with models of photosynthesis and stomatal conductance, and Bayesian approaches to estimating parameters in complex ecosystem simulation models. The paper concludes with some practical issues and thoughts on the direction of hierarchical Bayesian modeling in plant physiological and ecosystem ecology.

1 Introduction: Data–Model Synthesis

This paper is a review and prospectus of modern methods for integrating diverse data sources with process-based models for learning about plant physiological and ecosystem processes. In particular, we discuss how such data sources and models can be coupled within a hierarchical Bayesian modeling framework (Wikle 2003b; Clark 2005; Clark et al. 2005). This framework, however, has been underutilized in physiological and ecosystem ecology (Ellison 2004), despite its great potential for data–model integration in these areas. We summarize the use of Bayesian methods in ecological research and provide detailed examples highlighting uses of Bayesian and hierarchical Bayesian methods in plant physiological and ecosystem ecology. In the process, we provide a sort of tutorial on the use of hierarchical Bayesian methods for analyzing complex ecological problems. We conclude with some practical issues and thoughts on the direction of hierarchical Bayesian modeling in plant physiological and ecosystem ecology.

1.1 The Data

Plant physiological and ecosystem ecologists are proficient at collecting enormous amounts of data. This proficiency is partly due to advances in automated data collection devices (e.g. weather stations, soil moisture probes, eddy flux towers, continuous soil flux chambers, tunable diode lasers, sap flow sensors) that record at frequent intervals (e.g. every 30 min for multiple days, months, or years), and to sophisticated and expensive tools allowing hands-on field measurements (e.g. portable leaf-level gas exchange analyzers, stable isotopes, minirhizotrons). These diverse data represent ecological or physical processes that span different spatial scales (e.g. eddy flux and whole-ecosystem gas exchange vs soil chambers for measuring soil respiration), different temporal scales (e.g. carbon isotopes in leaf tissue and integrated water use efficiency vs leaf-level gas exchange and instantaneous water use efficiency), and different levels of biological organization (e.g. leaf temperature images from infrared cameras to infer patchy stomatal conductance vs sapflux and plant- or canopy-level transpiration). In addition, there is a plethora of information that already exists in the literature. How does one deal with such data? How can we extract information from different types of data to make inferences about complex, interrelated mechanisms governing plant and ecosystem processes?

1.2 The Process Model

Process models traditionally provide the means by which mechanisms operating at different temporal, spatial, and biological scales are formalized and explicitly linked. Development of a process model (or models) forces us to quantify our understanding of ecological systems, thus improving our heuristic and predictive understanding of plant physiological and ecosystem responses (e.g. Levin 1992; Canadell et al. 2000; Rastetter et al. 2003). Process models have been developed to describe and predict an array of physiological and ecosystem behaviors. Such models include, for example, the "Farquhar" biochemical-based model of leaf-level photosynthesis (Farquhar et al. 1980; Farquhar and Sharkey 1982; Farquhar and von Caemmerer 1982), the canopy-level photosynthesis and evaptranspiration model PnET (Aber and Federer 1992; Aber et al. 1996), and the plant–soil nutrient cycling model CENTURY (Parton et al. 1987). Components of such process models, including functional forms and parameter values, are often derived from empirical information, and these models are frequently used to learn about the real system through simulation experiments. However, a model of even moderate complexity is rarely fit to observed field or experimental data in a rigorous manner. Instead, data summaries (typically means) may be used for parameter values in a model. Or, a model may be "tuned" – parameters adjusted in a relatively ad-hoc manner – so that it fits a particular dataset or, more often, a summary thereof. Such approaches are often referred to as "forward modeling" where the primary focus is on understanding and predicting model and system behavior via simulation – running the model "forward" given a set of parameter values – with comparatively little emphasis on quantifying parameter uncertainty, or more generally, on rigorous data–model integration.

1.3 Data–Model Integration

So, we have varied and large datasets that are generated by complex, interrelated ecological processes, and we have process models designed specifically to learn about these complexities. When faced with several types of data that may inform the same process (or a set of processes), the standard approach is to analyze data in a piece-wise fashion – each dataset analyzed independently of others, largely ignoring the interconnectedness of the data. Instead of a separate analysis approach or a relatively ad-hoc tuning approach to process simulation, we argue a need for a more rigorous method that allows all data and prior knowledge to inform process parameters without requiring undue process model simplifications that could compromise the integrity of the inherent complexity of the ecological system embodied by the model. Ultimately, this data–model integration should allow exploration of ecological questions that are otherwise difficult to address via traditional approaches.

Recent developments in computing technology and Bayesian statistical methodology now allow for such integration of complex process models with large and diverse datasets within a unified statistical inferential framework. The Bayesian approach is emerging as a useful tool in population and community ecology (Clark 2003; Ellison 2004; Clark 2005; Clark and Gelfand 2006), but it has received comparatively little attention in plant physiological and ecosystem ecology (e.g. Ellison 2004; Van Oijen et al. 2005). The perspective of this paper is similar to that of Clark and Gelfand (2006), our intention here being to raise awareness of hierarchical Bayesian methods for coupling process models and data in plant physiological and ecosystem ecology.

In Section 2, we give a brief review of existing applications of Bayesian and hierarchical Bayesian methodologies in plant physiological and ecosystem ecology and closely related fields. Section 3 presents basic probability results, including hierarchical probability model specification and a statement of Bayes theorem, that are fundamental to hierarchical Bayesian modeling of complex processes. Section 4 presents detailed examples that illustrate how these results may be applied to relatively complex data–model systems in plant physiological and ecosystem ecology. In Section 5, we conclude with some practical suggestions and offer a few remarks on the future of hierarchical Bayesian modeling in these fields.

2 Bayesian Applications in Plant and Ecosystem Ecology

Bayesian statistical methods are relatively new to ecologists (Ellison 2004; Clark 2005). Ellison (2004) surveyed articles published in major ecological journals between 1996 and 2003 and found only 62 articles that used Bayesian data analysis tools. The ecological topics of these papers centered on population and community ecology, and plant physiological and ecosystem studies lagged behind in their use of Bayesian methods. Given the recent rapid increase in Bayesian applications in ecology, we expanded the scope of Ellison (2004) to include top plant physiological and ecosystem journals for the period 1996 to 2006. The results from our survey are summarized in Table 1 and Fig. 1.

Bayesian applications in population and community ecology continue to grow (Fig. 1), with only 20–25% of these studies addressing problems of a botanical nature. It should be noted that Bayesian inference in phylogenetics (Yang and Rannala 1997; Larget and Simon 1999; Mau et al. 1999; Huelsenbeck and Ronquist 2001) has become popular due to statistical and computational demands associated with analyzing enormous molecular datasets within the context of complex evolutionary models (e.g. Huelsenbeck et al. 2001). A large number of papers – 58 for the search criteria in Table 1 and 1411 for unrestricted dates and journals – employed Bayesian methods to construct phylogenies to infer, for example, evolutionary relationships among species (e.g. Miller et al. 2004; McKown et al. 2005). Despite a comparable level of complexity in plant physiological and ecosystem ecology, Bayesian methods remain underutilized in these areas (Fig. 1, Table 1). Once the plant physiological and ecosystem community becomes more aware of these methods, we anticipate a rapid increase in the application of Bayesian methodologies.

There are, however, a few notable examples of Bayesian applications in plant physiological and ecosystem ecology and closely related fields. For example, global

Table 1 Survey of ecological studies using Bayesian methods published between 1996 and 2006 in the following major ecological journals: *American Journal of Botany*, *American Naturalist*, *Conservation Biology*, *Ecological Applications*, *Ecological Monographs*, *Ecology*, *Ecology Letters*, *Ecosystems*, *Functional Ecology*, *Global Change Biology*, *Journal of Applied Ecology*, *Journal of Animal Ecology*, *Journal of Ecology*, *New Phytologist*, *Oecologia*, *Oikos*, *Plant Cell and Environment*, and *Tree Physiology*. See Ellison (2004) for a similar survey, based on a subset of these journals, for papers published between 1996 and 2003. *Note*: we used the ISI Web of Science search engine to find papers in the focal journals with "Bayes" or "Bayesian" in the title, abstract, or subject. Thus, this is not a comprehensive survey of all ecology-related papers using Bayesian methods, as there are surely other papers in journals that we did not consider and there are likely papers using Bayesian methods that do not have the term in the title, abstract, or subject. Papers related to population ecology were split into two categories, those focusing on population-related and life-history aspects of single species and those focusing on interacting species (e.g. predator–prey dynamics). Our rules for classifying articles may have differed from Ellison (2004), and thus our numbers are not in perfect agreement with Ellison's. We also tallied papers that presented a problem with a clear animal or plant focus. The numbers of plant and animal papers do not always add up to the total number of papers because some papers were classified as both (e.g. plant–animal interactions), and some were not classified as either (e.g. they may have focused on, for example, soil biogeochemistry). A full bibliography of these studies can be obtained from the authors

Topic	Number of papers (1996–2006)	Primary focus		Selected examples
		Plant	Animal	
Population ecology (single species)	73	10	63	O'Hara et al. (2002), Clark et al. (2003a, 2005), Wikle (2003a)
Population ecology (interacting species)	12	3	11	Dixon et al. (2005), Ovaskainen and Laine (2006)
Community ecology	24	12	14	Crome et al. (1996), Cottingham and Schindler (2000), Etienne and Olff (2005)
Physiological ecology	1	1	0	Ogle et al. (2004)
Ecosystem ecology	9	6	1	Carpenter et al. (1996), Rains et al. (2004), Braswell et al. (2005)
Review/comment	9			Ellison (1996, 2004), Clark (2005)
Total	128	32	89	

change scientists have embraced Bayesian inversion methods that couple various CO_2 flux and concentration data with atmospheric transport and/or terrestrial biosphere models to predict regional, continental, and global sources and sinks of atmospheric CO_2 (Enting et al. 1995; Dargaville et al. 2002; Law et al. 2004). Bayesian applications are surfacing in smaller-scale studies related to plant physiological ecology (e.g. Ogle et al. 2004) and ecosystem carbon, water, and nutrient dynamics (e.g. Braswell et al. 2005; Hong et al. 2005), and the ecosystem-related studies in Section 4.2 serve as good examples of how to incorporate informative prior information. Examples of Bayesian applications in ecosystem-level studies include: partitioning of ecosystem carbon fluxes into photosynthesis and respiration (Ogee et al. 2004), inverting a carbon cycle model to infer soil carbon loss and transfer between different litter, microbial, and soil organic matter pools (Xu et al. 2006), and parameterizing process-based models of tree growth and forest productivity

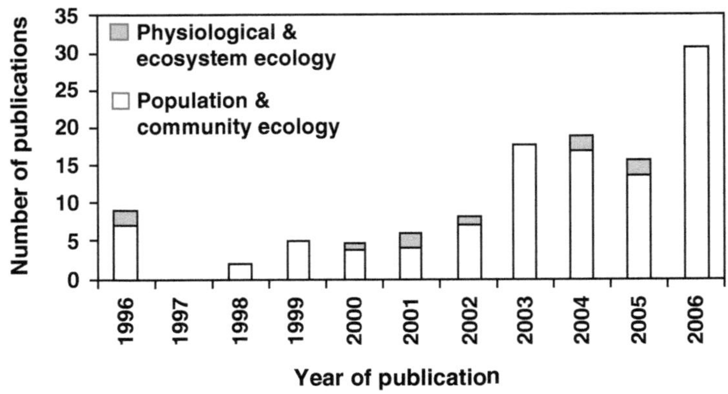

Fig. 1 Number of publications related to physiological and ecosystem ecology (*gray bars*) vs population and community ecology (*open bars*) that used Bayesian methods during the period 1996–2006. Details of the literature survey are given in Table 1 and discussed in Section 2.

(Gertner et al. 1999; Radtke et al. 2002; Van Oijen et al. 2005). These studies and those listed in Table 1 are prime examples of how Bayesian statistical methods can be used to integrate complex ecological models and data related to plant physiological and ecosystem processes.

3 Overview of the Bayesian Approach

In this section, we give a necessarily broad overview of hierarchical Bayesian modeling. Our intent is not to extol the virtues or to criticize the weaknesses of statistical paradigms, Bayesian or otherwise – we would have little more to offer to the philosophical debates among statistical schools of thought. Instead, we attempt to focus the discussion toward current developments of hierarchical statistical modeling of complex ecological systems. We start by presenting basic probability rules that are useful for a subsequent discussion of hierarchical probability modeling and for a statement of Bayes theorem. We conclude this section with what is loosely becoming known as the "process sandwich", a largely conceptual way to view the incorporation of process models into a hierarchical probability framework.

3.1 Hierarchical Probability Models

We begin with some basic probability rules. Let $\mathbf{y} = (y_1, y_2, \ldots, y_n)'$ denote a vector of n random variables, with $p(\mathbf{y})$ denoting a mathematical formula describing their probabilistic behavior. That is, $\mathbf{y}$ is a random (vector) variable with (joint probability) distribution $p(\mathbf{y})$. For example, $\mathbf{y}$ may be a collection of n observed leaf transpiration rate values, and $p(\mathbf{y})$ may be a multivariate Gaussian (or "normal") probability density function. We use $\mathbf{y} \sim p(\mathbf{y})$ to denote the distribution of a random variable – i.e., $\mathbf{y}$ is "distributed as" $p(\mathbf{y})$.

Often, we wish to learn about the joint behavior of two random variables, for example, $\mathbf{x}$ and $\mathbf{y}$, together, and we write $(\mathbf{x}',\mathbf{y}')' \sim p(\mathbf{x},\mathbf{y})$ to denote the joint distribution of $\mathbf{x}$ and $\mathbf{y}$. In this context, we now refer to $p(\mathbf{x})$ and $p(\mathbf{y})$ as marginal distributions where $p(\mathbf{y})=\int p(x,y)d\mathbf{x}$ or $p(\mathbf{y})=\sum_{\mathbf{x}} p(\mathbf{x},\mathbf{y})$ if $\mathbf{x}$ is discrete-valued.

In a modeling situation, the marginal distributions $p(\mathbf{x})$ or $p(\mathbf{y})$ may be relatively easy to specify, but the joint nature of the relationship between $\mathbf{x}$ and $\mathbf{y}$ may make direct specification of $p(\mathbf{x}, \mathbf{y})$ more difficult. Also, if given a value of, say, $\mathbf{x}$, then the problem of specifying a distribution for $\mathbf{y}$ may become easier. That is, the data analyst or modeler often finds the conditional distribution of $\mathbf{y}$ given $\mathbf{x}$ easier to specify than $p(\mathbf{y})$. We use $p(\mathbf{y}|\mathbf{x})$ to denote the dependence of the distribution of $\mathbf{y}$ on the value of $\mathbf{x}$, and we read "$\mathbf{y}|\mathbf{x}$" as $\mathbf{y}$ "given" or "conditional upon" $\mathbf{x}$. As we discuss below, the idea of conditioning is fundamental to building complex joint probability models from relatively simple "pieces."

Perhaps the most familiar example of conditional model specification occurs with regression analysis or with analysis of variance (ANOVA) whereby the distribution – typically Gaussian – of the response $\mathbf{y}$ is specified conditionally on the value of the covariate $\mathbf{x}$ usually through the mean of $p(\mathbf{y}|\mathbf{x})$, i.e., through the "regression function." In many circumstances, depending on the assumptions for $\mathbf{x}$ (e.g. $\mathbf{x}$ may be assumed "fixed" or measured without error), it suffices to specify only $p(\mathbf{y}|\mathbf{x})$ in order to "best" learn about the parameters (e.g. the intercept and slope coefficient) in the regression function. Other cases lead to specification of a distribution for $\mathbf{x}$, $p(\mathbf{x})$, which, incidentally, is done when modeling "measurement error" or "error in variables" (e.g. Stefanski 2000). If we consider both $\mathbf{y}$ and $\mathbf{x}$ as random quantities, then, by basic probability rules, $p(\mathbf{y}|\mathbf{x})$ and $p(\mathbf{x})$ specify a joint probability distribution for $\mathbf{x}$ and $\mathbf{y}$ together,

$$p(\mathbf{x},\mathbf{y}) = p(\mathbf{y}|\mathbf{x}) \cdot p(\mathbf{x}). \tag{1}$$

As is frequently the case, if we assume independence across observations, then we may work component-wise and use the multiplication rule for multiplying distributions of independent quantities to get the joint distribution,

$$\begin{aligned} p(\mathbf{x},\mathbf{y}) &= \prod_i p(x_i, y_i) \\ &= \prod_i p(y_i|x_i) \cdot p(x_i) \end{aligned} \tag{2}$$

To illustrate this simple hierarchical probability model (Eqs. 1, 2), consider the problem of measuring leaf-level transpiration rates (T) and stomatal conductance to water vapor (g), and let T_i and g_i represent the ith observation of each. Let us also assume that the mean transpiration rate, denoted as $E(T)$, is given by Fick's law of diffusion, $E(T) = g \cdot \Delta W$, where ΔW is the gradient in water vapor from inside the leaf to the atmosphere (Pearcy et al. 1989; Nobel 1999). For simplicity, we treat ΔW as a fixed covariate and assume that the mean conductance value is $E(g) = \mu_g$, with $p(T_i \mid g_i)$ and $p(g_i)$ given by Gaussian probability densities with variances σ_T^2 and σ_g^2. That is:

$$T_i \mid g_i \sim No(g_i \cdot \Delta W, \sigma_T^2) \tag{3}$$

and:

$$g_i \sim No(\mu_{g_i}, \sigma_g^2). \tag{4}$$

We use Eq. 1 to combine the conditional and marginal distributions in Eqs. 3 and 4 to obtain a bivariate normal density for the joint distribution of T_i and g_i:

$$\begin{pmatrix} T_i \\ g_i \end{pmatrix} \sim MNo\left(\begin{pmatrix} \mu_{g_i} \cdot \Delta W \\ \mu_{g_i} \end{pmatrix}, \begin{pmatrix} (\Delta W)^2 \cdot \sigma_g^2 + \sigma_T^2 & (\Delta W)^2 \cdot \sigma_g^2 \\ (\Delta W)^2 \cdot \sigma_g^2 & \sigma_g^2 \end{pmatrix} \right) \tag{5}$$

where the unconditional mean of T_i is $\mu_{gi} \cdot \Delta W$, the unconditional variance of T_i is $(\Delta W)^2 \cdot \sigma_g^2 + \sigma_T^2$, the covariance between T_i and g_i is $(\Delta W)^2 \cdot \sigma_g^2$, and μ_{gi} and σ_g^2 are as given above. In this example, the conditional and marginal distributions for $T_i|g_i$ and g_i are relatively straightforward to specify, but it is less obvious how to specify the joint distribution in Eq. 5 directly, without the relationships in Eqs. 3 and 4. Assuming independence across $i = 1,\ldots, n$ observations gives:

$$p(\mathbf{T},\mathbf{g}) = \prod_{i=1}^{n} p(T_i, g_i) \tag{6}$$

where $p(T_i, g_i)$ is given by Eq. 5.

Equation 1 is a particular case of a completely general result for obtaining joint distributions from the product of conditional and marginal distributions. For a generic random vector $\mathbf{z} = (z_1, z_2,\ldots, z_n)'$, basic rules of probability allow us to write the joint distribution of $\mathbf{z}$, $p(\mathbf{z}) = p(z_1)\times p(z_2|z_1)\times p(z_3|z_2,z_1)\times\ldots\times p(z_n| z_{n-1},\ldots,z_1)$, where the components z_i may themselves be sub-vectors of $\mathbf{z}$, with the particular sub-vectors and sequencing of conditioning being arbitrary. For example, with $\mathbf{z} = (\mathbf{z}_1', \mathbf{z}_2')'$, we might choose to model $p(\mathbf{z}) = p(\mathbf{z}_1|\mathbf{z}_2)\times p(\mathbf{z}_2)$ or, instead, $p(\mathbf{z}) = p(\mathbf{z}_2|\mathbf{z}_1)\times p(\mathbf{z}_1)$. It should be noted that when building a joint $p(\mathbf{z})$ from marginals and conditionals, the resulting $p(\mathbf{z})$ may differ, depending on the specified marginal and conditional distributions. Whichever way we choose to condition, we get a joint distribution $p(\mathbf{z})$ in the end. For example, we may have specified a conditional distribution for $\mathbf{g}|\mathbf{T}$ and marginal for $\mathbf{T}$, but the resulting joint distribution would generally be different from that in Eq. 5. Again, the usefulness of this simple result is that each conditional distribution $p(\mathbf{z}_j|\mathbf{z}_{j-1},\ldots,\mathbf{z}_1)$ is likely easier to specify, whereas the task of specifying a model for a joint distribution $p(\mathbf{z})$ directly is often beyond our ability – it is simply easier to think about the smaller dimensions of $\mathbf{z}_1$, $\mathbf{z}_2$, through $\mathbf{z}_n$ than to think at the dimensionality of $\mathbf{z}$. The conditional approach allows us to work with relatively simple pieces to arrive at a complex joint model that would otherwise be difficult or impossible to specify directly. This approach is sometimes loosely referred to as "modeling locally."

Initially, the flexibility of the conditional modeling approach may seem bewildering. Which conditional specification should we choose? But, the particular specification used in practice is often motivated and aided by consideration of the problem at hand, including the available data and ecological or biophysical theory. In other words, complex models of data, quantitative theory, and conceptual understanding can often be accommodated via the flexibility of conditional specification. It is the specification of a joint model via a sequence of conditional models that is generally referred to as hierarchical modeling. Some of the examples that we consider later illustrate how the conditional specification is key to modeling complex ecological processes. Other tools such as graphical models can facilitate the derivation of the corresponding probabilistic model by helping to identify relationships among components, which ultimately informs the conditional specification (see Brooks 1998, and references therein), and which may be particularly useful to ecologists with a conceptual model in mind (e.g. Clark and Gelfand 2006). For example, ecologists often use "systems diagrams" (see Kitching 1983), a form of graphical modeling, to help conceptualize the ecological system, and thus they should find graphical models familiar and helpful in building probability models.

3.2 *Bayes Theorem and Bayesian Inference*

Of course, we want to put the above discussion on hierarchical probability modeling into a Bayesian statistical framework, and we start with a statement of Bayes theorem, at the foundation of Bayesian statistical inference. The theorem has existed in various forms since the eighteenth century when introduced by Thomas Bayes (Barnard 1958; Malakoff 1999). The result of the theorem is a straightforward consequence of fundamental probability rules. Bayes theorem (or Bayes rule) is without controversy, existing separate from notions of statistical inference:

$$p(\mathbf{x}|\mathbf{y}) = \frac{p(\mathbf{y}|\mathbf{x}) \cdot p(\mathbf{x})}{p(\mathbf{y})}. \tag{7}$$

To place this into a Bayesian statistical context, consider a probability distribution function representing the distribution of data, **y**, and depending on other quantities, θ (generally a vector), called the parameter(s) of the distribution. Such dependence is often made explicit via notation such as $p(\mathbf{y}|\theta)$ or $p(\mathbf{y}|\mathbf{x}, \theta)$ if considering covariates **x**. Continuing the transpiration–conductance example (Eqs. 3–6), the data are $\mathbf{y} = (\mathbf{T},\mathbf{g})$, the parameters are $\theta = (\mu_g, \sigma_T^2, \sigma_g^2)$, and the covariates are $\mathbf{x} = (\Delta\mathbf{W})$. The distribution $p(\mathbf{y}|\theta)$ is often called the data distribution, the sampling distribution, or, when viewed as a function of θ with **y** fixed at observed data values, it is called the likelihood (of the data). Of course, θ is almost always unknown to some degree, and the objective is to infer about θ, to predict unobserved **y** values, or, more generally, to infer about some function of these quantities.

To enable inference about the parameter θ, we introduce a prior probability distribution $p(\theta)$ – an explicit summary, in the form of a probability distribution, of what we know about θ prior to observing the data. So, with $p(\mathbf{y}|\theta)$ as our sampling distribution and $p(\theta)$ as our prior, then, by the basic probability rule in Eq. 1, we have a "full" probability model for data and parameters,

$$p(\mathbf{y},\theta) = p(\mathbf{y}|\theta) \cdot p(\theta). \tag{8}$$

Bayes theorem is used to actuate the common (but not universally followed) statistical principle of conditioning on what is known ($\mathbf{y}$, data) in order to learn about what is unknown (θ, parameters). In other words, once the probability model is specified, Bayes theorem provides a complete prescription for statistical inference. It is often said that Bayes theorem "updates" prior knowledge of θ as quantified by $p(\theta)$ to reflect the information in the data, arriving at the posterior (conditional) distribution of θ (given the data $\mathbf{y}$),

$$p(\theta|\mathbf{y}) = \frac{p(\mathbf{y}|\theta) \cdot p(\theta)}{p(\mathbf{y})} \tag{9}$$

We provide a detailed example in Section 4.1, similar to the transpiration–conductance problem, of how to arrive at the posterior distribution of unknown quantities.

Usually, only in the simplest of cases are we able to use analytic techniques to obtain a closed form solution to $p(\theta|\mathbf{y})$, and often we must resort to numerical or stochastic techniques to approximate the distribution $p(\theta|\mathbf{y})$ or to obtain a sample of θ values from $p(\theta|\mathbf{y})$. Widespread use of Bayesian methodology occurred with the development of computing power (Malakoff 1999) and concomitant developments of algorithms for sampling from distributions (e.g. Gelfand et al. 1990; Gilks et al. 1996; Gamerman and Lopes 2006). These sampling methods are generally referred to as Markov chain Monte Carlo (MCMC) methods (Gilks et al. 1996; Brooks 1998; Cappe and Robert 2000; Robert and Casella 2004; Gamerman and Lopes 2006), including Gibbs (Gelfand et al. 1990; Casella and George 1992), Metropolis (Metropolis et al. 1953; Robert and Casella 2004), and Metropolis–Hastings algorithms (Hastings 1970; Chib and Greenberg 1995). Although these methods are not inherently Bayesian, their usefulness for sampling from the posterior distribution has made them standard fare in Bayesian methodology. Together, MCMC methodology and Bayesian inference are becoming a very useful combination for complex models (e.g. Clark and Gelfand 2006). We do not cover MCMC for Bayesian inference but refer the reader to the above references, especially the texts by Gamerman and Lopes (2006), Gilks et al. (1996), and Robert and Casella (2004).

Upon obtaining a sample of θ values from $p(\theta|\mathbf{y})$, inference about θ then proceeds via ordinary computation of means, medians, quantiles, credible regions, etc. (e.g. Smith and Gelfand 1992), exactly analogous to summarizing a sample of data. Furthermore, the (marginal) posterior distribution and summaries for any function $f(\theta)$, or more generally, for any function $f(\theta, \mathbf{y}, \mathbf{x})$, may then be obtained in a straightforward manner as well. For example, for each value of θ, $f(\theta)$ is computed, yielding a sample of $f(\theta)$ values, which can be summarized in the same way as θ.

All of the most popular computational methods for obtaining samples from posterior distributions depend only on knowing $p(\theta|\mathbf{y})$ up to some multiplicative constant, a factor that does not depend on θ. When the data are given, $p(\mathbf{y})$ is constant with respect to θ, and we often see an alternative form of Bayes theorem:

$$p(\theta|\mathbf{y}) \propto p(\mathbf{y}|\theta) \cdot p(\theta), \tag{10}$$

and we often hear the chant "posterior is proportional to likelihood times prior."

3.3 *Hierarchical Bayesian Modeling*

Recalling the previous discussion (Section 3.1) of modeling joint distributions via conditional specifications, we see that the right-hand side of Bayes theorem as expressed in Eq. 10 is the joint distribution $p(\mathbf{y},\theta)$. Thus, the enterprise of Bayesian statistical modeling can be seen as the development of a full probability model $p(\mathbf{y},\theta)$ for the data and unknown parameters, inference being fully prescribed, in principle, via Bayes theorem once the model is specified. Once again it is important to note that we may avail ourselves of the flexibility of the hierarchical specification of joint distributions, which is especially important for complex modeling problems – modeling locally and, now, inferring globally so-to-speak.

Before proceeding to discuss hierarchical Bayesian modeling in more detail, we remind the reader that hierarchical statistical modeling is, of course, not exclusively Bayesian. In fact, we suspect that the vast majority of literature on hierarchical modeling remains nonBayesian despite recently increasing activity in hierarchical Bayesian modeling. The literature on hierarchical statistical modeling is extensive and diverse. We provide only a short, incomplete and somewhat idiosyncratic set of references that is, nonetheless, sufficient to allow the interested reader to explore this literature more fully.

Hierarchical statistical modeling, especially its nonBayesian versions, tends to fall under the guises of repeated measurement models (Lindsey 1993), longitudinal data models (e.g. Diggle et al. 2002), multilevel models (Goldstein 2002) or, more generally, mixed effects models. Laird and Ware (1982) are frequently cited for their work on linear mixed effects (LME) models, and Lindley and Smith (1972) introduce the hierarchical Bayesian linear model. Lindstrom and Bates (1990) discuss the nonlinear mixed effects model (NLME), with Gelfand et al. (1990) providing a Bayesian perspective. The extension of hierarchical models to nonGaussian data is often attributed to Breslow and Clayton (1993) for their treatment of generalized linear mixed models (GLMM). The text by Searle et al. (1992) is a comprehensive review of LME and variance component models. Davidian and Giltinan (1995) present a nice discussion of linear and nonlinear hierarchal models, and Gelman et al.

(2004) provides a general introduction to Bayesian methodology with chapters on hierarchical modeling. Popular software for fitting nonBayesian hierarchical models includes the R (R Development Core Team 2006) package by Pinheiro et al. (2006) for LME and NLME models (see also Pinheiro and Bates 2000), the SAS MIXED procedure for LME models, and SAS %NLINMIX macro for NLME models (SAS Institute 2001; see also Little et al. 1996). The SAS GLIMMIX procedure is designed for fitting GLMM models (SAS Institute 2006). WinBUGS (Lunn et al. 2000) is a general purpose software package for implementing Bayesian models, including hierarchical Bayesian models.

So, hierarchical statistical modeling is neither new nor exclusively Bayesian. Why focus on hierarchical Bayesian models? Hierarchical Bayesian modeling is proving to be the primary methodology to study complex relationships embodied in process models, data, and prior information, and it is the only method that allows for explicit incorporation of prior information. Berliner (1996) is often credited with introducing what has become loosely referred to as the "process sandwich" as an idealized way of thinking about how to incorporate the information contained in a process model into a Bayesian statistical framework along with (i.e., "between") the data and prior information. Similar to Eq. 10, the process sandwich is:

$$p(\text{process, parameters}\,|\,\text{data}) \propto$$

$$p(\text{data}\,|\,\text{process, parameters}) \cdot p(\text{process}\,|\,\text{parameters}) \cdot p(\text{parameters}), \tag{11}$$

and Bayes theorem and (usually) MCMC methodology are applied to learn about the (unknown or latent) process and parameters via the posterior *p*(process, parameters|data). In this context, the process is now modeled as a stochastic quantity with probability distribution *p*(process|parameters) with a mean that is usually specified by a deterministic process model – what we have been calling the "process model" up to this point. Now, we may refer to the process (probability) distribution model or to its mean, the process model. The process model may describe a theoretical physical process in the form of, say, partial differential equations, but in its most general form may have some combination of mechanistic, semi-mechanistic, phenomenological, empirical, or stochastic components that represent, for example, real underlying physiological, ecological, or biogeochemical dynamics.

The process sandwich approach is espoused and illustrated by Wikle and others in, for example, Wikle et al. (1998), Wikle et al. (2001), and Wikle (2003b); with much of this work having a spatial or spatial-temporal statistical bent within a geophysical application. Wikle (2003a) applies similar methodology in modeling the spread of the house finch (*Carpodacus mexicanus*) using a reaction–diffusion partial differential equation, and Clark and colleagues have applied the approach to a variety of population and community ecology problems (e.g. Clark 2003; Clark et al. 2003b, 2005). We return to hierarchical Bayesian modeling and the process sandwich in our examples below.

4 Example Applications

In this section, we provide some specific examples of Bayesian and hierarchical Bayesian applications in plant physiological and ecosystem studies. None of the physiological- and ecosystem-related papers in the literature survey (Table 1, Fig. 1) employed a complete hierarchical Bayesian framework as defined by Eq. 11. That is, although many incorporated complex ecological process models, none explicitly modeled process error – the process is treated as deterministic rather than stochastic. Thus, in Section 4.1, we begin by providing our own detailed example of a hierarchical Bayesian modeling problem using process models of photosynthesis and stomatal conductance with leaf-level gas exchange data. In Section 4.2, we then highlight three published examples that use (nonhierarchical) Bayesian methods with informative priors to parameterize complex ecosystem simulation (process) models.

4.1 Example 1: Leaf-Level Gas Exchange

4.1.1 The Problem

This example builds on a study by Ogle and Reynolds (2002) that focused on *Larrea tridentata* (creosote bush), a common evergreen shrub of the hot deserts of North America (Barbour et al. 1977). *Larrea* experiences large year-to-year, seasonal, and diurnal fluctuations in temperature and water availability, yet it remains metabolically active year-round (e.g. Oechel et al. 1972) due to its ability to tolerate severe temperature and water stress. Thus, *Larrea* makes an interesting candidate for studying plant physiological strategies for coping with environmental stress in field settings. A primary goal of this study was to learn how temperature acclimation and stomatal control of photosynthesis in *Larrea* contribute to its ability to tolerate severe water and temperature stress. Ogle and Reynolds (2002) combined extensive field data with physiological process models to tease apart potential mechanisms underlying *Larrea*'s ability to tolerate harsh arid conditions.

4.1.2 The Data

Ogle and Reynolds (2002) used an open-system infrared gas exchange analyzer to measure leaf-level, instantaneous net photosynthesis (*Anet*), stomatal conductance to water vapor (*g*), and leaf-internal CO_2 concentration (*Ci*). Diurnal gas-exchange measurements were made on eight shrubs on each of 16 days throughout the spring and summer seasons in the Chihuahuan desert of New Mexico. Additional stomatal conductance measurements were made with a steady-state porometer on 16 shrubs (including the eight above) and 25 days

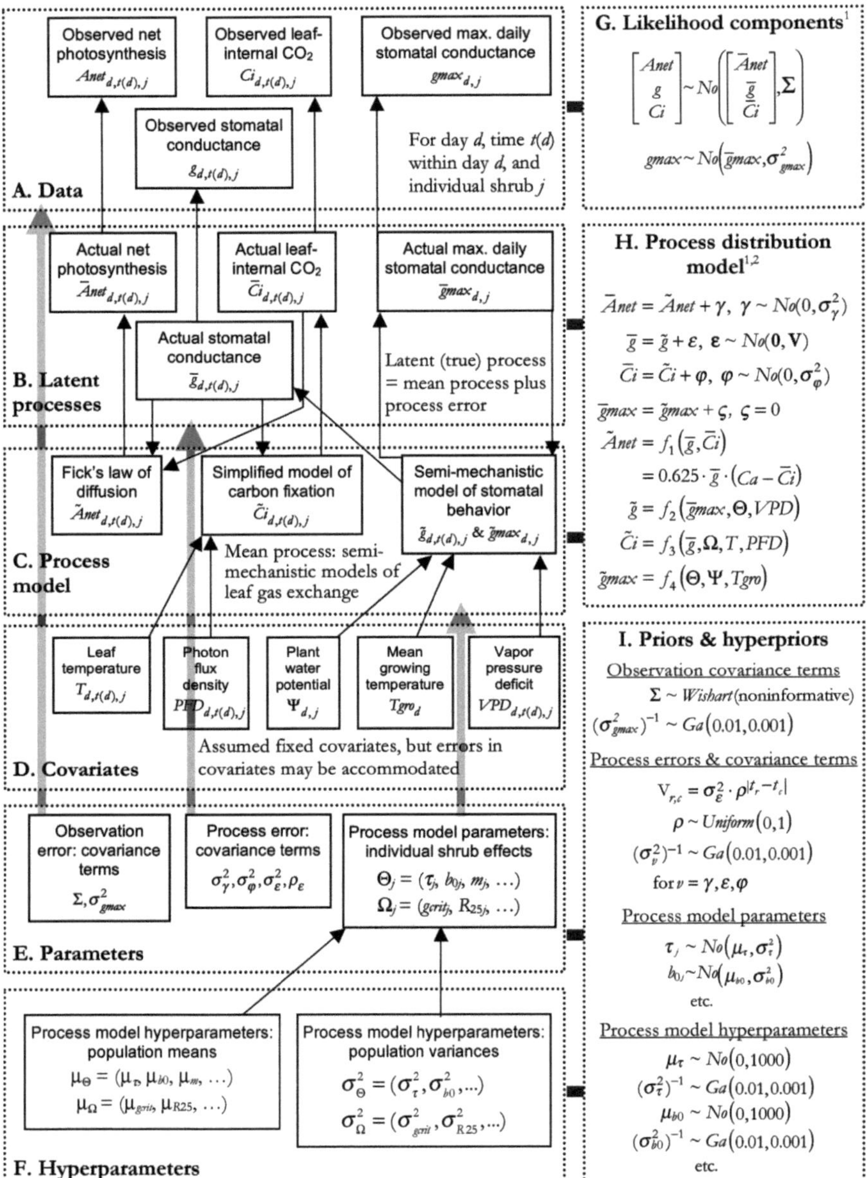

[1]For ease of presentation, we omit the subscripts d, $t(d)$, and j.

[2]Process errors represent diurnal (time within day) effects, and exploratory analyses suggest that the $\bar{g}$ process errors are temporally autocorrelated, but the $\bar{Anet}$ and $\bar{Ci}$ process errors are independent with respect to time of day. The $\bar{g}$ process errors were nonidentifiable, thus we set them equal to zero. See Ogle & Reynolds (2002) for the explicit equations used to specify f_2, f_3, and f_4; note that f_1 is Fick's law.

Fig. 2 Graphical model (*Boxes A–F*) and corresponding distribution and process models (*Boxes G–H*) associated with the leaf-level gas exchange example in Section 4.1

(including the 16 days above) to estimate daily maximum conductance (*gmax*). Referring to Fig. 2 (Box A), *d* indexes the day of observation, *t*(*d*) refers to the time of measurement within day *d*, and *j* indexes the shrub. Using the notation introduced in Section 3.1, the dataset includes 934 observations of the triplet $(Anet_{d,t(d),j}, g_{d,t(d),j}, Ci_{d,t(d),j})'$, 218 observations of $gmax_{d,j}$, and various biotic and abiotic covariates that were measured at the same time (see Fig. 2, Box D). See Ogle and Reynolds (2002) for a detailed account of the field experiment and sampling design and methods.

4.1.3 The Process Model

Ogle and Reynolds (2002) modified the Hybrid model (Katul et al. 2000) to describe *Larrea*'s *Ci* dynamics. This model is derived from simplified versions of an A-Ci curve (e.g. see pp 17–19 in Lambers et al. 1998) and a biochemical model of CO_2 assimilation (Farquhar et al. 1980; Farquhar and Sharkey 1982; Farquhar and von Caemmerer 1982). Stomatal conductance is an important factor controlling *Ci*. Instantaneous *g* is modeled as a threshold-type function of *gmax* and vapor pressure deficit (*VPD*) in such a way that agrees with a biophysical model of how stomata regulate leaf water potential (see Oren et al. 1999; Ogle and Reynolds 2002); *gmax* is assumed to vary with plant water stress and growing temperature. Once the *Ci*, *g*, and *gmax* models are specified, than net assimilation (*Anet*) is simply given by Fick's law of diffusion. The coupling of *Ci*, *g*, *gmax*, and *Anet* is illustrated in Fig. 2 (Boxes B, C, H), and we cover more details in the next section.

4.1.4 Data–Model Integration

We develop a hierarchical Bayesian model to couple the data and process models discussed above. We provide a detailed discussion of how the conditional modeling framework established in Section 3 can be applied to the *Larrea* leaf-level gas exchange problem. Although this is a specific example, it is representative of the types of data and process models that are commonly encountered by plant ecophysiologists, and the steps that we discuss here are applicable to a variety of problems. Of course, in practice, a primary goal of implementing such a framework is to obtain the posterior distribution of all ecologically meaningful process parameters, latent variables, and covariance terms that describe errors in observation (i.e. measurement) and process (i.e. model), thereby updating our understanding of the underlying physiological system. However, our goal here is to emphasize the means by which to arrive at the posterior distribution.

Using notation similar to that in Eq. 11, we present the stages of the hierarchical model. We elaborate on Eq. 11 by (1) partitioning the "parameters" into "process parameters", "process hyperparameters", and "covariance parameters", (2) renaming "process" as "latent process" and introducing explicit dependence on the deterministic "process model", and (3) indicating explicit dependence on "covariates". Thus, the posterior distribution for all unknown parameters and latent processes:

$$p(\text{parameters, latent process} \mid \text{process model, data, covariates}) \cdot \quad (12)$$

is proportional to:

$$p(\text{data} \mid \text{latent process, covariance parameters}) \cdot \quad (12.1)$$

$$p(\text{latent process} \mid \text{process model, process parameters, covariates}) \cdot \quad (12.2)$$

$$p(\text{process parameters} \mid \text{process hyperparameters}) \cdot p(\text{covariance parameters}) \cdot \quad (12.3a)$$

$$p(\text{process hyperparameters}) \cdot \quad (12.3b)$$

The first stage (Eq. 12.1) is the likelihood of the data, which corresponds directly to *p*(data|process, parameters) in Eq. 11. The second stage (Eq. 12.2) is the stochastic (or probability) model for the "unobservable" latent processes, i.e. *p*(process|parameters) in Eq. 11. The third stage (Eqs. 12.3a, b) is *p*(parameters) in Eq. 11 and is modeled, using an assumption of independence of process-related parameters and covariance parameters and applying the hierarchical specification in Eq. 1 to *p*(process parameters, process hyperparameters). Notice that the joint prior is specified hierarchically, and we may want to distinguish different levels of the joint prior as the "prior" (Eq. 12.3a) and the "hyperprior" (Eq. 12.3b).

In the following sections, we discuss each of the above stages (Eqs. 12.1–12.3) in the context of the leaf gas exchange problem. Following Clark (2005) and Clark and Gelfand (2006), we provide a graphical model that illustrates these stages (Fig. 2, Boxes A–F), and we also provide a summary of the probability distributions associated with each stage (Fig. 2, Boxes G–I).

Stage 1

The first stage is represented by Boxes A and G (Fig. 2), which denote the likelihood components of individual observations of $(Anet_{d,t(d),j}, g_{d,t(d),j}, Ci_{d,t(d),j})'$ and $gmax_{d,j}$. The gas-exchange instrument software computes "observed" values of *Ci* from "observed" photosynthesis rates (*Anet*), transpiration rates, and stomatal conductance (*g*; LI-COR 2004). Similarly, *Anet* and *g* are computed from other measured quantities (e.g. mole fractions of water vapor and CO_2 and flow rates). Thus, it seems reasonable to consider the "observation" errors in *Ci*, *Anet*, and *g* to be correlated, and we assume a multivariate normal distribution for $(Anet_{d,t(d),j}, g_{d,t(d),j}, Ci_{d,t(d),j})'$, as given in Box G of Fig. 2 [subscripts *d*, *t*(*d*), and *j* are omitted for ease of presentation]. (In fact, the posterior distribution of Σ supports this assumption.) Given that a different instrument was used to measure *gmax*, we assume independence of *gmax* observations and use a univariate normal distribution to describe the likelihood of $gmax_{d,j}$ (Fig. 2, Box G). The likelihood of all data (i.e. **Anet**, **g**, **Ci**, **gmax**) is obtained by multiplying the likelihood components (Fig. 2, Box G) across all days, shrubs, and times, giving:

$$p(\text{data} = \{\mathbf{Anet}, \mathbf{g}, \mathbf{Ci}, \mathbf{gmax}\} | \text{latent process, covariance parameters}) =$$

$$\prod_d \prod_j No\left(gmax_{d,j} \,\middle|\, \overline{g}max_{d,j}, \sigma^2_{gmax}\right) \left(\prod_{t(d)} No\left(\begin{bmatrix} Anet_{d,t(d),j} \\ g_{d,t(d),j} \\ Ci_{d,t(d),j} \end{bmatrix} \middle| \begin{bmatrix} \overline{A}net_{d,t(d),j} \\ \overline{g}_{d,t(d),j} \\ \overline{C}i_{d,t(d),j} \end{bmatrix}, \Sigma \right) \right) \quad (13)$$

(see examples discussed with Eqs. 2, 6). Note that when the data are conditioned on the true or latent processes (i.e. $\overline{\mathbf{A}}\boldsymbol{net}$, $\overline{\mathbf{g}}$, $\overline{\mathbf{C}}\boldsymbol{i}$, $\overline{\mathbf{g}}\boldsymbol{max}$), as indicated in Eq. 13, then any remaining unexplained variability should be attributable to measurement error, and we assume that these errors are uncorrelated across shrubs and time. Although this is a common assumption, it may not always be appropriate depending on the measurement instrument used.

Stage 2

The second stage (Eq. 12.2) is depicted by Boxes B, C, D, and H in Fig. 2. Stage 2 embodies the specification of the stochastic and deterministic process models (Fig. 2, Boxes B, C, H), and we incorporate process error into the stochastic process model (Box H). The process error accounts for the fact that we are approximating the true photosynthetic and stomatal dynamics with relatively simple equations that cannot reproduce the truth exactly. Thus, we model the latent process as the mean process plus some random process error (see Fig. 2, Box H).

One approach to modeling process errors is to assume that they are independent. However, if observation errors are also assumed independent, then this may lead to identifiability problems – i.e. process and observation variance components may not be disentangled – unless a tight prior is specified for the variance components of the observation error. Instrument manuals often provide estimates of instrument precision, which can be used to construct tight (or informative) priors for observation error variances. In the absence of informative (or informative) priors, identifiability of process error and observation error variance terms may be facilitated by incorporation of different error structures for process and observation errors. We cannot possibly measure and account for all factors affecting the true process, so we cannot model the true process perfectly via the process mean alone and are led to consider the structure of the process error. These unobserved factors are almost invariably temporal, spatial, or biological in nature and usually have little impact on measurement error. Thus, we may want to incorporate temporal, spatial, or biological structure into the process errors.

In the leaf-gas exchange problem, we did not have sufficient information to construct tight priors for the observation covariance parameters. Thus, for illustration, we opted to incorporated temporal structure into the *Anet*, *g*, and *Ci* process errors

(Fig. 2, γ, ε, and φ in Box H). Based on previous experience (Ogle and Reynolds 2002), we assumed diurnal effects such that each time period within a day gets a particular random effect (error) that applies to all days. For example, the diurnal effect at, say, noon is the same for all days – effects do not vary by day. Exploratory analysis with different temporal error structures suggested the *Anet* and *Ci* diurnal effects were uncorrelated with respect to time. However, the *g* diurnal effects appeared autocorrelated, and we used an exponential covariance function – defined by the variance parameter σ_ε^2 and the correlation coefficient ρ – to describe this correlation (see Fig. 2, Boxes H, I). The shrub itself is where the physiological processes are taking place, but shrubs differ in their genetic composition, age, and growth history. Thus, rather than incorporating additive shrub effects into the process error, we allow the process parameters themselves (i.e. Θ_j and Ω_j in Box E) to vary by shrub, with distributions specified in Stage 3. Specification of the deterministic process model is highly dependent on existing ecological theory, the understanding of the specific system of interest, and the particular objectives of the study. Process models have a long history in ecology, and we do not cover details of process model development. However, the process model is key because it contains important information about the ecological system, and we illustrate how it is incorporated into the hierarchical Bayesian modeling framework. In this example, each shrub *j* gets a unique process mean (i.e. Fig. 2, $\widetilde{Anet}$, $\tilde{g}$, $\widetilde{Ci}$, $\widetilde{gmax}$ in Boxes C, H) because the deterministic process models (Box C) depend on the shrub-specific process parameters (Θ_j and Ω_j in Box E). The specific deterministic process models that we employ in this example are discussed in Section 4.1.3 and their equations are outlined in Box H (Fig. 2). The shrub-specific process model parameters (Boxes E, I) include, for example, maximum potential conductance (τ_j), sensitivity of stomata to changes in *VPD* (represented by b_{0j} and m_j), leaf dark respiration rate at 25 °C (R_{25j}), and so forth (for a list of parameters, see Ogle and Reynolds 2002). The process models also incorporate the effects of important plant and environmental covariates (see Box D for specific covariates, Box H for model dependencies on covariates). In this example, we assume that the covariates are measured without error, which is probably a reasonable assumption because there is likely relatively little error associated with measuring these quantities compared to "measuring" *Anet*, *g*, *Ci*, and *gmax*. However, it is fairly straightforward to incorporate errors in covariates into a hierarchical modeling framework (Clark et al. 2003b).

Stage 3

Finally, Stage 3 (Eq. 12.3) involves defining the priors for all parameters (process parameters, covariance parameters, process hyperparameters; Fig. 2, Boxes E, F, I). As mentioned above, we lack information on the instrument measurement error and thus use fairly noninformative (relatively diffuse) priors for the observation covariance parameters (see *Wishart* for Σ and gamma for precision term σ^2_{gmax} in Box I). Similarly, we specified noninformative priors for the process error variance parameters [see gammas for $(\sigma_\gamma^2)^{-1}$, $(\sigma_\varepsilon^2)^{-1}$, and $(\sigma_\varphi^2)^{-1}$ in Box I] and a uniform (flat) prior for the correlation parameter ρ (Box I). All distribution parameterizations follow Gelman et al. (2004).

As noted in Stage 2, we assume that the process parameters (Θ_j and Ω_j, Box E) vary by shrub. The shrub effects are viewed as coming from population distributions with means μ_Θ and μ_Ω and variances σ_Θ^2 and σ_Ω^2 (see Box E). For example, Θ_j represents the shrub-specific process parameters related to the stomatal conductance models (see f_2 and f_4 in Box H), and let us consider one element of Θ_j, maximum potential conductance (τ_j). We first note that τ_j is a positive-valued quantity, and thus we model $\log(\tau_j)$ – which can be positive- or negative-valued – as coming from a population distribution, with mean μ_τ and variance σ_τ^2, that describes shrub-to-shrub variability in τ_j,

$$\log(\tau_j) \sim No\,(\mu_\tau, \sigma_\tau^2) \tag{14}$$

We assume a priori that the elements of Θ_j and Ω_j are independent given μ_Θ, μ_Ω, σ_Θ^2, and σ_Ω^2, and thus priors similar to Eq. 14 are specified for each element of Θ_j and Ω_j (Box I). We assume independence of the covariance parameters and the shrub-specific parameters so that the complete prior in Eq. 12.3a is given by the multiplication of the marginal priors specified in first three sections of Box I (Fig. 2), analogous to Stage 1, Eq. 13. Note that the simplifying modeling assumption of a priori independence does not necessarily imply a posteriori independence.

Finally, we specify the hyperprior (Eq. 12.3b) for the hyperparameters (i.e. μ_Θ, μ_Ω, σ_Θ^2, σ_Ω^2) describing the population distributions of shrub effects. Continuing to illustrate using τ_j, we specify independent diffuse priors for μ_τ and σ_τ^2 (process model hyperparameters, Box I), with similar a priori independent priors for the remaining hyperparameters, and the full hyperprior (Eq. 12.3b) following again by multiplication of distributions. Note that in most cases, we are actually most interested in learning about such hyperparameters (population-level parameters); for example, here, μ_Θ and μ_Ω tell us about the physiological behavior of *Larrea* in general (we may not be interested in the behavior of a particular shrub).

Now that we have described the various components making up the different stages of the hierarchical model, some in more detail than others, they may be combined to yield the full conditional specification in Eq. 12. The resulting model is rather cumbersome, thus we omit the technical details. Moreover, we implemented the model in WinBUGS, and the WinBUGS environment does not require one to specify the fully probability model in Eq. 12, but only requires one to specify each of the conditional components, just as we have done here.

4.1.5 Comments

We do not discuss specific results of the analysis because this is not the objective of this example, but we would like to address the question: now that we have samples from the posterior, what to we do with them? We can summarize the posterior results by computing posterior statistics such as means, variances, and percentiles (a measure of "uncertainty") for each parameter of interest. We may also want to estimate correlations betweens parameters or to learn about the distributional form of the marginal posterior distributions. For example, many marginals are often normal-like, but some can exhibit interesting behavior such as a high degree of skewness or multiple modes. Multiple modes may indicate indentifiability problems that may be addressed via tighter priors, reparameterization, or restructuring of related model components. In principle, we can obtain the posterior distributions for most any ecologically relevant functions of the parameters, data, or covariates. For example, we could compute daily-integrated photosynthesis from predictions of diurnal photosynthesis (e.g. $\bar{Aday}_{d,j} = \Sigma_{t(d)} \bar{Anet}_{d,t(d),j}$), yielding a posterior distribution for $\bar{Aday}_{d,j}$. We could also use the joint posterior distribution of the parameters within a forward modeling context to learn how uncertainty in the photosynthesis and conductance parameters are propagated to model output related to leaf-level gas exchange under, for example, different scenarios of temperature and water stress. We discuss the use of the posterior in forward modeling simulations in Section 4.2. How does the hierarchical Bayesian modeling approach compare with other, more traditional approaches? We note that Ogle and Reynolds (2002) fitted the aforementioned process models to the gas exchange using a piece-wise approach that proceeded as follows: (1) fit *gmax* observations to the *gmax* process model using nonlinear least squares, (2) fit the *g* observations to the *g* process model using a nonlinear mixed effects (NLME) approach, with the *gmax* process parameters fixed at the point estimates obtained in (1), and (3) fit *Anet* observations to the *Anet* process model using a NLME approach, with *gmax* and *g* parameters fixed at the point estimates obtained in (1) and (2). The NLME approach was considered a fairly rigorous and state-of-the-art approach at the time (e.g. Ogle and Reynolds 2002; Peek et al. 2002).

In hindsight, this piece-wise approach is unsatisfactory in several ways. First, standard NLME packages cannot accommodate a multivariate likelihood, thus compelling an assumption of independence of the *Anet*, *g*, and *Ci* observations, but the posterior distribution of Σ (Fig. 2, Box G) suggested strong dependence. Second, uncertainty in parameter estimates could not be propagated from one model fit to the next. And, third, there were convergence issues with the NLME routines such that some of the parameters could not be estimated and thus were fixed at "realistic" values to ensure convergence of the estimates of the other parameters. The hierarchical Bayesian analysis was not subject to these problems, and more importantly, it simultaneously integrates all data sources and the interrelated process models.

4.2 Example 2: Use of Prior Information with Process Models

4.2.1 The Problem

In this example, we highlight three studies that discuss and illustrate the utility of Bayesian methods for parameterizing relatively detailed ecosystem process models. These models include the coupled hydrologic-nitrogen cycle model SINIC

(Hong et al. 2005), the ecosystem-level photosynthesis and evapotranspiration model SIPNET (Braswell et al. 2005), and the forest carbon and nitrogen simulation model BASFOR (Van Oijen et al. 2005). The primary goal of each study was to demonstrate how rigorous parameter estimation could be accomplished within a simple Bayesian framework that incorporates different types of data and prior information. Similar across all studies is the specification of priors based on knowledge embodied in the scientific literature. Each study also addresses the questions: how "well constrained" are the parameters; how well does the data update prior knowledge and uncertainty about the parameters; and, how does parameter uncertainty propagate to model output uncertainty?

4.2.2 The Data

The three studies vary in the types and amount of data that they use to construct the likelihood. Hong et al. (2005) used long-term (1964–1994) measurements of annual streamflow and annual streamflow nitrate flux from a watershed located in the Hubbard Brook Experimental Forest in New Hampshire. Braswell et al. (2005) used net ecosystem CO_2 exchange (NEE) data for daily daytime and nighttime intervals for the period 1992–2002. NEE values were calculated from eddy flux measurements made at the Harvard Forest Long-Term Ecological Research site in Massachusetts. Van Oijen et al. (2005) utilized forest inventories from Skogaby, Sweden that provided measurements of, for example, tree heights, carbon and nitrogen contents of different plant and soil pools, leaf area index, and net primary productivity (data from Schulze 2000). In all studies, the data are directly related to outputs generated by the process models.

4.2.3 The Process Model

Although the three models were constructed for dissimilar purposes and represent different ecological systems, they are similar in their level of complexity. Each model contains a series of interrelated equations describing the daily dynamics of various ecosystem processes of interest, and the process models require a moderately large number of parameters. For example, SINIC simulates a variety of hydrologic processes (e.g. evapotranspiration, vertical and horizontal soil water flux, runoff, groundwater flow) and nitrogen cycling processes (e.g. atmospheric deposition, mineralization, plant uptake, nitrification, denitrification, horizontal and vertical fluxes in the soil, discharge to streams; Hong et al. 2005). Twenty parameters are needed for the hydrologic submodel and 17 for the nitrogen submodel. SIPNET is derived from the canopy photosynthesis and evapotranspiration model PnET-Day (Aber et al. 1996), and it models NEE as the difference between respiration and gross primary productivity. These flux processes depend on 25 parameters related to the initial pool values, photosynthesis, autotrophic and heterotrophic respiration, and plant–soil water relations (Braswell et al. 2005). BASFOR is a forest simulator that contains 11 state variables representing pools of carbon and nitrogen in different tree, soil, and litter compartments (Van Oijen et al. 2005). The model produces a variety of outputs including, but not limited to, time series of the state variables, net primary productivity, and leaf area index. BASFOR has 39 parameters related to, for example, initial pool values, carbon allocation, tree morphology, tissue turnover rates, decomposition, and environmental response functions. All models require similar climate-related input data such as precipitation, light, and temperature.

4.2.4 Data–Model Integration

Unlike Example 1 (Section 4.1), none of the three studies couch the Bayesian data–model integration within a hierarchical framework. The process modeling stage is based on the deterministic output of the simulation models, and there are no random effects associated with, for example, different trees, time periods, or locations. An important element of all three studies, however, is the use of prior information obtained from the scientific literature. Braswell et al. (2005) and Van Oijen et al. (2005) implemented relatively wide uniform priors for all stochastic parameters. The upper and lower bounds were primarily derived from the literature, but some were based on conventional knowledge or ranges of values used in other similar ecosystem models (e.g. Braswell et al. 2005). Hong et al. (2005) implemented a set of rules for choosing the type of distribution and the level of prior uncertainty. Most priors were bounded either with a uniform distribution (for parameters with high prior uncertainty) or a beta distribution (for parameters with intermediate prior uncertainty). Normal distributions were used for those parameters where direct data were available, and the mean and variance were computed from available data. Hong et al. (2005) also incorporated logical constraints that eliminated physically or biologically impossible parameter values. For example, "day of leaf out" must be less than "day of leaf fall" and "soil water content at field capacity" must be less than "saturated soil water content".

4.2.5 Comments

We applaud Braswell et al. (2005), Hong et al. (2005), and Van Oijen et al. (2005) for introducing and demonstrating the utility of Bayesian data–model integration methods to their respective communities. The Bayesian parameter estimation approach that they employed could be easily adapted to a hierarchical modeling framework, and Braswell et al. (2005) acknowledge that the approach can be extended to include "structural failures in the model" – or process error as defined herein. For example, the SIPNET model had difficulty capturing interannaul patterns of NEE (Braswell et al. 2005), indicating that the incorporation of temporal process error may be appropriate here.

These studies provide valuable insight into the identification of parameters that were constrained by the observed data. For example, Braswell et al. (2005) found that approximately half of the parameters were poorly constrained (little updating of the prior), or their marginal posterior distributions were concentrated at one of the edges of the uniform prior, suggesting discordance between prior and data. The other half were well constrained by the NEE data such that their marginal posteriors were concentrated inside the uniform prior. Hong et al. (2005) combined a sensitivity analysis of the SINIC model with a rigorous evaluation of the posterior distribution to identify key process parameters controling streamflow and steamflow nitrate flux. Perhaps not surprisingly, they found that, in general, those parameters having the most influence on the output of the SINIC model were also those that were most well constrained by the data. They also note that, because the SINIC model is relatively insensitive to some of the other parameters, improved (less uncertain) estimates of such parameters may not necessarily improve predictions of steam nitrate flux.

How can the posterior distributions of poorly constrained parameters be improved? Aside from certain types of parameters like, for example, the individual shrub random effects Θ_j and Ω_j of the *Larrea* example of Section 4.1, more data is often the answer. But, what kind of data and how much? To answer this question, one could conduct a statistical experiment similar to that carried out by Van Oijen et al. (2005). They systematically left out different data sources, reran the Bayesian analysis, and evaluated the resulting posteriors. This allowed them to identify the most useful data – for example, tree height growth data were most informative for the BASFOR model – providing a guide for future data collection efforts. Although Braswell et al. (2005) constructed the likelihood function based on only one data source (i.e. computed NEE values), they noted that the incorporation of other data sources – such as water vapor fluxes, soil respiration measurements, and carbon stock assessments – would likely improve estimates of poorly constrained parameters. Because the model parameters are associated with a variety of different, yet interrelated processes, the incorporation of diverse datasets reflecting these different processes is expected to refine our understanding of the process parameters and underlying ecological mechanisms. A primary goal of Bayesian parameter estimation is to produce "good" estimates of the process model parameters that are informed by information-rich data sources. Given these parameter estimates, one may wish to use the process model in a series of simulation experiments to explore, for example, potential effects of environmental perturbations (e.g. altered precipitation, elevated CO_2, increased nitrogen deposition) on model behavior, with implications for real system responses. As mentioned earlier, such simulations are often referred to as "forward modeling," but the incorporation of the Bayesian parameter estimates adds a new and improved (i.e. data- and prior-informed) dimension to this predictive modeling approach. For example, we could carry out Monte Carlo simulations whereby process parameters are drawn from the joint posterior distribution, input to the process model, a process model simulation is initiated, and the resulting output is stored. Repeated simulations are conducted and distributions of simulation outputs – induced by the posterior distribution of the process parameters – can be evaluated. Drawing parameter values from the joint posterior distribution preserves important correlations between parameters, and Monte Carlo simulations that draw parameters independently (e.g. from the priors) are likely to draw "biologically unrealistic" combinations and greatly over-estimate model output uncertainty (see Van Oijen et al. 2005).

5 Final Remarks

We conclude with three sections on various issues that we consider important to our treatment of hierarchical Bayesian statistical modeling of complex ecological systems. Section 5.1 discusses the relevance of this modeling framework to the issue of scaling in plant physiological and ecosystem ecology. Section 5.2 provides some practical computational suggestions and a brief discussion of convergence diagnostics and model checking. The final section (5.3) offers some direction on statistical education and practice.

5.1 Scaling Plant Physiological and Ecosystem Ecology

One particularly exciting aspect of the hierarchical Bayesian modeling framework is the potential to link plant and ecosystem processes that span a range of temporal, spatial, and biological scales. For example, we may want to learn about the underlying mechanisms controlling carbon fluxes and stocks at different scales – spanning leaves, canopies, soils, and whole-ecosystems – and how mechanisms operating at one scale may feed back to another scale. Process models are very useful tools for linking mechanisms across scales (Rastetter et al. 2003), but a rigorous coupling of such models with data that also span a range of scales is generally lacking. Importantly, the process models contain ecologically and biophysically meaningful parameters related to scale-dependent mechanisms, and some parameters may even be important to scaling from one level to another. A hierarchical Bayesian integration of process models and datasets spanning different scales provides a method for obtaining "good" estimates of these process parameters.

For example, consider linking leaf-level gas exchange (fast, small scales), ecosystem carbon fluxes (fast, larger scales), and ecosystem carbon stocks (slow, larger scales). We may have data related to each of these processes, such as measurements of leaf-level photosynthesis, canopy-level gas exchange, soil respiration, net ecosystem exchange, and soil carbon stocks. We could use the conditional specification introduced earlier to write, for example, the likelihood of the carbon stock data given latent carbon stocks (predicted by an ecosystem carbon cycle model; part of Stage 1, Eq. 12.1), the distribution of latent carbon stocks given latent soil respiration, latent net ecosystem exchange, and latent initial carbon stocks (part of Stage 2, Eq. 12.2), the likelihood of the net ecosystem exchange data given latent net ecosystem exchange (part of Stage 1), and so forth.

Note that data related to a specific process are conditioned on the associated latent process, as described in Section 4.1 (Fig. 2, Box G). Here, we may assume that the different types of observations (data) are conditionally independent given the latent processes such that Stage 1 can be written as:

$$\begin{gathered} p(\text{data}\,|\,\text{latent process}, ...)= \\ p(\text{carbon stocks data}\,|\,\text{latent catbon stocks}, ...)\times \\ p(\text{net ecosystem exchange data}\,|\,\text{latent net ecosystem exchange}, ...)\times \\ ...\times \\ p(\text{photosynthesis data}\,|\,\text{latent photosynthesis}, ...), \end{gathered} \tag{15}$$

where "…" refers to "covariance parameters." Likewise, the explicit linkage of processes operating at different scales is achieved in Stage 2 (Eq. 12.2), and we can also use the conditional specification ("modeling locally") to decompose the latent process model in Eq. 12.2 into a hierarchy of conditional latent process models, for example:

$$\begin{gathered} p(\text{latent process}\,|\,...)= \\ p(\text{laent carbon stocks}|\text{latent soil respiration, latent net ecosystems} \\ \text{exchange, latent initial stocks}, ...)\times \\ p(\text{latent net ecosystem exchange}|\text{latent soil respiration, latent} \\ \text{photosynthesis}, ...)\times \\ ...\times \\ p(\text{latent photosynthesis}\,|\,...), \end{gathered} \tag{16}$$

where, here, "…" is short-hand for "process model(s), process parameters, covariates." Note, the latent processes depend on ecological process models and process model parameters, in addition to other covariance parameters associated with process error. The final product is a full probability model from which we get the posterior distribution of all unknown quantities, including latent processes and process parameters, allowing inference about mechanisms operating across different scales and governing whole-system behavior.

5.2 *Implementation Issues*

Many hierarchical models may be "fit" using specialized software packages as mentioned in Section 3.3. For Bayesian computations, we recommend using WinBUGS when possible because the language is relatively transparent and development time is usually short enough to justify the longer run-times typical of a high-level, specialized language. One of the advantages of WinBUGS is that it does not require specification of the details of sampling from the posterior. However, the complexity of the problem may not be amenable to implementation in WinBUGS (e.g. the simulation models discussed in Section 4.2 cannot be accommodated by WinBUGS) or run-time may be prohibitive and other options may be more suitable. For example, R and MATLAB (The MathWorks) are less specialized, but still convenient enough to keep production time reasonable. If run-time is still prohibitive, then a language like C, C++, or Fortran may be worth the extra development time usually required of these lower-level, but generally faster languages. Often a model may be coded in a higher-level language first, followed by a faster running lower-level version. This dual approach can provide results with a relatively small initial development investment and can serve as a means to check code and results between the two implementations.

There are other important issues that we have avoided thus far but deserve some discussion. In particular, we should mention something about convergence and model checking. Convergence refers to iterative algorithms used to sample from the posterior distribution. Many of these algorithms do not immediately obtain parameter values from the posterior, requiring some number – often many thousands – of iterations before samples can be reasonably assumed to come from the posterior distribution. Visual and other standard diagnostics are available to assess convergence (e.g. Cowles and Carlin 1996; Brooks and Gelman 1998). Sometimes, apparent convergence issues may be due to nonidentifiability problems that may be resolved by re-parameterization or by the introduction of more informative priors. Also, the efficiency with which the posterior parameter space is explored does vary. One solution to a poorly efficient sampling algorithm is simply to let the sampling algorithm continue for a large number of iterations – which may take days, weeks, or months! When run-time is of the essence, substantial energy and experience may be required to devise more efficient sampling strategies (see Gamerman and Lopes 2006).

In principle, once the probability model is specified, inference is fully prescribed via Bayes theorem. This is the strength of Bayesian inference. But this is completely model dependent. So, how can we be sure that we have a "good" model? Of course, "model" refers to any component of a hierarchical model including the likelihood, prior, and any intervening hierarchical structures, including, for example, the process model. Addressing this question requires "model checking" tools, including, for example, evaluating the sensitivity of the posterior to the prior specification, constructing plots of observed vs predicted (e.g. $\bar{A}net_{d,\,t(d),j}$ vs $\bar{A}net_{d,t(d),j}$ in Section 4.1) to evaluate model goodness-of-fit (e.g. Ogle et al. 2006), comparing posterior predictive abilities of competing models (e.g. using DIC; Spiegelhalter et al. 2002), and evaluating parameter identifiability by comparing prior vs posterior distributions (Braswell et al. 2005, Hong et al. 2005). Many standard model-checking methods are described by Gelman et al. (2004). Finally, we mention the use of synthetic data generated from models with known parameter values, which can be compared with the posterior distribution obtained by subjecting the synthetic data to the hierarchical Bayesian modeling framework (see e.g. Ogle et al. 2004, Braswell et al. 2005).

5.3 *Thoughts on Statistical Education and Practice*

In his commentary on the Bayes/frequentist "roadmap" for the future of statistics as a discipline, Little (2006) concludes with three primary suggestions on future directions for statistics education and practice. Although aimed at statisticians, we believe Little's suggestions are more broadly relevant to include statistics as taught to and practiced by ecologists, including the hierarchical framework emphasized in this paper.

1. Given the current prominence of Bayesian statistics in science, Bayesian statistics must be taught, beginning at the undergraduate level. Our experience is that the vast majority of statisticians have little to no formal training or professional interest in Bayesian statistics and, consequently, have been slow to incorporate Bayesian methods into the statistics curriculum. Ecologists can help to change this by encouraging their statistician colleagues to teach Bayesian statistical methods.
2. More emphasis should be given to statistical modeling over the methods. We suggest that this include instruction on development of graphical models and their associated probability models, something we only briefly mention in Section 3.1 (for more discussion on graphical and hierarchical modeling, see Clark and Gelfand 2006). We need to provide students and researchers with the tools to allow complex problems to be solved rather instructing them on how to design a study to fit a relatively restrictive analysis framework.
3. More attention should be devoted to assessing model fit. Model-checking typically involves comparisons of predictions to observed data, i.e. to empirical evidence, and in the case of a Bayesian model, includes comparison of posterior quantities to our current knowledge about the parameters. In as much as the models that we discuss here serve as scientific hypotheses, model-checking is then vital to their use in the inherently empirical enterprise of the scientific method. We believe this to be especially important for the complex models discussed in this paper.

Acknowledgements This material is based on work supported by the National Science Foundation (under a Biological Informatics Postdoctoral Fellowship awarded to K.O. in 2003 and Biological Informatics Starter Grant 0630119 awarded to K.O. in 2006) and the Department of Energy National Institute for Climatic Change Research (grant awarded to K.O. in 2006). The authors would like to thank Kimberly Garvie for conducting the literature searches associated with Table 1 and Fig. 1.

References

Aber JD, Federer CA (1992) A generalized, lumped-parameter model of photosynthesis, evapotranspiration and net primary production in temperate and boreal forest ecosystems. Oecologia 92:463–474

Aber JD, Reich PB, Goulden ML (1996) Extrapolating leaf CO_2 exchange to the canopy: A generalized model of forest photosynthesis compared with measurements by eddy correlation. Oecologia 106:257–265

Barbour MG, MacMahon JA, Bamberg SA, Ludwig JA (1977) The structure and distribution of *Larrea* communities. In: Mabry TJ, Hunziker JH, Difeo DR (eds) Creosote bush: biology and chemistry of *Larrea* in New World deserts. (US/IBP synthesis series 6) Halsted, New York, pp 227–251

Barnard GA (1958) Studies in the history of probability and statistics. IX. Thomas Bayes's essay towards solving a problem in the doctrine of changes. Biometrika 45:293–295

Berliner LM (1996) Hierarchical Bayesian time series models. In: Hanson KM, Silver RN (eds) Maximum entropy and Bayesian methods. Kluwer, Dordrecht, pp 15–22
Braswell BH, Sacks WJ, Linder E, Schimel DS (2005) Estimating diurnal to annual ecosystem parameters by synthesis of a carbon flux model with eddy covariance net ecosystem exchange observations. Global Change Biol 11:335–355
Breslow NE, Clayton DG (1993) Approximate inference in generalized linear mixed models. J Am Stat Assoc 88:9–25
Brooks SP (1998) Markov chain Monte Carlo method and its application. J R Stat Soc D 47:69–100
Brooks SP, Gelman A (1998) General methods for monitoring convergence of iterative simulations. J Comput Graph Stat 7:434–455
Canadell JG, Mooney HA, Baldocchi DD, Berry JA, Ehleringer JR, Field CB, Gower ST, Hollinger DY, Hunt JE, Jackson RB, Running SW, Shaver GR, Steffen W, Trumbore SE, Valentini R, Bond BY (2000) Carbon metabolism of the terrestrial biosphere: a multitechnique approach for improved understanding. Ecosystems 3:115–130
Cappe O, Robert CP (2000) Markov chain Monte Carlo: 10 years and still running! J Am Stat Assoc 95:1282–1286
Carpenter SR, Kitchell JF, Cottingham KL, Schindler DE, Christensen DL, Post DM, Voichick N (1996) Chlorophyll variability, nutrient input, and grazing: evidence from whole-lake experiments. Ecology 77:725–735
Casella G, George EI (1992) Explaining the Gibbs sampler. Am Stat 46:167–174
Chib S, Greenberg E (1995) Understanding the Metropolis–Hastings algorithm. Am Stat 49:327–335
Clark JS (2003) Uncertainty and variability in demography and population growth: a hierarchical approach. Ecology 84:1370–1381
Clark JS (2005) Why environmental scientists are becoming Bayesians. Ecol Lett 8:2–14
Clark JS, Gelfand AE (2006) A future for models and data in environmental science. Trends Evol Ecol 21:375–380
Clark JS, Lewis M, McLachlan JS, Hiller IS, Lambers J (2003a) Estimating population spread: what can we forecast and how well? Ecology 84:1979–1988
Clark JS, Mohan J, Dietze M, Ibanez I (2003b) Coexistence: how to identify trophic trade-offs. Ecology 84:17–31
Clark JS, Ferraz G, Oguge N, Hays H, DiCostanzo J (2005) Hierarchical Bayes for structured, variable populations: from recapture data to life-history prediction. Ecology 86:2232–2244
Cottingham KL, Schindler DE (2000) Effects of grazer community structure on phytoplankton response to nutrient pulses. Ecology 81:183–200
Cowles MK, Carlin BP (1996) Markov chain Monte Carlo convergence diagnostics: a comparative review. J Am Stat Assoc 91:883–904
Crome FHJ, Thomas MR, Moore LA (1996) A novel Bayesian approach to assessing impacts of rain forest logging. Ecol Appl 6:1104–1123
Dargaville R, McGuire AD, Rayner P (2002) Estimates of large-scale fluxes in high latitudes from terrestrial biosphere models and an inversion of atmospheric CO_2 measurements. Clim Change 55:273–285
Davidian M, Giltinan DM (1995) Nonlinear models for repeated measurement data. Chapman and Hall/CRC, Boca Raton
Diggle PJ, Heagerty P, Liange KY, Zeger SL (2002) Analysis of longitudinal data. Oxford University Press, New York
Dixon PM, Ellison AM, Gotelli NJ (2005) Improving the precision of estimates of the frequency of rare events. Ecology 86:1114–1123
Ellison AM (1996) An introduction to Bayesian inference for ecological research and environmental decision-making. Ecol Appl 6:1036–1046
Ellison AM (2004) Bayesian inference in ecology. Ecol Lett 7:509–520
Enting IG, Trudinger CM, Francey RJ (1995) A synthesis inversion of the concentration and $\delta^{13}C$ of atmospheric CO_2. Tellus B Chem Phys Meteorol 47:35–52

Etienne RS, Olff H (2005) Confronting different models of community structure to species-abundance data: a Bayesian model comparison. Ecol Lett 8:493–504
Farquhar GD, Caemmerer S von (1982) Modelling of photosynthetic response to environmental conditions. In: Lange OL, Nobel PS, Osmond CB, Ziegler H (eds) Water relations and carbon assimilation. (Encyclopedia of plant physiology. Physiological plant ecology II) Springer, Berlin Heidelberg New York, pp 549–587
Farquhar GD, Sharkey TD (1982) Stomatal conductance and photosynthesis. Annu Rev Plant Physiol Plant Mol Biol 33:317–345
Farquhar GD, Caemmerer S von, Berry JA (1980) A biochemical model of photosynthetic CO_2 assimilation in leaves of C_3 species. Planta 149:78–90
Gamerman D, Lopes HF (2006) Markov chain Monte Carlo. Chapman and Hall/CRC, Boca Raton
Gelfand AE, Hills SE, Racinepoon A, Smith AFM (1990) Illustration of Bayesian inference in normal data models using Gibbs sampling. J Am Stat Assoc 85:972–985
Gelman A, Carlin JB, Stern HS, Rubin DB (2004) Bayesian data analysis. Chapman and Hall/CRC, Boca Raton
Gertner GZ, Fang SF, Skovsgaard JP (1999) A Bayesian approach for estimating the parameters of a forest process model based on long-term growth data. Ecol Model 119:249–265
Gilks WR, Richardson S, Spiegelhalter DJ (1996) Markov chain Monte Carlo in practice. Chapman and Hall/CRC, Boca Raton
Goldstein H (2002) Mulitlevel statistical models. Hodder Arnold, London
Hastings WK (1970) Monte Carlo sampling methods using Markov chains and their applications. Biometrika 57:97–109
Hong B, Stawrman RI, Swaney DP, Weinstein DA (2005) Bayesian estimation of input parameters of a nitrogen cycle model applied to a forested reference watershed, Hubbard Brook Watershed six. Water Resour Res 41. DOI 10.1029/2004WR003551
Huelsenbeck JP, Ronquist F (2001) MRBAYES: Bayesian inference of phylogenetic trees. Bioinformatics 17:754–755
Huelsenbeck JP, Ronquist F, Nielsen R, Bollback JP (2001) Bayesian inference of phylogeny and its impact on evolutionary biology. Science 294:2310–2314
Katul GG, Ellsworth DS, Lai CT (2000) Modelling assimilation and intercellular CO_2 from measured conductance: a synthesis of approaches. Plant Cell Environ 23:1313–1328
Kitching RL (1983) Systems ecology: an introduction to ecological modelling. University of Queensland Press, St. Lucia
Laird NM, Ware JH (1982) Random-effects models for longitudinal data. Biometrics 38:963–974
Lambers H, Chapin FS, Pons TL (1998) Plant physiological ecology. Springer, Berlin Heidelberg New York
Larget B, Simon DL (1999) Markov chain Monte Carlo algorithms for the Bayesian analysis of phylogenetic trees. Mol Biol Evol 16:750–759
Law RM, Rayner PJ, Wang YP (2004) Inversion of diurnally varying synthetic CO_2: network optimization for an Australian test case. Global Biogeochem Cycles 18:GB1044. DOI 1010.1029/2003GB002136
Levin SA (1992) The problem of pattern and scale in ecology. Ecology 73:1943–1967
LI-COR (2004) Using the LI–6400: Portable Photosynthesis System, ver 5. LI-COR Biosciences, Lincoln
Lindley DV, Smith AFM (1972) Bayes estimates for the linear model. J R Stat Soc B Stat Method 34:1–41
Lindsey JK (1993) Models for repeated measurements. Oxford University Press, New York
Lindstrom MJ, Bates DM (1990) Nonlinear mixed effects models for repeated measures data. Biometrics 46:673–687
Little RC, Milliken GA, Stroup WW, Wolfinger RD (1996) SAS system for mixed models. SAS Institute, Cary
Little RJ (2006) Calibrated Bayes: a Bayes/frequentist roadmap. Am Stat 60:213–223

Lunn DJ, Thomas A, Best N, Spiegelhalter DJ (2000) WinBUGS – a Bayesian modelling framework: concepts, structure, and extensibility. Stat Comput 10:325–337

Malakoff D (1999) Bayes offers a 'new' way to make sense of numbers. Science 286:1460–1464

Mau B, Newton MA, Larget B (1999) Bayesian phylogenetic inference via Markov chain Monte Carlo methods. Biometrics 55:1–12

McKown AD, Moncalvo JM, Dengler NG (2005) Phylogeny of *Flaveria* (Asteraceae) and inference of C_4 photosynthesis evolution. Am J Bot 92:1911–1928

Metropolis N, Rosenbluth AW, Rosenbluth MN, Teller AH, Teller E (1953) Equation of state calculations by fast computing machines. J Chem Phys 21:1087–1092

Miller RE, McDonald JA, Manos PS (2004) Systematics of *Ipomoea* subgenus *quamoclit* (Convolvulaceae) based on its sequence data and a Bayesian phylogenetic analysis. Am J Bot 91:1208–1218

Nobel PS (1999) Physicochemical and environmental plant physiology. Academic, San Diego

O'Hara RB, Arjas E, Toivonen H, Hanski I (2002) Bayesian analysis of metapopulation data. Ecology 83:2408–2415

Oechel WC, Odening WR, Strain BR (1972) Photosynthetic rates of a desert shrub, *Larrea divaricata* Cav., under field conditions. Photosynthetica 6:183–188

Ogee J, Peylin P, Cuntz M, Bariac T, Brunet Y, Berbigier P, Richard P, Ciais P (2004) Partitioning net ecosystem carbon exchange into net assimilation and respiration with canopy-scale isotopic measurements: an error propagation analysis with $^{13}CO_2$ and $CO^{18}O$ dGB2019ata. Global Biogeochem Cycles 18:GB2019. DOI 10.1029/2003GB002166

Ogle K, Reynolds JF (2002) Desert dogma revisited: coupling of stomatal conductance and photosynthesis in the desert shrub, *Larrea tridentata*. Plant Cell Environ 25:909–921

Ogle K, Wolpert RL, Reynolds JF (2004) Reconstructing plant root area and water uptake profiles. Ecology 85:1967–1978

Ogle K, Uriarte M, Thompson J, Johnstone J, Jones A, Lin Y, McIntire EJB, Zimmerman JK (2006) Implications of vulnerability to hurricane damage for long-term survival of tropical tree species: a Bayesian hierarchical analysis. In: Clark JS, Gelfand AE (eds) Hierarchical modelling for the environmental sciences: statistical methods and applications. Oxford University Press, New York, pp 98–117

Oren R, Sperry JS, Katul GG, Pataki DE, Ewers BE, Phillips N, Schafer KVR (1999) Survey and synthesis of intra- and interspecific variation in stomatal sensitivity to vapour pressure deficit. Plant Cell Environ 22:1515–1526

Ovaskainen O, Laine AL (2006) Inferring evolutionary signals from ecological data in a plant–pathogen metapopulation. Ecology 87:880–891

Parton WJ, Schimel DS, Cole CV, Ojima DS (1987) Analysis of factors controlling soil organic matter levels in Great Plains grasslands. Soil Sci Soc Am J 51:1173–1179

Pearcy RW, Schulze ED, Zimmermann R (1989) Measurement of transpiration and leaf conductance. In: Pearcy RW, Ehleringer J, Mooney HA, Rundel PW (eds) Plant physiological ecology. Chapman and Hall, London, pp 137–160

Peek MS, Russek-Cohen E, Wait DA, Forseth IN (2002) Physiological response curve analysis using nonlinear mixed models. Oecologia 132:175–180

Pinheiro JC, Bates DM (2000) Mixed-effects models in S and S-PLUS. Springer, Berlin Heidelberg New York

Pinheiro JC, Bates DM, DebRoy S, Sarkar D (2006) Linear and non-linear mixed effects models: R package version 3.1–73. R Foundation for Statistical Computing, Vienna

R Development Core Team (2006) R: a language and environment for statistical computing. R Foundation for Statistical Computing, Vienna

Radtke PJ, Burk TE, Bolstad PV (2002) Bayesian melding of a forest ecosystem model with correlated inputs. For Sci 48:701–711

Rains MC, Mount JE, Larsen EW (2004) Simulated changes in shallow groundwater and vegetation distributions under different reservoir operations scenarios. Ecol Appl 14:192–207

Rastetter EB, Aber JD, Peters DPC, Ojima DS, Burke IC (2003) Using mechanistic models to scale ecological processes across space and time. Bioscience 53:68–76
Robert CP, Casella G (2004) Monte Carlo statistical methods. Springer, Berlin Heidelberg New York
SAS Institute (2001) The MIXED procedure and %NLINMIX macro. SAS Institute, Cary. Available at http://www.sas.com/statistics/doc.html)
SAS Institute (2006) The GLIMMIX procedure, June 2006. SAS Institute, Cary. Available at http://www.sas.com/statistics/doc.html)
Schulze ED (2000) Carbon and nitrogen cycling in european forest ecosystems. Springer, Berlin Heidelberg New York
Searle SR, Casella G, McCulloch CE (1992) Variance components. Wiley, New York
Smith AFM, Gelfand AE (1992) Bayesian statistics without tears: a sampling-resampling perspective. Am Stat 46:84–88
Spiegelhalter DJ, Best NG, Carlin BR, Linde A van der (2002) Bayesian measures of model complexity and fit. J R Stat Soc B Stat Method 64:583–616
Stefanski LA (2000) Measurement error models. J Am Stat Assoc 95:1353–1358
Van Oijen M, Rougier J, Smith R (2005) Bayesian calibration of process-based forest models: bridging the gap between models and data. Tree Physiol 25:915–927
Wikle CK (2003a) Hierarchical Bayesian models for predicting the spread of ecological processes. Ecology 84:1382–1394
Wikle CK (2003b) Hierarchical models in environmental science. Int Stat Rev 71:181–199
Wikle CK, Berliner LM, Cressie N (1998) Hierarchical Bayesian space-time models. Environ Ecol Stat 5:117–154
Wikle CK, Milliff RF, Nychka D, Berliner LM (2001) Spatiotemporal hierarchical Bayesian modeling: tropical ocean surface winds. J Am Stat Assoc 96:382–397
Xu T, White L, Hui DF, Luo YQ (2006) Probabilistic inversion of a terrestrial ecosystem model: analysis of uncertainty in parameter estimation and model prediction. Global Biogeochem Cycles 20:GB2007. DOI 10.1029/2005GB002468
Yang ZH, Rannala B (1997) Bayesian phylogenetic inference using DNA sequences: a Markov chain Monte Carlo method. Mol Biol Evol 14:717–724

Quaternary Palaeoecology: Africa and its Surroundings

Burkhard Frenzel

Contents

Abstract The remarkable richness in biodiversity of various African mountain systems poses several problems, which are difficult to solve in terms of Quaternary vegetation history and palaeoecology because, regrettably, good archives which go far back into the past are repeatedly lacking there. These archives should inform us about: (a) how climate and vegetation might have changed there during the Quaternary; (b) whether palaeoclimatic changes had simultaneously happened everywhere in Africa and, if so, whether these changes had always followed identical tendencies; (c) and – last but not least – at which time and with what consequences a remarkable impact of man on his environment began. For solving these problems the biological and sedimentological contents of deep-sea cores are used, as well as all available palaeoecological tracers on land. It will be seen that – in general – a better knowledge exists only for certain phases of the Last Glaciation and for those of the Holocene, and that still only little information exists on the timing of the first remarkable human impact on his environment in different regions, although the history of mankind in Africa goes back into the distant past.

1 African Biodiversity

Mutke and Barthlott (2005) showed that the African biodiversity centres of mosses and ferns are situated in the Cape Region of South Africa and in eastern Madagascar, whereas according to Barthlott et al. (2005) the African centres of vascular plants –

Burkhard Frenzel
Institut für Botanik und Botanischer Garten 210, Universität Hohenheim, D-70593 Stuttgart, Germany
e-mail: bfrenzel@uni-hohenheim.de

U. Lüttge et al. (eds.), *Progress in Botany 69.*

in general – are found in Cameroon–Guinea, with the Mt. Cameroon and the Guinean sub-centres, within the Lake Albert Rift Valley, in Maputuland and Pondoland, in the Cape Region and in Madagascar as well. The Quaternary histories of climate, vegetation and human impact on his environment in these centres will soon be intensively studied. In contrast, McClean et al. (2005) investigated the interrelations between plant biodiversity and climate change, concentrating their efforts on African plant species, which occur to the south of 27° N. It is suggested that the modern distribution patterns of the 5197 species studied are caused by natural processes only. Regrettably, this means that the history of human influences on these distribution patterns is not considered. Yet it is calculated how these distribution patterns will change up to 2085, if the atmospheric concentrations of important greenhouse gases increase. Kier et al. (2005) studied changes in the global patterns of plant biodiversity, which might depend on density changes of human populations, on man-made land transformation, electric power and transport infrastructures. Yet in eight out of 14 biomes no distinct interrelations between "human footprints" and plant species richness could be observed. Knowing what happened in Europe since the beginning of agriculture, this is rather astonishing. Finally Küper et al. (2004) investigated the African "hotspots" of biodiversity, defining hotspots as regions with a high plant biodiversity and a strong human impact. However, what is a strong human impact? From the following it can be seen that, in the past, primitive agricultural practises and overgrazing by probably small herds had repeatedly a strong and disastrous influence on the ecological situation in various regions of Africa.

2 Deep-Sea Cores as Archives of African Vegetation History and of Palaeoecology

In Africa the number of suitable archives of Quaternary vegetation history is still restricted. This holds for the number of sites studied as well as for the time span which could be investigated until now. Thus, it is quite reasonable that deep-sea cores, obtained off the coasts of Africa, have become very interesting for Quaternary palaeoecology. However, sporomorphs, which have been transported by winds or rivers from the continent into the ocean will not be immediately deposited on the sea-floor but they will probably be dislocated and also to some extent be changed by the ocean's currents. Thus, it becomes important to see what is known about the history of these currents and moreover about the distribution of suitable marine sites, where pollen and spores might be deposited. Bylinskaya and Golovina (1990) described the Quaternary stratigraphy of 11 cores, obtained between the Cape Verde Islands and the Mid-Atlantic Ridge. In general the cores seem to be rich in floral and faunal remains. The dating was done by palaeomagnetic investigations and Blazhchishin (1978) produced an interesting map of Holocene sediments of the tropical and subtropical Atlantic Ocean. In view of the African palaeoecology, the Cape Verde Islands are of great interest. Pushcharovskij et al. (1989) showed that deep sea cores are strongly influenced by post-sedimentary disturbances, caused both by tectonics, sliding on steep submarine slopes as well as by carbonate solution. Thus the biological remains of 11 deep-sea cores there are badly distorted. Gorshkov et al. (1977) published very interesting maps of deep-sea currents of the Atlantic and the Indian Ocean, which might become important for a correct interpretation of marine palynological data. All these efforts are concentrated on modern times. However, Upper Quaternary changes of these marine currents have also repeatedly been investigated. For the tropical eastern Atlantic Ocean see Höll and Kemle von Mücke (2000), for the southeastern Atlantic Ocean off the coasts of Namibia see Dingle et al. (1996), Schmiedl and Mackensen (1997) and Kirst et al. (1999). Based on the analysis of the presently prevailing plankton foraminifera communities, Little et al. (1997) described changes of marine currents during the past 160 000 years, including the upwelling phenomenon off the coasts of Namibia. It is assumed that communities of the past, which were dominated by only one species or subspecies, had the same ecological requirements as related modern communities. But is it true that the ecological "requirements" of these taxa have remained constant, if the composition of the communities might have changed during the course of time?

Hooghiemstra et al. (2006) comprehensively analysed the Upper and Middle Quaternary West African vegetation history, based on pollen analyses of several deep sea cores and on the present-day distribution patterns of sporomorphs on land and in the bottom surfaces of the East Atlantic Ocean. Additionally, wind systems at various elevations above ground and ocean currents at various depths were duly taken into consideration. It turns out that, on land and in the marine bottom surfaces, the modern distribution patterns of the Sudanian, the Sudano-Guinean and the Guinean pollen taxa correspond quite well with each other, while the European Mediterranean, the Saharan and the Sahelian taxa are transported much farther to the south in the ocean than on land. In this respect it is noteworthy that, according to Rognon and Coudé-Gaussen (1992, 1996) the near-surface wind systems of various northwest African regions and those of the atmosphere's upper levels have followed quite different patterns since the last glacial maximum (LGM), which of course, significantly influences pollen transport there. In northwest Africa, strong changes in the geographical positioning of the boundaries between main plant geographical entities of the Middle and Upper Quaternary periods were intensively analysed by Hooghiemstra (1988, 1996), Dupont (1993), Dupont et al. (1996) and Hooghiemstra et al. (2006), although the dating quality of interesting stratigraphical levels sometimes is not very convincing (Hooghiemstra 1988). When the climate became drier during Miocene and Pliocene times, the wooded West African savannah replaced the evergreen forests there. Dupont (1993) and Hooghiemstra (1996) as well as Frédoux (1994) and Hooghiemstra et al. (2006) could show that the tropical rain forest seems to have existed in mountain systems of western Africa even during stadials of earlier glaciations, yet that this vegetation type did not survive the last glaciation there. In contrast, *Podocarpus*-dominated vegetation types, which are now characteristic elements of high-mountain vegetation belts in central Western Africa, moved strongly downwards from northern Angola to the Guinean mountains at the end of the last but one glaciation within the region and again during the stadials of the beginning of the last glaciation, but that during the LGM they were restricted to the mountains of the Congo basin and Cameroon (Dupont et al. 1996). This points to the possibility of a different species composition of exacting plant communities within glacial refuge areas during various cold-climate periods of the Quaternary. Regrettably, up to now, this problem has not been analysed, but it is interesting that, in the relevant pollen diagrams by Jahns (1996), the following modern tropical rain forest taxa could not be found in marine sediments of the Congo fan, dating from stadials of the preceding glaciation and the LGM: *Cestis*, *Diospyros*, *Manilkaro* type, *Trichilia*, ± *Sesbania*, *Rauvolfia*, ± *Lophira*, *Elaeis*, ± *Allophyllus*, *Flacourtia*, *Tapinanthus* and *Crudia*. The same holds for plants of the transitional and dry forest, like *Burkea*, *Brachystegia*, *Sauvagesia* and *Daniellia*. The problem is whether this situation is only caused by poor pollen production and insufficient pollen transport, or whether these taxa were missing in the existing vegetation types during that time. To answer this question many more analyses like those of Lézine and Vergnaud Grazzini (1993) and Dupont and Weinelt (1996) are required. Nevertheless, it is questionable whether Lézine and Vergnaud Grazzini (1993) are right in suggesting that the high numbers of pteridophyte spores in sediments of the LGM off the coast of Ghana point to the former existence of forests on the nearby coasts, since an over-representation of fern spores, together with a small total number of pollen counted very often points to a secondary enrichment of these very resistant microfossils. It is interesting to note that evidently the Niger River sedimented much more terrestrial material on its fan during stadials of the last two glaciations than during warm-climate periods, because vegetation covered much more of the land surfaces during the latter than during stadials (Zabel et al. 2001).

Shi and Dupont (1997), Shi et al. (2000) and Dupont and Behling (2006) made a pollen analysis study of the Upper Quaternary vegetation history of southwestern Africa, together with the history of the adjacent ocean currents from deep-sea cores. Shi and Dupont (1997) contributed a wealth of most interesting information, partially given in maps, showing important steps of vegetation history during the past 300 000 years. Regrettably, zonation and dating of the pollen diagrams only follows that of the $\delta^{18}O$ curves of the ocean's plankton foraminifera. It may be questioned, whether this type of dating is reliable enough since, most of all, vegetation history on land is not always directly connected with only climate history, if complicated migrations of the involved taxa took place. Further, Shi and Dupont (1997) suggested that the Afromontane Forest was intensively cleared by man during the Upper Holocene. This problem of the onset of a remarkable human impact on the African vegetation will be discussed below. In contrast, Shi et al. (2000) made a pollen analysis investigation of a deep-sea core which had been obtained in front of the mouth of the Cunene River, northwestern-most Namibia. It spans the past 21 000 years (nine ^{14}C AMS data). Evidently, in this region there were two important phases of aridity: 21 000 to 17 500 cal BP, and 14 300 to 12 600 cal BP. Both of these phases were characterized by a strong upwelling of the Benguela Current. Finally, there was a third phase of arid climate, from 11 000 to 8900 cal

BP, in which upwelling was reduced. It could be shown that the beginning of climate warming happened there at about 17 500 cal BP, i.e. much earlier than the Bølling-Allerød phase, and that the second phase of an extremely arid climate (14 300 to 12 600 cal BP) occurred much earlier than the beginning of the Younger *Dryas* period. From this it is concluded that there were strong regional variations in the influence of the ocean's thermohaline circulation. This was confirmed by Dupont and Behling (2006), who could show that, at least during late-glacial and early Holocene times, the vegetation history in western Angola and in northwestern Namibia differed remarkably from each other. All these observations fit very well into the picture of the Upper Quaternary palaeoecological changes in South Africa which will be discussed in Section 5.

3 Quaternary Palaeoecology of Northern Africa and of its Surroundings

Reconstructions of former vegetation patterns are meanwhile strongly improved by investigations on the modern pollen deposition in various parts of northern and central Africa. Lézine and Edorh (1991) studied the West African Sudan, showing that single surface samples taken at only one site can be rather misleading. A similar comprehensive study for southern Arabia is available from Lézine et al. (1998). Vincens et al. (1997) analysed a wealth of surface samples originating from the main plant geographical formations of western Uganda; and DeBusk (1997) analysed 54 surface samples from Lake Malawi, together with 12 samples from its surroundings. Again it could be shown that water transport within the lake seriously impedes the correct reconstruction of former vegetation types.

For the LGM, Anhuf (1997) and Anhuf et al. (2006) mapped the central and north African vegetation. It could be shown that, at that time, the tropical rain forest in west and east Africa had retreated to small areas of higher mountain systems and that these refuge areas were more extended than Frenzel et al. (1992) suggested. Some problems related to this question have already been discussed in Section 2. At any rate, during full-glacial times, the hyper-arid parts of the Sahara were much larger than they are now. Yet the Late-Glacial and early Holocene climate amelioration first began at higher elevations in the eastern Sahara (Pachur and Wünnemann 1996) and – somewhat later – in the northeastern part of the Sahara (Kuper and Kröpelin 2006). Comparable regional climate history differences are reported by several other authors working in north Africa, e.g. Salzmann et al. (2002), concerning northeast Nigeria, and Van der Veen (1995), who studied the agricultural history of Libya. This phase of moister conditions of the eastern Sahara lasted from about 12 000 BP to c. 5000 BP. More information about this is provided by Stevenson et al (1993) and Swezey et al. (1999). However, it should also be considered that already during various phases of the Last Glaciation groundwater had accumulated there in deep-lying aquifers (Sonntag, in Frenzel 1992; Sultan et al. 1997). More water was added by precipitation during Late-Glacial and Early Holocene times. Thus, the eastern Saharan exact vegetation of that time depended not only on the existing climate conditions, but also on much older waters. This has always to be considered when a quantitative climate reconstruction is attempted. In view of these problems, it is appreciated that Abell et al. (1996) and Ayliffe et al. (1996) used the stable carbon and oxygen isotopic compositions of subfossil gastropod shells for a reconstruction of palaeohydrological changes in various parts of the Sudan, independent from geomorphological observations. Churcher (1992), Brookes (1993), Iacumin et al. (1996), Lario et al. (1997), Bard (2003), Williams and Nottage (2006) and Williams et al. (2006) extensively dealt with changing palaeoecological conditions during Late-Glacial and Holocene times in eastern Egypt. This concerns flora and fauna, as well as man. Williams et al. (2006) stated that around 15 000 BP the White Nile began to drain Lake Victoria, thus causing strong floods in the vicinity of Khartoum, when the summer monsoon abruptly started to bring much moisture there. Studying the catastrophic inundations of the Blue and White Niles of August 1999 in the vicinity of Khartoum, Williams and Nottage (2006) demonstrated that this could have also been repeatedly caused by strong precipitation in the Ethiopian Highlands as a consequence of La Niña events. Nevertheless, the climatically induced moister period of the Early Holocene cannot be neglected here, as was convincingly shown by Kindermann et al. (2006) in the now hyper-arid surroundings of Djara, situated between Asyut and the Farfara Oasis. Much charcoal was found there in levels dating from 8060 BP to 7300 BP, containing remnants of *Acacia*, *Tamarix*, Capparaceae, together with fruits of *Zilla spinosa*, *Anastatica hierochuntica*, as well as animal remains of *Gazella dorcas*, *G. dama*, *Addax nasomaculatus*, *Oryx dammah* and *Caracal caracal*. Undoubtedly, this is a fauna of dry, but not hyper-arid regions. These observations were confirmed by Peters (1992), who analysed the mammalian fauna of the Central Sudanese and the Upper Atbara regions. A wealth of data was used for constructing palaeovegetation maps concerning the final Pleistocene (desert and semi-desert vegetation), the Early Holocene (low rainfall savannah moved far to the north), the Middle Holocene (deserts began to expand again) and the late Middle Holocene, when the latter tendency increased further. Comparable Early Holocene changes in vegetation and climate were investigated in northwestern Sudan by Jahns (1995) and a wealth of critically evaluated, most interesting botanical data was reported by Barakat (1995a, c), Dahlberg et al. (1995), Hather (1995), Mitka and Wasylikowa (1995), Wasylikowa et al. (1995), Dahlberg and Wasylikowa (1996), and Wasylikowa (1997) about Early Neolithic settlement layers at Nabta Playa to the west of the Abu Simbel Temple in the Egyptian Western Desert, dating from about 8600 BC to 7600 BC. The charcoal remains of *Tamarix* sp., *Acacia raddeana*, *A. ehrenbergiana*, *A. nilotica*, *Acacia* sp., *Cassia senna* and *Capparis decidua* point to relatively easily accessible groundwater. Evidently there was an oasis-like vegetation in a region in which the present-day precipitation rate is less than 1 mm/year. Yet Wasylikowa et al. (1995) quite correctly warn against a too simple reconstruction of climate, because all these plant remains had been collected by man, i.e. they do not necessarily reflect the correct former natural situation. Comparable, very comprehensive data were reported by Goodman et al. (1992) concerning the Holocene fauna found in a limestone cave in the Egyptian Eastern Desert. Here, the present-day precipitation rate is very low, yet the Early Holocene fauna was astonishingly rich. It is suggested that these animals lived there in suitable localities only, not in broad vegetation belts. The same situation was described by Ballouche and Neumann (1995) for the Holocene vegetation of the West African Sahel. This points to another geobotanical concept than is repeatedly used for climate modelling: in areas of climatic transition it seems appropriate not to assume belt-like vegetation types only, which moved to and fro in the past, but rather a mosaic of ecologically distinct plant and animal communities, which behaved differentially when climate changed. Yet the question remains unanswered whether the deterioration in the ecological setting of these sites leading to modern conditions was caused by climate change or by man. Based on the modern browsing and grazing preferences of the species found, the former vegetation was reconstructed and Dittmann (1990), by means of geomorphology and archaeology, investigated the palaeoecological evolution within the Egyptian Eastern Desert. It is interesting to note that observations are reported there which point to activities of hunters and gatherers from more than 30 000 BP onwards, already (see below). According to fossil remains from a Neolithic settlement 18 km to the north of Khartoum, on the right-hand bank of the River Nile, dating from between 5960 BP and 5030 BP, cattle, sheep, goats and dogs were held there by plant-gatherer communities, which preferably used wild grasses, like *Sorghum*, *Panicum*, *Setaria*, *Eragrostis*, *Digitaria*, *Eleusine* and *Andropogon*. According to charcoal determinations, a thorn-bush savannah with a lot of fire-resistant species existed there at that time. It is discussed whether man and his animals already intensively influenced the vegetation. Further, Van der Veen (1996) reported valuable observations done in a Roman settlement (Mons Claudianus) in the hyper-arid Egyptian Eastern Desert, where, due to the use of artificial wells, an astonishingly rich flora existed at that time. However, in general, the Holocene phases of increased moisture there were caused by changing patterns in the atmosphere's circulation. These processes were intensively studied in the Arabian Sea by Sirocko et al. (1991), Sirocko (1996), Reichart et al. (1997), Zonneveld et al. (1997), Andrews et al. (1998), Cayre and Bard (1999), Rad et al. (1999) and Preusser and Radies (2006). Here, the Upper Quaternary phase of extreme aridity during the last stadial of the Last Glaciation was followed since about 14 700 cal BP by much moister conditions, which were caused there by the summer monsoon. Since about 5500 cal BP this influence decreased again, finally leading in an oscillating way to modern climatic conditions. During this phase of a moister climate lakes existed there within several dune valleys, surrounded by a sparse vegetation, the tree taxa of which were of a modern Somalia–Masai type: *Acacia*, *Commiphora*, *Maerua*, *Salvadora persica* and the Afromontane type: *Olea*, *Juniperus*, *Myrica*, *Podocarpus* (long-distance transport of its pollen grains?), *Ephedra*, *Dodonaea viscosa*, *Rhus* and *Celtis*. All these taxa, with the exception of *Podocarpus*, presently occur in Yemen. Thus, the geobotanical differences from modern conditions were obviously not too strong (Lézine et al. 1998, 2006). At that time (8500–6000 cal. BP) a grass savannah, dominated by C_3 plants, seems to have been the dominant vegetation type (Parker et al. 2006; Pickens Davies 2006), so that Neolithic cultures could exist and develop there since at least 8000 cal BP. However it is even possible that man already lived in southeastern Arabia since late Middle-Pleistocene times, having come there from eastern Africa (Parker et al. 2006). It should also be mentioned that, at Abu Hureyra, 35° 52' N in the Euphrates valley, cereals, most of all rye, were already cultivated by hunter–gatherer communities since at least 13 000 cal years BP (Hillman et al. 2001).

In the Ahaggar Massif, Central Sahara, an astonishingly diverse pollen flora was found some time ago, which (at about 4600 BP) seems to point to the occurrence there of both tropical (*Cordia*, *Cadaba*) and Central European plants (e.g. *Alnus*, *Fagus*, *Quercus*, etc.). When reinvestigating this site, Thinon et al. (1996) found that this pollen flora had evidently been brought there by birds of passage. This seems to be corroborated by Barakat (1995b), who investigated in the same mountain system droppings of *Procavia capensis*, which has no browsing nor grazing preferences. These excrements contained a rich fossil flora, dating from between 4910 BP and 4630 BP, which occurs there in approximately the same composition, yet without the just-mentioned taxa. Further, Jousse and Escarguel (2006) used reports of other research groups on the Holocene occurrence of 50 bovid taxa (without domestic animals) for reconstructing major Holocene changes of the West African vegetation in a very generalized way.

Of course, the considerable Upper Quaternary differences in available moisture within the Sudan influenced plant biomasses there, too. Lioubimtseva et al. (1996) tried to quantify these changes, assuming that the plant's dry matter consists of approx. 45% C. It was estimated that, at 18 000 BP, the Sudanian phytomass amounted to about 0.65–1.5 Gt C, at 9000–8000 BP to ca. 4.30–7.70 Gt C and at present to ca. 2.5 Gt C. According to Lézine (1989), before 13 000 BP the vegetation in the surroundings of Lake Chad was a very open grassland of Saharian type, which was at that time replaced by a Sahelian grassland; at ca. 10 000 BP Sudanian open forests and Guinean-type forests immigrated. Approximately at the same time thunderstorms seem to have been important elements of the North Nigerian weather (Sponholz et al. 1993). Yet at about 6500 BP the Early Holocene forest vegetation in the surroundings of Lake Chad began to retreat (Lézine 1989). At the same time also a remarkable human influence in the spontaneous vegetation could be felt there (Ballouche and Neumann 1995). Comparable changes in the Late-Glacial and Holocene vegetation of the Sahelian and Sudanian zones of Northeast Nigeria were intensively studied by Salzmann (1996) and Salzmann et al. (2002). Though these changes followed comparable patterns, they did not seem to have happened simultaneously everywhere. Thus the question remains whether these differences were caused by man or only by climate. However, clear traces pointing to an overall influence of man in this region could not be found (Salzmann 1996). Thus, Street-Perrott et al. (2000) interpreted the changing dust deposition in the West African Sahel solely in terms of climate change. But how can this be separated from human impact? The simultaneously acting huge Holocene changes in the surface area of Lake Chad were mapped by Leblanc et al. (2006), using space shuttle images. It was shown that these were part of a strongly oscillating process, which influenced human activities, too. Klee et al. (2004) investigated the plant remains of settlements there, dating from the beginning of the second millennium BC, to about 800 BC. At those times there a Saharo-Sudanian savannah had existed in the vicinity of the study sites, in which forests had repeatedly occurred. Thus, a relatively rich flora could be collected there by the former farmers and it seems that the problem of an early, ecologically important human impact on the environment should be investigated much more intensively in these regions. Of course, this influence became very strong at about 1660 BP when, in the north of Timbuktu, iron ores were smelted (Rolando and Raimbault 1992). During the Holocene, the palaeoecological situation near the coast of Senegal was strongly influenced by marine transgressions, whose influence reached far into the land (Lézine 1988). At about the middle of the first millennium BC, man settled in northern Tenerife, but it seems that a strong human influence on flora and vegetation occurred only since the fifteenth century AD (Machado Yanes et al. 1997).

4 Quaternary Palaeoecology of Central Africa

4.1 Palaeoecology of Glacial Times

According to isotope studies on carbonates of the Olduvai Gorge, northern Tanzania, the climate between 2 200 000 and 1 300 000 years BP seems to have been moister and about 5–7 °C colder than at present. Evidently, this caused a remarkable reduction in the amount of C_4 plants at that time (Cerling and Hay 1986). The rodent fauna (Middle Bed I) at first points to a relatively dense woodland, which changed into some type of savannah (Upper Bed I) and, finally, at about 1 795 000 BP to a vegetation composed of palms, sedges, grasses and dicots (phytoliths; see Bamford et al. 2006). For the early hominids and the coeval fauna of small mammals there, see Denys et al. (1996) and Deocampo et al. (2002). It seems that, at that time, Mt. Kenya had already experienced two mountain glaciations (between about 2 000 000 and 1 670 000 years BP), yet nothing is known about the vegetation thriving there simultaneously (Mahaney 1991). It is interesting to note that stone artefacts (Levallois flakes) on the western beach of Lake Bavingo, to the northeast of Lake Victoria, date from about 200 000 to 250 000 years BP, although again nothing is known about man's impact on his environment at that time (Tryon and McBrearty 2006).

The reconstruction of the Last Glacial vegetation of Central Africa poses some problems, because the number of suitable sites is strongly restricted. At any rate, during full-glacial times of the last two glaciations, the River Niger transported much more terrigenous material into the ocean than during warm-climate periods. Thus, vegetation must have been much more open during the glaciations (Zabel et al. 2001). At about 16 300 years cal BP the meteorite crater of Lake Bosumtwi in southern Ghana had a 60 m lower lake level than at present. Some time later, at about 14 500 years cal BP, the lake level began to rise, to +16 m above its modern position. Thus, climate must have become moister at that time. Regrettably nothing is reported about the coeval vegetation (Shanahan et al. 2006). The Shum Lake Cave, western Cameroon, 1650 m a.s.l., was surrounded after 30 300 BP by a vegetation composed of *Protea* sp., *Kigelia* cf. *africana*, *Drypetes* sp., *Triumphetta* sp. and *Hypericum* sp. Evidently, this was a montane forest with several open patches, pointing to drier conditions than nowadays (Moeyersons 1997). According to $\delta^{13}C$ values of organic matter, a mosaic of forest stands and Poaceae/Cyperaceae savannah dominated West Cameroon (Lake Barombi Mbo, 301 m a.s.l.), where at present dense forests prevail spontaneously (Giresse et al. 1994). Traces of this dry period were also found at other sites: Congo (Elenga et al. 1994; Mercader et al. 2000), the surroundings of the Central Rift Valley, Kenya (Taylor 1990; Bonnefille et al. 1991; Maitima 1991; Jansen et al. 1995) and at the northern part of Lake Tanganyika (Vincens 1993; Verschuren et al. 1996). According to palaeopedological investigations on Mt. Kilimanjaro (Zech 2006), vegetation belts seem to have been at least 1000 m lower than at present. Yet the question cannot be answered how the vegetation was composed at these lower levels: did the unchanged vegetation belts just move downhill, or at that time did a mixture of plants exist, which presently occurs in different vegetation belts? The same problem exists in higher latitudes of the northern and southern hemispheres. For a long time, it was suggested that Lake Victoria did not exist during full-glacial times of the Last Glaciation. Yet in this lake more than 500 endemic "species" of cichlids are reported (Johnson et al. 2005): Did all these species develop there only during the past 15 000 years, when the lake was formed again? Further, based on sedimentological investigations, Garcin et al. (2006) suggested that the climate of the LGM and the Younger *Dryas* time was moister in southwestern Tanzania than it is at present. Generally, during Late-Glacial and Early Holocene times, climate became much moister in Equatorial Africa than it is at present, causing a fast spread of various forest types [for Southwest Cameroon see Wirrmann et al.(2001); for Ethiopia see Vincens (1989); Telford and Lamb (1999); Lamb et al. (2000); for Equatorial East Africa see Casanova and Hillaire-Marcel (1992); Jolly and Bonnefille (1992); Mäckel (1992); Taylor (1993); Jolly et al. (1994); Marchant and Taylor (1998); Barker et al. (2000); Thevenon et al. (2003)]. In Central Madagascar, at about 1800 m a.s.l., a crater lake was surrounded by an Ericaceae-dominated vegetation from about 40 800 to ca. 22 300 years cal BP. The lake dried up during the LGM. The Ericaceae rapidly retreated between 16 900 and 16 500 cal BP. After a short cold spell from 16 000 to 15 100 BP, this tendency continued and tree taxa of middle and lower vegetation belts immigrated, together with plants of a savannah vegetation (Gasse and van Campo 1998).

4.2 Holocene Palaeoecology

During Holocene times, considerable changes in vegetation, repeatedly happened in Central Africa, most of all an opening of the forest vegetation and an expansion of various types of savannah. In general this is attributed to climate change [for Western Central Africa see Giresse et al. (1994); Wirrmann et al. (2001); for Ethiopia see Bonnefille and Mohammed (1994); Machado et al. (1998); Vincens et al. (1998); Lamb et al. (2000); for Central East Africa see Stager (1988); Maitima (1991); Bonnefille et al. (1991); Casanova and Hillaire-Marcel (1992); Jolly and Bonnefille (1992); Barker et al. (2000); Peyron et al. (2000); Lamb et al. (2003)]. However, Telford and Lamb (1999) pointed to the fact that, at least in volcanic areas like those of Ethiopian crater lakes, groundwater input, which is caused by geothermal processes and which is independent of climatic change has also to be taken into consideration. Further, man already worked there for long times in various regions.

Mercader et al. (2000) reported that man settled in Ituri, Congo, at 18 800 ± 100 BP and, after a long interruption, began there again at 10 530±50 BP in Matangai Turu; and Mworia-Maitima (1997) wrote that man already influenced the East African vegetation by artificially setting fires since ca. 12 000 BP. However in general man is discussed as an ecological factor only for much later times there [for Western Cameroon see Moeyersons (1997); for Northern Ethiopia see Darbyshire et al. (2003); for Equatorial East Africa see Vincens (1989); Taylor (1990, 1993); Bonnefille et al. (1991); Jolly and Bonnefille (1992); Jolly et al. (1994); Marchant and Taylor (1998); Taylor et al. (2000), Thevenon et al. (2003)]. Thus, it seems that this problem should be much more intensively investigated than in the past, since there is a discrepancy between the very old traces of mankind in Africa and his allegedly late impact on the ecological situation. Perhaps it might be good to look not only for the pollen of cultivated plants but also for those selected by grazing domesticated animals, as we have done studying the history of human impact on the Tibetan vegetation (Frenzel and Adamczyk 2004). It is well-known that in Madagascar 17 species of mammals, reptiles and birds became extinct during the Upper Holocene. It is assumed that this was caused by man, who seems to have settled on this island by about 1200–1000 years BP (Burney 1987a, b; Clarke et al. 2006). Evidently man soon strongly changed the spontaneous natural vegetation by the use of fire (Burney 1987a), yet intensive fires already occurred there between about 36 000 BP and 35 000 BP. Their intensity strongly diminished during full-glacial times, but increased again during the Holocene (Burney 1987b). This might have influenced the living conditions of some animal species, which soon became extinct there. However at least in the case of the extinction of the elephant bird (*Aepyornis*), a direct connection between man's activities and the extinction of this bird cannot be proven, though it happened at the same time as man's impact became strong (Clarke et al. 2006).

5 Quaternary Palaeoecology of Southern Africa

According to a fossil fauna near Johannesburg (Makapansgat Limeworks Cave) and dating from about 3.2–2.5 million years BP, the vegetation seems to have been of a bush and woodland type at that time (Schubert et al. 2006), with a relatively rich mammalian fauna (*Parmularius rugosus*, *Pelorovis* sp. (*antiquus*), *Hippotragus gigas*, *Antidorcas recki*, *Redunca arundinum*, *Rabaticeras arambourgi*), pointing to a comparable vegetation that lived there mostly at about 990 000 BP (Herries et al. 2006).

At least during Middle and Upper Quaternary times, the climate history of Southern Africa in general seems to have followed other pathways than it did on the Northern Hemisphere, thus causing moist periods there even during the LGM (for Southwestern Kalahari, Etosha Pan, Okavango and Makgadikgadi in Botswana see Heine 1992). This was corroborated by Brook et al. (1992, 1996), Klein and Cruz-Uribe (1996), Partridge et al. (1997) and Huntsman-Mapila et al. (2006). Information on a very rich mammalian fauna, which lived there during stadial phases of Upper Quaternary glaciations is given by Klein et al. (1999). Nevertheless there were also quite remarkable regional differences (O'Connor and Thomas 1999; Thomas et al. 1997). Further, the timing and quality of climate change in northwesternmost Namibia strongly resembles that of Central and Northern Africa (Holmgren and Karlén 1995; Eitel and Zöller 1996; Gingele 1996; Srivastava et al. 2006; and perhaps Avery 1992; yet in Avery's paper the dating quality is not very convincing).

As stated in Section 1, one of the most important global biodiversity centres is situated in South Africa. Here, Pleistocene history of the Fynbos vegetation, which is so extremely rich in Ericaceae, Restionaceae and Asteraceae, could be followed at least in some major steps by Carr et al. (2006a). It turned out that, during the LGM, this vegetation moved down to lower elevations, even to the Agulhas Plain, when the sea level sank considerably, thus enabling in this plain the formation of dune systems at drier sites, besides the Fynbos at moister habitats (Carr et al. 2006b). Yet at present, it is unknown how the Fynbos associations might have looked like during the LGM. Were they of the same composition as they are now, or had they mixed with plant species which generally occur at lower levels? If so, the ecological "requirements" of at least some Fynbos plants might have changed, too. This would be important for extrapolating the climatological conditions of that time from their occurrence. Even near sea level (Klaarfontein Springs, Western Cape) the Fynbos vegetation seems to have experienced some changes in composition during Middle to Late Holocene times. Yet it is unknown whether man caused some of these changes by his cattle grazing since about 1900 BP or whether they were caused by climate only (Meadows and Baxter 2001). At any rate, sea surface temperatures there were approximately 5 °C lower during the LGM than they are now (Hutson 1980). Sea-level changes of the Upper Pleistocene (Ramsay and Cooper 2002) and Holocene (Compton 2001, 2006) were analysed in various regions of the South African coasts. These results are important for better understanding the history of the coastal vegetation. Simultaneously there were climate changes which, at least during the past 900 years, were comparable in timing and quality with those which happened at the same time in Europe (Holmgren et al. 1999). These Holocene climate changes also influenced plant community composition in the eastern Orange Free State and in Lesotho, as could be reconstructed from charcoal layers in rock shelters (Esterhuysen 1992; Esterhuysen and Mitchell 1996), as well as in some sites near Windhoek and near Tsumeb (Scott et al. 1991).

6 Conclusions

During the past decades considerable progress has been achieved in the study of Quaternary palaeoecological changes in Africa. This holds for the drier regions as well as for the area of the modern tropical rain forest. Yet it seems that our knowledge in palaeobotany is much better than in palaeozoology. Concerning palaeoecological changes, it is very difficult to differentiate between climate-dependent processes and the influence of man, although in Africa man competed with vegetation and fauna for a much longer time than in other regions of the Earth. It may be that this situation is caused by the modern way of research, which in general focuses on possible interrelations between climate and biosphere, but it should duly be taken into consideration that man has always been a very independent and an astonishingly effective palaeoecological factor. All these investigations might be substantially improved, if better and more precisely acting dating methods could be employed.

References

Abell P, Hoelzmann P, Pachur H-J (1996) Stable isotope ratios of gastropod shells and carbonate sediments of NW Sudan as palaeoclimatic indicator. Palaeoecol Africa 24:33–52

Andrews JE, Singhvi AK, Kailath AJ, Kuhn R, Dennis PF, Tandon SK, Dhir RP (1998) Do stable isotope data from calcrete record Late Pleistocene monsoonal climatic variation in the Thar Desert of India? Quat Res 50:240–251

Anhuf D (1997) Paleovegetation in West Africa for 18 000 and 8500 B.P. Eiszeitalter Geg 47:112–119

Anhuf D, Ledru M-P, Behling H, Da Cruz FW Jr, Cordeira RC, Hammen T van der, Karmann I, Marengo JA, Oliveira PE de, Pessenda L, Siffedine A, Albuquerque AL, Silva Dias PL da (2006) Palaeogeog Palaeoclim Palaeoecol 239:510–527

Avery DM (1992) The environment of early modern humans at Border Cave, South Africa: micromammalian evidence. Palaeogeog Palaeoclim Palaeoecol 91:71–87

Ayliffe D, Williams MAJ, Sheldon F (1996) Stable carbon and oxygen isotopic composition of Early Holocene gastropods from Wadi Mansurab, north-central Sudan. Holocene 6:157–169

Ballouche A, Neumann K (1995) A new contribution to the Holocene vegetation history of the West African Sahel: pollen from Oursi, Burkina Faso and charcoal from three sites in northeast Nigeria. Veg Hist Archaeobot 4:31–39

Bamford MK, Albert RM, Cabanes D (2006) Plio-Pleistocene macroplant fossil remains and phytoliths from Lowermost Bed II in the eastern palaeolake margin of Olduvai Gorge, Tanzania. Quat Int 148:95–112

Barakat HN (1995a) Middle Holocene vegetation and human impact in Central Sudan: charcoal from the Neolithic site at Kadero. Veg Hist Archaeobot 4:101–108

Barakat HN (1995b) Plant macrofossils from Z'bib n Elias: a subfossil middens from, a prehistoric cave in the Hoggar, Central Sahara. Acta Palaeobot 35:99–103

Barakat HN (1995c) Charcoals from Neolithic site at Nabta Playa (E-75-6), Egypt. Acta Palaeobot 35:163–166

Bard KA (2003) Radiocarbon dates from Halfiah Gibli (Abadiyeh), a predynastic settlement in Upper Egypt. Radiocarbon 45:123–130

Barker PH, Telford R, Merdaci OU, Williamson D, Taieb M, Vincens A, Gibert E (2000) The sensitivity of a Tanzanian crater lake to catastrophic tephra input and four millennia of climate change. Holocene 10:303–310
Barthlott W, Mutke J, Rafiqpoor D, Kier G, Kreft H (2005) Global centers of vascular plant diversity. Nova Acta Leopold 92:61–83
Blazhchishin AI (1978) Thickness and volume of Holocene sediments of the Atlantic Ocean (in Russian). Doklady Akad Nauk SSSR 242:161–167
Bonnefille R, Mohammed U (1994) Pollen-inferred climatic fluctuations in Ethiopia during the last 3000 years. Palaeogeog Palaeoclim Palaeoecol 109:331–343
Bonnefille R, Riollet G, Buchet G (1991) Nouvelle séquence pollinique d'une tourbière de la crête Zaire-Nil (Burundi). Rev Palaeobot Palynol 67:315–330
Brook GA, Haberyan KA, Filippis ST de (1992) Evidence of a shallow lake at Tsodilo Hills, Botswana, 17 500 to 15 000 yr BP: further confirmation of a widespread Late Pleistocene humid period in the Kalahari Desert. Palaeoecol Afr 23:165–175
Brook GA, Cowart JB, Marais E (1996) Wet and dry periods in the southern African summer rainfall zone during the last 300 Kyr from speleothem, tufa and sand dune age data. Palaeoecol Afr 24:147–158
Brookes IA (1993) Geomorphology and quaternary geology of the Dakhla Oasis region, Egypt. Quat Sci Rev 12:529–552
Burney DA (1987a) Late Holocene vegetational change in Central Madagascar. Quat Res 28:130–143
Burney DA (1987b) Late Quaternary stratigraphic charcoal records from Madagascar. Quat Res 28:274–280
Bylinskaya ME, Golovina LA (1990) Stratigraphy of Quaternary sediments of the tropical Atlantic Ocean according to their foraminifera fauna and nanoplankton (in Russian). Byull Kom Izuch Chetvertichn Perioda Akad Nauk SSSR 59:117–128
Carr AS, Thomas DSG, Bateman MD (2006a) Climatic and sea level controls on Late Quaternary eolian activity on the Agulhas Plain, South Africa. Quat Res 65:252–263
Carr AS, Thomas DSG, Bateman MD, Meadows ME, Chase B (2006b) Late Quaternary palaeoenvironments of the winter-rainfall zone of southern Africa: palynological and sedimentological evidence from the Agulhas Plain. Palaeogeog Palaeoclim Palaeoecol 239:147–165
Casanova J, Hillaire-Marcel C (1992) Late Holocene hydrological history of Lake Tanganyika, East Africa, from isotope data on fossil stromatolites. Palaeogeog Palaeoclim Palaeoecol 91:35–48
Cayre O, Bard E (1999) Planktonic foraminiferal and alkenone records of the last deglaciation from the Eastern Arabian Sea. Quat Res 52:337–342
Cerling ThE, Hay RL (1986) An isotopic study of paleosol carbonates from Olduvai Gorge. Quat Res 25:63–78
Churcher CS (1992) Ostrich bones from the Neolithic of Dakhleh Oasis, Western Desert of Egypt. Palaeoecol Afr 23:67–71
Clarke SJ, Miller GH, Fogel ML, Chivas AR, Murray-Wallace CV (2006) The amino acid and stable isotope biogeochemistry of elephant bird (*Aepyornis*) eggshells from southern Madagascar. Quat Sci Rev 25:2343–2356
Compton JS (2001) Holocene sea-level fluctuations inferred from the evolution of depositional environments of the southern Langebaan Lagoon salt marsh, South Africa. Holocene 11:395–405
Compton JS (2006) The mid-Holocene sea-level highstand at Bogenfels Pan on the southwest coast of Namibia. Quat Res 66:303–310
Dahlberg JA, Wasylikowa K (1996) Image and statistical analyses of early *Sorghum* remains (8000 B.P.) from the Nabta Playa archaeological site in the Western Desert, southern Egypt. Veg Hist Archaeobot 5:293–299
Dahlberg JA, Evans J, Johnson E, Hyman M, Biehl E, Wendorf F (1995) Attempts to identify 8000-BP *Sorghum* using image-analysis, infrared spectroscopy, and biotechnological procedures. Acta Palaeobot 35:167–173
Darbyshire I, Lamb H, Umer M (2003) Forest clearance and regrowth in northern Ethiopia during the last 3000 years. Holocene 13:537–546

DeBusk GHjr. (1997) The distribution of pollen in the surface sediments of Lake Malaŵi, Africa, and the transport of pollen in large lakes. Rev Palaeobot Palynol 97:123–153

Denys C, Williams CT, Dauphin Y, Andrews P, Fernandez-Jalvo Y (1996) Diagenetical changes in Pleistocene small mammal bones from Olduvai Bed I. Palaeogeog Palaeoclim Palaeoecol 126:121–134

Deocampo DM, Blumenschine RJ, Ashley GM (2002) Wetland diagenesis and traces of early hominids, Olduvai Gorge, Tanzania. Quat Res 57:271–281

Dingle RV, Bremner JM, Giraudeau J, Buhmann D (1996) Modern and palaeo-oceanographic environments under Benguela upwelling cells off southern Namibia. Palaeogeog Palaeoclim Palaeoecol 123:85–105

Dittmann A (1990) Die Kombination geomorphologischer und prähistorischer Arbeitsmethoden bei der Lösung paläogeographischer Fragen in der Eastern Desert Ägyptens. Eiszeitalter Geg 40:139–147

Dupont LM (1993) Vegetation zones in NW Africa during the Brunhes Chron reconstructed from marine palynological data. Quat Sci Rev 12:189–202

Dupont L, Behling H (2006) Land-sea linkages during deglaciation: high-resolution records from the eastern Atlantic off the coast of Namibia and Angola (ODP site 1078). Quat Int 148:19–28

Dupont LM, Weinelt M (1996) Vegetation history of the savanna corridor between the Guinean and the Congolian rain forest during the last 150 000 years. Veg Hist Archaeobot 5:273–292

Dupont LM, Shi N, Jahns S, Marret F (1996) *Podocarpus* in West Africa during the Late Pleistocene. Palaeoecol Africa 24:85–101

Eitel B, Zöller L (1996) Soils and sediments in the basin of Dieprivier-Uitskot (Khorixas District, Namibia): age, geomorphic and sedimentological investigation, palaeoclimatic interpretation. Palaeoecol Afr 24:159–172

Elenga H, Schwartz D, Vincens A (1994) Pollen evidence of Late Quaternary vegetation and inferred climate changes in Congo. Palaeogeog Palaeoclim Palaeoecol 109:345–356

Esterhuysen AB (1992) Charcoal from Leliehoek shelter, eastern Orange Free State, South Africa. Palaeoecol Afr 23:193–201

Esterhuysen AB, Mitchell PJ (1996) Palaeoenvironmental and archaeological implications of charcoal assemblages from Holocene sites in western Lesotho, southern Africa. Palaeoecol Afr 24:203–232

Frédoux A (1994) Pollenanalysis of a deep-sea core in the Gulf of Guinea: vegetation and climatic changes during the last 225 000 years B.P. Palaeogeog Palaeoclim Palaeoecol 109:317–330

Frenzel B (1992) Projektgruppe “Terrestrische Paläoklimatologie” (im Rahmen des nationalen Klimaforschungsprogramms der Bundesregierung) Akad Wiss Lit Mainz Jahrb 1991:211–227

Frenzel B, Adamczyk S (2004) Über den pollenanalytischen Nachweis einer prähistorischen Weidewirtschaft auf dem Tibetischen Plateau. H Obermaier Ges Erforsch Eiszeitalt Steinz Tag Greifswald Abstr 2004:13–14

Frenzel B, Pécsi M, Velichko AA (1992) Atlas of paleoclimates and paleoenvironments of the Northern Hemisphere: Late Pleistocene – Holocene. Fischer, Stuttgart, pp 53

Garcin Y, Williamson D, Taieb M, Vincens A, Mathé P-E, Majule A (2006) Centennial to millennial changes in maar-lake deposition during the last 45 000 years in tropical Southern Africa (Lake Masoko, Tanzania). Palaeogeog Palaeoclim Palaeoecol 239:334–354

Gasse F, Campo E van (1998) A 40 000-year pollen and diatom record from Lake Tritrivakely, Madagascar, in the Southern Tropics. Quat Res 49:299–311

Gingele FX (1996) Holocene climatic optimum in Southwest Africa – evidence from the marine clay mineral record. Palaeogeog Paleoclim Palaeoecol 122:77–87

Giresse P, Maley J, Brenac P (1994) Late Quaternary palaeoenvironments in the Lake Barombi Mo (West Cameroon) deduced from pollen and carbon isotopes of organic matter. Palaeogeog Palaeoclim Palaeoecol 107:65–78

Goodman StM, Hobbs JJ, Brewer DJ (1992) Nimir Cave: Mophology and fauna of a cave in the Egyptian Eastern Desert. Palaeoecol Afr 23:73–90

Gorshkov SG, Alekseev VN, Rassokho AI, Faleev VI, Khaetov GM, Seregin MP (1977) Atlas of the oceans – the Atlantic and Indian Oceans (in Russian). Ministerstvo Oborony SSSR, Voenno-Morskoy Flot, Moskva
Hather J (1995) Parenchymatous tissues from the Early Neolithic site E75-6 at Nabta Playa, Western Desert, South Egypt. Preliminary report. Acta Palaeobot 35:157–162
Heine K (1992) On the ages of humid Late Quaternary phases in southern African arid areas (Namibia, Botswana). Palaeoecol Afr 23:149–164
Herries AIR, Reed KE, Kuykendall KL, Latham AG (2006) Speleology and magnetostratigraphic chronology of the Buffalo Cave fossil site, Makapansgat, South Africa. Quat Res 66:233–245
Hillman G, Hedges R, Moore A, Colledge S, Pettitt P (2001) New evidence of Lateglacial cereal cultivation at Abu Hureyra on the Euphrates. Holocene 11:383–393
Höll CHR, Kemle-von Mücke S (2000) Late Quaternary upwelling variations in the eastern equatorial Atlantic Ocean as inferred from dinoflagellate cysts, planktonic foraminifera, and organic carbon content. Quat Res 64:58–67
Holmgren K, Karlén W (1995) Paleoclimatic significance of the stable isotopic composition and petrology of a Late Pleistocene stalagmite from Botswana. Quat Res 43:320–328
Holmgren K, Karlén W, Lauritzen SE, Lee-Thorp JA, Partridge TC, Piketh S, Repinski P, Stevenson C, Svariered O, Tyson PD (1999) A 3000-year high-resolution stalagmite-based record of palaeoclimate for northeastern South Africa. Holocene 9:295–309
Hooghiemstra H (1988) Changes of major wind belts and vegetation zones in NW Africa 20 000 – 5 000 yr B.P., as deduced from a marine pollen record near Cape Blanc. Rev Palaeobot Palynol 55:101–140
Hooghiemstra H (1996) Aspects of Neogene – Quaternary environmental and climatic change in Equatorial and Saharan Africa. Palaeoecol Afr 24:115–132
Hooghiemstra H, Lézine A-M, Leroy SAG, Dupont L, Marret F (2006) Late Quaternary palynology in marine sediments: a synthesis of the understanding of pollen distribution patterns in the NW African setting. Quat Int 148:29–44
Huntsman-Mapila P, Ringrose S, Markay AW, Downey WS, Modisi M, Coetzee SH, Tiercelin J-J, Kampunzu AB, Vanderpost C (2006) Use of geochemical and biological sedimentary record in establishing palaeoenvironments and climate change in the Lake Ngami Basin, NW Botswana. Quat Int 148:51–64
Hutson WH (1980) The Agulhas Current during the Late Pleistocene: analysis of modern faunal analogs. Science 207:64–66
Iacumin P, Bocherens H, Mariotti A, Lenginelli A (1996) An isotopic palaeoenvironmental study of human skeletal remains from the Nile Valley. Palaeogeog Palaeoclim Palaeoecol 126:15–30
Jahns S (1995) A Holocene pollen diagram from El Atrun, northern Sudan. Veg Hist Archaeobot 4:23–30
Jahns S (1996) Vegetation history and climate changes in West-Equatorial Africa during the Late Pleistocene and Holocene, based on a marine pollen diagram from the Congo fan. Veg Hist Archaeobot 5:207–213
Jansen F, Ufkes E, Bey Khelifa L (1995) The younger *Dryas* in equatorial and southern Africa and in the southeast Atlantic Ocean. – In: Troelstra SR, Van Hinte JE, Ganssen GM (eds) The younger *Dryas*. K Ned Akad Wet Verhandl Afd Natuurk 44:141–147
Johnson TC, Talbot MR, Odada EQ (2005) A successful decade of East African lake research. Global Change News Lett 64:7–9
Jolly D, Bonnefille R (1992) Histoire et dynamique du marécage tropical de Ndurumu (Burundi), données polliniques. Rev Palaeobot Palynol 75:133–151
Jolly D, Bonnefille R, Roux M (1994) Numerical interpretation of a high resolution Holocene pollen record from Burundi. Palaeogeog Palaeoclim Palaeoecol 109:357–370
Jousse D, Escarguel G (2006) The use of Holocene bovid fossils to infer palaeoenvironment in Africa. Quat Sci Rev 25:763–783
Kier G, Mutke J, Dinerstein E, Ricketts TH, Küper W, Kreft H, Barthlott W (2005) Global patterns of plant diversity and floristic knowledge. J Biogeog 32:1107–1116

Kindermann K, Bubenzer O, Nussbaum St, Riemer H, Darius F, Pöllath N, Smettan U (2006) Palaeoenvironment and Holocene land use of Djara, Western Desert of Egypt. Quat Sci Rev 25:1619–1637
Kirst GJ, Schneider RR, Müller P, von Storch I, Wefer G (1999) Late Quaternary temperature variability in the Benguela Current system derived from alkenones. Quat Res 52:92–103
Klee M, Zach B, Stika H-P (2004) Four thousand years of plant exploitation in the Lake Chad Basin (Nigeria), part III: plant impressions in potsherds from the Final Stone Age Gajiganna Culture. Veg Hist Archaeobot 13:131–142
Klein RG, Cruż-Uribe K (1996) Size variation in the rock hyrax (*Procavia capensis*) and Late Quaternary climatic change in South Africa. Quat Res 46:193–207
Klein RG, Cruż-Uribe K, Halkett D, Hart T, Parkington JE (1999) Palaeoenvironmental and human behavioral implications of the Boegoeberg 1 Late Pleistocene Hyena Den, Northern Cape Province, South Africa. Quat Res 52:393–403
Kuper R, Kröpelin ST (2006) Climate-controlled Holocene occupation in the Sahara: motor of Africa's evolution. Science 313:803–807
Küper W, Sommer JH, Lovett JC, Mutke J, Linder HP, Beentje HJ, Van Rompaey RSAL, Chatelaine C, Sosef M, Barthlott W (2004) Africa's hotspots of biodiversity redefined. Ann Mo Bot Gard 91:525–535
Lamb AL, Leng MJ, Lamb HF, Mohammed MU (2000) A 9000-year oxygen and carbon isotope record of hydrological change in a small Ethiopian crater lake. Holocene 10:167–177
Lamb H, Darbyshire I, Verschuren D (2003) Vegetation response to rainfall variation and human impact in Central Kenya during the past 1100 years. Holocene 13:285–292
Lario J, Sanchez-Moral S, Fernandez V, Jimeno A, Menendez M (1997) Paleoenvironmental evolution of the Blue Nile (Central Sudan) during the Early and Mid-Holocene (Mesolithic – Neolithic transition). Quat Sci Rev 16:583–588
Leblanc M, Farreau G, Maley J, Nazoumou Y, Leduc CHR, Stagnitti F, Van Oevelen PJ, Delclaux F, Lemoalle J (2006) Reconstruction of Megalake Chad using shuttle radar topographic mission data. Palaeogeog Palaeoclim Palaeoecol 239:16–27
Lézine A-M (1988) New pollen data from the Sahel, Senegal. Rev Palaeobot Palynol 55:141–154
Lézine A-M (1989) Late Quaternary vegetation and climate of the Sahel. Quat Res 32:317–334
Lézine A-M, Edorh TM (1991) Modern pollen deposition in West African Sudanian environments. Rev Palaeobot Palynol 67:41–58
Lézine A-M, Vergnaud Grazzini C (1993) Evidence of forest extension in West Africa since 22 000 BP: a pollen record from the eastern Tropical Atlantic. Quat Sci Rev 12:203–210
Lézine A-M, Saliège J-K, Robert C, Wertz F, Inizan M-L (1998) Holocene lakes from Ramlat as Sab'atayn (Yemen) illustrate the impact of monsoon activity in Southern Arabia. Quat Res 50:290–299
Lézine A-M, Robert C, Saliège J-F, Tiercelin J-J (2006) Sediment, pollen and isotope evidence for an Early to Middle Holocene humid period in the desert of Yemen. PAGES News 14:30–32
Lioubimtseva E, Faure H, Faure-Denard L, Page N, Wickens GE (1996) Sudan biomass changes since 18 000: a test-area for tropical Africa. Palaeoecol Afr 24:71–84
Little MG, Schneider RR, Kroon D, Price B, Bickert T, Wefer G (1997) Rapid palaeooceanographic changes in the Benguella upwelling system for the last 160 000 years as indicated abundances of planktonic foraminifera. Palaeogeog Palaeoclim Palaeoecol 130:135–161
Machado MJ, Pérez-González A, Benito G (1998) Paleoenvironmental changes during the last 4000 yr in the Tigray, Northern Ethiopia. Quat Res 49:312–321
Machado Yanes M del C, Del Arco Aguilar M del C, Vernet J-L, Ourcival J-M (1997) Man and vegetation in northern Tenerife (Canary Islands, Spain) during the Prehispanic Period based on charcoal analyses. Veg Hist Archaeobot 6:187–195
Mäckel R (1992) New ^{14}C data of Kenya and their geomorphological and palaeoecological interpretation. Palaeoecol Africa 23:117–128

Mahaney WC (1991) Geomorphologic and lithologic evidence for an expanded mountain ice sheet on Mount Kenya during the Early Quaternary. Eiszeitalter Geg 41:132–140

Maitima JM (1991) Vegetation response to climatic change in Central Rift Valley, Kenya. Quat Res 35:234–245

Marchant R, Taylor D (1998) Dynamics of montane forest in Central Africa during the Late Holocene: a pollen-based record from Western Uganda. Holocene 8:375–381

McClean CJ, Lovett JC, Küper W, Hannah L, Sommer JH, Barthlott W, Termansen M, Smith GF, Tokumine S, Taplin JRD (2005) African plant diversity and climate change. Ann Mo Bot Gard 92:139–152

Meadows ME, Baxter AJ (2001) Holocene vegetation history and palaeoenvironments at Klaarfontein Springs, Western Cape, South Africa. Holocene 11:699–706

Mercader J, Runge F, Vrydaghs L, Doutrelepont H, Ewongo CEN, Juan-Tresseras J (2000) Phytoliths from archaeological sites in the tropical forest of Ituri, Democratic Republic of Congo. Quat Res 54:102–112

Mitka J, Wasylikowa K (1995) Numerical analysis of charred seeds and fruits from an 8000 years old site at Nabta Playa, Western Desert, South Egypt. Acta Palaeobot 35:175–184

Moeyersons J (1997) Geomorphological processes and their palaeoenvironmental significance at the Shum Lake Cave (Bamenda, Western Cameroon). Palaeogeog Palaeoclim Palaeoecol 133:103–116

Mutke J, Barthlott W (2005) Patterns of vascular plant diversity at continental to global scale. Biol Skr 55:521–531

Mworia-Maitima J (1997) Prehistoric fires and land-cover change in western Kenya: evidences from pollen, charcoal, grass cuticles and grass phytoliths. Holocene 7:405–417

O'Connor PW, Thomas DSG (1999) The timing and environmental significance of Late Quaternary linear dune development in Western Zambia. Quat Res 52:44–55

Pachur H-J, Wünnemann B (1996) Reconstruction of the palaeoclimate along 30° E in the eastern Sahara during the Pleistocene/Holocene transition. Palaeoecol Afr 24:11–32

Parker AG, Goudie AS, Stokes ST, White K, Hodson MI, Manning M, Kennet D (2006) A record of Holocene climate change from lake geochemical analyses in southeastern Arabia. Quat Res 66:465–476

Partridge TC, Demenocal PB, Lorentz SA, Paiker MJ, Vogel JC (1997) Orbital forcing of climate over South Africa: A 200 000-year rainfall record from the Pretoria Saltpan. Quat Sci Rev 16:1125–1133

Peters J (1992) Late Quaternary mammalian remains from Central and Eastern Sudan and their palaeoenvironmental significance. Palaeoecol Afr 23:91–115

Peyron O, Jolly D, Bonnefille R, Vincens A, Guiot J (2000) Climate of East Africa 6000 ^{14}C years B.P. as inferred from pollen data. Quat Res 54:90–101

Pickens Davies C (2006) Holocene paleoclimates of southern Arabia from lacustrine deposits of the Dhomar Highlands, Yemen. Quat Res 66:454–460

Preusser F, Radies D (2006) Quaternary climate change in south-eastern Arabia. PAGES News 14:38–39

Pushcharovskij YM, Pejve AA (eds) (1989) Stroenie zony razloma Zelyonogo Mysa, Central'naya Atlantika (Structure of the Cape Verde fracture zone, Central Atlantic; in Russian). (Acad Nauk SSSR/Geol Inst Trudy 439) Nauka, Moskva, pp 187

Rad U von, Schaaf M, Michels KH, Schulz H, Berger WH, Sirocko F (1999) A 5000-yr record of climate change in varved sediments from the oxygen minimum zone off Pakistan, Northeastern Arabian Sea. Quat Res 51:39–53

Ramsay PJ, Cooper JAG (2002) Late Quaternary sea-level change in South Africa. Quat Res 57:82–90

Reichart GJ, Dulk M den, Visser HJ, Weijden CH van der, Zachariasse WJ (1997) A 225 Kyr record of dust supply, paleoproductivity and the oxygen minimum zone from the Murray Ridge (northern Arabian Sea). Palaeogeog Palaeoclim Palaeoecol 134:149–169

Rognon P, Coudé-Gaussen G (1992) Reconstitution des circulations atmosphériques du Pléistocène terminal et de l'Holocène au large de l'Afrique entre 15 et 35°N. Palaeoecol Afr 23:1–25

Rognon P, Coudé-Gaussen G (1996) Paleoclimates off northwest Africa (28° – 35° N) about 18 000yr B.P. based on continental eolian deposits. Quat Res 46:118–126
Rolando C, Raimbault M (1992) Vegetation associated with the protohistorical mound of "Mouyssam II" (KNTZ) in the Malian Sahel: a reconstruction. Palaeoecol Afr 23:57–66
Salzmann U (1996) Holocene vegetation history of the Sahelian zone of NE-Nigeria: preliminary results. Palaeoecol Afr 24:103–114
Salzmann U, Hoelzmann Ph, Morczinek I (2002) Late Quaternary climate and vegetation of the Sudanian zone of Northeast Nigeria. Quat Res 58:73–83
Schmiedl G, Mackensen A (1997) Late Quaternary paleoproductivity and deep water circulation in the eastern South Atlantic Ocean: evidence from benthic foraminifera: Palaeogeog Palaeoclim Palaeoecol 130:43–80
Schubert BW, Ungar RS, Sponheimer M, Reed KE (2006) Microwear evidence for Plio-Pleistocene bovid diets from Makapansgat Limeworks Cave, South Africa. Palaeogeog Palaeoclim Palaeoecol 241:301–319
Scott L, Cooremans B, Wet JS de, Vogel JC (1991) Holocene environmental changes in Namibia inferred from pollen analysis of swamp and lake deposits. Holocene 1:8–13
Shanahan TM, Overpeck JT, Wheeler CW, Beck JW, Pigati JS, Talbot MR, Scholz CHA, Peck J, King JW (2006) Paleoclimatic variations in West Africa from a record of Late Pleistocene and Holocene lake level stands of Lake Bosumtwi, Ghana. Palaeogeog Palaeoclim Palaeoecol 22:287–302
Shi N, Dupont LM (1997) Vegetation and climate history of Southwest Africa: a marine palynological record of the last 300 000 years. Veg Hist Archaeobot 6:117–131
Shi N, Dupont LM, Beug H-J, Schneider R (2000) Correlation between vegetation in southwestern Africa and oceanic upwelling in the past 21 000 years. Quat Res 54:72–80
Sirocko F (1996) The evolution of the monsoon climate over the Arabian Sea during the last 24 000 years. Palaeoecol Afr 24:53–69
Sirocko F, Sarnthein M, Lange H, Erlenkeuser H (1991) Atmospheric summer circulation and coastal upwelling in the Arabian Sea during the Holocene and the Last Glaciation. Quat Res 36:72–93
Sponholz B, Baumhauer R, Felix-Henningsen P (1993) Fulgurites in the southern Central Sahara, Republic of Niger, and their palaeoenvironmental significance. Holocene 3:97–104
Srivastava P, Brock GA, Marais E, Morthekai P, Singhvi AK (2006) Depositional environment and OSL chronology of the Homeb Silt deposits, Kuiseb River, Namibia. Quat Res 65:478–491
Stager JC (1988) Environmental changes at lake Cheshi, Zambia, since 40 000 years B.P. Quat Res 29:54–65
Stevenson AC, Phethean SJ, Robinson JE (1993) The palaeosalinity and vegetational history of Garaet el Ichkeul, Northwest Tunisia. Holocene 3:201–210
Street-Perrott FA, Holmes JA, Waller MP, Allen MJ, Barber NGH, Fothergill PA, Harkness DD, Ivanovich M, Kroon D, Perrott RA (2000) Drought and dust deposition in the West African Sahel: A 5500-year record from Kajemarum Oasis, north eastern Nigeria. Holocene 10:293–302
Sultan M, Sturchio N, Hassan FA, Rahman Hamdan MA, Mahmood AM, Alfy Z el, Stein T (1997) Precipitation source inferred from stable isotopic composition of Pleistocene groundwater and carbonate deposits in the Western Desert of Egypt. Quat Res 48:29–37
Swezey C, Lancaster N, Kocurek G, Deynoux M, Blum M, Price D, Pion J-C (1999) Response of aeolian systems to Holocene climatic and hydrologic changes in the northern margin of the Sahara: a high-resolution record from the Chott Rharsa Basin, Tunisia. Holocene 9:141–147
Taylor DM (1990) Late Quaternary pollen records from two Ugandan mires: evidence for environmental change in the Rukiga Highlands of Southwest Uganda. Palaeogeog Palaeoclim Palaeoecol 80:283–300
Taylor DM (1993) Environmental change in montane Southwest Uganda: a pollen record for the Holocene from Ahakagyezi Swamp. Holocene 3:324–332
Taylor D, Robertshaw P, Marchant RA (2000) Environmental change and political-economic upheaval in precolonial western Uganda. Holocene 10:527–536

Telford RJ, Lamb HF (1999) Groundwater-mediated response to Holocene climatic change recorded by the diatom stratigraphy of an Ethiopian crater lake. Quat Res 52:63–75

Thevenon F, Williamson D, Vincens A, Taieb M, Merdaci OU, Decobert H, Buchet G (2003) A Late-Holocene charcoal record from Lake Masoko, SW Tanzania: climatic and anthropologic implications. Holocene 13:785–792

Thinon M, Ballouche A, Reille M (1996) Holocene vegetation of the Central Saharan mountains: the end of a myth. Holocene 6:457–462

Thomas DSG, Stokes ST, Shaw PA (1997) Holocene aeolian activity in the southwestern Kalahari Desert, southern Africa: significance and relationships to Late-Pleistocene dune-building events. Holocene 7:273–281

Tryon CA, McBrearty S (2006) Tephrostratigraphy of the Bedded Tuff Member (Kapthurin Formation, Kenya) and the nature of archaeological change in the Later Middle Pleistocene. Quat Res 65:492–507

Van der Veen M (1995) Ancient agriculture in Libya: a review of the evidence. Acta Palaeobot 35:85–98

Van der Veen M (1996) The plant remains from *Mons Claudianus*, a Roman quarry settlement in the Eastern Desert of Egypt – an interim report. Veg Hist Archaeobot 5:137–141

Verschuren D, Dumont H, Armengol-Diaz J (1996) Utilisation de cladocères et chironomides fossiles pour réconstruire l'évolution hydrologique de leur habitat marécageux dans la tourbière de Kashiru (Burundi) depuis 40 000 ans BP. Palaeoecol Afr 24:133–145

Vincens A (1989) Paléoenvironnements du Bassin Nord Tanganyika (Zaïre, Burundi, Tanzania) au cours des 13 derniers mille ans: apport de la palynologie. Rev Palaeobot Palynol 61:69–88

Vincens A (1993) Nouvelle séquence pollinique du Lac Tanganyika: 30 000 ans d'histoire botanique et climatique du bassin nord. Rev Palaeobot Palynol 78:381–394

Vincens A, Ssemmanda I, Roux M, Jolly D (1997) Study of the modern pollen rain in western Uganda with a numerical approach. Rev Palaeobot Palynol 96:145–168

Vincens A, Schwartz D, Bertaux J, Elenga H, Namur C de (1998) Late Holocene climatic changes in western Equatorial Africa inferred from pollen from Lake Sinuda, Southern Congo. Quat Res 50:34–45

Wasylikowa K (1997) Flora of the 8000 years old archaeological site E-75-6 at Nabta Playa, Western Desert, Southern Egypt. Acta Palaeobot 37:99–205

Wasylikowa K, Schild R, Wendorf F, Królik H, Kubiak-Martens L, Harlan JR (1995) Archaeobotany of the Early Neolithic site E-75-6 at Nabta Playa, Western Desert, South Egypt. Acta Palaeobot 35:133–155

Williams M, Nottage J (2006) Impact of extreme rainfall in the central Sudan during 1999 as a partial analogue for reconstructing Early Holocene prehistoric environments. Quat Int 150:82–94

Williams M, Talbot M, Aharon P, Abdl Salaam Y, Williams F, Brendeland KI (2006) Abrupt return of the summer monsoon 15 000 years ago: new supporting evidence from the lower White Nile valley and Lake Albert. Quat Sci Rev 25:2651–2665

Wirrmann D, Bertaux J, Kossoni A (2001) Late Holocene paleoclimatic changes in western Central Africa inferred from mineral abundance in dated sediments from Lake Ossa (Southwest Cameroon). Quat Res 56:275–287

Zabel M, Schneider RR, Wagner T, Adegbie AT, Vries U de, Kolonic S (2001) Late Quaternary climate changes in Central Africa as inferred from terrigenous input to the Niger Fan. Quat Res 56:207–217

Zech M (2006) Evidence for Late Pleistocene climate changes from buried soils on the southern slope of Mt. Kilimanjaro, Tanzania. Palaeogeog Palaeoclim Palaeoecol 242:303–312

Zonneveld KAF, Ganssen G, Troelstra S, Versteegh GJM, Visscher H (1997) Mechanisms forcing abrupt fluctuations of the Indian Ocean summer monsoon during the last deglaciation. Quat Sci Rev 16:187–201

The Application of Novel Optical Sensors (Optodes) in Experimental Plant Ecology

Progress and Perspectives in Non-Invasive Bioprocess Analysis and Biogeochemical Interaction

Dirk Gansert (✉) and Stephan Blossfeld

Contents

Abstract The state of the art of optical chemical sensors, optodes, as an enabling technology for non- and minimal invasive bioprocess analysis in plant ecological research is presented. The measuring principle of optodes is briefly described and discussed in respect of their performance, physical properties and analytical efficiency for parameters, such as O_2, CO_2, temperature or pH, used as measures of bioprocesses. Independence of measurement from the state of aggregation is a major technical advantage of optodes that overcomes present limitations of conventional optical sensor techniques. Novel hybrid optodes that provide simultaneous information of two parameters, such as O_2 and pH, at the same spot are introduced. A survey of the application of optodes for in vivo bioprocess analysis during the past decade is presented, particularly the use of oxygen optodes in plant ecophysiological research.

Dirk Gansert
Department of Ecology & Ecosystem Research, University of Göttingen, Untere Karspüle 2, D-37073 Göttingen, Germany
e-mail: dganser@gwdg.de

U. Lüttge et al. (eds.), *Progress in Botany 69*.

1 Introduction

Scientific progress in elucidating and understanding life processes from the subcellular level to organisms and ecosystems requires methods and techniques that provide non-invasive access to bioprocesses on a high-resolution scale in space and time. Hence, there is a great need for analytical tools that adequately satisfy the requirements of miniaturization, high analyte sensitivity and analyte specifity, excellent measuring performance, longevity and versatility. Further, adaptive measuring systems at low operating costs are of increasing significance for future analytical tools that can be applied for in vivo investigations of bioprocesses under highly variable physico-chemical conditions, particularly in natural environments. At present, however, progress in non-invasive bioprocess analysis is hampered by a severe lack of such analytical tools, even for quantities such as oxygen, carbon dioxide, pH or temperature used as basic quantitative measures of metabolism and the physiological conditions of heterotrophous and autotrophous organisms and their respective compartments. With regard to gas analysis, conventional systems used for investigations of bioprocesses, such as photosynthesis, oxidative respiration, fermentation, denitrification or methanogenesis, are technically restricted to measure either in the gaseous or in the aqueous phase. However, the phase transition of gases from the aqueous to the gaseous phase, such as CO_2 release from respiring tissues to the atmosphere or vice versa, CO_2 uptake by photosynthesizing leaves, addressing autotrophous plants either as carbon sources or sinks, plays a major role in the exploration of matter flux across biotic–abiotic interfaces and the underlying major pathways for gas transport. Hence, there is a strong need for sensor technologies qualified for measuring in the gaseous phase *and* in the aqueous phase without restricting the analyte-specific sensing performance.

Generally, oxygen and carbon dioxide dissolved in the aqueous phase are electrochemically measured using Clark-type or Severinghaus electrodes (Gansert et al. 2006). In the Clark-type electrode, molecular oxygen is reduced to hydroxy anions at the cathode, while the electrons consumed by the reduction reaction stem from the oxidation of silver at the anode, where silver chloride is produced (Von Willert et al. 1995). The resulting current intensity is proportional to the number of O_2 molecules consumed by this redox process. Miniaturized Clark-type oxygen electrodes are now available down to 5 µm in diameter at the needle tip (Armstrong et al. 2000; Buerk 2004). The electrochemical measurement of CO_2 is based on measuring pH changes as a function of CO_2 concentration via the formation of carbonic acid in the aqueous phase of the electrolyte solution of the electrode (McGuire and Teskey 2002). In order to accelerate the relatively slow formation of carbonic acid, and thereby to reduce the response time of the electrode to changes of the CO_2 concentration in the sample, the system was modified by additon of the enzyme carbonic anhydrase in the electrolyte solution (Hanstein et al. 2001). Such an electrochemical CO_2 biosensor can be minimized to 2 µm at the sensor tip and capable of CO_2 resolution down to 50 ppm. These examples also illustrate that analyte consumption, here O_2 and CO_2, by the electrochemical measuring principle affects the accuracy of bioprocess analysis due to an artificial gas sink being formed during the invasive measuring process.

In the gas phase, CO_2 measurement is excellently performed by high-resolution infrared gas analyzers (IRGAs). Using the infrared light absorption characteristics of the CO_2 molecule (main absorption band at 4.26 μm wavelength), the measuring principle is based on the decrease in quantum flow density of infrared light as a quantitative measure of CO_2. Even though the application of the IRGA technique as a proper tool for measuring CO_2 gas exchange of organisms reaches back to the 1940s, the major progress in understanding the dynamics of carbon and water balance of higher plants was made in the 1960s and 1970s, when the IRGA technique became fully developed for simultaneous measurements of CO_2 and H_2O gas exchange rates through the stomata of leaves under natural field conditions (Lösch and Larcher 1998; Lösch 2001). However, the technical advantage of parallel measurement of CO_2 and H_2O by infrared light absorption in the same device dismissed the physiological fact that, in contrast to O_2 consumption, CO_2 production is no accurate measure for oxidative respiration. Consequently, as outlined below, flaws in carbon balance and its modeling from the plant to the ecosystem level are largely due to a misinterpretation of CO_2 release from plants as respiration and the calculation of its temperature dependence derived from these measurements. Oxygen in the gas phase can be most accurately measured by analyzers which use the paramagnetic character of oxygen as the measuring principle (Gansert et al. 2006; Kovacich et al. 2006). Even though measuring techniques for CO_2 and O_2 in the gas phase are free from analyte-consumption flaws, these techniques require flow systems equipped with precision logic and control units to generate constant gas flows through cuvettes that house the samples, tubings and analyzers. In order to minimize the boundary layer resistance for gas exchange between plant and atmosphere, especially that of leaves, strong ventilation of the cuvettes creates conditions considerably different from those encountered in the natural environment (Von Willert et al. 1995; Lösch 2001 and literature cited therein; Lange 2002). Therefore, similar to the insufficiencies of electrodes for gas measurement in the aqueous phase, flow systems for gas measurement in the gas phase also show methodical limitations for in vivo quantification of bioprocesses involved in plant gas exchange, especially with respect to aqueous–gaseous phase transitions and boundary layers.

Optodes represent a novel category of optical chemical sensors that are capable of continuously recording a physical parameter or a chemical compound, an analyte, based on the fluorescence properties of an analyte-specific indicator dye (Wolfbeis 2005). Fluorescence lifetime appears to be a superior optical principle for the measurement of a physico-chemical parameter, such as temperature (Klimant et al. 1997; Liebsch et al. 1999), CO_2 (Neurauter et al. 2000), O_2 (Klimant et al. 1995, 1997; Holst et al. 1997; Campbell and Uttamchandani 2004; Arain et al. 2005; Roche et al. 2006; Tengberg et al. 2006), pH value (Werner et al. 1997; Kosch et al. 1999; Liebsch et al. 2001; Hulth et al. 2002; Hazneci et al. 2004; Zhu et al. 2005; Li et al. 2006; Safavi et al. 2006) or H_2O_2 (Wolfbeis et al. 2002). Regarding ion-specific analysis of membrane transport processes, heavy metal toxicity or nutritional physiology, the development of optical chemical ion sensors has made great progress during the past five years – e.g. for NH_4^+ (Absalan et al. 2004), Cl^- (Pimenta et al. 2004; Radu and Bakker 2005), Na^+ and Ca^{2+} (Xu et al. 2005), Pb^{2+}, Hg^{2+}, Cd^{2+} (Meng et al. 2005) and Ni^{2+} (Amini et al. 2004).

Concerning the measurement of gases, optodes are distinguished by their excellent measuring performance in both gaseous and aqueous phase. In optodes the use of light as the carrier of information allows decoupling between the analyte-specific sensor and the detector, a prerequisite for non-invasive investigations of bioprocesses. Real-time measurement of bioprocesses can thus be performed without metrological disturbance and/or contamination of sterile media. Moreover, the incorporation of an analyte-specific indicator dye in a polymer matrix allows the individual design of optodes to the specific purpose of bioprocess analysis; as fibrous microsensors of only few micrometers in diameter or as planar sensors, i.e. sensor foils, from square millimeters to square decimeters in size (Gansert et al. 2006). Their flexible design provides the potential for multipurpose measurements of biological parameters over a wide range of dimension, from microspots to large-scale two- and three-dimensional real-time bioprocess analysis. Recently, the use of an oxygen-sensitive phosphorescent indicator dye encapsulted in polystyrene microbeads led to a breakthrough in plant science, because for the first time, intracellular oxygen concentration of green photosynthetically active cells could be measured in vivo (Schmälzlin et al. 2005). During the past decade, optical nanosensors for intracellular measurement of analytes reached the status of an enabling technology in life sciences, because the nanosensor matrix imparts two key benefits: (a) protection of the sensing component from interfering species within the intracellular environment and (b) protection of the intracellular environment from any toxic effects of the sensing component (Aylott 2003).

Hybrid optodes represent the latest state of optical chemical sensor technology. These sensors are able to simultaneously measure two parameters, e.g. oxygen and temperature (Stehning and Holst 2004; Borisov et al. 2006a) or pH and oxygen (Arain et al. 2006; Vasylevska et al. 2006), at the same sensor spot (Gansert et al. 2006). Hence, for the first time, physiological processes can be non-invasively measured in the effective physico-chemical condition of the micro-environments wherein the bioprocesses take place. This chapter provides basic information on the principles of optode sensor functioning and outlines the potential of optodes as a novel optical tool for bioprocess analysis in plant ecological research.

2 Measuring Principles of Optical Sensors

In an optical sensor the optical properties of a given physico-chemical parameter are analyzed and monitored spectroscopically over a certain distance. Typically, optical properties such as absorbance, reflectance and luminescence are monitored (Wolfbeis 2005). Using light as the carrier of information, optical sensors provide the possibility of non-invasive measurement because no physical contact is required between the sensor material, which is placed in the medium to be analyzed, and the detection system.

Optical sensing is usually performed in three different ways (for details see Gansert et al. 2006).

1. An analyte in a given medium is measured directly by its specific optical properties. For example, carbon dioxide can be quantified by its infrared light absorption characteristics.
2. An indicator is added to the medium where it reacts with the analyte. The change of the optical properties of the indicator derived from this reaction can then be measured. This type of optical sensing is broadly used, e.g. pH measurement via absorption spectroscopy.
3. Analyte sensing is performed by an indicator which is embedded in a matrix. The analyte diffuses from the medium into the matrix, and chemically or photophysically changes the optical properties of the indicator. The interaction should be fully reversible and the indicator should return to its initial state when the analyte is removed. This principle is technically realized in optodes (see below).

In optical sensor technology fluorescence spectroscopy is broadly used because of its extremely high sensitivity, where even single photons can be detected. Moreover, several optical parameters can be measured which include luminescence intensity, fluorescence decay time, polarization, quenching efficiency, radiative and non-radiative energy transfer as well as combinations of these parameters.

Fluorescence decay time can be measured either by the time domain or by the frequency domain method. In the time domain method fluorescence emission is measured as a function of time after a short excitation light pulse. In the frequency domain method a sample is excited by a sinusoidally modulated light. Luminescent light, that comprises fluorescent and phosphorescent light, has the same waveform but is modulated and phase-shifted from the excitation light. Thus, the decay time can be determined by measuring the phase shift or demodulation. In technical terms, this time delay is the phase angle between the excitation light signal and the light emitted from the indicator dye. The phase angle is shifted as a function of the analyte concentration: the higher the analyte concentration the lower the phase angle. It therefore represents a quantitative measure of the substance – molecules or ions – to be optically analyzed (Gansert et al. 2006). This phase modulation technique allows precise decay time measurement in micro- and millisecond ranges.

Some analytes, such as molecular oxygen, show proportionality between the analyte concentration and the fluorescence decay time of the corresponding indicator. Thus, optical measurement of molecular oxygen is based on the effect of dynamic luminescence quenching by oxygen as the quenching molecule (Fig. 1). The collision between the O_2-sensitive luminescent indicator molecule (O_2 luminophore) in its excited state and the quencher (molecular oxygen) results in radiationless deactivation, and this is called collisional or dynamic quenching. After collision, energy transfer takes place from the excited indicator molecule to oxygen which consequently

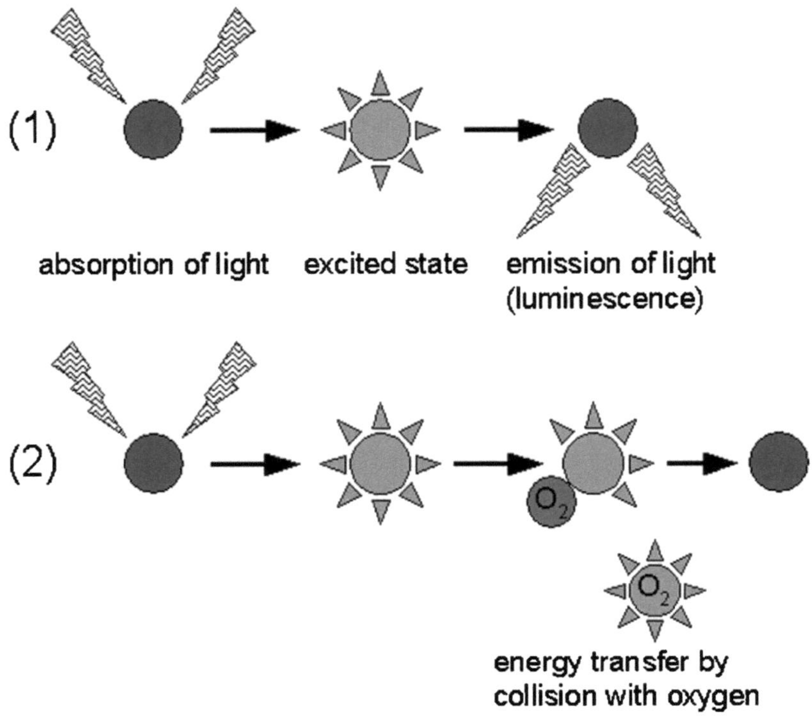

Fig. 1 Principle of dynamic luminescence quenching of an excited oxygen-sensitive indicator dye (luminophore) by molecular oxygen. (1) Luminescence of the excited luminophore in absence of oxygen. (2) Luminescence quenching by energy transfer from the excited luminophore to oxygen as the quenching molecule (Gansert et al. 2006)

is transferred from its ground state (triplet state) to its excited singlet state. There is a non-linear decrease in luminescence with increasing concentration of molecular oxygen. The signal can be linearized by means of the Stern–Volmer equation (Stern and Volmer 1919) which describes dynamic fluorescence quenching by molecular oxygen (Klimant et al. 1995).

For some analytes like pH and CO_2 no indicators exist that change their fluorescence decay time in the range of microseconds but in the nanosecond time scale. Therefore, a new measuring principle was introduced: the dual lifetime referencing (DLR) method (Huber et al. 2001; Klimant et al. 2001; Liebsch et al. 2001). This method uses a couple of luminophores with different decay times but similar excitation spectra. In principle, an analyte-insensitive, long-lived phosphorescent luminophore (decay time $>10^{-7}$ s) is used for referencing of the analyte-sensitive, short-lived (decay time 10^{-9}–10^{-8} s) fluorescent indicator (fluorophore) and fluorescence intensity is converted into a phase shift. The phase shift Φ_m of the overall luminescence obtained at a single frequency depends on the ratio of intensities of the reference luminophore and the indicator dye. Φ_m

can be represented as the superposition of the single sine wave signals of the indicator and the reference luminophore. The reference luminophore gives a constant background signal while the fluorescence signal of the indicator depends on the analyte concentration. The average phase shift Φ_m directly reflects the intensity of the indicator dye and, consequently, the analyte concentration. The modulation frequency is adjusted to the decay time of the reference dye. Ideally, the two luminophores have overlapping excitation and emission spectra so that they can be excited at the same wavelength and their fluorescence can be detected using the same emission window and photodetector.

The principle of fluorescence decay time measurement used in optodes proved to be superior to conventional fluorescence intensity measurement applied in other optical sensors. This is due to four major reasons:

1. The decay time does not depend on fluctuations of the excitation light intensity and the sensitivity of the detector.
2. The decay time is not influenced by signal loss caused by fiber bending or by intensity changes caused by changes in the geometry of the sensor.
3. The decay time is, to a great extent, independent of the concentration of the indicator in the sensitive layer. Thus, photobleaching and leaching of the indicator dye has less influence on the measuring signal.
4. The decay time is not influenced by variations in the optical properties of the sample including turbidity, refractive index and coloration.

2.1 Optodes

Optodes offer the advantage of a broad range of application for in vivo investigations of metabolic processes in cells, tissues, organs, whole organisms and the corresponding environments. Free shaping of optodes is due to the process of manufacture where analyte-specific luminescent microparticles (luminophores) are incorporated in a specifically designed polymer matrix. The matrix not only defines the shape of the sensor spot but also tunes sensor properties such as sensitivity, response times and especially cross-sensitivity to other analytes. It further determines physical and mechanical qualities of an optode sensor. Thus, the properties of a polymer matrix can vary considerably for different sensor types. For example, the polymer matrix of a pH optode sensor must be permeable to protons, while for an oxygen sensor ion permeability can be a serious drawback as far as ions can cause undesirable quenching of the oxygen indicator (Gansert et al. 2006). Therefore, luminophores and matrices form functional units that characterize the optode sensors.

Hybrid optodes, the present frontier in optode sensor technology, provide independent information on the concentration or the amount of two analytes at the same location and at the same time. Hitherto, analyte pairs of (a) oxygen and temperature, (b) oxygen and pH, (c) oxygen and carbon dioxide, (d) temperature and carbon dioxide, and (e) oxygen and chlorophyll have been developed for biological application. Ongoing investigations on sensor development have directed the attention to combinations of nitrous oxide (N_2O, laughing gas), nitric oxide (NO) or hydrogen peroxide (H_2O_2).

In hybrid optodes one of the major problems to overcome is to avoid cross-sensitivity of the different indicators. For example, cross-sensitivity of an indicator "ind 1" to the analyte "ana 2" can arise from poor separation of the light emission spectra of the indicators. Two different approaches, the spectral technique and the multifrequency differentiation technique, can be applied to separate the light-encoded information about the concentrations or the amount of two analytes using the frequency domain method.

In the spectral approach, absorption and emission spectra of the indicator "ind 1" should be shifted significantly from the spectra of the indicator "ind 2". This technique requires separate excitation light sources for both indicators, but also different filter set-ups for separation of the fluorescence light emitted from the indicators (Fig. 2a). The spectral approach has been successfully realized in the oxygen–temperature hybrid optode (Stehning and Holst 2004; Gansert et al. 2006).

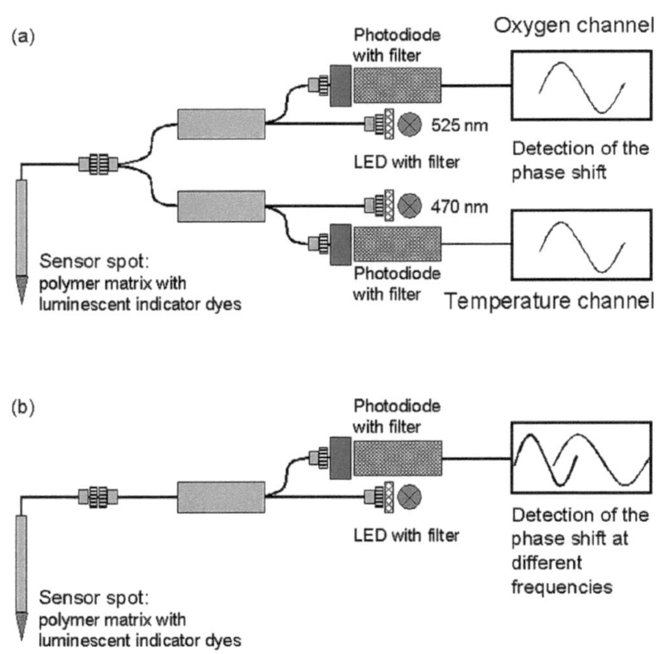

Fig. 2 Optical system components for spectral (**a**), and multifrequency (**b**) measuring principles applied in hybrid optical chemical sensor technology (Neurauter et al. 2000; Gansert et al. 2006). The spectral approach is exemplarily shown for the oxygen–temperature hybrid optode (for further details refer to the text)

In the multifrequency approach, which has been applied for the oxygen–pH hybrid optode, both indicators should have overlapping absorption and emission spectra. Thereby, simultaneous excitation of both indicator dyes with one light source is possible. One emission filter is chosen to pass the emission light from both indicators. According to this principle, an extended DLR sensor was designed (Borisov et al. 2006b). The only difference from the DLR principle described above is that each reference dye is now sensitive to one of the two analytes. Thus, for separation of information derived from both luminophores, phase shift measurements are performed at two or more different modulation frequencies and mathematical models are used for data separation. Signal discrimination is possible because the dyes, having different decay times, are modulated differently. The major advantage of multifrequency measurement lies in a simplified, and therefore much cheaper optical set-up (Fig. 2b).

Optodes and hybrid optodes offer two substantial advantages over other optical sensors. First, their optical measuring principle of fluorescence lifetime is independent of the physical aggregation state, which allows measurements in the gaseous and aqueous phase without technical modifications. Second, they can be used for non-invasive measurements. If an optically transparent window is present a sensor spot can be positioned inside a system to be investigated, and the selected parameters can be optically measured from outside through this window. This measuring principle provides a broad spectrum of applications in plant ecological research, which were not possible before. For example, visualization of the spatial and temporal dynamics of growth and differentiation of plant organs by high-resolution image processing (Nagel et al. 2006; Schurr et al. 2006; Walter et al. 2007) can now be linked with non-invasive visualization of biologically functional parameters, such as CO_2, O_2 or pH, and their growth-related change. In the root–rhizosphere–soil interface the combination of biometric and functional parameters, both derived from non-invasive measuring techniques, will pave the way for real-time and uninfluenced bioprocess analysis, a prerequisite for the progress in our understanding of biogeochemical interactions between plant and soil (Buscot et al. 2000; Bonkowski 2004; Hinsinger et al. 2005). High-resolution two-dimensional imaging of physico-chemical parameters such as pH (Zhu et al. 2005; Blossfeld and Gansert 2007), gas partial pressures, e.g. O_2 (Frederiksen and Glud 2006), CO_2 or N_2O as well as temperature by planar optodes will provide new insights in metabolite transport processes and signal transfer between microorganisms, mycorrhizal fungi and fine roots in the rhizosphere of higher plants. By the use of planar hybrid optodes disturbance-free two-dimensional visualization of two parameters, being simultaneously measured at the same spot (Stehning and Holst 2004; Arain et al. 2006), allows in vivo quantification of the dynamics of metabolic processes in terms of educts, products and physiological information derived from it. For example, the intensity of oxidative respiration of plant cells, tissues or organs can be quantified by simultaneous measurement of rates of O_2 consumption and CO_2 production at the same spot, and the respiratory quotient derived from it provides information on the kind of substrate respired.

3 Application of Optodes in Biogeochemical Process Analysis – State of the Art

3.1 The Use of Oxygen Optodes for Process Studies in Aquatic Ecosystems

During the past decade, optodes have received increasing appreciation as a high-performance optoanalytical tool for studies of biogeochemical process dynamics. From 2002 to 2006 inclusive, a total of 30 papers on investigations of oxygen dynamics of organisms and their environments based on optode sensor technology were published in peer-review journals. More than half the papers (16) address marine animal physiology, ecophysiology, and marine environments, especially sediments (cited below). Six papers from bio-medicinal science focus on animal and human oxygen physiology of the blood and oxygen supply to certain tissues (Bassnett and McNulty 2003; Menzel et al. 2003; Nwaigwe et al. 2003; McKenzie et al 2004; Pearce et al. 2005; Shui et al. 2006). In relation to this emphasis of 22 papers on animal and marine research, the remaining eight papers on oxygen-related plant ecophysiology indicate some deficits in use of this optoanalytical tool in plant research. Up to now, there is only a small spectrum of applications where oxygen optodes have been used for in plant research. This includes: (a) in situ investigation of oxygen gradients in the rhizosphere of wetland plants (*Carex rostrata*: Mainiero and Kazda 2005; *Juncus effusus*: Blossfeld and Gansert 2007) and aquatic plants (*Zostera marina*: Jensen et al. 2005; Frederiksen and Glud 2006), and (b) oxygen balance of the wood-body of trees (Del Hierro et al. 2002; Gansert 2003; Spicer and Holbrook 2005; Sorz and Hietz 2006). The following examples selected from the different fields of research where oxygen optodes have been used so far address the metrological potential for in vivo bioprocess analysis and interaction between biological and geochemical processes under experimental laboratory conditions as well as in natural environments.

In aquatic ecosystems, investigations of oxygen gradients in undisturbed, biologically active sediment layers, especially those in marine environments, improved our understanding of the habitat conditions of the aerobic and anaerobic benthic biota, and of the mineralization processes of organic matter (Tengberg et al. 2003; Viollier et al. 2003; Precht et al. 2004; Glud et al. 2005; König et al. 2005; Polerecky et al. 2005; Franke et al. 2006). In their comprehensive review of existing in situ technologies for the study of benthic biogeochemistry dynamics and related scientific applications, Viollier et al. (2003) summarize that rapid progress in the technology of in situ sensors and benthic chambers yielded major breakthroughs in the scientific understanding of benthic biogeochemistry. Among others the use of planar optodes for two-dimensional process analysis proved that the interaction of sediment topography and oscillating boundary flow between sediment and water determines the spatial and temporal oxygen distribution in the upper sediment layer. A simultaneously occurring inflow of oxygenated water into and anoxic pore water coming out of the sediment creates undulating oxic–anoxic boundaries and patchiness of oxic and anoxic microhabitats in bioturbated sediments of quite different physico-chemical conditions (Precht et al. 2004; Glud et al. 2005). Measurements of respiratory oxygen consumption further proved that bacterial degradation of detritus is spatially decoupled from the final degradation of dissolved organic matter by pore-water flow, demonstrating the complex interaction between biologically and physically driven processes in sediments (Franke et al. 2006).

Due to the metrological independence of the state of aggregation oxygen optodes provided first access to the oxygen dynamics between the liquid–solid phase transition of water during sea ice formation (Mock et al. 2002, 2003). Microprofiles of oxygen concentration through the ice–water interface were continuously recorded by the use of fibrous microoptodes. The profiles revealed that different diffusive boundary layers exist in the ice–water interface, related to a small-scale patchiness of algae. This quantitative approach allowed the differentiation between physically and biologically driven oxygen shifts in the ice–water interface. Herein, the photoautotrophic, oxygen-producing ice diatom *Fragilariopsis cylindrus* played a key role as a biological source for oxygen (Mock et al. 2003).

In marine zoology optodes have been successfully used for the in vivo characterization of the oxygen status and metabolic activity of animal tissues and whole organisms (Gatti et al. 2002; Sartoris et al. 2003; Davenport and Irwin 2003; Thuesen et al. 2005a, b; Irwin and Davenport 2006).

3.2 Oxygen Gradients in the Root–Rhizosphere–Soil Interface of Aquatic and Wetland Plants

In aquatic plant ecophysiology oxygen microelectrodes and planar oxygen optodes were used to map the microdistribution of oxygen and radial oxygen loss (ROL) from roots of common seagrass (*Zostera marina* L.) kept in natural sediment (Jensen et al. 2005). It could be demonstrated that oxygen release from the roots to the rhizosphere was restricted to the root tip and extended only up to 8 mm along the root (Frederiksen and Glud 2006). Thus, a strong oxygen gradient existed along the roots of *Z. marina*, from 80% air saturation at the root tip down to 5% air saturation or even anoxia at positions up to 6 mm behind the root apex (Jensen et al. 2005). Radial oxygen diffusion off the root surface created a dynamic oxic microzone about three times as wide as the root diameter (Jensen et al. 2005). Moreover, it could be proven that oxygen release from the roots of *Z. marina* was clearly related with photosynthetic oxygen production in the leaves. The volume of oxic sediment around the roots increased linearily with irradiance up to 250 μmol photons $m^{-2}\ s^{-1}$, but the oxygen release rate saturated at maximum irradiance of 500 μmol photons $m^{-2}\ s^{-1}$ (Frederiksen and Glud 2006). Roots of *Juncus effusus* L. also revealed a great oxygenation potential under strong hypoxic conditions of submerged soil (Blossfeld and Gansert 2007). Similar to *Z. marina*, high oxygen concentrations up to 26% air saturation (75 μmol $O_2\ l^{-1}\ H_2O$) were measured close to the root surface, which was 80-fold higher than in the hypoxic bulk soil (0.9 µmol O_2 l^{-1} H_2O). However, it was not the root tip but the elongation zone of the roots of *J. effusus* that showed the highest oxygen release. In contrast to *Z. marina*, *J. effusus* did not respond to rising photosynthetic photon flux density by an increase in ROL, confirming previous findings derived from investigations in hydroponic systems (Wießner et al. 2002). Under natural field conditions Mainiero and Kazda (2005) showed that the presence of *Carex rostrata* Stok. significantly increased the oxygen content in the soil (56%), relative to a control plot without vegetation (26.6%). The high portion of aerenchyma tissue of nearly half the total shoot area (48%) is the morphological prerequisite that enables *C. rostrata* to thrive well in anaerobic soil conditions (Visser et al. 2000; Busch 2001). This is the more striking when one takes into account that, in anoxic conditions, rhizomes of *C. rostrata* are not able to survive longer than four days when separated from the above-ground organs, indicating poor physiological tolerance to anoxia (Barclay and Crawford 1982). Thus, avoidance of anoxia by sufficient oxygen supply of the below-ground rhizomes and roots via gaseous diffusion through a wide-luminous aerenchyma is the key for survival of *C. rostrata* in anoxic soils.

Although it seems obvious that the key factors for oxygen distribution patterns in roots are species-specific root growth rate, aerenchyma morphology and the formation of anatomical barriers in the outer cortex that maintain efficient oxygen conductance to compartments of high metabolic activity (Colmer and Bloom 1998), the interaction between structural and functional parameters that determine the dynamics of oxygen supply to the rhizosphere is still poorly understood. Non-invasive in vivo measurement techniques will considerably contribute to the understanding of the character and strength of oxygen sources and sinks in the rhizosphere and surrounding soil texture by undisturbed, simultaneous quantitative visualization of the spatial and temporal variation of oxygen concentration, root biometry and other functional physico-chemical parameters like pH.

3.3 Optical Analyses of O_2 and CO_2 in Tree Stems Substantiate the Effect of Sap Flow on Respiration

It has long been recognized that the transpiration stream plays an important role in the aeration of living tissues surrounded by the cambium of woody plants (Bailey 1913; Haberlandt 1914). Bailey clearly recognized the importance of xylem sap flow for oxygen supply to and carbon dioxide removal from respiring parenchymatous tissues of the sapwood when he stated that the xylem of arborescent plants has an important function for transport of gases dissolved in water from the roots to the cambium and leaves. During the past century, from the first analyses of CO_2 and other gases in trees (Bushong 1907; MacDougal and Working 1933; Chase 1934) onward, a conspicuous prevalence on measuring CO_2 in tree stems can be found in the literature (Levy et al. 1999; McGuire and Teskey 2002 and references therein; Pruyn et al. 2002a, b; Lavigne et al. 2004; Gansert and Burgdorf 2005; Teskey and

McGuire 2005; Maier and Clinton 2006). This prevalence of interest in CO_2 is mainly due to the fact that CO_2 concentration ([CO_2]) of the wood-body surrounded by the cambium (2–26%; McGuire and Teskey 2002) surmounts the atmospheric concentration (0.038%) by up to three orders of magnitude, irrespectively of tree species, tree age or the measuring technique applied. Hence from the first measurements the conclusion was drawn that a considerable amount of CO_2 efflux from woody plant parts to the atmosphere originates from a strong diffusion gradient, rather than from the temperature-dependent respiration of wood parenchyma, primarily the cortex, which caused continuous doubts on CO_2 efflux rates as a suitable measure of oxidative respiration. Laboratory and field experiments on direct manipulation of the CO_2 concentration of the xylem sap produced rapid and reversible changes in CO_2 efflux from the stems of different tree species (Negisi 1979; Bowman et al. 2005; Teskey and McGuire 2002, 2005). For example, the CO_2 efflux from the stems of saplings of sycamore (*Platanus occidentalis* L.) and sweetgum (*Liquidambar styraciflua* L.) was directly proportional to xylem [CO_2] and explained about 75% of its variation (Teskey and McGuire 2005). Being in line with previous findings these results substantiated the authors' conclusion that CO_2 transported in the xylem may confound measurements of respiration based on CO_2 efflux to the atmosphere. The interference of sap flow-mediated CO_2 transport in the xylem with radial CO_2 efflux from the stem was also shown for *Pinus taeda* seedlings (Martin et al. 1994), *Combretum micranthum* G. shrub (Levy et al. 1999) and mature *Betula pendula* Roth. (Gansert and Burgdorf 2005).

Investigations carried out on mature sycamore trees showed that stem CO_2 efflux to the atmosphere consisted of 45% CO_2 transported in the xylem and 55% CO_2 derived from local respiring cells, i.e. about one-half of the CO_2 diffusing from the stem to the atmosphere was not due to the respiration of local cells (Teskey and McGuire 2007). This finding is in line with that derived from mature *Betula pendula* (Gansert and Burgdorf 2005). On the experimental condition of preventing a temperature gradient between atmosphere and outer sapwood, during daytime, CO_2 efflux decreased by up to 40% of gross CO_2 release calculated from the nighttime temperature dependence of CO_2 efflux during the same day. This diurnal pattern of hysteresis between temperature and CO_2 efflux did not occur when sap flow was small or even ceased on warm days with low insolation. Then, CO_2 efflux showed the typical exponential increase with temperature, as can also be measured during the leafless state of deciduous trees in spring or autumn. For young loblolly pine (*Pinus taeda*) trees in spring, Maier and Clinton (2006) reported the same reduction of stem CO_2 efflux between 18% and 40% during daytime. However, for this conifer tree species the hysteresis of CO_2 efflux was not due to the retention of CO_2 in the sapwood by xylem sap flow, but seemed to be related with diffusion gradients from the cambium and inner bark to the atmosphere, caused by high respiratory activity during rapid circumferential growth at this time of the year. In consideration of different wood anatomy of coniferous and deciduous tree species, these seemingly contrary findings between CO_2 efflux and xylem sap flow provide evidence of the superposition and interrelation between aqueous and gaseous pathways for CO_2 transport through the wood-body via xylem sap flow and gaseous diffusion through the intercellular continuum, respectively, and substantiate the need for a reassessment of woody-tissue respiration quantified in terms of CO_2 efflux rates. Misinterpretation of woody-tissue respiration not only affects models of carbon balance, that commonly calculate respiration as an exponential function of temperature, but it also brings into question current concepts of growth and maintenance respiration of woody plants (Lavigne et al. 1996; Ryan et al. 1996; Bosc et al. 2003).

Surprisingly, oxidative respiration of woody plant parts is commonly not expressed in terms of oxygen consumption nor, from the even more accurate physiological viewpoint, quantified by simultaneous measurement of O_2 consumption and CO_2 production. To our knowledge, only two investigations carried out on mature trees of Norway spruce, *Picea abies* (L.) Karst. (Eklund 1990), English oak, *Quercus robur* L., and Norway maple, *Acer platanoides* L. (Eklund 1993) considered the seasonal variation of both CO_2 and O_2 gas concentrations in the outer sapwood. Although these investigations did not address the respiratory activity of sapwood parenchyma, only the restriction of secondary growth induced by O_2 limitation of cambial activity, the relation between the two gas concentrations indicated a greater oxygen deficit than could be deduced from respiratory CO_2 production. Optoanalytical measurements of CO_2 and O_2 gas exchange rates between atmosphere and stems of mountain birch saplings (*Betula pubescens* Ehr.) confirmed the existence of an enhanced oxygen deficit, that strongly increased as soon as the xylem sap flow ceased (Gansert 2003). This imbalance between CO_2 and O_2 in the gas phase was due both to the rate of oxygen supply to the sapwood and to the withdrawal of dissolved CO_2 by xylem sap flow, which may suggest an underestimation of sapwood respiration. The reversible interruption of xylem sap flow in mountain birch saplings provided quantitative evidence that xylem sap flow contributed about 60% to the total oxygen supply to the sapwood. Investigations on oxygen supply to the sapwood of olive saplings *Olea europaea* L. indicated that up to 80% of the oxygen concentration in the sapwood was delivered by xylem sap flow (Mancuso and Marras 2003). Similar results were obtained from recurring flooding and drainage of the root system of a potted 21-year-old *Laurus nobilis* L. tree, that caused a reversible decrease in oxygen concentration of the aboveground sapwood by 6% (Del Hierro et al. 2002).

For mountain birch it could further be proven that sap flow not only affects the oxygen status of the sapwood but also has an effect on radial oxygen transport between stem and atmosphere (Gansert 2003). Estimates of radial gaseous diffusion accounted for 0.2–0.5 nmol O_2 l^{-1} s^{-1}, while sap flow accounted for a specific rate of oxygen supply of 2 nmol O_2 l^{-1} s^{-1} ($mmol^{-1}$ H_2O m^{-2} leaf area s^{-1} transpired), which is one order of magnitude higher than the diffusion component of oxygen transport (Gansert 2003). The oxygen relations between stem and atmosphere and in the sapwood substantiate the importance of sap flow for oxygen supply to the sapwood. Moreover, the results support the concept of a dual transport system that supplies wood parenchyma of trees with oxygen via radial gas diffusion and axial flow of oxygen dissolved in the xylem sap, as suggested earlier (Hook et al. 1972; Gansert et al. 2001). During daytime, xylem sap flow is the major path for oxygen supply while at nighttime, when sap flow approximates zero, the gaseous path for oxygen transport prevails, driven by diffusion gradients radially through intercellular gas spaces.

Interestingly, oxygen depletion in the sapwood of mountain birch saplings down to 77 µmol O_2 l^{-1} (27% of air saturation at T = 20 °C), that was caused by experimental interruption of xylem sap flow, was similar to the low oxygen concentration of 42 µmol O_2 l^{-1}, which is only 15% of air saturation at T = 20 °C, measured in mature *Betula pendula* stems during bud break, when sap flow is not yet resumed (Gansert et al. 2001). The phenological state of bud break marked the upper limit of oxygen depletion in the sapwood of this tree species. During the period of intensive circumferential growth, a second maximum of oxygen depletion in the sapwood occurs (70 µmol O_2 l^{-1}). However, due to daily recovery from oxygen deficiency by oxygen supply via sap flow this period of oxygen depletion may be less significant than hypoxia during bud break (Gansert et al. 2001).

Radial oxygen profiles from the cambium to the pith of different tree species should provide information on: (a) the existence of a consistent pattern of radial oxygen decline, (b) the strength and period of potentially hypoxic conditions, and (c) its effect on parenchyma cells in the inner stem (Spicer and Holbrook 2005). A novel system was developed for in situ measurement of O_2 concentration at different depths from 0.5 cm to 9.0 cm inside the cambium of stems of different tree species (*Acer rubrum* L., *Fraxinus americana* L., *Tsuga canadensis* (L.) Carr., *Quercus rubra* L.). Stainless steel tube assemblies equipped with fibrous oxygen optodes and thermocouples were installed that provided continuous records of O_2 partial pressure in equilibrium with the local internal gas composition in the stems. In summer, radial oxygen gradients were observed in *F. americana*, *T. canadensis* and *Q. rubra* that ranged from 10% gaseous mole fraction on average down to minima around 3–5% (corresponding to approx. 15–25% air O_2 content) in the innermost sapwood. Before the spring leaf flush, the oxygen content in the outer sapwood was reduced in *Q. rubra* and *T. canadensis* relative to summer, and it was occasionally even lower than in the inner sapwood. Both findings are in line with the observation of strongest O_2 depletion in the sapwood of *B. pendula* during bud break (Gansert et al. 2001). From their investigation of radial oxygen gradients Spicer and Holbrook (2005) deduced that parenchyma respiration is not severly limited by oxygen deficiency or may induce parenchyma cell death during sapwood senscence. Within-stem O_2 levels may instead be most relevant to metabolism in the cambial zone and phloem, for which sapwood could serve as a significant source of oxygen. Recent measurements of gas diffusion through wood of different anatomy (coniferous, ring-, diffuse-porous) showed that radial diffusion in water-saturated sapwood can be too low to ensure the supply of respiring sapwood parenchyma with sufficient oxygen (Sorz and Hietz 2006). Even on the assumption of a relatively low respiration rate of 100 µmol CO_2 m^{-3} sapwood s^{-1} (Edwards and Hanson 1996; Lavigne et al. 1996), Sorz and Hietz (2006) estimated that the total storage capacity of water-satured sapwood for oxygen (0.21 mol O_2 m^{-3}) would last for only 35 min at zero sap flow.

Recently, a novel non-invasive single sensor laser heat pulse technique was introduced for the direct detection of xylem and phloem sap movement in woody plants of small stem size (Helfter et al. 2007). A near-infrared laser was used for application of a heat pulse as a tracer for sap flow, and heat propagation was monitored externally by means of an infrared camera. Heat pulse velocities were calculated from an equation requiring only one spatial measurement point, as opposed to two in the case of heat compensation techniques, and were related to mass flow rate. The method was successfully applied in measuring xylem and phloem flow rates in branches of about 4 mm in diameter, and in saplings of *Betula pendula* and *Quercus robur* to represent diffuse- and ring-porous wood-types, respectively.

In summary, the application of optical (chemical) sensors, especially oxygen optodes, in tree ecophysiology considerably contributes to the quantitative analysis and understanding of the role of xylem sap flow for endogenous source–sink relations of O_2 and CO_2 in trees and gas exchange between woody plant parts and the atmosphere. The findings significantly substantiate the early hypothesis that the water transport through the bole of trees plays a vital role in aeration of xylary tissues surrounded by a relatively impervious sheath of cambial cells, as was last emphasized 35 years ago in a comprehensive review on the aeration of trees (Hook et al. 1972). The progress in our understanding of these relations will certainly depend on success in quantitative differentiation between: (a) biogenous processes, such as respiration, growth or bark photosynthesis, and (b) physico-chemical processes, such as diffusion, solubility, dissociation or flow in the gaseous and aqueous phase, in woody plants. For example, the effect of diurnal and seasonal variation of xylem sap pH, the most important parameter of the aqueous phase for CO_2 solubility and dissociation of carbonic acid, on the equilibrium between CO_2 partial pressure in the intercellular gas phase and the CO_2 concentration in the aqueous phase of the apoplast has yet to be explored. Moreover, uptake of soil water into the apoplast of a plant implies uptake of dissolved CO_2 that originates from root systems of different neighboring plant individuals, metabolic activity of the edaphon in the rooted soil compartments, and in particular, from the organism network of the rhizosphere (Fig. 3). The amount of pedogenous CO_2 uptake into the xylem strongly depends on the pH-mediated balance between dissolved CO_2 and bicarbonate. At pH<5 the amount of HCO_3^- in the soil water is negligible, so that considerable amounts of CO_2 can enter the plant body. This is because the plasma membrane at the endodermis is six orders of magnitude less permeable for HCO_3^- than for CO_2 (HCO_3^-: $3E^{-12}$ m s^{-1}; CO_2: $1E^{-6}$ m s^{-1}; Maberly and Madsen 1998). Therefore, miniaturized high-performance pH sensors suitable for minimal invasive insertion in woody tissues are required that accurately measure small variations of xylem sap pH at the same spot where CO_2 is measured at the same time. Underwater access to the sapwood by the use of pressurized, degassed and sterilized water, that drastically reduces the risk of embolism in tracheids and vessels, provides a worthwhile method for the insertion of a variety of sensors and microanalytical tools in the sapwood without detrimental effects on the natural dynamics of xylem sap flow (Gansert et al. 2001; Gansert 2006). This approach examplarily proves that the combination of miniaturized, non-invasive optoanalytical sensor technology and elaborate methods for its in situ application will pave the way for our progress in biological and physico-chemical process analysis of endogenous gas transport and metabolism of woody plants.

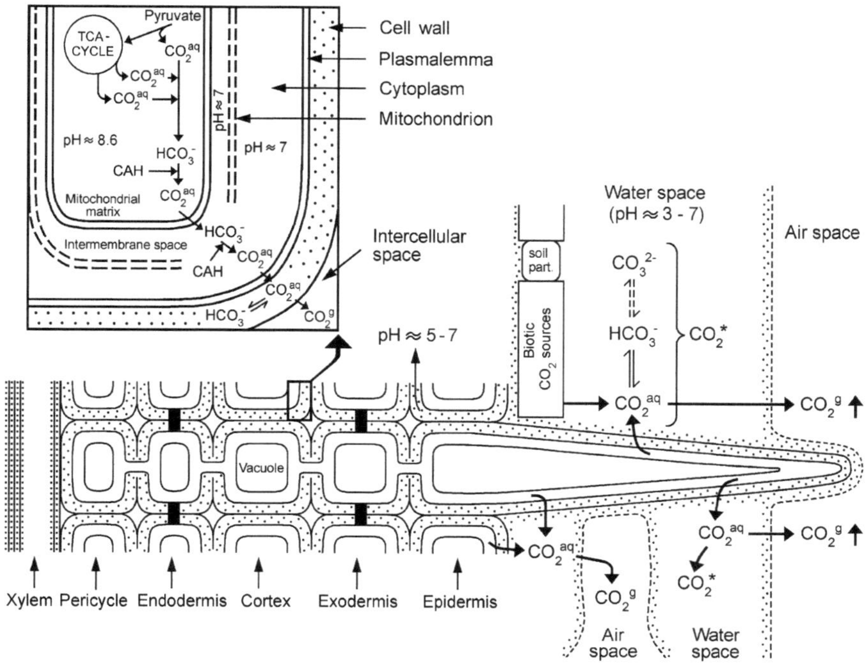

Fig. 3 Scheme of CO_2 exchange processes between plant roots (cross-section with root hair indicated) and surrounding pore water (*dotted area*) and pore air of soil. The pore water forms an aqueous continuum with the apoplast. Ranges of pH are indicated for each aqueous compartment. CO_2* indicates the sum of dissolved carbon dioxide (CO_2^{aq}), bicarbonate (HCO_3^-) and carbonate (CO_3^{2-}). The CO_2 source of the rhizosphere and the edaphon is indicated as *Biotic CO_2 sources*. Respiratory CO_2 production by the root is shown enlarged. The enzyme carbonic anhydrase (CAH) mediates symplastic CO_2 transport from the mitochondrial matrix across the plasmalemma into the apoplast. Depending on the condition of xylem sap flow, soil water enriched with pedogenous CO_2 is taken up by the roots

3.4 Planar pH Optodes as a Novel Tool for Non-Invasive Visualization of pH Dynamics in the Rhizosphere

Recently, a novel non-invasive optical method using planar pH optodes for quantitative visualization of pH dynamics in the rhizosphere was introduced that overcomes present limitations of conventional pH measurement. Quantitative analysis of the spatial and temporal micropattern of pH in the root–soil interface of growing roots without any disturbance of the biological and physico-chemical conditions is now possible (Blossfeld and Gansert 2007). Long-term records at 4 s intervals over more than eight weeks revealed considerable pH dynamics in the rhizosphere of *Juncus effusus* L. of about 10% of the pH scale. Root-induced acidification of the rhizosphere in alkaline submerged soil by about one pH unit (pH 7.5–8.5) was observed during daytime, while at nighttime the intensity of acidification decreased. Interestingly, conspicuous acidification of the rhizosphere was measured shortly after the start of illumination, indicated by a drop in pH by 0.5 units within only 30min. The long-term investigation revealed a pronounced heterogeneity of the acidification patterns of main and lateral roots. Young lateral roots showed a strong potential of acidification that gradually decreased with age. The planar pH optodes used for visualization of pH dynamics in the rhizosphere proved the existence of a significant diurnal variation and spatial heterogeneity of pH that is closely related to root type (main or lateral), root age and root growth rate of the wetland plant species *J. effusus*. In accordance with other findings (Marschner 1995; Bloom et al. 2003), the results also confirmed that acidification during daytime was due to ammonia uptake and equivalent proton release from the roots, such as the prevailing physiological pathway for nitrogen aquisition by wetland plants under reduced conditions of submerged soils. The quantitative analysis of plant-induced alteration of pH in the root–rhizosphere–soil interface is of major importance for the understanding of: (a) the physico-chemical conditions of nutrient aquisition and uptake, (b) the microbial network and food webs of the rhizosphere, and (c) the biogenic production of the greenhouse gases methane and nitrous oxide in wetland soils (Blossfeld and Gansert 2007).

The optical chemical measurement of pH by fibrous microoptodes and their minimal invasive insertion in roots, stems or branches (Gansert et al. 2001; Gansert 2006) provides an innovative approach for online measurement of the spatial and temporal dynamics of xylem sap pH that overcomes conventional discontinuous pH measurements of extracted samples of xylem sap (Gollan et al. 1992; Schurr 1998; Wilkinson et al. 1998; Heizmann et al. 2001; Alves et al. 2004; Teskey and McGuire 2007). Simultaneous measurement of pH and CO_2 in the xylem sap provides the basis for quantification of the amount of dissolved inorganic carbon (DIC) in the xylem, calculated as the sum of dissolved [CO_2], [HCO_3^-] and [CO_3^{2-}] (Butler 1991; Stumm and Morgan 1996; McGuire and Teskey 2002; Kirk 2005). The knowledge of plant endogenous inorganic carbon pools and fluxes through the apoplast driven by the transpiration stream from the rooted soil towards the leaves, and its effect on gaseous CO_2 efflux to the atmosphere, will considerably improve our understanding of the function of plants as periodical carbon sinks and sources. One of the open questions is: how much of dissolved carbon that is released from respiratory assimilate catabolism in the symplast of the roots and from the rhizosphere will be displaced in the apoplast and translocated towards the leaves by xylem sap flow? Thus, a rapid carbon cycling within the cormus has to be taken into account that intersects with carbon allocation to the soil, its biogeochemical processing, and fate via aqueous and gaseous pathways.

4 Outlook

Hybrid optodes, able to measure two physico-chemical parameters in one sensor, represent a coming optical chemical sensor technology for progress in non- and minimal invasive quantitative process analysis in natural, life and environmental sciences. The multipurpose design of hybrid optodes over orders of magnitude of dimension, from microbeads to foils, and their high sensing performance in the gaseous and liquid phase also represent a methodical step forward in plant ecology, comparable with the introduction of the IRGA and PAM technologies for measurement of carbon assimilation, transpiration and the energetic states of the photosystems of photoautotrophic plants in their natural environment.

For example, simultaneous measurement of oxygen and pH is of major importance for any investigation of metabolic processes in living systems, because these parameters not only determine the oxygenation state and enzyme activity in living compartments but also affect the redox potential, and thus, the overall energy state in these compartments. Hypoxia and anaerobiosis concurrently result in a low energy charge of cells and a decrease in the cytoplasmic pH (Crawford 1992, 2003; Crawford and Braendle 1996; Schlüter and Crawford 2001; Kirk 2005). Acidification of the cytoplasm not only causes detrimental effects on enzyme activity but also induces a rise in redox potential, which in turn affects electron transport processes across plasma membranes. Hence, the physiological interplay between pH and oxygen becomes more important the more plants have to cope with periodic or enduring limitation of oxygen supply. For aerenchymal plants oxygen supply to the submerged roots is much better understood than the dynamics of pH across the root–rhizosphere–soil interface, and along oxidation gradients caused by oxygen release from the roots (Armstrong et al. 1994, 1996; Colmer 2003). Oxygen–pH hybrid optodes (Arain et al. 2006) are now available to simultaneously quantify pH and oxygen gradients along roots and from the root surface towards the bulk soil. The non-invasive visualization of plant-induced spatial and temporal patterns of physico-chemical conditions will provide novel information on the biogeochemistry of inorganic and organic matter fluxes and turnover in the root–mycorrhizal–microbial food web (Bezbaruah and Zhang 2004; Bonkowski 2004; Booth et al. 2005; Fitter et al. 2005; Hinsinger et al. 2005; Patra et al. 2005; Chapman et al. 2006).

Two major strands of progression should be sketched for future optical sensing technology:

1. Multiple sensors with novel analyte-specific indicators will widen the spectrum of analytes towards the establishment of different analyte families relevant to life and environmental science such as biogenic gases (CH_4, NO, NO_x, N_2O, H_2O_2, etc.), inorganic cations and anions, organic molecules (acids, hormones, energy equivalents, redox compounds, etc.) or metabolites (sugars, amino acids, fatty acids, etc.).
2. Concurrent with the extension of the analyte spectrum, miniaturization will be a continuous challenge for optical sensor technology. Both miniaturization and highly specific analyte detection is essential for progress in biogeochemical process analysis on the microscale. For example, metabolite transport and chemical signal transmission within cells and between cells in tissues and organs is a prerequisite for understanding the communication of cells in living organisms that can be traced by use of optical chemical micro- and nanosensors

(Aylott 2003). The suitability of optodes for non-invasive measurement of several parameters at the same time opens up the perspective of a multifactorial in vivo communication process analysis in and between organisms.

Further, two- and three-dimensional hybrid imaging will provide an innovative and powerful tool for the elucidation of coordinated process dynamics, such as species-specific growth, the differentiation and senescence of tissues and organ- or organismic interrelations, such as symbioses (e.g. mycorrhiza) or parasitism. The high-resolution non-invasive visualization of biogeochemical process dynamics in combination with anatomical and morphological image processing will provide a future platform for new insights in matter fluxes, organismic interactions and source–sink relations during the formation, metabolism and mineralization of biomass.

References

Absalan G, Soleimani M, Asadi M, Ahmadi MB (2004) Construction of a new optical sensor for monitoring ammonia in water samples using bis (acetylacetoneethylenediamine)-tributylphosphin cobalt(III) tetraphenylborate complex-coated triacetylcellulose. Anal Sci 20:1433–1436

Alves G, Ameglio T, Guilliot A, Fleurat-Lessard P, Lacointe A, Sakr S, Petel G, Julien J-L (2004) Winter variation in xylem sap pH of walnut trees: involvement of plasma membrane H^+-ATPase of vessel-associated cells. Tree Physiol 24:99–105

Amini MK, Momeni-Isfahani T, Khorasani JH, Pourhossein M (2004) Development of an optical chemical sensor based on 2-(5-bromo-2-pyridylazo)-5-(diethylamino)phenol in Nafion for determination of nickel ion. Talanta 63:713–720

Arain S, Weiss S, Heinzle E, John GT, Krause C, Klimant I (2005) Gas sensing in microplates with optodes: influence of oxygen exchange between sample, air, and plate material. Biotechnol Bioeng 90:271–280

Arain S, John GT, Krause C, Gerlach J, Wolfbeis OS, Klimant I (2006) Characterization of microtiterplates with integrated optical sensors for oxygen and pH, and their applications to enzyme activity screening, respirometry, and toxicological assays. Sensor Actuat B Chem 113:639–648

Armstrong W, Brändle R, Jackson MB (1994) Mechanisms of flood tolerance in plants. Acta Bot Neerl 43:307–358

Armstrong J, Armstrong W, Beckett PM, Halder JE, Lythe S, Holt R, Sinclair A (1996) Pathways of aeration and the mechanisms and beneficial effects of humidity- and Venturi-induced convections in *Phragmites australis* (Cav.) Trin. Ex Steud. Aquat Bot 54:177–197

Armstrong W, Cousins D, Armstrong J, Turner DW, Beckett PM (2000) Oxygen distribution in wetland plant roots and permeability barriers to gas-exchange with the rhizosphere: a microelectrode and modelling study with *Phragmites australis*. Ann Bot Lond 86:687–703

Aylott JW (2003) Optical nanosensors – an enabling technology for intracellular measurements. Analyst 128:309–312

Bailey IW (1913) The preservation treatment of wood. II. The structure of the pit membranes in the tracheids of conifers and their relation to the penetration of gases, liquids, and finely divided solids into green and seasoned wood. Forest Q 11:12–20

Barclay AM, Crawford RMM (1982) Plant growth and survival under strict anaerobiosis. J Exp Bot 33:541–549

Bassnett S, McNulty R (2003) The effect of elevated intraocular oxygen on organelle degradation in embryonic chicken lens. J Exp Biol 206:4353–4361

Bezbaruah AN, Zhang TC (2004) pH, redox, and oxygen microprofiles in the rhizosphere of bulrush (*Scirpus validus*) in a constructed wetland treating municiple wastewater. Biotechnol Bioeng 88:60–70
Bloom AJ, Meyerhoff PA, Taylor AR, Rost TL (2003) Root development and absorption of ammonium and nitrate from the rhizosphere. J Plant Growth Regul 21:416–431
Blossfeld S, Gansert D (2007) A novel non-invasive optical method for quantitative visualization of pH dynamics in the rhizosphere of plants. Plant Cell Environ 30:176–186
Bonkowski M (2004) Protozoa and plant growth: the microbial loop in soil revisited. New Phytol 162:617–631
Booth MS, Stark JM, Rastetter E (2005) Controls on nitrogen cycling in terrestrial ecosystems: a synthetic analysis of literature data. Ecol Monogr 75:139–157
Borisov SM, Vasylevska AS, Krause Ch, Wolfbeis OS (2006a) Composite luminescent material for dual sensing of oxygen and temperature. Adv Funct Mater 16:1536–1542
Borisov SM, Neurauter G, Schroeder C, Klimant I, Wolfbeis OS (2006b) A modified dual lifetime referencing method for simultaneous optical determination and sensing of two analytes. Appl Spectrosc 60:1167–1173
Bosc A, Grandcourt A de, Loustau D (2003) Variability of stem and branch maintenance respiration in a *Pinus pinaster* tree. Tree Physiol 23:227–236
Bowman WP, Barbour MM, Turnbull MH, Tissue DT, Whitehead D, Griffin KL (2005) Sap flow rates and sapwood density are critical factors in within- and between-tree variation in CO_2 efflux from stems of mature *Dacrydium cupressinum* trees. New Phytol 167: 815–828
Buerk DG (2004) Measuring tissue pO_2 with microelectrodes. Methods Enzymol 38:665–690
Busch J (2001) Characteristic values of key ecophysiological parameters in the genus *Carex*. Flora 196:1–26
Buscot F, Munch JC, Charcosset JY, Gardes M, Nehls U, Hampp R (2000) Recent advances in exploring physiology and biodiversity of ectomycorrhizas highlight the functioning of these symbioses in ecosystems. FEMS Miocrobiol Rev 24:601–614
Bushong FW (1907) Composition of gas from cottonwood trees. Kans Acad Sci Trans 21:53
Butler JN (1991) Carbon dioxide equilibria and their applications. Lewis, Chelsea
Campbell A, Uttamchandani D (2004) Optical dissolved oxygen lifetime sensor based on sol–gel immobilisation. IEE Proc Sci Meas Technol 151:291–297
Chapman SK, Langley JA, Hart SC, Koch GW (2006) Plants actively control nitrogen cycling: uncorking the microbial bottleneck. New Phytol 169:27–34
Chase WW (1934) The composition, quantity, and physiological significance of gases in tree stems. (Minn Agric Exp Stn Tech Bull 99) University of Minnesota, Minneapolis
Colmer TD (2003) Long-distance transport of gases in plants: a perspective on internal aeration and radial oxygen loss from roots. Plant Cell Environ 26:17–36
Colmer TD, Bloom (1998) A comparison of NH_4^+ and NO_3^- net fluxes along roots of rice and maize. Plant Cell Environ 21:240–246
Crawford RMM (1992) Oxygen availability as an ecological limit to plant distribution. Adv Ecol Res 23:93–185
Crawford RMM (2003) Seasonal differences in plant responses to flooding and anoxia. Can J Bot 81:1224–1246
Crawford RMM, Braendle R (1996) Oxygen deprivation stress in a changing environment. J Exp Bot 47:145–159
Davenport J, Irwin S (2003) Hypoxic life of intertidal acorn barnacles. Mar Biol 143:555–563
Del Hierro AM, Kronberger W, Hietz P, Offenthaler I, Richter H (2002) A new method to determine the oxygen concentration inside the sapwood of trees. J Exp Bot 53:559–563
Edwards NT, Hanson PJ (1996) Stem respiration in a closed-canopy upland oak forest. Tree Physiol 16:433–439
Eklund L (1990) Endogenous levels of oxygen, carbon dioxide and ethylene in stems of Norway spruce trees during one growing season. Trees 4:150–154

Eklund L (1993) Seasonal variations of O_2, CO_2, and ethylene in oak and maple stems. Can J For Res 23:2608–2610
Fitter AH, Gilligan CA, Hollingworth K, Kleczkowski A, Twyman RM, Pitchford JW (2005) Biodiversity and ecosystem function in soil. Funct Ecol 19:369–377
Franke U, Polerecky L, Precht E, Huettel M (2006) Wave tank study of particulate organic matter degradation in permeable sediments. Limnol Oceanogr 51:1084–1096
Frederiksen MS, Glud RN (2006) Oxygen dynamics in the rhizosphere of *Zostera marina*: a two-dimensional planar optode study. Limnol Oceanogr 51:1072–1083
Gansert D (2003) Xylem sap flow as a major pathway for oxygen supply to the sapwood of birch (*Betula pubescens* Ehr.). Plant Cell Environ 26:1803–1814
Gansert D (2006) Miniaturisierte und multifunktionale optosensorische Meß- und Regelschleuse für die minimalinvasive Applikation von Mikrosensoren und Mikrowerkzeugen in lebenden und abiotischen Systemen zur quantitativen räumlich und zeitlich hochauflösenden Prozessanalyse, Präparation und experimentellen Manipulation unter Freiland- und Laborbedingungen. Patent DE 102005 018845 A1
Gansert D, Burgdorf M (2005) Effects of xylem sap flow on carbon dioxide efflux from stems of birch (*Betula pendula* Roth). Flora 200:444–445
Gansert D, Burgdorf M, Lösch R (2001) A novel approach to the in situ measurement of oxygen concentrations in the sapwood of woody plants. Plant Cell Environ 24:1055–1064
Gansert D, Arnold M, Borisov S, Krause C, Müller A, Stangelmayer A, Wolfbeis OS (2006) Hybrid optodes (HYBOP). In: Popp J, Strehle M (eds) Biophotonics: visions for better health care. Wiley–VCH, Weinheim, pp 477–518
Gatti S, Brey T, Muller WEG, Heilmayer O, Holst G (2002) Oxygen microoptodes: a new tool for oxygen measurements in aquatic animal ecology. Mar Biol 140:1075–1085
Glud RN, Wenzhofer F, Tengberg A, Middelboe M, Oguri K, Kitazato H (2005) Distribution of oxygen in surface sediments from central Sagami Bay, Japan: in situ measurements by microelectrodes and planar optodes. Deep Sea Res Part I Oceanogr Res Pap 52:1974–1987
Gollan T, Schurr U, Schulze E-D (1992) Stomatal response to drying soil in relation to changes in the xylem sap composition of *Helianthus annuus*. I. The concentration of cations, anions amino acids in, and pH of the xylem sap. Plant Cell Environ 15:551–559
Haberlandt G (1914) Physiological plant anatomy. Macmillan, London
Hanstein S, Beer D de, Felle HH (2001) Miniaturised carbon dioxide sensor designed for measurements within plant leaves. Sensor Actuat B Chem 81:107–114
Hazneci C, Ertekin K, Yenigul B, Cetinkaya E (2004) Optical pH sensor based on spectral response of newly synthesized Schiff bases. Dyes Pigments 62:35–41
Heizmann U, Kreuzwieser J, Schnitzler JP, Brüggemann N, Rennenberg H (2001) Assimilate transport in the xylem sap of pedunculate oak (*Quercus robur*) saplings. Plant Biol 3: 132–138
Helfter C, Shephard JD, Martinez-Vilalta J, Mencuccini M, Hand DP (2007) A noninvasive optical system for the measurement of xylem and phloem sap flow in woody plants of small stem size. Tree Physiol 27:169–179
Hinsinger P, Gobran GR, Gregory PJ, Wenzel WW (2005) Rhizosphere geometry and heterogeneity arising from root-mediated physical and chemical processes. New Phytol 168:293–303
Holst G, Glud RN, Kühl M, Klimant I (1997) A microoptode array for fine-scale measurement of oxygen distribution. Sensor Actuat B Chem:122–129
Hook DD, Brown CL, Wetmore RH (1972) Aeration in trees. Bot Gaz 133:443–454
Huber C, Klimant I, Krause C, Wolfbeis OS (2001) Dual lifetime referencing as applied to a chloride optical sensor. Anal Chem 73:2097–2103
Hulth S, Aller RC, Engstrom P, Selander E (2002) A pH plate fluorosensor (optode) for early diagenetic studies of marine sediments. Limnol Oceanogr 47:212–220
Irwin S, Davenport J (2006) Implications of water flow and oxygen gradients for molluscan oxygen uptake and respirometric measurements. J Mar Biol Assoc UK 86:401–402

Jensen SI, Kuhl M, Glud RN, Jorgensen LB, Prieme A (2005) Oxic microzones and radial oxygen loss from roots of *Zostera marina*. Mar Ecol Prog Ser 293:49–58
Kirk G (2005) The biogeochemistry of submerged soils. Wiley, Chichester
Klimant I, Meyer V, Kühl M (1995) Fiber-optic oxygen microsensors, a new tool in aquatic biology. Limnol Oceanogr 40:1159–1165
Klimant I, Kühl M, Glud RN, Holst G (1997) Optical measurement of oxygen and temperature in microscale: strategies and biological applications. Sensor Actuat B Chem 38:29–37
Klimant I, Huber C, Liebsch G, Neurauter G, Stangelmayer A, Wolfbeis OS (2001) Dual lifetime referencing (DLR) – a new scheme for converting fluorescence intensity into a frequency-domain or time-domain information. In: Valeur B, Brochon JC (eds) New trends in fluorescence spectroscopy: application to chemical and life sciences. Springer, Berlin Heidelberg New York, pp 257–275
König B, Kohls O, Holst G, Glud RN, Kuhl M (2005) Fabrication and test of sol–gel based planar oxygen optodes for use in aquatic sediments. Mar Chem 97:262–276
Kosch U, Klimant I, Wolfbeis OS (1999) Long-lifetime based pH micro-optodes without oxygen interference. Fresen J Anal Chem 364:48–53
Kovacich RP, Martin NA, Clift MG, Stocks C, Gaskin I (2006) Highly accurate measurement of oxygen using a paramagnetic gas sensor. Meas Sci Technol 17:1579–1585
Lange OL (2002) Photosynthetic productivity of the epilithic lichen *Lecanora muralis*: long-term field monitoring of CO_2 exchange and its physiological interpretation. I. Dependence of photosynthesis on water content, light, temperature, and CO_2 concentration from laboratory measurements. Flora 197:1–17
Lavigne MB, Franklin SE, Hunt ER (1996) Estimating stem maintenance respiration rates of dissimilar balsam fir stands. Tree Physiol 16:687–696
Lavigne MB, Little CHA, Riding RT (2004) Changes in stem respiration rate during cambial reactivation can be used to refine estimates of growth and maintenance respiration. New Phytol 162:81–93
Levy PE, Meir P, Allen SJ, Jarvis PG (1999) The effect of aqueous transport of CO_2 in xylem sap on gas exchange. Tree Physiol 19:53–58
Li CY, Zhang XB, Han ZX, Akermark B, Sun LC, Shen GL, Yu RQ (2006) A wide pH range optical sensing system based on a sol–gel encapsulated amino-functionalised corrole. Analyst 131:388–393
Liebsch G, Klimant I, Wolfbeis OS (1999) Luminescence lifetime temperature sensing based on sol–gels and poly(acrylonitrile)s dyed with ruthenium metal-ligand complexes. Adv Mater 11:1296
Liebsch G, Klimant I, Krause C, Wolfbeis OS (2001) Fluorescent imaging of pH with optical sensors using time domain dual lifetime referencing. Anal Chem 73:4354–4363
Lösch R (2001) Wasserhaushalt der Pflanzen. Quelle & Meyer, Wiebelsheim
Lösch R, Larcher W (1998) Entwicklung und Trends der Ökophysiologie im 20. Jahrhundert. Wetter Leben 50:291–336
Maberly SC, Madsen TV (1998) Affinity for CO_2 in relation to the ability of freshwater macrophytes to use HCO_3^-. Funct Ecol 12:99–106
MacDougal DT, Working EB (1933) The pneumatic system of plants, especially trees. (Publication 441) Carnegie Institute of Washington, Washington, D.C.
Maier CA, Clinton BD (2006) Relationship between stem CO_2 efflux, stem sap velocity and xylem CO_2 concentration in young loblolly pine trees. Plant Cell Environ 29:1471–1483
Mainiero R, Kazda M (2005) Effects of *Carex rostrata* on soil oxygen in relation to soil moisture. Plant Soil 270:311–320
Mancuso S, Marras AM (2003) Different pathways of the oxygen supply in the sapwood of young *Olea europaea* trees. Planta 216:1028–1033
Marschner H (1995) Mineral nutrition of higher plants. Academic, London
Martin TA, Teskey RO, Dougherty PM (1994) Movement of respiratory CO_2 in stems of loblolly pine (*Pinus taeda* L.) seedlings. Tree Physiol 14:481–495

McGuire MA, Teskey RO (2002) Microelectrode technique for in situ measurement of carbon dioxide concentrations in xylem sap of trees. Tree Physiol 22:807–811

McKenzie DJ, Wong S, Randall DJ, Egginton S, Taylor EW, Farrell AP (2004) The effects of sustained exercise and hypoxia upon oxygen tensions in the red muscle of rainbow trout. J Exp Biol 207:3629–3637

Meng XM, Liu L, Guo QX (2005) Progress in fluorescence sensors for Pb(II), Hg(II) and Cd(II) cations. Prog Chem 17:45–54

Menzel M, Soukup J, Henze D, Engelbrecht K, Senderreck M, Scharf A, Rieger A, Grond S (2003) Experiences with continuous intra-arterial blood gas monitoring: precision and drift of a pure optode-system. Intensive Care Med 29:2180–2186

Mock T, Dieckmann GS, Haas C, Krell A, Tison JL, Belem AL, Papadimitriou S, Thomas DN (2002) Micro-optodes in sea ice: a new approach to investigate oxygen dynamics during sea ice formation. Aquat Microb Ecol 29:297–306

Mock T, Kruse M, Dieckmann GS (2003) A new microcosm to investigate oxygen dynamics at the sea ice water interface. Aquat Microb Ecol 30:197–205

Nagel KA, Schurr U, Walter A (2006) Dynamics of root growth stimulation in *Nicotiana tabacum* in increasing light intensity. Plant Cell Environ 29:1936–1945

Negisi K (1979) Bark respiration rate in stem segments detached from young *Pinus densiflora* trees in relation to velocity of artificial sap flow. J Jpn For Soc 61:88–93

Neurauter G, Klimant I, Wolfbeis OS (2000) Fiber-optic microsensor for high resolution pCO_2 sensing in marine environment. Fresen J Anal Chem 366:481–487

Nwaigwe CI, Roche MA, Grinberg O, Dunn JF (2003) Brain tissue and sagittal sinus pO_2 measurement using the lifetimes of oxygen-quenched luminescence of a ruthenium compound. Oxygen transport to tissue XXIV Adv Exp Med Biol 530:101–111

Patra AK, Abbadie L, Clays-Josserand A, Degrange V, Grayston SJ, Loiseau P, Louault F, Mahmood S, Nazaret S, Philippot L, Poly F, Prosser JI, Richaume A, Le Raux X (2005) Effects of grazing on microbial functional groups involved in soil N dynamics. Ecol Monogr 75:65–80

Pearce FJ, Waasdorp C, Hufnagel H, Burtris D, DeFeo J, Soballe P, Drucker WR (2005) Subcutaneous pO_2 as an index of the physiological limits for hemodilution in the rat. J Appl Physiol 99:814–821

Pimenta AM, Araujo AN, Conceicao M, Montenegro BSM, Pasquini C, Rohwedder JJR, Raimundoo IM (2004) Chloride-selective membrane electrodes and optodes based on an indium(III) porphyrin for the determination of chloride in a sequential injection analysis system. J Pharmaceut Biomed 36:49–55

Polerecky L, Franke U, Werner U, Grunwald B, Beer D de (2005) High spatial resolution measurement of oxygen consumption rates in permeable sediments. Limnol Oceanogr Methods 3:75–85

Precht E, Franke U, Polerecky L, Huettel M (2004) Oxygen dynamics in permeable sediments with wave-driven pore water exchange. Limnol Oceanogr 49:693–705

Pruyn ML, Gartner BL, Harmon ME (2002a) Within-stem variation of respiration in *Pseudotsuga menziesii* (Douglas-fir) trees. New Phytol 154:359–372

Pruyn ML, Gartner BL, Harmon ME (2002b) Respiratory potential in sapwood of old versus young ponderosa pine trees in the Pacific Northwest. Tree Physiol 22:105–116

Radu A, Bakker E (2005) Shifting the measuring range of chloride selective electrodes and optodes based on the anticrown ionophore [9]mercuracarborand-3 by the addition of 1-decanethiol. Chem Anal Warsaw 50:71–83

Roche P, Al-Jowder R, Narayanaswamy R, Young J, Scully P (2006) A novel luminescent lifetime-based optrode for the detection of gaseous and dissolved oxygen utilising a mixed ormosil matrix containing ruthenium (4,7-diphenly-1,10-phenanthroline) (3) Cl-2 (Ru.dpp). Anal Bioanal Chem 386:1245–1257

Ryan MG, Hubbard RM, Pongracic S, Raison RJ, McMurtrie RE (1996) Foliage, fine-root, woody-tissue and stand respiration in *Pinus radiata* in relation to nitrogen status. Tree Physiol 16:333–343

Safavi A, Rostamzadeh A, Maesum S (2006) Wide range pH measurements using a single H^+-selective chromoionophore and a time-based flow method. Talanta 68:1469–1473

Sartoris FJ, Bock C, Serendero I, Lannig G, Portner HO (2003) Temperature-dependent changes in energy metabolism, intracellular pH and blood oxygen tension in the Atlantic cod. J Fish Biol 62:1239–1253

Schlüter U, Crawford RMM (2001) Long-term anoxia tolerance in leaves of *Acorus calamus* L. and *Iris pseudacorus* L. J Exp Bot 52:2213–2225

Schmälzlin E, Dongen JT van, Klimant I, Marmodée B, Steup M, Fisahn J, Geigenberger P, Löhmannsröben H-G (2005) An optical multifrequency phase-modulation method using microbeads for measuring intracellular oxygen concentrations in plants. Biophys J 89:1339–1345

Schurr U (1998) Xylem sap sampling – new approaches to an old topic. Trends Plant Sci 3:293–298

Schurr U, Walter A, Rascher U (2006) Functional dynamics of plant growth and photosynthesis – from steady-state to dynamics – from homogeneity to heterogeneity. Plant Cell Environ 29:340–352

Shui YB, Fu JJ, Garcia C, Dattilo LK, Rajagopal R, McMilllan S, Mak G, Holekamp NM, Lewis A, Beebe DC (2006) Oxygen distribution in the rabbit eye and oxygen consumption by the lens. Invest Ophthalmol Vision Sci 47:1571–1580

Sorz J, Hietz P (2006) Gas diffusion through wood: implications for oxygen supply. Trees 20:34–41

Spicer R, Holbrook NM (2005) Within-stem oxygen concentration and sap flow in four temperate tree species: does long-lived xylem parenchyma experience hypoxia? Plant Cell Environ 28:192–201

Stehning C, Holst GA (2004) Addressing multiple indicators on a single optical fiber – digital signal processing approaches for temperature compensated oxygen sensing. IEEE Sens J 4:153–159

Stern O, Volmer M (1919) Über die Abklingzeit der Fluoreszenz. Phys Z 20:183–188

Stumm W, Morgan JJ (1996) Aquatic chemistry: chemical equilibria and rates in natural waters. Wiley, New York

Tengberg A, Hovdenes J, Barranger D, Brocandel O, Diaz R, Sarkkula J, Huber C, Stangelmayer A (2003) Optodes to measure oxygen in the aquatic environment. Sea Technol 44:10–15

Tengberg A, Hovdenes J, Andersson HJ, Brocandel O, Diaz R, Hebert D, Arnerich T, Huber C, Kortzinger A, Khripounoff A, Rey F, Ronning C, Schimanski J, Sommer S, Stangelmayer A (2006) Evaluation of a lifetime-based optode to measure oxygen in aquatic systems. Limnol Oceanogr Methods 4:7–17

Teskey RO, McGuire MA (2002) Carbon dioxide transport in xylem causes errors in estimation of rates of respiration in stems and branches of trees. Plant Cell Environ 25:1571–1577

Teskey RO, McGuire MA (2005) CO_2 transported in xylem sap affects CO_2 efflux from *Liquidambar styraciflua* and *Platanus occidentalis* stems, and contributes to observed wound respiration phenomena. Trees 19:357–362

Teskey RO, McGuire MA (2007) Measurement of stem respiration of sycamore (*Platanus occidentalis* L.) trees involves internal and external fluxes of CO_2 and possible transport of CO_2 from roots. Plant Cell Environ 30:529–669

Thuesen EV, Rutherford LD, Brommer PL, Garrison K, Gutowska MA, Towanda T (2005a) Intragel oxygen promotes hypoxia tolerance of scyphomedusae. J Exp Biol 208: 2475–2482

Thuesen EV, Rutherford LD, Brornmer PL (2005b) The role of aerobic metabolism and intragel oxygen in hypoxia tolerance of three ctenophores: *Pleurobrachia bachel*, *Bolinopsis infundibulum* and *Mnemiopsis leidyi*. J Mar Biol Assoc UK 85:627–633

Vasylevska AS, Borisov SM, Krause Ch, Wolfbeis OS (2006) Indicator-loaded and permeation-selective microparticles for use in fiber optic simultaneous sensing of pH and dissolved Oxygen. Chem Mater **18**:4609–4616

Viollier E, Rabouille C, Apitz SE, Breuer E, Chaillou G, Dedieu K, Furukawa Y, Grenz C, Hall P, Janssen F, Morford JL, Poggiale JC, Roberts S, Shimmield T, Taillefert M, Tengberg A, Wenzhofer F, Witte U (2003) Benthic biogeochemistry: state of the art technologies and guidelines for the future of in situ survey. J Exp Mar Biol Ecol 285:5–31

Visser EJW, Bögemann GM, Van de Steeg HM, Pierik R, Blom WPM (2000) Flooding tolerance of *Carex* species in relation to field distribution and aerenchyma formation. New Phytol 148:93–103

Von Willert DJ, Matyssek R, Herppich W (1995) Experimentelle Pflanzenökologie. Thieme, Stuttgart

Walter A, Scharr H, Gilmer F, Zierer R, Nagel KA, Ernst M, Wiese A, Virnich O, Christ MM, Uhlig B, Jünger S, Schurr U (2007) Dynamics of seedling growth acclimation towards altered light conditions can be quantified via GROWSCREEN: a setup and procedure designed for rapid optical phenotyping of different plant species. New Phytol. DOI 10.1111/j.1469-8137.2007.02002.x

Werner T, Huber Ch, Heinl S, Kollmannsberger M, Daub J, Wolfbeis OS (1997) Novel optical pH-sensor based on a boradiaza-indacene derivative. Fresen J Anal Chem 359:150–154

Wießner A, Kuschk P, Stottmeister U (2002) Oxygen release by roots of *Typha latifolia* and *Juncus effusus* in laboratory hydroponic systems. Acta Biotechnol 22:209–216

Wilkinson S, Corlett JE, Oger L, Davies WJ (1998) Effects of xylem pH on transpiration from wild-type and *flacca* tomato leaves. Plant Physiol 117:703–709

Wolfbeis OS (2005) Materials for fluorescence-based optical chemical sensors. J Mater Chem 15:2657–2669

Wolfbeis OS, Dürkop A, Wu M, Lin Z (2002) A europium-ion-based luminescent sensing probe for hydrogen peroxide. Angew Chem Int Ed 41:4495–4498

Xu C, Wygladacz A, Qin Y, Retter R, Bell M, Bakker E (2005) Microsphere optical ion sensors based on doped silica gel templates. Anal Chim Acta 537:135–143

Zhu QZ, Aller RC, Fan YZ (2005) High-performance planar pH fluorosensor for two-dimensional pH measurements in marine sediment and water. Environ Sci Technol 39:8906–8911

Indirect Defence – Recent Developments and Open Questions

Martin Heil

Contents

Abstract Many plants defend themselves indirectly against herbivores via the attraction of carnivores. Carnivores can be provided with resources such as cellular food bodies, domatia, or extrafloral nectar (EFN), and they can be attracted by herbivore-induced volatile organic compounds (HI-VOCs) that signal the presence of prey. Both HI-VOCs and EFN are usually induced in response to herbivore attack. HI-VOCs have been intensively studied with respect to their (bio)chemistry and the signalling events leading to their induction, but field studies on the role of HI-VOCs in the plant's indirect defence via tritrophic interactions are scarce. In contrast, EFN has been described for hundreds of species, while we lack detailed knowledge on its chemical ecology and the physiology of EFN secretion. Food bodies and domatia have been almost ignored by researchers in the past ten years. Tritrophic interactions

Martin Heil
Departamento de Ingeniería Genética, CINVESTAV–Irapuato, Km. 9.6 Libramiento Norte, CP 36821, Irapuato, Guanajuato, Mexico
e-mail: mheil@ira.cinvestav.mx

U. Lüttge et al. (eds.), *Progress in Botany 69.*

are increasingly being discussed as environmentally friendly tool in crop protection, but current research is highly unbalanced in focusing on entirely different aspects of the different defensive traits. Future studies should consider what is already known on other traits in order to gain a more complete understanding of a plant's possibilities to defend itself by establishing mutualisms with the third trophic level.

1 Introduction

An important invention of life is mutualism: instead of featuring all traits required by them on their own, organisms are engaged in mutually beneficial interactions to exchange resources, which are derived from syndromes of traits the respective partner species could hardly ever evolve on its own. Usually, one partner provides a service in order to receive a reward from the other partner (Bronstein 1994). For instance, plants lack mobility and they are therefore engaged in mutualisms with animals in which the partner's service consists of transport, such as transport of pollen (→ pollination) or offspring (→ dispersal). Similarly, plants with few exceptions form part of the first trophic level and lack traits that would allow them to prey on other organisms. In order to get rid of animals that feed on them (i.e., herbivores), many plant species are engaged in mutualisms with those who are born for this task: the enemies of their enemies. By attracting carnivores, plants can actively increase top-down control of herbivores and thereby defend themselves indirectly (Price et al. 1980).

2 Phenomenon and Definitions

2.1 Introduction

Plants have evolved a bewildering array of chemical and mechanical strategies for protection from herbivores and pathogens (Walling 2000). This chapter is devoted to one particular group of strategies, the 'indirect defences'. Most textbook examples of plant defence consider toxins, repellents and mechanical barriers. Yet, research during recent decades discovered the relevance of indirect defences: plants can attract or even house predators that act as defending agents (Arimura et al. 2005; Dicke 1999; Karban and Baldwin 1997).

'Direct defence' comprises all plant traits that exhibit a bottom-up control of herbivores, i.e., plant traits that directly affect herbivores and as a consequence reduce the damage inflicted by them. Herbivores, however, are controlled not only by the availability and quality of food, but also by their natural enemies (top-down control). Every trait that increases the presence of members of the third trophic level (carnivores) and thereby decreases the abundance of herbivores on a plant thus has the potential to positively affect plant fitness by reducing herbivory. It therefore can be termed a defence, though an indirect one, since there is no immediate negative effect of plant traits on the herbivore. Historically, this idea was introduced by Price et al. (1980), but Dicke and Sabelis (1988) apparently were the first to introduce the terms 'direct' and 'indirect' defence (M. Dicke, personal communication).

In order to be more attractive to carnivores, plants can provide food, shelter, or simply the information that they are attacked. Plant traits that function in these interactions are herbivore-induced volatile organic compounds (HI-VOCs, which usually are released upon herbivore damage, see Section 3.1), extrafloral nectar (EFN, nectar that is not involved in pollination but in the attraction and nutrition of predators, see Section 3.2), food bodies (FBs, cellular structures that serve the attraction and nutrition of predators, see Section 3.3) and domatia (structures that provide predators with shelter, see Section 3.4). Finally, plants may also make use of the biosynthetic capacities of fungi and become less palatable to herbivores due to substances that are not synthesized by themselves, but rather by endophytic fungi (see Section 3.5).

Research has greatly augmented our knowledge on signalling cascades and genes involved in the expression of induced defences. Many excellent reviews are available on, e.g., jasmonates and the octadecanoid signalling cascade (Blechert et al. 1995; Creelman and Mullet 1997a; Farmer et al. 2003; Mithöfer et al. 2005a; Schilmiller and Howe 2005; Wasternack and Parthier 1997), and on the involvement of ethylene in these responses (Glazebrook et al. 2003). Microarray techniques have led to an exploding knowledge on genes that are up- or down-regulated during resistance induction (Kant et al. 2004; Mercke et al. 2004; Pauw and Memelink 2005; Ralph et al. 2006; Reymond and Farmer 1998; Reymond et al. 2000; Schenk et al. 2000; Voelckel and Baldwin 2004). The present chapter therefore concentrates on the chemical ecology of indirect defences.

2.2 Phenotypic Plasticity of Indirect Defence

Based on their phenotypic plasticity, defence traits can be characterized as 'constitutive', i.e., being expressed all the time or at least hardly dependent on environmental conditions, or as 'induced', i.e., being expressed in response to a first, eliciting event, which usually is feeding by an enemy (Karban and Baldwin 1997) but even might be 'only' egg deposition by herbivorous insects (Hilker and Meiners 2006). Besides VOCs, also EFN can be induced by herbivore damage.

Why do plants wait for a first attack before they express defence, though induced strategies are hampered by the intrinsic problem of a time-lag between attack and resistance action? Mainly two explanations are given for the evolution of induced defences: signal reliability and fitness costs.

2.2.1 Signal Reliability in Tritrophic Interactions

HI-VOCs are no resource per se but advertize the presence of prey. Signal reliability thus is an important aspect in the plant–predator mutualism, which would be evolutionarily instable if plants would 'cheat' by attracting carnivores in the absence of herbivores. Parasitoids in particular learn quickly which signals to use for prey localization (De Boer et al. 2005; Petitt et al. 1992; Turlings et al. 1990) and avoid plant signals that do not reliably indicate the presence of their prey. EFN, in contrast, contains water, carbohydrates, and amino acids, and thus represents a valuable resource on its own. Still, plants increase EFN secretion rates dramatically when they are damaged (see Section 3.2.2). In these cases, fitness costs are the explanation for the evolution of inducible defence traits: The production of defence traits is assumed to be metabolically or ecologically costly and is therefore restricted to time spans that actually require an active defence.

2.2.2 Fitness Costs of Indirect Defence

Defence indeed can be costly. *Arabidopsis* and wheat plants that were treated with chemical elicitors to induce their resistance to pathogens or herbivores responded with a reduced growth and seed set when cultivated under enemy-free conditions that prevented the resistance from benefiting the plants (Cipollini 2002, 2007; Dietrich et al. 2005; Heil et al. 2000b). Inducing herbivore resistance also led to significant fitness costs in tobacco (Baldwin 1998), *Lepidium* (Agrawal 2000a), or wild radish (Agrawal et al. 1999).

While there is now overwhelming evidence for the existence of costs of induced direct resistances (Cipollini et al. 2003; Heil 2002; Heil and Baldwin 2002) much less is known on costs of indirect defences. Physiological costs of HI-VOCs are assumed to be low (Dicke and Sabelis 1989; Hoballah et al. 2004), and allocation costs of EFN secretion have not been studied. Food bodies (see Section 3.3), in contrast, can represent some 10 % of aboveground tissue construction costs (Heil et al. 1997). Besides these direct physiological costs, particularly those resistance traits that include the exchange of information can also be used by other than the mutualistic arthropods and then cause ecological costs (Dicke 1999; Heil 2002).

In short, costs of indirect defences can vary greatly in quantity and quality, and evidence for their occurrence and relevance is fairly anecdotal. Since defence traits vary phenotypically and on the level of populations, much more research is required to understand whether (and which types of) costs indeed can serve as explanation for the induced expression of indirect defences.

3 Basic Principles: Exchange of Information and Resources

3.1 Volatiles (HI-VOCs)

Flower scents are the most prominent plant volatiles, yet in fact all above-ground parts of plants and even roots release VOCs. Upon damage, the quantity and quality of VOCs in a plant's headspace usually changes dramatically (Arimura et al. 2005;

Dicke 1994; Dicke and Vet 1999; Farmer 2001; Paré and Tumlinson 1997b; Tumlinson et al. 1999; Turlings et al. 1995), a response at least in part mediated by the plant hormone, jasmonic acid (JA; Boland et al. 1995). The number of volatile compounds released by plants exceeds 1000, and more substances are likely to be detected in the future (Pichersky et al. 2006a).

Substances that are immediately released after damage cause the characteristic odour of mowed pastures and are therefore called green-leaf volatiles (GLVs). The majority of these substances are C_6 volatiles, i.e., isomers of hexenol, hexenal or hexenyl acetate. The release of (3*Z*)-hex-3-enal has been observed within 20 s after the disruption of *Arabidopsis* leaf tissues (Matsui et al. 2000). Though enzymatic activity appears generally required for GLV synthesis, the extremely rapid release upon damage (Turlings et al. 1998) points to the presence of precursors. Some GLVs may even be completely present in the undamaged leaves and just released passively due to disruption of the cuticle. The passive nature of GLV release is underlined by the observation that it was quantitatively unchanged in maize plants that were infected by pathogenic fungi, while the release of other HI-VOCs was reduced to almost the half of what is released by healthy plants (Rostas et al. 2006).

The majority of HI-VOCs are synthesized completely de novo after damage (Dudareva et al. 2004; Paré and Tumlinson 1997b; Pichersky et al. 2006a). Esters such as methyl salicylate and methyl jasmonate, monoterpenes such as limonene, linalool, or ocimene, and sesquiterpenes such as bergamotene, caryophyllene and farnesene are released in general as several isomers, typically starting 24 h after attack (Arimura et al. 2005; Halitschke et al. 2000; Turlings et al. 1998). Many terpenoids are synthesized by a family of terpene synthases, a prominent member of which is TPS10 from corn plants, forming (E)-β-farnesene and (E)-α-bergamotene from farnesyl diphosphate (Schnee et al. 2006). However, experimental evidence for the biosynthetic origin of most plant volatiles is still missing (Pichersky et al. 2006a), and I therefore name all volatile organic compounds *HI-VOCs* that are released at significantly higher rates in response to herbivore damage.

HI-VOCs are induced systemically, i.e., they are released also from yet undamaged organs of a damaged plant (Paré and Tumlinson 1999). There is now even compelling evidence for aboveground–belowground signalling, as attack of roots by belowground herbivores can induce the release of HI-VOCs from shoots (Bezemer and van Dam 2005). Chemical signals were isolated such as cell wall constituents that originate from the feeding event (Creelman and Mullet 1997b; Doares et al. 1995) and elicitors that are derived from the herbivores' saliva or that are even resulting from a conjugation of plant- and herbivore-derived precursors. These are transported within the plant or elicit the synthesis of hormones such as JA and thereby induce the systemic release of HI-VOCs (Alborn et al. 1997; Mattiacci et al. 1995; Paré et al. 1998; Ryan and Pearce 1998; Schaller and Ryan 1995; Turlings and Tumlinson 1992). Recent studies, however, demonstrated that mere mechanical damage can be sufficient to elicit the majority of JA-dependent genes or defence traits (Heil et al. 2001, 2004c; Major and Constabel 2006, 2007; Mithöfer et al. 2005b).

3.1.1 HI-VOCs-Mediated Indirect Defence

Predator–prey interactions are a classical field in ecological and behavioural research. Still, it took almost ten years after the review by Price et al. (1980) until the first hints pointed to a role of damaged plants in the prey-searching behaviour of carnivores (Dicke 1986). This idea was rapidly confirmed by follow-up experimental studies (Dicke and Sabelis 1988; Dicke et al. 1990; Turlings et al. 1990). Since then, many studies demonstrated the defensive effects of HI-VOCs. It is now generally accepted that HI-VOCs can attract predatory arthropods (De Moraes et al. 1998; Dicke and Dijkman 1992; Dicke et al. 1993; Kessler and Baldwin 2001; Thaler 1999; Turlings et al. 1995) and/or repel herbivores (Birkett et al. 2000; De Moraes et al. 2001; Dicke and Dijkman 1992; Heil 2004a) and thus serve as a means of plant resistance (Karban and Baldwin 1997; Paré and Tumlinson 1999; Tumlinson et al. 1999). Though single compounds can have significant attractive effects (see examples in Table 1), blends as released from damaged plants usually elicit stronger responses (Wei and Kang 2006).

Classic examples are the attraction of the blind predatory mite *Phytoseiulus persimilis* that uses odours released by Lima bean (*Phaseolus lunatus*) plants to localize its prey mite, *Tetranychus urticae* (Dicke and Sabelis 1988, 1989; Dicke et al. 1993), and the parasitoidic wasp *Cotesia marginiventris* that uses odours of damaged maize (*Zea mays*) plants to localize its host caterpillar, *Spodoptera exigua* (Turlings and Tumlinson 1992; Turlings et al. 1990, 1995). Studies into these two systems demonstrated that carnivores discriminate: (a) among damaged and undamaged plants, (b) among plants infested by different herbivore species, and (c) among different plants infested by the same herbivore, and that (d) the odours released by the herbivore itself are much less important as a host localization cue than the odours released by the infested plants (Dicke 1994). Ongoing research then demonstrated that HI-VOCs might even be induced already during egg deposition by herbivorous insects (for a comprehensive review, see Hilker and Meiners 2006) and then usually serve the attraction of egg parasitoids (Colazza et al. 2004a; Meiners and Hilker 2000). HI-VOCs induced by egg deposition on various bean species were, for instance, (E)-β-caryophyllene, DMNT and TMTT, i.e., compounds that are commonly involved in various plant–arthropod interactions (Colazza et al. 2004b; see also Table 1).

The taxonomic list of plants for which a volatile-mediated attraction of carnivores in response to herbivore damage has been demonstrated is surprisingly short: the most recent overviews comprised fewer than 30 species, namely apple (*Malus domestica*), broad bean (*Vicia faba*), common bean (*Phaseolus vulgaris*), Lima bean (*Phaseolus lunatus*), cabbage (*Brassica oleracea*), cassava (*Manihot esculentum*), *Chrysanthemum* sp., corn (*Zea mays*), cotton (*Gossypium hirsutum*), cowpea (*Vigna unguiculata*), cucumber (*Cucumis sativus*), *Gerbera jamesonii*, ground ivy (*Glechoma hederacea*), hawthorn (*Crataegus* sp.), lettuce (*Lactuca sativa*), nasturtium (*Tropaeolum majus*), rose (*Rosa X hybrida*), pear (*Pyrus communis*), potato (*Solanum tuberosum*), sesame (*Sesamum indicum*), tomato (*Lycopersicon esculentum*), wheat (*Triticum aestivum*), and yellow cress

Table 1 Multiple roles of VOCs in indirect defence and pollination. Comprehensive overviews have been published both on volatiles released from flowers (Knudsen et al. 1993, 2006) and on HI-VOCs. I therefore present only a few examples for which at least anecdotal knowledge exists on the ecological functions and which are taxonomically widespread compounds of floral scent according to Knudsen et al. (2006). Stereochemistry is not reported in many, particularly behavioural, studies and is therefore not indicated

Substance	Function as HI-VOC in tritrophic interactions	Function in flowering
E-β-Ocimene, β-3,7-dimethyl-1,3,6-octatriene 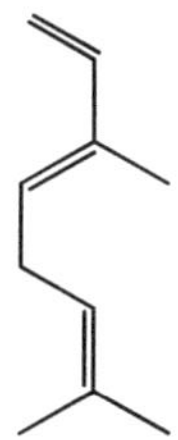	Synthesized de novo in herbivore-damaged cotton leaves (Paré and Tumlinson 1997a) and common HI-VOC of other plants such as, e.g., Lima bean (Kost and Heil 2006). Induced in weevil-damaged strawberry flowers and eliciting EAG responses in these insects (Bichão et al. 2005)	Component of flower odours in more than 50% of families investigated so far (Knudsen et al. 2006). Released from stigmas of carob tree flowers (Custódio et al. 2006) and from different parts of *Mirabilis jalapa* flowers (Effmert et al. 2005). Major component in the flower odour of *Wulffia baccala* (Zoghbi et al. 2000)
β-Caryophyllene, 8-methylene-4,11,11-trimethyl-bicyclo [7.2.0] undec-4-ene 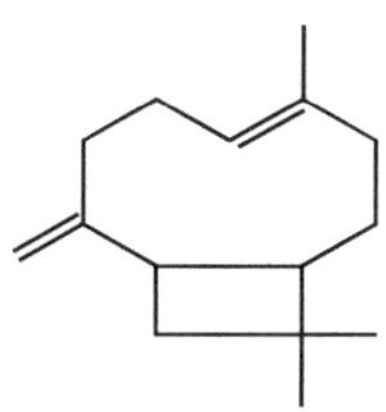	Released by Ginkgo in response to JA treatment (Van den Boom et al. 2004). Released by maize roots in response to insect damage and attractive to an entomopathogenic nematode (Rasmann et al. 2005). Induced in weevil-damaged strawberry flowers and eliciting EAG responses in these insects (Bichão et al. 2005). Released by damaged bean plants and putatively attractive to the egg parasitoid *Trissolcus basalis* (Colazza et al. 2004b)	Component of flower odours in more than 50% of families investigated so far (Knudsen et al. 2006). Major compound of flower odour of fly-pollinated *Alocasia* sp. (Miyake and Yafuso 2005). Released from *Arabidopsis* flowers (Tholl et al. 2005). Emitted in diurnal rthythms from *Hoya* flowers (Altenburger and Matile 1990)

(continued)

Table 1 (continued)

Substance	Function as HI-VOC in tritrophic interactions	Function in flowering
DMNT, 4,8-dimethyl-1,3,7-nonatriene 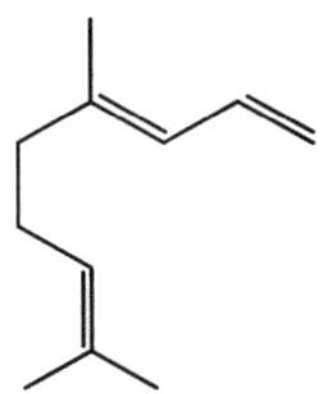	Synthesized de novo in herbivore-damaged cotton leaves (Paré and Tumlinson 1997a). Released by herbivore-infested apple, cabbage, corn, cowpea, cucumber, *Eucalyptus* sp., Lima bean, *Sedum* sp. and soybean (Dicke 1994; Kost and Heil 2006). Released from damaged *Magnolia* leaves (Azuma et al. 1997) and from JA-treated grape vine (Hampel et al. 2005). Attracts predatory arthropods, responses by several parasitic wasps (Gouinguené et al. 2005). Induced in strawberry plants due to weevil feeding and elicits EAG responses in these insects (Bichão et al. 2005)	Major compound of flower odour of fly-pollinated *Alocasia* sp. (Miyake and Yafuso 2005) and *Magnolia* sp. (Azuma et al. 1997). Found in flower odour of the moth-pollinated species, *Escobedia grandiflora* (Knudsen and Tollsten 1993)
TMTT, (E,E)-4,8,12,-trimethyl-1,3,7,11,-tridecatetraene 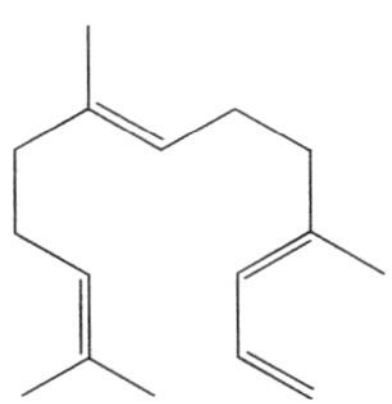	Synthesized de novo in herbivore-damaged cotton leaves (Paré and Tumlinson 1997a). Released by Lima bean in response to feeding by mites and beetles (Kost and Heil 2006). Used by predatory mites to discriminate among plants infested by prey and non-prey herbivores (De Boer et al. 2004). Elicits strong EAG responses by the parasitic wasp, *Opius dissitus*, and attracts it in the Y-tube (Wei and Kang 2006). Released by damaged bean plants and putatively attractive to the egg parasitoid *Trissolcus basalis* (Colazza et al. 2004b)	Widespread compound of floral scent (Knudsen et al. 2006). Found for instance in flower-odour of several moth-pollinated species (Knudsen and Tollsten 1993), *Hoya carnosa* (Altenburger and Matile 1990), several Fabaceae, Cactaceae and Orchidaceae (Kaiser 1991). Emitted in high amounts from butterfly-pollinated flowers of *Centrathus ruber* (Andersson et al. 2002)

(E,E)-α-Farnesene, 3,7,11-trimethyl-1,3,6,10 dodeca-tetraene 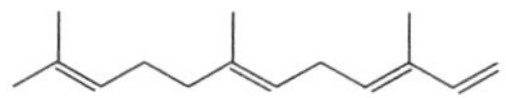	Synthesized de novo in herbivore-damaged cotton leaves (Paré and Tumlinson 1997a). Released constitutively by Ginkgo (Van den Boom et al. 2004). Released by damaged bean plants and putatively attractive to the egg parasitoid *Trissolcus basalis* (Colazza et al. 2004b)	Released from carob tree flowers (Custódio et al. 2006) and major compounds of floral scent of mistletoe (Bungert et al. 2002) and *Lonicera japonica* (Schlotzhauer et al. 1996). Minor compound of flower odour of fly-pollinated *Alocasia* sp. (Miyake and Yafuso 2005)
(Z)-3-Hexen-1-ol 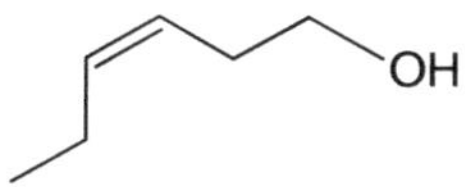	GLV that induces and primes resistance in neighbouring plants (Farag et al. 2005; Ruther and Kleier 2005). Increased herbivore egg predation rate on native tobacco (Kessler and Baldwin 2001)	Important component in floral scent of *Silene* sp. (Jürgens 2004) and *Lonicera japonica* (Schlotzhauer et al. 1996). Emitted from flowers of several butterfly-pollinated species (Andersson et al. 2002)
(E)-3-Hexenyl acetate acetic acid (E)-3-hexen-1-yl ester 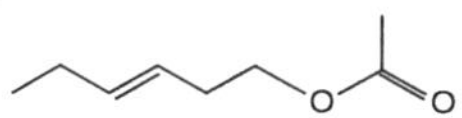	Common GLV that attracts predatory arthropods (James 2003)	Widespread compound of floral scent (Knudsen et al. 2006). Released from flowering *Mirabilis jalapa* plants (Effmert et al. 2005)
(Z)-3-Hexenyl acetate, acetic acid (Z)-3-hexen-1-yl ester 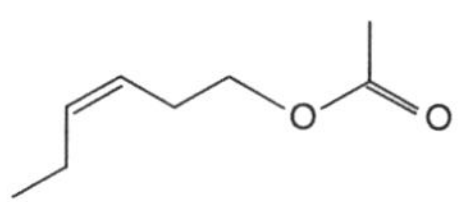	Common GLV that induces and primes resistance in neighbouring plants (Bate and Rothstein 1998; Kost and Heil 2006). Attracts predatory arthropods when used on artificial baits in the field (James 2003). Elicits strong EAG responses by several parasitic wasps (Gouinguené et al. 2005), e.g., by *C. chlorideae* and *O. dissitus*, and attracts them in the Y-tube (Wei and Kang 2006; Yan and Wang 2006). Induced in weevil-damaged strawberry flowers and eliciting EAG responses in these insects (Bichão et al. 2005)	Widespread compound of floral scent (Knudsen et al. 2006), e.g., in *Silene* species (Jürgens 2004). Emitted from flowers of many butterfly-pollinated species (Andersson et al. 2002)

(continued)

Table 1 (continued)

Substance	Function as HI-VOC in tritrophic interactions	Function in flowering
Linalool, 3,7-dimethyl-1,6-octadien-3-ol 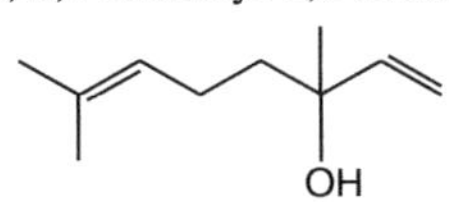	Taxonomically widespread, herbivore- and JA-responsive monoterpene (Miller et al. 2005; Mumm and Hilker 2006). Synthesized de novo in herbivore-damaged cotton leaves (Paré and Tumlinson 1997a). Elicits EAG responses in the parasitoid *C. chlorideae* and attracts it in the Y-tube (Yan and Wang 2006). Increased herbivore egg predation rate on native tobacco while having a repellent effect on herbivores (Kessler and Baldwin 2001). Elicits strong EAG responses by several parasitic wasps (Gouinguené et al. 2005; Wei and Kang 2006)	Component of flower odours in more than 50% of families investigated so far (Knudsen et al. 2006). Important component in floral scent of *Silene* sp. (Jürgens 2004) and *Lonicera japonica* (Schlotzhauer et al. 1996). Emitted in diurnal rthythms from *Hoya* flowers (Altenburger and Matile 1990) and released as several derivatives from carob tree flowers (Custódio et al. 2006). Emitted at high amounts from flowers of *Cirsium arvense* and *Buddleja davidii* and elicits strong antennal responses in pollinating butterflies (Andersson 2003)
Limonene, 1-methyl-4-prop-1-en-2-yl-cyclohexene 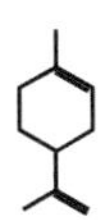	Attractive to parasitoid *Cotesia* sp. in Y-tube olfactometer (Ibrahim et al. 2005)	Component of flower odours in more than 50% of families investigated so far (Knudsen et al. 2006). Major component in the flower odour of *Wulffia baccala* (Zoghbi et al. 2000)

Methyl salicylate, methyl-2-hydroxybenzoate 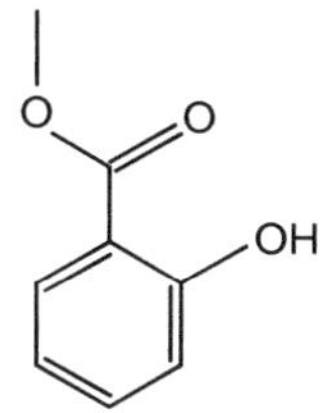	Released by many plants in response to herbivore attack (particularly mites) and pathogen infection, important elicitor of resistance responses directed towards pathogens and herbivores (Schenk et al. 2000). Attracts predatory arthropods when used on artificial baits (James 2003). Used by predatory mites to discriminate among plants infested by prey and non-prey herbivores (De Boer et al. 2004). Induced in strawberry plants due to weevil feeding and eliciting EAG responses in these insects (Bichão et al. 2005). Induced in aphid-infested soybean plants and attractive to the aphid predator *Coccinella septempunctata*, eliciting EAG responses in this beetle (Zhu and Park 2005)	Component of flower odours in more than 50% of families investigated so far (Knudsen et al. 2006). Major compound of flower odour of fly-pollinated *Alocasia* spp (Miyake and Yafuso 2005). Released from male cones of insect-pollinated cycads (Pellmyr et al. 1991). Compound of *Lonicera japonica* floral scent (Schlotzhauer et al. 1996)

(*Rorippa indica*; Dicke 1999). Since 1999, HI-VOC-mediated tritrophic interactions have been described for additional plant species, such as leek (*Allium porrum*; Dugravot and Thibout 2006), *Silphium laciniatum* (Tooker and Hanks 2006), spruce (Hulcr et al. 2006; Pettersson and Boland 2003; Ralph et al. 2006), strawberry (Himanen et al. 2005), and rice (Lou et al. 2005). However, these species represent a broad spectrum of life histories, ranging from annual plants to trees, and they belong to taxonomically widespread families (Asteraceae, Brassicaeae, Cucurbitaceae, Euphorbiaceae, Fabaceae, Liliaceae, Malvaceae, Poaceae, Pinaceae, Rosaceae, Scrophulariaceae, Solanaceae, Tropaeolaceae). It is therefore most likely that the low species number just indicates the current concentration of research on model plants and the resulting lack of taxonomic broadness in the scientific efforts.

3.1.2 Ecological Relevance of HI-VOCs

Decades of work resulted in detailed knowledge on the chemistry and synthesis of HI-VOCs (Dudareva et al. 2006; Pichersky et al. 2006a). However, the vast majority of these studies were conducted under laboratory conditions on a few selected model organisms. HI-VOCs vary with plant genotype (Fritzsche-Hoballah et al. 2002; Halitschke et al. 2000; Loughrin et al. 1995), species and developmental stage of the attacking herbivore (Dicke 1994, 1999; Ozawa et al. 2000; Takabayashi and Dicke 1996; Takabayashi et al. 1995) and abiotic factors such as temperature, light intensity, rainfall, or nutrient availability (Gouinguené and Turlings 2002; Vallat et al. 2005). A few studies investigated HI-VOCs release with respect to time of day and generally revealed marked diurnal patterns, with release rates dropping dramatically during night even in response to continuous damage (Arimura et al. 2005; Kunert et al. 2002; Loughrin et al. 1994). Nothing is known whether this presents a functional adaptation or is rather a result of physiological constrains.

The previous experience of carnivores affects their behaviour towards particular VOCs (De Boer et al. 2005; Krips et al. 2001; Petitt et al. 1992; Turlings et al. 1990). Moreover, plants naturally grow in mixed stands, and carnivores must be able to distinguish their blends from the general background, an aspect overlooked in almost all laboratory studies (Dicke et al. 2003). Instead of deterring them, HI-VOCs also can attract herbivores (Bolter et al. 1997; Carroll et al. 2006; Dicke 1999; Heil 2004a; Horiuchi et al. 2003; Kalberer et al. 2001; Loughrin et al. 1996). Finally, the various carnivore species present, with different responses to HI-VOCs, may show complex interactions under natural conditions, and even a successful increase of parasitation rates of herbivores does not necessarily benefit the parasitoid-attracting plant. The question therefore remained open whether or not the release of HI-VOCs really benefits plants due to the attraction of carnivores and a resulting reduction of herbivory under natural conditions.

It was only during recent years that investigation focused on such questions. It was demonstrated in field trials that the specialist parasitic wasp *Cardiochiles nigriceps* preferred plants infested by its host, *Heliothis virescens*, over plants infested by the non-host *Heliothis zea*, and that this capacity was based on plant odours (De Moraes et al. 1998). One year later, Thaler (1999) demonstrated that caterpillars suffered higher parasitation rates when caged near tomato plants that were induced with JA to release VOCs. Parasitation of herbivores indeed can increase fitness of the plants on which the herbivores feed (Fritzsche-Hoballah and Turlings 2001; Van Loon et al. 2000). Artificially applied (*Z*)-3-hexene-1-ol, linalool, and (*Z*)-α-bergamotene increased rates at which *Manduca sexta* eggs fixed on *Nicotiana attenuata* plants at their natural growing site were predated by the generalist *Geocoris pallens* (Kessler and Baldwin 2001). Lima bean plants treated repeatedly with JA in nature released more HI-VOCs and suffered less herbivory than controls, and they produced more leaves, flowers and fruits (Heil 2004b).

Lima bean, however, responds to JA treatment with the combined increase of both VOCs release and EFN secretion (see Section 3.2). The latter study thus failed to nail down the observed defence effects to HI-VOCs. Higher parasitation or egg predation rates of herbivores do not necessarily benefit plants, depending on the intensity of these effects and the competition of herbivores for suitable hosts. 'Induced defence' should be used exclusively for cases in which a beneficial effect on plant fitness has been demonstrated (Karban and Baldwin 1997). Since plants can compensate for material loss to herbivores to some degree (Agrawal 2000b), demonstrations of increased predation pressure on herbivores (De Moraes et al. 1998; Kessler and Baldwin 2001; Thaler 1999) due to HI-VOCs, or the repellent function of HI-VOCs to specialized herbivores (De Moraes et al. 2001; Heil 2004a; Kessler and Baldwin 2001) are not sufficient proof of the defensive function of HI-VOCs: beneficial effects of HI-VOCs release on plant fitness in nature must be demonstrated before this response can be termed induced defence *sensu* Karban and Baldwin (1997).

3.1.3 HI-VOCs-Mediated Priming and Plant–Plant Communication

Volatiles form openly presented information on the status of attack of a plant, and they can be used by neighbouring plants to adjust their defensive phenotype accordingly. Since the first report of 'plant–plant communication' (Baldwin and Schultz 1983), it has been debated controversially whether this phenomenon indeed plays a role in nature (Baldwin et al. 2002, 2006; Dicke and Bruin 2001).

Originally, Baldwin and Schultz (1983) reported that undamaged plants sharing the same air with damaged plants responded with increased levels of chemical anti-herbivore defence. This observation was made under laboratory conditions, as was a behavioural study demonstrating that undamaged cotton seedlings become more attractive to predatory mites and less attractive to herbivorous mites when exposed to air from infested conspecific plantlets (Bruin et al. 1992). Later, field studies revealed that wild tobacco plants growing next to clipped sagebrush had increased levels of the defence enzyme, polyphenol oxidase (Karban et al. 2000), and defoliation of alder trees increased the herbivore resistance of neighbouring trees (Dolch and Tscharntke 2000).

Studies aimed at a mechanistic understanding of these phenomena reported changes in the expression of defence-related genes (Arimura et al. 2000; Farag et al. 2005; Paschold et al. 2006) or increased production rates of JA and defensive compounds (Baldwin and Schultz 1983; Engelberth et al. 2004; Farmer and Ryan 1990; Ruther and Kleier 2005) in volatile-exposed plants. HI-VOCs can even induce resistance to fungal pathogens (Kishimoto et al. 2005). Several other studies failed to demonstrate resistance induction by particular HI-VOCs at natural concentrations and thus doubted the ecological relevance of such observations (Preston et al. 2004). This apparent contradiction might in part be due to the fact that HI-VOCs at low concentrations prime rather than induce resistance traits in exposed plants.

Primed plants respond stronger once they are attacked or infected themselves, yet they do not show detectable expression of resistance traits before damage occurs (Conrath et al. 2006; Zimmerli et al. 2000). Exposing undamaged corn plants to HI-VOCs caused them to produce JA and terpenes more intensively and/or rapidly in response to caterpillar-caused damage than plants that were damaged without this pre-treatment (Engelberth et al. 2004). Primed corn plants are significantly more attractive to parasitic *Cotesia marginiventris* wasps, HI-VOCs-induced priming thus in fact leads to enhanced levels of direct and indirect resistance against insect attack (Ton et al. 2007). Priming is also involved in the signalling between sagebrush and tobacco: Transcriptional responses in tobacco exposed to clipped sagebrush foliage were not followed by a direct production of defensive chemicals or proteins, while the synthesis of proteinase inhibitors after feeding by *Manduca sexta* caterpillars on those plants was higher than in controls that had not been exposed to sagebrush foliage (Kessler et al. 2006). Studying the effects of VOCs on genetic or biochemical changes without exposing plants to enemy attack misses such priming effects, which however can play an important role under more natural growing conditions.

Yet, why do plants warn their neighbours at all, and why do they even warn plants belonging to other species (Karban 2001; Karban et al. 2000)? Plants usually compete with their neighbours. Receiving information on the status of attack of a neighbour and adjusting its phenotype accordingly would benefit the receiver at the cost of the emitter. How can such a signal evolve? One explanation would be that VOCs serve also plant-internal functions and for instance mediate signalling among different parts of the same plant individual (Farmer 2001; Orians 2005).

In fact, airflow from damaged to undamaged parts appeared necessary for systemic resistance induction among different branches of individual sagebrushes (Karban et al. 2006; Shiojiri and Karban 2006). The same observation was most recently made for VOCs-induced EFN secretion (see Section 3.2.3) and in fact might be simply an overseen general phenomenon, since studies focusing on systemic defence induction usually do not control air flow among the different plant parts. Neighbouring plants probably only 'eavesdrop' on what is within-plant signalling 'worn on the outside' rather than true plant–plant communication.

HI-VOCs identified so far to cause priming or induction of defence in undamaged plants include (3*Z*)-hex-3-enyl acetate (Kost and Heil 2006) and several structurally related C_6 volatiles (Bate and Rothstein 1998; Engelberth et al. 2004; Farag et al. 2005; Kishimoto et al. 2005; Ruther and Kleier 2005), i.e., the substances involved in this phenomenon are typical GLVs and thus released extremely rapidly after damage (see Section 3.1). HI-VOCs-mediated within-plant signalling might thus be faster in eliciting a systemic response than any signal that is transported as a classic hormone in phloem or xylem.

3.2 *Extrafloral Nectar*

Extrafloral nectar (EFN) can be secreted on the shoots, the leaves and even the inflorescences of plants. It is functionally not involved in pollination, but nevertheless is taxonomically widespread: more than 70 families of plants comprise EFN-producing species (Elias 1983; Koptur 1992). Plants bearing extrafloral nectaries can make up ecologically relevant proportions of the vegetation and are particularly common in tropical and subtropical habitats (Blüthgen et al. 2000; Fiala and Linsenmair 1995; Koptur 1992). EFN consequently has been the target of many field studies on ant–plant interactions. Alas, research into its chemical ecology and the regulation of its secretion is still in its infancy.

3.2.1 Indirect Defence via EFN

During the first half of the past century it was heavily debated whether extrafloral nectaries serve a 'physiological' role in secreting 'excess carbohydrates' or whether they indeed have an ecological function and attract ants as 'pugnacious bodyguards' (Bentley 1977). In response to the 'physiological' theory, an ever-increasing number of studies demonstrated convincingly that the increased presence of ants, wasps, mites and spiders that results from the secretion of EFN can reduce herbivory rates and thereby benefit plant fitness (De la Fuente and Marquis 1999; delClaro et al. 1996; Heil and McKey 2003; Oliveira et al. 1999; Sobrinho et al. 2002). Most recently, an artificial EFN mimick was applied to Lima bean plants to demonstrate that the presence of EFN alone can benefit plants in nature (Kost and Heil 2005).

The vast majority of studies on EFN focused on ants and their role as defenders. However, other arthropods from the Araneae, Diptera, Orthoptera, Hemiptera, Coleoptera, Dermaptera, Hymenoptera and Lepidoptera visit extrafloral nectaries (Koptur 1992). For instance, mites and ladybird beetles regularly visit extrafloral nectaries of many plants (Pemberton 1993; Pemberton and Vandenberg 1993; Van Rijn and Tanigoshi 1999) and thus might compete with ants, as do stingless bees (O'Dowd 1979) and certain flies (Heil et al. 2004d). Other visitors include ichneumonid and braconid wasps (Bugg et al. 1989; Cuautle and Rico-Gray 2003; Stapel et al. 1997), jumping spiders (Ruhren and Handel 1999), mosquitoes (Foster 1995) and neuropterans (Limburg and Rosenheim 2001). Some of these EFN consumers can protect plants against herbivores via increased parasitation rates (Pemberton and Lee 1996) or by preying on them (Cuautle and Rico-Gray 2003; Ruhren and Handel 1999).

Much less attention was paid to the influence that environmental and/or physiological conditions have on EFN secretion. Studies conducted so far made it likely that EFN secretion depends on water and light availability. Secretion also depends on the rate of nectar removal (Heil et al. 2000a, 2001, 2004). However, the underlying physiological mechanisms are completely unknown. A few studies also elucidated diurnal rhythms in the secretion rates of EFN (Heil et al. 2000a; Raine et al. 2002; Wickers 1997), yet without finding any general pattern.

Extrafloral nectar contains mainly mono- and disaccharides and free amino acids dissolved in water (Heil et al. 2000a; Koptur 1992), but research in our laboratory now demonstrates that more compound classes are certainly present. Though water per se can be an important resource (Ruffner and Clark 1986), most ants prefer EFNs that are rich in sugars and amino acids (Baker et al. 1978; Blüthgen and Fiedler 2004; Koptur 1994; Lanza 1988; Ruffner and Clark 1986; Smith et al. 1990). Alas, very little scientific efforts have been spent on unravelling the exact chemical composition of EFN and the role of further substances. In fact, the chemical ecology of EFN can be sophisticated. EFN of specialized *Acacia* myrmecophytes (i.e., plants that are obligately inhabited by specific, symbiotic ant colonies; see Heil and McKey 2003) has invertase activity to keep it free of sucrose, a disaccharide being generally attractive to hymenopterans. This EFN thus is unattractive to non-symbiotic ants that might compete with the plants' obligate *Pseudomyrmex* ant partners. The specialized ant inhabitants of these plants, in turn, lack invertase activity in their digestive tracts and thus depend on the 'predigested' EFN of their hosts (Heil et al. 2005).

The invertase in the *Acacia* EFN was the first described case of an enzyme regulating the carbohydrate composition of nectar, and just the second example of nectar enzyme being functionally characterized at all. The first group of nectar enzymes that have ever been described are nectarins from floral nectar of tobacco, which serve the nectars' antibacterial defence (Carter and Thornburg 2004; Thornburg et al. 2004). Antibacterial enzymes appear in *Acacia* EFN, too (Svatos and Heil, unpublished data). From behavioural studies it is now clear that EFN releases odours, which facilitate the orientation of carnivores (Röse et al. 2006). These odours, however, remain to be chemically characterized. Only recently were the first behaviourally relevant volatile compounds from floral nectar of tobacco characterized; and they were found to serve both repellent (nicotine) and attractant (benzylacetone) functions (Kessler and Baldwin 2007). In short, the detailed composition of EFN plays a crucial role in its ecological functions, yet researchers have so far virtually ignored whole compound classes, as they did also for floral nectar.

3.2.2 EFN as an Induced Mechanism

Already being hypothesized in the 1980s (Koptur 1989; Stephenson 1982), it was discovered a few years ago that EFN secretion is inducible. The Euphorbiaceae *Macaranga tanarius* responds to leaf damage with dramatically increased rates of

EFN secretion. This increases the numbers of ants showing up on the plant, which in turn reduces herbivore pressure. Experiments based on both the exogenous application of JA and the application of an inhibitor of endogenous JA synthesis demonstrated that the transient increase in endogenous JA that can be observed after damaging the plants is both required and sufficient to increase EFN secretion. EFN thus appears basically regulated via the same signalling pathway that controls HI-VOCs and many direct defences to herbivores: the octadecanoid signalling cascade (Heil et al. 2001). In consecutive studies, increases of EFN secretion upon herbivory and/or mechanical damage were demonstrated for cotton and castor (Wäckers et al. 2001), the Bignoniaceae *Catalpa bignonioides* (Ness 2003), and several species of the family Fabaceae, including species from both the Faboideae (*Phaseolus lunatus*; Heil 2004b) and the Mimosoideae (Heil et al. 2004c). Studies on cotton revealed that EFN is also involved in aboveground–belowground signalling, as herbivore damage to roots induces EFN secretion on leaves (Wäckers and Bezemer 2003).

All studies cited so far found increases of EFN secretion by already existing nectaries, though this can be secondarily changed to a constitutive secretion when required by the plant's specific life history (Heil et al. 2004c). Another species of the Fabaceae, *Vicia faba*, even increased the number of nectaries present on the plant in response to mechanically removing a part of the overall leaf surface (Mondor and Addicott 2003), a phenomenon caused by the production of more nectary-bearing stipules on the growing shoot (Mondor et al. 2006).

3.2.3 EFN in Plant–Plant Communication

Only in 2006, two groups reported independently that HI-VOCs induce EFN secretion by Lima bean, *Phaseolus lunatus*. Choh et al. (2006) used spider-mite-infested plants in laboratory studies and found that EFN secretion increased in uninfected beans kept in the same aerial space as the infested plants. In a field study at the plant's original growing site in Mexico, Kost and Heil (2006) demonstrated that undamaged plants respond to volatiles released by herbivore-damaged tendrils and to an artificial volatile blend mimicking what is released from a herbivore-damaged plant. Experiments using single volatile compounds demonstrated that the green-leaf volatile (Z)-3-hexenyl acetate alone elicits a significant increase in EFN secretion (Kost and Heil 2006).

Follow-up studies then confirmed that two further important phenomena discussed for HI-VOCs also apply to EFN: Lima bean plants that had been exposed to HI-VOCs responded with augmented EFN secretion when being mechanically damaged, i.e., EFN is sensitive to priming (Heil and Kost 2006). Most recently, we demonstrated that exposition to VOCs released by beetle-damaged Lima bean tendrils increased defence and growth of plants growing in their natural habitat, and we found that HI-VOCs released by damaged Lima bean leaves induce EFN secretion not only by neighbouring plants, but also by neighbouring leaves of the same individual plant (Heil and Silva Bueno 2007). This confirmed the observations on sagebrush (Karban et al. 2006) and underlines the putatively general role of HI-VOCs as a within-plant signal 'worn on the outside'.

3.3 Food Bodies

Food bodies (FBs) are cellular structures containing mainly carbohydrates, proteins and lipids. Charles Darwin (1877) apparently was the first to use the term 'food bodies', which he applied to the small cellular structures produced at the leaflet tips of Central American *Acacia* shrubs and on hairy pads located at the leaf bases of *Cecropia peltata*. These plants, like several other obligate myrmecophytes, produce FBs to nourish their obligate ant inhabitants (Janzen 1966, 1974), while many more plant species use FBs to attract ants facultatively from the vicinity (O'Dowd 1982). In both cases, FBs serve as food for carnivores and thus as a means of indirect plant defence.

FBs appear chemically adapted to their functions as ant food (Fischer et al. 2002; Heil et al. 1998, 2004a; Rickson 1971) and in plant defence (Folgarait and Davidson 1994, 1995), and FBs of several specialized plant species are induced by the presence of ants (Heil et al. 1997; Risch and Rickson 1981) or mere mechanical removal (Folgarait et al. 1994). Since ants are highly mobile and show adaptive patterns in their defensive efforts (Agrawal and Dubin-Thaler 1999; Agrawal and Rutter 1998; Heil et al. 2004b), FBs can be produced distantly to those organs that require the most intensive defence (Heil et al. 1997) and thereby circumvent physiological restrictions on the defence of young, growing plant organs (Herms and Mattson 1992).

Unfortunately, these promising beginnings did not elicit many follow-up studies on the chemical ecology, the evolution, or let alone the genetics of FB production. In recent years, little research has been conducted into FBs (Webber et al. 2007).

3.4 Domatia

Any resource that can be used by predatory arthropods increases their frequency on a particular plant and thus can serve as a means of indirect defence. While EFN and FBs nourish ants, domatia are hollow structures that can be used as nesting space. Ant domatia are apparently restricted to tropical plants. Yet, here they are described by an ever-increasing number of studies (Heil and McKey 2003; Huxley and Cutler 1991). Nesting space appears a generally crucial resource in tropical ecosystems and indeed can limit colony size (Fonseca 1993). Hence, the mere provision of hollow twigs serving as artificial domatia increased the abundance and diversity of ants on coffee plants (Philpott and Foster 2005).

The majority of domatia are caulinary, i.e., hollow stems and shoots (Heil and McKey 2003, and see references therein). However, ant domatia may be also localized in hollow thorns (Janzen 1966), in leaf pouches (Alvarez et al. 2001; Bizerril and Vieira 2002; Edwards et al. 2006), in leaf petioles (Clarke and Kitching 1995; Risch et al. 1977) and even on fruits (Kato et al. 2004). Domatia may serve not only ants but also smaller predators such as mites and bugs. These domatia are generally localized on leaves (O'Dowd and Pemberton 1994; O'Dowd and Willson 1991). In contrast to ant domatia, leaf domatia are known from tropical and temperate regions (Romero and Benson 2005; Walter 1996).

The evidence for the protective effect of ants housed in domatia is overwhelming and not repeated here (Beattie 1985; Bronstein 1998; Davidson and McKey 1993; Heil and McKey 2003; Huxley and Cutler 1991). Evidence for the defensive effects of other domatia-inhabiting carnivores is less abundant, yet still convincing. Removal of leaf domatia reduced the abundance of predatory mites on *Viburnum tinus* (Grostal and Odowd 1994), and experimental addition of leaf domatia to cotton plants significantly increased numbers of predatory thrips and bugs and enhanced plant performance in terms of fruits, the weight of the lint and the number of seeds produced (Agrawal and Karban 1997; Agrawal et al. 2000). Defensive effects of mites housed in leaf domatia have also been demonstrated in the natural ecosystem (Romero and Benson 2004).

Domatia appear to be simple morphological traits but in fact bear features that hint at an intensive coevolution among plant and inhabitants. The majority of cauli-nary or stipular ant domatia are closed, and ants must actively open them to gain access; yet, this access can be facilitated to specialist ants by pre-formed openings, called prostomata (Brouat et al. 2001; Federle et al. 2001; Yu and Davidson 1997). Ant plants may shed domatia in order to prevent defending ants from destroying flowers and fruits (Izzo and Vasconcelos 2002), or to sanction non-defending ants (Edwards et al. 2006). On the other hand, ants may induce the swelling required for domatia formation (Blüthgen and Wesenberg 2001). In spite of the obvious importance of ants as biocontrol agents and the importance of nesting space in this function (Philpott and Foster 2005), even less scientific effort has been spent on domatia than on FBs. Many more studies are required to understand their development and evolution, their relevance in natural ecosystems, and particularly their putative function in sustainable, low-cost biocontrol strategies (see Section 4).

3.5 Endophytic Fungi

Endophytic fungi are widespread and particularly common in grasses and woody host plants (Clay 1990). Microorganisms deprive their hosts of valuable nutrients and thus are usually regarded as parasites. In addition, infection in general activates a plant's systemic resistance to pathogens and thereby reduces its resistance to herbivores (Heil and Bostock 2002). However, fungi are potent producers of many toxic compounds such as alkaloids and thus have the potential to defend their host indirectly against herbivores, thereby turning the interaction from a parasitic into a mutualistic one. Indeed, grasses infected by fungi are often more resistant to vertebrate and insect herbivores, and sometimes also to pathogenic microorganisms. Endophytic fungi can even produce VOCs that have defensive effects against pathogenic fungi and bacteria (Strobel 2006). Similarly, endophytic bacteria are just being discovered and can have a wide range of effects on plant health, growth, nutrient assimilation and defence (Rosenblueth and Martínez-Romero 2006). Several excellent reviews are available (Clay 1990; Faeth 2002; Faeth and Bultman 2002) and this topic is thus not treated in more detail in the present chapter.

Anyway, much more knowledge will be required before endophyte-mediated defences are understood to a satisfying degree. Endophyte-infection of grasses can strongly alter food-web dynamics (Omacini et al. 2001), and the interactions among the different trophic levels are highly complex. To give one example, aphids feeding on grasses that are infected with endophytic fungi grew more slowly and had a reduced fecundity, most probably due to alkaloids produced by the fungus (Bultman et al. 2006). Such effects can be constant over several generations (Meister et al. 2006). However, the detailed outcomes of such interactions depend on the genotypes of plant, fungus and aphid (Clement et al. 2005) and on soil nutrient availability (Lehtonen et al. 2005). Adding even more complexity to the system, fungus-derived alkaloids are obviously transported through the food chain, since ladybird beetles predating on aphids living on endophyte-infected grasses also showed reduced growth and reproduction rates (Sassi et al. 2006). Similarly, caterpillars feeding on endophyte-infected grasses were more resistant to infection by an entomopathogenic nematode (Clay 1990; Faeth 2002; Faeth and Bultman 2002). It appears currently impossible to predict how such cascading effects influence the populations of predators and aphids and – finally – the plants.

4 Evolutionary Origins

The high diversity of HI-VOCs can be put down to duplication and following divergent evolution of genes that previously served other metabolic processes (Pichersky et al. 2006a). How (and why) did plants, however, start to produce odours at all, and how did it happen that nectar was first produced? Was the first nectar floral or extrafloral? The following section is intended to present some – necessarily highly speculative – ideas.

4.1 Evolutionary Origins of HI-VOCs

Plant volatiles are involved in indirect defence but also function in direct protection against herbivores and microorganisms and in protection from abiotic stress, as for instance isoprene protects the photosystem from thermal damage and oxidative stress (Loreto and Velikova 2001). HI-VOCs serve further internal roles as volatile plant hormones (see Sections 3.1.3, 3.2.3). Yet, since the majority of plant volatiles still lack a clearly defined ecological or physiological function, it was suggested that they are by-products of other metabolic pathways, whose loss simply cannot be avoided by plants, which at best 'make use of the unavoidable' (Peñuelas and Llusiá 2004). However, ecological functions are described for an ever increasing number of compounds, and a lack of *known* function should not be wrongly taken for a lack of *existing* function (Pichersky et al. 2006b).

The majority of HI-VOCs are also components of flower odours (Dudareva et al. 2006; Pichersky and Gershenzon 2002; Pichersky et al. 2006a; see Table 1 for examples), an observation that had been made already in the earliest studies (Dicke et al. 1990). JA not only regulates direct and indirect plant responses to herbivores but also serves central roles in flower development and fruit ripening (Peña-Cortés et al. 2005). Though also flowers can induce volatile release when damaged (Bichão et al. 2005), identical volatile compounds can be released from damaged leaves and undamaged flowers of the same species (Azuma et al. 1997). Unfortunately, few studies nail the attractive effect of flower scents down to a pollinator-attracting effect of isolated compounds (Pichersky and Gershenzon 2002), and detailed stereochemical information is often lacking in behavioural studies. The identity of an HI-VOC with a component of a flower odour therefore must be tested critically. Yet, several enzymes involved in the synthesis in volatiles have been discovered serving both as floral scents and in indirect defence (Pichersky and Gershenzon 2002).

Cheekily speaking, one could conclude that plants use the hormones that are involved in the regulation of flowering to make leaves more 'flower-ish' and thereby attract pollinating hymenopterans to leaves, where they behave as plant defenders. The wide overlap in chemical composition and regulation phenomena makes it likely that floral volatiles and HI-VOCs have a common evolutionary origin, but can we decide whether the history suggested by this statement is correct (i.e., flower odours evolved first and their function was then broadened to indirect defence)? Flower odours are generally exclusive to animal-pollinated flowering plants, i.e., to angiosperms (Knudsen et al. 1993, 2006), though scattered evidence exists on the release of odours from insect pollinated cycads (Pellmyr et al. 1991; Thien et al. 2000) and from flowers of some conifers (Borg-Karlson et al. 1985). Concordantly, insect pollination per se is mostly restricted to angiosperms, though some extinct pteridosperms and cycads were apparently insect-pollinated as well (Labandeira 1997). The herbivore-induced or JA-mediated release of VOCs, in contrast, is described for angiosperms and also for gymnosperms, such as conifers (Miller et al. 2005; Mumm and Hilker 2006) and Ginkgo (Boland et al. 1995; Van den Boom et al. 2004), and even for the fern, *Dryopteris filix-mas* (Boland et al. 1995).

A phylogenetic reconstruction based on parsimony principles thus concludes that flower odours (appearing as autapomorphy of the monophyletic Angiospermae, or more conservatively as autapomorphy of seed plants) are the derived trait in relation to the plesiomorphic HI-VOCs. Following Pichersky et al. (2006a) in that 'convergent evolution is often responsible for the ability of distally related species to synthesize the same volatile' weakens any phylogenetic reconstruction, and floral scents in general turned out to be evolutionarily too labile to be useful for phylogenetic reconstructions (Knudsen et al. 2006) when used on a single-compound level. I do also not see any chance that reliable fossil records of indirect defence by volatiles can ever be obtained (but see Section 4.2).

However, the attack of plants by herbivores dates back to the Paleozoic, with supporting gut contents of insects being preserved, e.g., from the Carboniferous

(Labandeira 1997) – the need of a functioning protection from herbivores thus clearly preceded the need to adapt to pollinators. This is in line with the fact that the common HI-VOC, (*E,E*)-α-farnesene, is released constitutively from Ginkgo leaves (Van den Boom et al. 2004), and that volatiles released from male cones of insect-pollinated cycads are potent herbivore deterrents (Pellmyr et al. 1991), as are most compounds released from flowers of wind-pollinated conifers (Borg-Karlson et al. 1985).

In summary, plant volatiles serve multiple functions as flower odours, fruit scents and as both direct and indirect plant defences against pathogens and herbivores (Dudareva et al. 2006; Pichersky 2004; Pichersky et al. 2006a). HI-VOcs are taxonomically more broadly distributed than floral scents, and the odours released from non-angiosperm flowers have chemical properties that support their defensive function. All these observations are most parsimoniously explained by an original function of plant volatiles in plant-internal processes or as metabolic by-products (Peñuelas and Llusiá 2004) and by their role in defence preceding their much better known role in pollination.

4.2 Evolutionary Origins of EFN

Being secreted in intact flowers and on damaged leaves, nectar bears the same parallels that have just been pointed out for VOCs. Interestingly, JA is involved in the induction of EFN secretion (Heil 2001, 2004b, c) and there are now hints that floral nectar secretion is also regulated by this hormone (Venkatesan, unpublished data). Nectar in the context of pollination is generally assumed to have appeared in the Jurassic, as inferred from the evolution of specialized nectar-sucking mouthparts in Diptera (Labandeira 1997; Ren 1998). The first reported cases of floral nectaries apparently date back to the Late Cretaceous (Schönenberger et al. 2001), while the earliest evidence for extrafloral nectaries comes from a 35-million-year-old *Poplar* species (Bronstein et al. 2006). The oldest known floral nectaries thus precede the oldest extrafloral nectaries. However, the fossil record is highly scattered, extrafloral nectaries may have been overseen due to the low awareness of botanists of the importance of these structures, and nectar-sucking mouthparts of insects can be interpreted as insect adaptations to extrafloral or floral nectar. Indeed, Labandeira (1997) mentions dipteran mouthparts suitable to feed on *extrafloral* nectar from the Late Triassic or Early Jurassic. A phylogenetic reconstruction based on the distribution of EFN in extant plants thus is likely to give the better information. As HI-VOCs, EFN is known from a taxonomically wide range of plant species including not only angiosperms but also non-flowering plants such as ferns (Rashbrook et al. 1992; Rumpf et al. 1994; Tempel 1983). Again the most parsimonious explanation is that nectar was first secreted on vegetative plant parts and that its role in pollination represents the derived state.

5 Potential Application in Agronomy

5.1 Tritrophic Interactions as Means of Sustainable Crop Protection

Indirect defences function in nature, are apparently cheap to induce, and are based on merely biological interactions, thereby being particularly environmentally friendly. An application in crop protection thus is an obvious and explicitly expressed goal of research in this direction (Degenhardt et al. 2003; Shiojiri et al. 2006; Turlings and Ton 2006; Walling 2001). The majority of plants for which a volatile-mediated role of the plant in tritrophic interactions has been demonstrated are crops (Dicke 1999). Putative defenders were attracted to artificially released HI-VOCs in a hop yard, and plots in grape vineyards and hop yards baited with methyl salicylate contained a nearly four times larger population of predators, while populations of spider mites were reduced (James 2003; James and Price 2004). Even planting an odorous grass into maize fields reduced herbivore damage. This grass, *Melinitis minutiflora*, constitutively emits a compound that is typically released by maize in response to caterpillar damage, and that attracted parasitoids (Khan et al. 1997). Predators were attracted to banana pseudostems damaged by their prey herbivores, the banana weevil, under field conditions (Tinzaara et al. 2005).

Breeding has decreased several but not all crops' indirect defences, as commercial cultivars of cotton released seven times less HI-VOCs than a naturalized line (Loughrin et al. 1995). Cultivated Lima bean released less HI-VOCs and less EFN than the wild form (Mendoza García and Heil, unpublished data), while Benrey et al. (1998) reported volatiles not to be reduced in certain *Brassica* and *Phaseolus* cultivars. For cultivated maize the overall release of volatiles from aboveground parts was quantitatively similar to its wild ancestor (Gouinguené et al. 2001). However, North American maize lines do not release (*E*)-β-caryophyllene from their roots in response to beetle damage and are not attractive to an entomopathogenic nematode that serves as a promising crop protection agent for caryophyllene-releasing European lines (Rasmann et al. 2005). Though changing the volatile bouquet of crops might severely interact with its functions in taste and perceived food quality (Goff and Klee 2006), efforts are being made to 'give plants back what they have lost during cultivation', and first plants now have been genetically engineered to alter their volatile release (Kappers et al. 2005; Schnee et al. 2006; Shiojiri et al. 2006).

Interestingly, these seemingly innovative developments have in fact been preceded by a century-long history of using ants as biocontrol agents. In China, artificial domatia have been used for centuries in *Citrus* plantations, and ant nests are actively brought to cacao and other plantations in different tropical countries (examples from Philpott and Foster 2005). Leaf domatia are a characteristic of some important crop plants and can exhibit relevant effects on the production of fruits and lint by cotton plants (Agrawal et al. 2000). Defending ants are also successfully attracted to extrafloral nectaries of cultivated cashew (Rickson and Rickson 1998).

5.2 Problems and Putative Solutions

Unfortunately, no study has so far clearly demonstrated that attracting carnivores by HI-VOCs indeed can work in agronomy. Net effects on crop plants were not investigated by James (2003), nor by Rasmann et al. (2005), and the reduced herbivory in the intercropping experiment of Khan et al. (1997) could also have been caused by a repellent effect on the most important maize pest rather than an attraction of its parasitoids. The presumed positive effect of VOCs-mediated tritropic interactions on crop productivity therefore remains to be proven. How can this be explained in the presence of literally thousands of studies on HI-VOCs? In fact, several problems are connected to indirect defences which need intensive further investigation before tritrophic interactions can be used as tool in crop protection.

1. *Ecological costs*. As mentioned above (Sections 2.2, 3.1.2), HI-VOCs may attract certain herbivores. Many specialist herbivores use secondary compounds serving as defence against generalists to identify their host plant and thus turn the plant's weapon against it. Since VOC-releasing plants stand out from the background vegetation and thus are easier to localize (Dicke 1999), the constitutive release of HI-VOCs will have highly detrimental consequences in crop cultivation when the selected compounds are eventually used for host localization by specialist herbivores.
2. *Mutualist availability*. Another putative restriction is the shortage of suitable carnivores in modern agriculture: Though specifically managed field margins might provide important ecological services such as, e.g., food and refuge sites for carnivores (see Olson and Wäckers 2007, and references therein), monocultures usually lack the diverse fauna required for functioning tritrophic interactions. It is perhaps no coincidence that studies pointing to successful usage of parasitoids or ants (Khan et al. 1997; Philpott and Foster 2005; Rickson and Rickson 1998; Tinzaara et al. 2005) have been conducted in tropical, third-world agricultural systems.
3. *Signal reliability*. An evolutionarily highly important feature of HI-VOCs has already been discussed in the 'costs' section: signal reliability (Section 2.2.1). If parasitoids or other carnivores are attracted to a healthy plant that is genetically engineered to constitutively release HI-VOCs, searching behaviour will not be successful, and starving parasitoids will learn or evolve not to trust in these signals (Turlings and Ton 2006). This problem is less severe when using EFN or domatia, which represent a resource on their own, and it might again be no coincidence that traditional, experience-based crop protection techniques that successfully *do* make use of tritrophic interactions are entirely based on ants rather than VOC-attracted parasitoids.
4. *Food quality*. Crop plants have reduced direct and indirect defences, and these have not been lost just accidentally. Resistance traits cause allocation costs, and their expression by crop plants reduces growth rates and, finally, yield (Heil and Baldwin 2002). Probably even more importantly, many direct defensive compounds and volatiles are pharmaceutically active, and they cause a fruit's or

plant's characteristic aroma (Goff and Klee 2006; Pichersky 2004). Bringing our crop plants entirely back to the defensive level of their wild ancestors will bring them back also to the wild plants' growth rates, tastes and general suitability in human nutrition – probably not really a desirable target.

6 Conclusions and Outlook

Promising attempts to make use of tritrophic interactions need a more complete understanding than is currently available of the abiotic factors affecting VOCs and EFN secretion and the ecological backgrounds and restrictions of these interactions. If we are going to genetically tamper with volatiles, we at least should know their effects on the crop plant itself, its neighbours and the animal and plant species in the environment. In general, induced HI-VOCs release will minimize putative ecological and health-related risks as compared with a constitutive release (Turlings and Ton 2006). Then, beyond doubt, natural parasitoid populations will by far be too small to have effects of the desired magnitude in modern monocultures, and the active release of suitable parasitoids thus should accompany the induction of volatiles, a measure that also would minimize the risk that natural populations rapidly lose their affinity to HI-VOCs. Using HI-VOCs with direct repellent effects on herbivores will make the interaction less dependent on carnivores and hence more stable.

Research on HI-VOCs is still in its infancy (Dudareva et al. 2006; Pichersky et al. 2006a, b), but this statement is even more true for EFN and the other types of indirect defences discussed in this chapter. EFN has been described for many more plant species than HI-VOCs though it receives much less scientific attention. In contrast, HI-VOCs are genetically and physiologically much better investigated yet are described only for a short list of species (see Section 3.1.1). This imbalance is particularly regrettable since both types of indirect defences do interact with each other (see Section 3.2.3) and are likely to co-occur on many more plant species than Lima bean (Arimura et al. 2005).

In fact, there are hints on functioning crop protection by ants and other predators that are provided with 'honest' resources, though much less scientific effort has been invested in ant–plant interactions than in volatile-mediated defence. Populations of ants and predatory mites might also be too low in large annual monocultures, but they can serve effectively as a crop protection agent in shrubs and trees. To my knowledge, no study has ever tried to investigate putative crop protection by EFN, though this trait is known from many crops such as cotton, beans, pea, cashew, cherry, apple, etc. Future research should move beyond the focus on HI-VOCs and study all traits that serve in indirect defence under ecologically and agronomically realistic conditions in order to understand the interactions among these different types of defence and their manifold interactions with plant metabolism and the environment.

Acknowledgements I thank W. Beyschlag for the kind offer to contribute to this book, and M. Dicke, W. Boland and M. d'Alessandro for help with literature searches and many fruitful discussions. Financial support by the Deutsche Forschungsgemeinschaft (He 3169/4-2) is gratefully acknowledged.

References

Agrawal AA (2000a) Benefits and costs of induced plant defense for *Lepidium virginicum* (Brassicaceae). Ecology 81:1804–1813

Agrawal AA (2000b) Overcompensation of plants in response to herbivory and the by-product benefits of mutualism. Trends Plant Sci 5:309–313

Agrawal AA, Dubin-Thaler BJ (1999) Induced responses to herbivory in the Neotropical ant–plant association between *Azteca* ants and *Cecropia* trees: response of ants to potential inducing cues. Behav Ecol Sociobiol 45:47–54

Agrawal AA, Karban R (1997) Domatia mediate plant-arthropod mutualism. Nature 387:562–563

Agrawal AA, Rutter MT (1998) Dynamic anti-herbivore defense in ant-plants: the role of induced responses. Oikos 83:227–236

Agrawal AA, Strauss SY, Stout MJ (1999) Costs of induced responses and tolerance to herbivory in male and female fitness components of wild radish. Evolution 53:1093–1104

Agrawal AA, Karban R, Colfer RG (2000) How leaf domatia and induced plant resistance affect herbivores, natural enemies and plant performance. Oikos 89:70–80

Alborn T, Turlings TCJ, Jones TH, Stenhagen G, Loughrin JH, Tumlinson JH (1997) An elicitor of plant volatiles from beet armyworm oral secretion. Science 276:945–949

Altenburger R, Matile P (1990) Further observations on rhythmic emission of fragrance in flowers. Planta 180:194–197

Alvarez G, Armbrecht I, Jiménez E, Armbrecht H, Ulloa-Chacón P (2001) Ant-plant association in two *Tococa* species from a primary rain forest of Colombian *Choco* (Hymenoptera: Formicidae). Sociobiology 38:585–602

Andersson S (2003) Antennal responses to floral scents in the butterflies *Inachis io*, *Aglais urticae* (Nymphalidae), and *Gonepteryx rhamni* (Pieridae). Chemoecology 13:13–20

Andersson S, Nilsson LA, Groth I, Bergstrom G (2002) Floral scents in butterfly-pollinated plants: possible convergence in chemical composition. Bot J Linn Soc 140:129–153

Arimura G-i, Ozawa R, Shimoda T, Nishioka T, Boland W, Takabayashi J (2000) Herbivory-induced volatiles elicit defence genes in lima bean leaves. Nature 406:512–515

Arimura G, Kost C, Boland W (2005) Herbivore-induced, indirect plant defences. Biochim Biophys Acta 1734:91–111

Azuma H, Thien LB, Toyota M, Asakawa Y, Kawano S (1997) Distribution and differential expression of (E)-4,8-dimethyl-1,3,7-nonatriene in leaf and floral volatiles of *Magnolia* and *Liriodendron taxa*. J Chem Ecol 23:2467–2478

Baker HG, Opler PA, Baker I (1978) A comparison of the amino acid complements of floral and extrafloral nectars. Bot Gaz 139:322–332

Baldwin IT (1998) Jasmonate-induced responses are costly but benefit plants under attack in native populations. Proc Natl Acad Sci USA 95:8113–8118

Baldwin IT, Schultz JC (1983) Rapid changes in tree leaf chemistry induced by damage: evidence for communication between plants. Science 221:277–279

Baldwin IT, Kessler A, Halitschke R (2002) Volatile signaling in plant-plant-herbivore interactions: what is real? Curr Opin Plant Biol 5:351–354

Baldwin IT, Halitschke R, Paschold A, Dahl CC von, Preston CA (2006) Volatile signaling in plant-plant interactions: "talking trees" in the genomics era. Science 311:812–815

Bate NJ, Rothstein SJ (1998) C_6-volatiles derived from the lipoxygenase pathway induce a subset of defense-related genes. Plant J 16:561–569

Beattie AJ (1985) The evolutionary ecology of ant-plant mutualisms. Cambridge University Press, Cambridge
Benrey B, Callejas A, Rios L, Oyama K, Denno RF (1998) The effects of domestication of *Brassica* and *Phaseolus* on the interaction between phytophagous insects and parasitoids. Biol Contr 11:130–140
Bentley BL (1977) Extrafloral nectaries and protection by pugnacious bodyguards. Annu Rev Ecol Syst 8:407–427
Bezemer TM, Dam NM van (2005) Linking aboveground and belowground interactions via induced plant defenses. Trends Ecol Evol 20:617–624
Bichão H, Borg-Karlson AK, Araújo J, Mustaparta H (2005) Five types of olfactory receptor neurons in the strawberry blossom weevil *Anthonomus rubi*: selective responses to inducible host-plant volatiles. Chem Senses 30:153–170
Birkett MA, Campbell CAM, Chamberlain K, Guerrieri E, Hick AJ, Martin JL, Matthes M, Napier JA, Pettersson J, Pickett JA, Poppy GM, Pow EM, Pye BJ, Smart LE, Wadhams GH, Wadhams LJ, Woodcock CM (2000) New roles for cis-jasmone as an insect semiochemical and in plant defense. Proc Natl Acad Sci USA 97:9329–9334
Bizerril MXA, Vieira EM (2002) *Azteca* ants as antiherbivore agents of *Tococa formicaria* (Melastomataceae) in Brazilian Cerrado. Stud Neotrop Fauna Environ 37:145–149
Blechert S, Brodschelm W, Hölder S, Kammerer L, Kutchan TM, Müller MJ, Xia ZQ, Zenk MH (1995) The octadecanoid pathway: signal molecules for the regulation of secondary pathways. Proc Natl Acad Sci USA 92:4099–4105
Blüthgen N, Fiedler K (2004) Preferences for sugars and amino acids and their conditionality in a diverse nectar-feeding ant community. J Anim Ecol 73:155–166
Blüthgen N, Wesenberg J (2001) Ants induce domatia in a rain forest tree (*Vochysia vismiaefolia*). Biotropica 33:637–642
Blüthgen N, Verhaagh M, Goitía W, Jaffé K, Morawetz W, Barthlott W (2000) How plants shape the ant community in the Amazonian rainforest canopy: the key role of extrafloral nectaries and homopteran honeydew. Oecologia 125:229–240
Boland W, Hopke J, Donath J, Nüske J, Bublitz F (1995) Jasmonic acid and coronatin induce odor production in plants. Angew Chem Int Ed Engl 34:1600–1602
Bolter CJ, Dicke M, vanLoon JJA, Visser JH, Posthumus MA (1997) Attraction of Colorado potato beetle to herbivore-damaged plants during herbivory and after its termination. J Chem Ecol 23:1003–1023
Borg-Karlson A-K, Eidmann HH, Lindstrom M, Norin T, Wiersma N (1985) Odoriferous compounds from the flowers of the conifers *Picea abies*, *Pinus sylvestris* and *Larix sibirica*. Phytochemistry 24:455
Bronstein JL (1994) Our current understanding of mutualism. Quart Rev Biol 69:31–51
Bronstein JL (1998) The contribution of ant-plant protection studies to our understanding of mutualism. Biotropica 30:150–161
Bronstein JL, Alarcon R, Geber M (2006) The evolution of plant-insect mutualisms. New Phytol 172:412–428
Brouat C, Garcia N, Andary C, McKey D (2001) Plant lock and ant key: pairwise coevolution of an exclusion filter in an ant–plant mutualism. Proc R Soc Lond B Biol Sci 268:2131–2141
Bruin J, Dicke M, Sabelis MW (1992) Plants are better protected against spider-mites after exposure to volatiles from infested conspecifics. Experientia 48:525–529
Bugg RL, Ellis RT, Carlson RW (1989) Ichneumonidae (Hymenoptera) using extrafloral nectar of faba bean (*Vicia faba* L., Fabaceae) in Massachusetts. Biol Agric Hortic 6:107–114
Bultman TL, Pulas C, Grant L, Bell G, Sullivan TJ (2006) Effects of fungal endophyte isolate on performance and preference of bird cherry–oat aphid. Environ Entomol 35:1690–1695
Bungert M, Thiel R, Goedings P, Becker H (2002) (E,E)-α-Farnesene the main substance of the volatiles of the flowers from European mistletoe (*Viscum album* L.). Z Naturforsch Teil C 57:205–207
Carroll MJ, Schmelz EA, Meagher RL, Teal PEA (2006) Attraction of *Spodoptera frugiperda* larvae to volatiles from herbivore-damaged maize seedlings. J Chem Ecol 32:1911–1924

Carter CJ, Thornburg RW (2004) Tobacco nectarin V is a flavin-containing berberine bridge enzyme-like protein with glucose oxidase activity. Plant Physiol 134:460–469

Choh Y, Kugimiya S, Takabayashi J (2006) Induced production of extrafloral nectar in intact lima bean plants in response to volatiles from spider mite-infested conspecific plants as a possible indirect defense against spider mites. Oecologia 147:455–460

Cipollini D (2002) Does competition magnify the fitness costs of induced resistance in *Arabidopsis thaliana*? A manipulative approach. Oecologia 131:514–520

Cipollini D (2007) Consequences of the overproduction of methyl jasmonate on seed production, tolerance to defoliation and competitive effect and response of *Arabidopsis thaliana*. New Phytol 173:146–153

Cipollini D, Purrington CB, Bergelson J (2003) Costs of induced responses in plants. Basic Appl Ecol 4:79–89

Clarke CM, Kitching RL (1995) Swimming antd and pitcher plants: a unique ant–plant interaction from Borneo. J Trop Ecol 11:589–602

Clay K (1990) Fungal endophytes of grasses. Annu Rev Ecol Syst 21:275–297

Clement SL, Elberson LR, Bosque-Perez NA, Schotzko DJ (2005) Detrimental and neutral effects of wild barley – *Neotyphodium* fungal endophyte associations on insect survival. Entomol Exp Appl 114:119–125

Colazza S, Fucarino A, Peri E, Salerno G, Conti E, Bin F (2004a) Insect oviposition induces volatile emission in herbaceous plants that attracts egg parasitoids. J Exp Biol 207:47–53

Colazza S, McElfresh JS, Millar JG (2004b) Identification of volatile synomones, induced by *Nezara viridula* feeding and oviposition on bean spp, that attract the egg parasitoid *Trissolcus basalis*. J Chem Ecol 30:945–964

Conrath U, Beckers GJM, Flors V, Garcia-Agustin P, Jakab G, Mauch F, Newman MA, Pieterse CMJ, Poinssot B, Pozo MJ, Pugin A, Schaffrath U, Ton J, Wendehenne D, Zimmerli L, Mauch-Mani B (2006) Priming: getting ready for battle. Mol Plant Microbe Interact 19:1062–1071

Creelman RA, Mullet JE (1997a) Biosynthesis and action of jasmonates in plants. Annu Rev Plant Physiol Plant Mol Biol 48:355–381

Creelman RA, Mullet JE (1997b) Oligosaccharins, brassinolides, and jasmonates: Nontraditional regulators of plant growth, development, and gene expression. Plant Cell 9:1211–1223

Cuautle M, Rico-Gray V (2003) The effect of wasps and ants on the reproductive success of the extrafloral nectaried plant *Turnera ulmifolia* (Turneraceae). Funct Ecol 17:417–423

Custódio L, Serra H, Nogueira JMF, Goncalves S, Romano A (2006) Analysis of the volatiles emitted by whole flowers and isolated flower organs of the carob tree using HS-SPME-GC/MS. J Chem Ecol 32:929–942

Darwin C (1877) On the glandular bodies on *Acacia sphaerocephala* and *Cecropia peltata* serving as food for ants. J Linn Soc Lond Bot 15:398–409

Davidson DW, McKey D (1993) The evolutionary ecology of symbiotic ant–plant relationships. J Hym Res 2:13–83

De Boer JG, Posthumus MA, Dicke M (2004) Identification of volatiles that are used in discrimination between plants infested with prey or nonprey herbivores by a predatory mite. J Chem Ecol 30:2215–2230

De Boer JG, Snoeren TAL, Dicke M (2005) Predatory mites learn to discriminate between plant volatiles induced by prey and nonprey herbivores. Anim Behav 69:869–879

De la Fuente MAS, Marquis RJ (1999) The role of ant-tended extrafloral nectaries in the protection and benefit of a Neotropical rainforest tree. Oecologia 118:192–202

De Moraes CM, Lewis WJ, Paré PW, Alborn HT, Tumlinson JH (1998) Herbivore-infested plants selectively attract parasitoids. Nature 393:570–573

De Moraes CM, Mescher MC, Tumlinson JH (2001) Caterpillar-induced nocturnal plant volatiles repel conspecific females. Nature 410:577–580

Degenhardt J, Gershenzon J, Baldwin IT, Kessler A (2003) Attracting friends to feast on foes: engineering terpene emission to make crop plants more attractive to herbivore enemies. Curr Opin Biotechnol 14:169–176

delClaro K, Berto V, Reu W (1996) Effect of herbivore deterrence by ants on the fruit set of an extrafloral nectary plant, *Qualea multiflora* (Vochysiaceae). J Trop Ecol 12:887–892
Dicke M (1986) Volatile spider-mite pheromone and host-plant kairomone, involved in spaced-out gregariousness in the spider mite *Tetranychus urticae*. Physiol Entomol 11:251–262
Dicke M (1994) Local and systemic production of volatile herbivore-induced terpenoids: their role in plant–carnivore mutualism. J Plant Physiol 143:465–472
Dicke M (1999) Evolution of induced indirect defense of plants. In: Tollrian R, Harvell CD (eds) The ecology and evolution of inducible defenses. Princeton University Press, Princeton, pp 62–88
Dicke M, Bruin J (2001) Chemical information transfer between plants: back to the future. Biochem Syst Ecol 29:981–994
Dicke M, Dijkman H (1992) Induced defence in detached uninfested plant-leaves – effects on behavior of herbivores and their predators. Oecologia 91:554–560
Dicke M, Sabelis MW (1988) How plants obtain predatory mites as bodyguards. Neth J Zool 38:148–165
Dicke M, Sabelis MW (1989) Does it pay plants to advertize for bodyguards? Towards a cost–benefit analysis of induced synonome production. In: Lambers H, Cambridge ML, Konings H, Pons TL (eds) Variation in growth rate and productivity of higher plants. SPB Academic, The Hague, pp 342–358
Dicke M, Vet LEM (1999) Plant–carnivore interactions: evolutionary and ecological consequences for plant, herbivore and carnivore. In: Drent RH (ed) Herbivores: between plants and predators. Blackwell, Oxford, pp 483–520
Dicke M, Beek TA van, Posthumus MA, Ben Dom N, Bokhoven H van, Groot AE de (1990) Isolation and identification of volatile kairomone that affects *Acarine* predator–prey interactions. Involvement of the host plant int its production. J Chem Ecol 16:381–395
Dicke M, Baarlen P van, Wessels R, Dijkman H (1993) Herbivory induces systemic production of plant volatiles that attract predators of the herbivore: extraction of endogenous elicitor. J Chem Ecol 19:581–599
Dicke M, Boer JG de, Höfte M, Rocha-Granados MC (2003) Mixed blends of herbivore-induced plant volatiles and foraging success of carnivorous arthropods. Oikos 101:38–48
Dietrich R, Ploss K, Heil M (2005) Growth responses and fitness costs after induction of pathogen resistance depend on environmental conditions. Plant Cell Environ 28:211–222
Doares SH, Syrovets T, Weiler EW, Ryan CA (1995) Oligogalacturonides and chitosan activate plant defensive genes through the octadecanoid pathway. Proc Natl Acad Sci USA 92:4095–4098
Dolch R, Tscharntke T (2000) Defoliation of alders (*Alnus glutinosa*) affects herbivory by leaf beetles on undamaged neighbours. Oecologia 125:504–511
Dudareva N, Pichersky E, Gershenzon J (2004) Biochemistry of plant volatiles. Plant Physiol 135:1893–1902
Dudareva N, Negre F, Nagegowda DA, Orlova I (2006) Plant volatiles: Recent advances and future perspectives. Crit Rev Plant Sci 25:417–440
Dugravot S, Thibout E (2006) Consequences for a specialist insect and its parasitoid of the response of *Allium porrum* to conspecific herbivore attack. Physiol Entomol 31:73–79
Edwards DP, Hassall M, Sutherland WJ, Yu DW (2006) Selection for protection in an ant-plant mutualism: host sanctions, host modularity and the principal-agent game. Proc R Soc Lond B Biol Sci 273:595–602
Effmert U, Grosse J, Rose USR, Ehrig F, Kagi R, Piechulla B (2005) Volatile composition, emission pattern, and localization of floral scent emission in *Mirabilis jalapa* (Nyctaginaceae). Am J Bot 92:2–12
Elias TS (1983) Extrafloral nectaries: their structure and distribution. In: Bentley B, Elias TS (eds) The biology of nectaries. Columbia University Press, New York, pp 174–203
Engelberth J, Alborn HT, Schmelz EA, Tumlinson JH (2004) Airborne signals prime plants against insect herbivore attack. Proc Natl Acad Sci USA 101:1781–1785
Faeth SH (2002) Are endophytic fungi defensive plant mutualists? Oikos 98:25–36

Faeth SH, Bultman TL (2002) Endopyhtic fungi and interactions among host plants, herbivores, and natural enemies. In: Tscharntke T, Hawkins BA (eds) Multitrophic level interactions. Cambridge University Press, Cambridge, pp 89–123
Farag MA, Fokar M, Zhang HA, Allen RD, Paré PW (2005) (Z)-3-Hexenol induces defense genes and downstream metabolites in maize. Planta 220:900–909
Farmer EE (2001) Surface-to-air signals. Nature 411:854–856
Farmer EE, Ryan CA (1990) Interplant communication: airborne methyl jasmonate induces synthesis of proteinase inhibitors in plant leaves. Proc Natl Acad Sci USA 87:7713–7716
Farmer EE, Alméras E, Krishnamurthy V (2003) Jasmonates and related oxylipins in plant responses to pathogenesis and herbivory. Curr Opin Plant Biol 6:372–378
Federle W, Fiala B, Zizka G, Maschwitz U (2001) Incident daylight as orientation cue for hole-boring ants: prostomata in *Macaranga* ant-plants. Insect Soc 48:165–177
Fiala B, Linsenmair KE (1995) Distribution and abundance of plants with extrafloral nectaries in the woody flora of a lowland primary forest in Malaysia. Biodiv Conserv 4:165–182
Fischer RC, Richter A, Wanek W, Mayer V (2002) Plants feed ants: food bodies of myrmecophytic *Piper* and their significance for the interaction with *Pheidole bicornis* ants. Oecologia 133:186–192
Folgarait PJ, Davidson DW (1994) Antiherbivore defenses of myrmecophytic *Cecropia* under different light regimes. Oikos 71:305–320
Folgarait PJ, Davidson DW (1995) Myrmecophytic *Cecropia*: antiherbivore defenses under different nutrient treatments. Oecologia 104:189–206
Folgarait PJ, Johnson HL, Davidson DW (1994) Responses of *Cecropia* to experimental removal of Müllerian bodies. Funct Ecol 8:22–28
Fonseca CR (1993) Nesting space limits colony size of the plant-ant *Pseudomyrmex concolor*. Oikos 67:473–482
Foster WA (1995) Mosquito sugar feeding and reproductive energetics. Annu Rev Entomol 40:443–474
Fritzsche Hoballah ME, Turlings TCJ (2001) Experimental evidence that plants under caterpillar attack may benefit from attracting parasitoids. Evol Ecol Res 3:553–565
Fritzsche-Hoballah ME, Tamo C, Turlings TCJ (2002) Differential attractiveness of induced odors emitted by eight maize varieties for the parasitoid *Cotesia marginiventris*: is quality or quantity important? J Chem Ecol 28:951–968
Glazebrook J, Chen WJ, Estes B, Chang HS, Nawrath C, Metraux JP, Zhu T, Katagiri F (2003) Topology of the network integrating salicylate and jasmonate signal transduction derived from global expression phenotyping. Plant J 34:217–228
Goff SA, Klee HJ (2006) Plant volatile compounds: sensory cues for health and nutritional value? Science 311:815–819
Gouinguené SP, Turlings TCJ (2002) The effects of abiotic factors on induced volatile emissions in corn plants. Plant Physiol 129:1296–1307
Gouinguené S, Degen T, Turlings TCJ (2001) Variability in herbivore-induced odour emissions among maize cultivars and their wild ancestors (teosinte). Chemoecology 11:9–16
Gouinguené S, Pickett JA, Wadhams LJ, Birkett MA, Turlings TCJ (2005) Antennal electrophysiological responses of three parasitic wasps to caterpillar-induced volatiles from maize (*Zea mays mays*), cotton (*Gossypium herbaceum*), and cowpea (*Vigna unguiculata*). J Chem Ecol 31:1023–1038
Grostal P, O'Dowd DJ (1994) Plants, mites and mutualism: leaf domatia and the abundance and reproduction of mites on *Viburnum tinus* (Caprifoliaceae). Oecologia 97:308–315
Halitschke R, Keßler A, Kahl J, Lorenz A, Baldwin IT (2000) Ecophysiological comparison of direct and indirect defenses in *Nicotiana attenuata*. Oecologia 124:408–417
Hampel D, Mosandl A, Wust M (2005) Induction of de novo volatile terpene biosynthesis via cytosolic and plastidial pathways by methyl jasmonate in foliage of *Vitis vinifera* L. J Agric Food Chem 53:2652–2657
Heil M (2002) Ecological costs of induced resistance. Curr Opin Plant Biol 5:345–350

Heil M (2004a) Direct defense or ecological costs? Responses of herbivorous beetles to volatiles released by wild Lima bean (*Phaseolus lunatus*). J Chem Ecol 30:1289–1295

Heil M (2004b) Induction of two indirect defences benefits Lima bean (*Phaseolus lunatus*, Fabaceae) in nature. J Ecol 92:527–536

Heil M, Baldwin IT (2002) Fitness costs of induced resistance: emerging experimental support for a slippery concept. Trends Plant Sci 7:61–67

Heil M, Bostock RM (2002) Induced systemic resistance (ISR) in the context of induced plant defences. Ann Bot 89:503–512

Heil M, Kost C (2006) Priming of indirect defences. Ecol Lett 9:813–817

Heil M, McKey D (2003) Protective ant–plant interactions as model systems in ecological and evolutionary research. Annu Rev Ecol Evol Syst 34:425–453

Heil M, Silva Bueno JC (2007) Within-plant signaling by volatiles leads to induction and priming of an indirect plant defense in nature. Proc Natl Acad Sci USA 104:5467–5472

Heil M, Fiala B, Linsenmair KE, Zotz G, Menke P, Maschwitz U (1997) Food body production in *Macaranga triloba* (Euphorbiaceae): a plant investment in anti-herbivore defence via mutualistic ant partners. J Ecol 85:847–861

Heil M, Fiala B, Kaiser W, Linsenmair KE (1998) Chemical contents of *Macaranga* food bodies: adaptations to their role in ant attraction and nutrition. Funct Ecol 12:117–122

Heil M, Fiala B, Baumann B, Linsenmair KE (2000a) Temporal, spatial and biotic variations in extrafloral nectar secretion by *Macaranga tanarius*. Funct Ecol 14:749–757

Heil M, Hilpert A, Kaiser W, Linsenmair KE (2000b) Reduced growth and seed set following chemical induction of pathogen defence: does systemic acquired resistance (SAR) incur allocation costs? J Ecol 88:645–654

Heil M, Koch T, Hilpert A, Fiala B, Boland W, Linsenmair KE (2001) Extrafloral nectar production of the ant-associated plant, *Macaranga tanarius*, is an induced, indirect, defensive response elicited by jasmonic acid. Proc Natl Acad Sci USA 98:1083–1088

Heil M, Baumann B, Krüger R, Linsenmair KE (2004a) Main nutrient compounds in food bodies of Mexican *Acacia* ant-plants. Chemoecol 14:45–52

Heil M, Feil D, Hilpert A, Linsenmair KE (2004b) Spatio-temporal patterns in indirect defence of a South-east Asian ant-plant support the optimal defence hypothesis. J Trop Ecol 20:573–580

Heil M, Greiner S, Meimberg H, Krüger R, Noyer J-L, Heubl. G, Linsenmair KE, Boland W (2004c) Evolutionary change from induced to constitutive expression of an indirect plant resistance. Nature 430:205–208

Heil M, Hilpert A, Krüger R, Linsenmair KE (2004d) Competition among visitors to extrafloral nectaries as a source of ecological costs of an indirect defence. J Trop Ecol 20:201–208

Heil M, Rattke J, Boland W (2005) Post-secretory hydrolysis of nectar sucrose and specialization in ant/plant mutualism. Science 308:560–563

Herms DA, Mattson WJ (1992) The dilemma of plants: to grow or to defend. Quart Rev Biol 67:283–335

Hilker M, Meiners T (2006) Early herbivore alert: Insect eggs induce plant defense. J Chem Ecol 32:1379–1397

Himanen S, Vuorinen T, Tuovinen T, Holopainen JK (2005) Effects of cyclamen mite (*Phytonemus pallidus*) and leaf beetle (*Galerucella tenella)* damage on volatile emission from strawberry (*Fragaria* x *ananassa* Duch.) plants and orientation of predatory mites (*Neoseiulus cucumeris, N. californicus*, and *Euseius finlandicus*). J Agric Food Chem 53:8624–8630

Hoballah ME, Köllner TG, Degenhardt J, Turlings TCJ (2004) Costs of induced volatile production in maize. Oikos 105:168–180

Horiuchi JI, Arimura GI, Ozawa R, Shimoda T, Dicke M, Takabayashi J, Nishioka T (2003) Lima bean leaves exposed to herbivore-induced conspecific plant volatiles attract herbivores in addition to carnivores. Appl Entomol Zool 38:365–368

Hulcr J, Ubik K, Vrkoc J (2006) The role of semiochemicals in tritrophic interactions between the spruce bark beetle *Ips typographus*, its predators and infested spruce. J Appl Entomol 130:275–283

Huxley CR, Cutler DF (eds) (1991) Ant–plant interactions. Oxford University Press, Oxford
Ibrahim MA, Nissinen A, Holopainen JK (2005) Response of *Plutella xylostella* and its parasitoid *Cotesia plutellae* to volatile compounds. J Chem Ecol 31:1969–1984
Izzo TJ, Vasconcelos HL (2002) Cheating the cheater: domatia loss minimizes the effects of ant castration in an Amazonian ant-plant. Oecologia 133:200–205
James DG (2003) Synthetic herbivore-induced plant volatiles as field attractants for beneficial insects. Environ Entomol 32:977–982
James DG, Price TS (2004) Field-testing of methyl salicylate for recruitment and retention of beneficial insects in grapes and hops. J Chem Ecol 30:1613–1628
Janzen DH (1966) Coevolution of mutualism between ants and acacias in Central America. Evolution 20:249–275
Janzen DH (1974) Swollen-thorn acacias of Central America. Smithsonian Institution, Washington, D.C.
Jürgens A (2004) Flower scent composition in diurnal *Silene species* (Caryophyllaceae): phylogenetic constraints or adaption to flower visitors? Biochem Syst Ecol 32:841–859
Kaiser R (1991) Trapping, investigation and reconstitution of flower secnts. In: Müller PM, Lamparsky D (eds) Perfumes: art, science and technology. Elsevier, London, pp 213–250
Kalberer NM, Turlings TCJ, Rahier M (2001) Attraction of a leaf beetle (*Oreina cacaliae*) to damaged host plants. J Chem Ecol 27:647–661
Kant MR, Ament K, Sabelis MW, Haring MA, Schuurink RC (2004) Differential timing of spider mite-induced direct and indirect defenses in tomato plants. Plant Physiol 135:483–495
Kappers IF, Aharoni A, van Herpen T, Luckerhoff LLP, Dicke M, Bouwmeester HJ (2005) Genetic engineering of terpenoid metabolism attracts, bodyguards to *Arabidopsis*. Science 309:2070–2072
Karban R (2001) Communication between sagebrush and wild tobacco in the field. Biochem Syst Ecol 29:995–1005
Karban R, Baldwin IT (1997) Induced responses to herbivory. University of Chicago Press, Chicago
Karban R, Baldwin I, Baxter K, Laue G, Felton G (2000) Communication between plants: induced resistance in wild tobacco plants following clipping of neighboring sagebrush. Oecologia 125:66–71
Karban R, Shiojiri K, Huntzinger M, McCall AC (2006) Damage-induced resistance in sagebrush: Volatiles are key to intra- and interplant communication. Ecology 87:922–930
Kato H, Yamane S, Phengklai C (2004) Ant-colonized domatia on fruits of *Mucuna interrupta* (Leguminosae). J Plant Res 117:319–321
Kessler A, Baldwin IT (2001) Defensive function of herbivore-induced plant volatile emissions in nature. Science 291:2141–2144
Kessler D, Baldwin IT (2007) Making sense of nectar scents: the effects of nectar secondary metabolites on floral visitors of *Nicotiana attenuata*. Plant J 49:840–854
Kessler A, Halitschke R, Diezel C, Baldwin IT (2006) Priming of plant defense responses in nature by airborne signaling between *Artemisia tridentata* and *Nicotiana attenuata*. Oecologia 148:280–292
Khan ZR, AmpongNyarko K, Chiliswa P, Hassanali A, Kimani S, Lwande W, Overholt WA, Pickett JA, Smart LE, Wadhams LJ, Woodcock CM (1997) Intercropping increases parasitism of pests. Nature 388:631–632
Kishimoto K, Matsui K, Ozawa R, Takabayashi J (2005) Volatile C6-aldehydes and allo-ocimene activate defense genes and induce resistance against *Botrytis cinerea* in *Arabidopsis thaliana*. Plant Cell Physiol 46:1093–1102
Knudsen JT, Tollsten L (1993) Trends in floral scent chemistry in pollination syndromes: floral scent composition in moth-pollinated taxa. Bot J Linn Soc 113:263–284
Knudsen JT, Tollsten L, Bergstrom LG (1993) Floral scents – a checklist of volatile compounds isolated by headspace techniques. Phytochemistry 33:253–280
Knudsen JT, Eriksson R, Gershenzon J, Stahl B (2006) Diversity and distribution of floral scent. Bot Rev 72:1–120

Koptur S (1989) Is extrafloral nectar an inducible defense? In: Bock JH, Linhart YB (eds) The evolutionary ecology of plants. Westview, Boulder, pp 323–329

Koptur S (1992) Extrafloral nectary-mediated interactions between insects and plants. In: Bernays EA (ed) Insect–plant interactions, vol IV. CRC, Boca Raton, pp 81–129

Koptur S (1994) Floral and extrafloral nectars of Costa Rican Inga trees: a comparison of their constituents and composition. Biotropica 26:276–284

Kost C, Heil M (2005) Increased availability of extrafloral nectar reduces herbivory in Lima bean plants (*Phaseolus lunatus*, Fabaceae). Basic Appl Ecol 6:237–248

Kost C, Heil M (2006) Herbivore-induced plant volatiles induce an indirect defence in neighbouring plants. J Ecol 94:619–628

Krips OE, Willems PEL, Gols R, Posthumus MA, Gort G, Dicke M (2001) Comparison of cultivars of ornamental crop Gerbera jamesonii on production of spider mite-induced volatiles, and their attractiveness to the predator, *Phytoseiulus persimilis*. J Chem Ecol 27:1355–1372

Kunert M, Biedermann A, Koch T, Boland W (2002) Ultrafast sampling and analysis of plant volatiles by a hand-held miniaturised GC with pre-concentration unit: kinetic and quantitative aspects of plant volatile production. J Sep Sci 25:677–684

Labandeira CC (1997) Insect mouthparts: ascertaining the paleobiology of insect feeding strategies. Annu Rev Ecol Syst 28:153–193

Lanza J (1988) Ant preferences for *Passiflora* nectar mimics that contain amino acids. Biotropica 20:341–344

Lehtonen P, Helander M, Saikkonen K (2005) Are endophyte-mediated effects on herbivores conditional on soil nutrients? Oecologia 142:38–45

Limburg DD, Rosenheim JA (2001) Extrafloral nectar consumption and its influence on survival and development of an omnivorous predator, larval *Chrysoperla plorabunda* (Neuroptera: Chrysopidae). Environ Entomol 30:595–604

Loreto F, Velikova V (2001) Isoprene produced by leaves protects the photosynthetic apparatus against ozone damage, quenches ozone products, and reduces lipid peroxidation of cellular membranes. Plant Physiol 127:1781–1787

Lou YG, Ma B, Cheng JA (2005) Attraction of the parasitoid *Anagrus nilaparvatae* to rice volatiles induced by the rice brown planthopper *Nilaparvata lugens*. J Chem Ecol 31:2357–2372

Loughrin JH, Manukian A, Heath RR, Turlings TCJ, Tumlinson JH (1994) Diurnal cycle of emission of induced volatile terpenoids by herbivore-injured cotton plants. Proc Natl Acad Sci USA 91:11836–11840

Loughrin JH, Manukian A, Heath RR, Tumlinson JH (1995) Volatiles emitted by different cotton varieties damaged by feeding beet armyworm larvae. J Chem Ecol 21:1217–1227

Loughrin JH, Potter DA, HamiltonKemp TR, Byers ME (1996) Role of feeding-induced plant volatiles in aggregative behavior of the Japanese beetle (Coleoptera: Scarabaeidae). Environ Entomol 25:1188–1191

Major IT, Constabel CP (2006) Molecular analysis of poplar defense against herbivory: comparison of wound- and insect elicitor-induced gene expression. New Phytol 172:617–635

Major IT, Constabel CP (2007) Insect regurgitant and wounding elicit similar defense responses in Poplar leaves. Plant Signal Behav 2:e1–e3

Matsui K, Kurishita S, Hisamitsu A, Kajiwara T (2000) A lipid-hydrolysing activity involved in hexenal formation. Biochem Soc Trans 28:857–860

Mattiacci L, Dicke M, Posthumus MA (1995) Beta-glucosidase – an elicitor of herbivore-induced plant odor that attracts hostsearching parasitic wasps. Proc Natl Acad Sci USA 92:2036–2040

Meiners T, Hilker M (2000) Induction of plant synomones by oviposition of a phytophagous insect. J Chem Ecol 26:221–232

Meister B, Krauss J, Harri SA, Schneider MV, Muller CB (2006) Fungal endosymbionts affect aphid population size by reduction of adult life span and fecundity. Basic Appl Ecol 7:244–252

Mercke P, Kappers IF, Verstappen FWA, Vorst O, Dicke M, Bouwmeester HJ (2004) Combined transcript and metabolite analysis reveals genes involved in spider mite induced volatile formation in cucumber plants. Plant Physiol 135:2012–2024

Miller B, Madilao LL, Ralph S, Bohlmann J (2005) Insect-induced conifer defense. White pine weevil and methyl jasmonate induce traumatic resinosis, de novo formed volatile emissions, and accumulation of terpenoid synthase and putative octadecanoid pathway transcripts in Sitka spruce. Plant Physiol 137:369–382

Mithöfer A, Maitrejean M, Boland W (2005a) Structural and biological diversity of cyclic octadecanoids, jasmonates, and mimetics. J Plant Growth Regul 23:170–178

Mithöfer A, Wanner G, Boland W (2005b) Effects of feeding spodoptera littoralis on Lima bean leaves. II. Continuous mechanical wounding resembling insect feeding is sufficient to elicit herbivory-related volatile emission. Plant Physiol 137:1160–1168

Miyake T, Yafuso M (2005) Pollination of *Alocasia cucullata* (Araceae) by two *Colocasiomyia* flies known to be specific pollinators for *Alocasia odora*. Plant Spec Biol 20:201–208

Mondor EB, Addicott JF (2003) Conspicuous extra-floral nectaries are inducible in *Vicia faba*. Ecol Lett 6:495–497

Mondor EB, Tremblay MN, Messing RH (2006) Extrafloral nectary phenotypic plasticity is damage and resource-dependent in *Vicia faba*. Biol Lett 2:583–585

Mumm R, Hilker M (2006) Direct and indirect chemical defence of pine against folivorous insects. Trends Plant Sci 11:351–358

Ness JH (2003) *Catalpa bignonioides* alters extrafloral nectar production after herbivory and attracts ant bodyguards. Oecologia 134:210–218

O'Dowd DJ (1979) Foliar nectar production and ant activity on a neotropical tree, *Ochroma pyramidale*. Oecologia 43:233–248

O'Dowd DJ (1982) Pearl bodies as ant food: an ecological role for some leaf emergences of tropical plants. Biotropica 14:40–49

O'Dowd DJ, Pemberton RW (1994) Leaf domatia in Korean plants: floristics, frequency, and biogeography. Vegetatio 114:137–148

O'Dowd D, Willson MF (1991) Associations between mites and leaf domatia. Trends Ecol Evol 6:179–182

Oliveira PS, Rico-Gray V, Diaz-Castelazo C, Castillo-Guevara C (1999) Interaction between ants, extrafloral nectaries and insect herbivores in Neotropical coastal sand dunes: herbivore deterrence by visiting ants increases fruit set in *Opuntia stricta* (Cactaceae). Funct Ecol 13:623–631

Olson DM, Wäckers FL (2007) Management of field margins to maximize multiple ecological services. J Appl Ecol 44:13–21

Omacini M, Chaneton EJ, Ghersa CM, Muller CB (2001) Symbiotic fungal endophytes control insect host–parasite interaction webs. Nature 409:78–81

Orians C (2005) Herbivores, vascular pathways, and systemic induction: facts and artifacts. J Chem Ecol 31:2231–2242

Ozawa R, Arimura G, Takabayashi J, Shimoda T, Nishioka T (2000) Involvement of jasmonate- and salicylate-related signaling pathways for the production of specific herbivore-induced volatiles in plants. Plant Cell Physiol 41:391–398

Paré PW, Tumlinson JH (1997a) De novo biosynthesis of volatiles induced by insect herbivory in cotton plants. Plant Physiol 114:1161–1167

Paré PW, Tumlinson JH (1997b) Induced synthesis of plant volatiles. Nature 385:30–31

Paré PW, Tumlinson JH (1999) Plant volatiles as a defense against insect herbivores. Plant Physiol 121:325–331

Paré PW, Alborn HT, Tumlinson JH (1998) Concerted biosynthesis of an insect elicitor of plant volatiles. Proc Natl Acad Sci USA 95:13971–13975

Paschold A, Halitschke R, Baldwin IT (2006) Using 'mute' plants to translate volatile signals. Plant J 45:275–291

Pauw B, Memelink J (2005) Jasmonate-Responsive Gene Expression. J Plant Growth Regul 23:200–210

Pellmyr O, Tang W, Groth I, Bergstrom G, Thiens LB (1991) Cycad cone and angiosperm floral volatiles: Inferences for the evolution of insect pollination. Biochem Syst Ecol 19:623

Pemberton RW (1993) Observations of extrafloral nectar feeding by predaceous and fungivorous mites. Proc Entomol Soc Wash 95:642–643

Pemberton RW, Lee JH (1996) The influence of extrafloral nectaries on parasitism of an insect herbivore. Am J Bot 83:1187–1194

Pemberton RW, Vandenberg NJ (1993) Extrafloral nectar feeding by ladybird beetles (Coleoptera: Coccinellidae). Proc Entomol Soc Wash 95:139–151

Peña-Cortes H, Barrios P, Dorta F, Polanco V, Sánchez C, Sánchez E, Ramírez I (2005) Involvement of jasmonic acid and derivatives in plant responses to pathogens and insects and in fruit ripening. J Plant Growth Regul 23:246–260

Peñuelas J, Llusiá J (2004) Plant VOC emissions: making use of the unavoidable. Trends Ecol Evol 19:402–404

Petitt FL, Turlings TCJ, Wolf SP (1992) Adult experience modifies attraction of the leafminer parasitoid *Opius dissitus* (Hymenoptera: Braconidae) to volatile semiochemicals. J Insect Behav 5:623–634

Pettersson EM, Boland W (2003) Potential parasitoid attractants, volatile composition throughout a bark beetle attack. Chemoecology 13:27–37

Philpott SM, Foster PF (2005) Nest-site limitation in coffee agroecosystems: Artificial nests maintain diversity of arboreal ants. Ecol Appl 15:1478–1485

Pichersky E (2004) Plant scents – what we perceive as a fragrant perfume is actually a sophisticated tool used by plants to entice pollinators, discourage microbes and fend off predators. Am Sci 92:514–521

Pichersky E, Gershenzon J (2002) The formation and function of plant volatiles: perfumes for pollinator attraction and defense. Curr Opin Plant Biol 5:237–243

Pichersky E, Noel JP, Dudareva N (2006a) Biosynthesis of plant volatiles: Nature's diversity and ingenuity. Science 311:808–811

Pichersky E, Sharkey TD, Gershenzon J (2006b) Plant volatiles: a lack of function or a lack of knowledge? Trends Plant Sci 11:421–421

Preston CA, Laue G, Baldwin IT (2004) Plant-plant signaling: Application of *trans*- or *cis*-methyl jasmonate equivalent to sagebrush releases does not elicit direct defenses in native tobacco. J Chem Ecol 30:2193–2214

Price PW, Bouton CE, Gross P, McPheron BA, Thompson JN, Weis AE (1980) Interactions among three trophic levels: influence of plants on interactions between insect herbivores and natural enemies. Annu Rev Ecol Syst 11:41–65

Raine NE, Willmer P, Stone GN (2002) Spatial structuring and floral avoidance behavior prevent ant–pollinator conflict in a Mexican ant-acacia. Ecology 83:3086–3096

Ralph SG, Yueh H, Friedmann M, Aeschliman D, Zeznik JA, Nelson CC, Butterfield YSN, Kirkpatrick R, Liu J, Jones SJM, Marra MA, Douglas CJ, Ritland K, Bohlmann J (2006) Conifer defence against insects: microarray gene expression profiling of Sitka spruce (*Picea sitchensis*) induced by mechanical wounding or feeding by spruce budworms (*Choristoneura occidentalis*) or white pine weevils (*Pissodes strobi*) reveals large-scale changes of the host transcriptome. Plant Cell Environ 29:1545–1570

Rashbrook VK, Compton SG, Lawton JH (1992) Ant–herbivore interactions: reasons for the absence of benefits to a fern with foliar nectaries. Ecology 73:2167–2174

Rasmann S, Kollner TG, Degenhardt J, Hiltpold I, Toepfer S, Kuhlmann U, Gershenzon J, Turlings TCJ (2005) Recruitment of entomopathogenic nematodes by insect-damaged maize roots. Nature 434:732–737

Ren D (1998) Flower-associated Brachycera flies as fossil evidence for Jurassic angiosperm origins. Science 280:85–88

Reymond P, Farmer EE (1998) Jasmonate and salicylate as global signals for defense gene expression. Curr Opin Plant Biol 1:404–411

Reymond P, Weber H, Damond M, Farmer EE (2000) Differential gene expression in response to mechanical wounding and insect feeding in Arabidopsis. Plant Cell 12:707–719

Rickson FR (1971) Glycogen plastids in Müllerian body cells of *Cecropia peltata* – a higher green plant. Science 173:344–347

Rickson FR, Rickson MM (1998) The cashew nut, *Anacardium occidentale* (Anacardiaceae), and its perennial association with ants: extrafloral nectary location and the potential for ant defense. Am J Bot 85:835–849

Risch S, Rickson F (1981) Mutualism in which ants must be present before plants produce food bodies. Nature 291:149–150
Risch S, McClure M, Vandermeer J, Waltz S (1977) Mutualism between three species of tropical *Piper* (Piperaceae) and their ant inhabitants. Am Midl Nat 98:433–444
Romero GQ, Benson WW (2004) Leaf domatia mediate mutualism between mites and a tropical tree. Oecologia 140:609–616
Romero GQ, Benson WW (2005) Biotic interactions of mites, plants and leaf domatia. Curr Opin Plant Biol 8:436–440
Röse USR, Lewis J, Tumlinson JH (2006) Extrafloral nectar from cotton (*Gossypium hirsutum*) as a food source for parasitic wasps. Funct Ecol 20:67–74
Rosenblueth M, Martínez-Romero E (2006) Bacterial endophytes and their interactions with hosts. Mol Plant Microbe Interact 19:827–837
Rostas M, Ton J, Mauch-Mani B, Turlings TCJ (2006) Fungal infection reduces herbivore-induced plant volatiles of maize but does not affect naive parasitoids. J Chem Ecol 32:1897–1909
Ruffner GA, Clark WD (1986) Extrafloral nectar of *Ferocactus acanthodes* (Cactaceae): composition and its importance to ants. Am J Bot 73:185–189
Ruhren S, Handel SN (1999) Jumping spiders (Salticidae) enhance the seed production of a plant with extrafloral nectaries. Oecologia 119:227–230
Rumpf S, Cromey M, Webb CJ (1994) Ultrastructure and function of the nectaries of New Zealand bracken (*Pteridium esculentum* (Forst. f.) Cockayne). N Z J Bot 32:487–496
Ruther J, Kleier S (2005) Plant-plant signaling: Ethylene synergizes volatile emission in *Zea mays* induced by exposure to (Z)-3-Hexen-1-ol. J Chem Ecol 31:2217–2222
Ryan CA, Pearce G (1998) Systemin: A polypeptide signal for plant defensive genes. Annu Rev Cell Dev Biol 14:1–17
Sassi C, Muller CB, Krauss J (2006) Fungal plant endosymbionts alter life history and reproductive success of aphid predators. Proc R Soc Lond B Biol Sci 273:1301–1306
Schaller A, Ryan CA (1995) Systemin – a polypeptide defense signal in plants. BioEssays 18:27–33
Schenk PM, Kazan K, Wilson I, Anderson J, Richmond T, Somerville SC, Manners JM (2000) Coordinated plant defense responses in *Arabidopsis* revealed by microarray analysis. Proc Natl Acad Sci USA 97:11655–11660
Schilmiller AL, Howe GA (2005) Systemic signaling in the wound response. Curr Opin Plant Biol 8:369–377
Schlotzhauer WS, Pair SD, Horvat RJ (1996) Volatile constituents from the flowers of Japanese honeysuckle (*Lonicera japonica*). J Agric Food Chem 44:206–209
Schnee C, Köllner TG, Held M, Turlings TCJ, Gershenzon J (2006) The products of a single maize sesquiterpene synthase form a volatile defense signal that attracts natural enemies of maize herbivores. Proc Natl Acad Sci USA 103:1129–1134
Schönenberger J, Friis EM, Matthews ML, Endress PK (2001) Cunoniaceae in the Cretaceous of Europe: Evidence from fossil flowers. Ann Bot 88:423–437
Shiojiri K, Karban R (2006) Plant age, communication, and resistance to herbivores: young sagebrush plants are better emitters and receivers. Oecologia 149:214–220
Shiojiri K, Kishimoto K, Ozawa R, Kugimiya S, Urashimo S, Arimura G, Horiuchi J, Nishioka T, Matsui K, Takabayashi J (2006) Changing green leaf volatile biosynthesis in plants: An approach for improving plant resistance against both herbivores and pathogens. Proc Natl Acad Sci USA 103:16672–16676
Smith LL, Lanza J, Smith GS (1990) Amino acid concentrations in extrafloral nectar of *Impatiens sultani* increase after simulated herbivory. Ecology 71:107–115
Sobrinho TG, Schoereder JH, Rodrigues LL, Collevatti RG (2002) Ant visitation (Hymenoptera: Formicidae) to extrafloral nectaries increases seed set and seed viability in the tropical weed *Triumfetta semitriloba*. Sociobiology 39:353–368
Stapel JO, Cortesero AM, DeMoraes CM, Tumlinson JH, Lewis WJ (1997) Extrafloral nectar, honeydew, and sucrose effects on searching behavior and efficiency of *Microplitis croceipes* (Hymenoptera: Braconidae) in cotton. Environ Entomol 26:617–623

Stephenson AG (1982) The role of the extrafloral nectaries of *Catalpa speciosa* in limiting herbivory and increasing fruit production. Ecology 63:663–669
Strobel G (2006) Harnessing endophytes for industrial microbiology. Curr Opin Microbiol 9:240–244
Takabayashi J, Dicke M (1996) Plant-carnivore mutualism through herbivore-induced carnivore attractants. Trends Plant Sci 1:109–113
Takabayashi J, Takahashi S, Dicke M, Posthumus MA (1995) Developmental stage of herbivore *Pseudaletia separata* affets production of herbivore-induced synonome by corn plants. J Chem Ecol 21:273–287
Tempel AS (1983) Bracken fern (*Pteridium aquilinum*) and nectar-feeding ants: a nonmutualistic interaction. Ecology 64:1411–1422
Thaler JS (1999) Jasmonate-inducible plant defences cause increased parasitism of herbivores. Nature 399:686–688
Thien LB, Azuma H, Kawano S (2000) New perspectives on the pollination biology of basal angiosperms. Int J Plant Sci 161:S225–S235
Tholl D, Chen F, Petri J, Gershenzon J, Pichersky E (2005) Two sesquiterpene synthases are responsible for the complex mixture of sesquiterpenes emitted from Arabidopsis flowers. Plant J 42:757–771
Thornburg RW, Carter C, Powell A, Mittler R, Rizhsky L, Horner HT (2004) A major function of the tobacco floral nectary is defense against microbial attack. Plant Syst Evol 238:211–218
Tinzaara W, Gold CS, Dicke M, Van Huis A (2005) Olfactory responses of banana weevil predators to volatiles from banana pseudostem tissue and synthetic pheromone. J Chem Ecol 31:1537–1553
Ton J, D'Allesandro M, Jourdie V, Jakab G, Karlen D, Held M, Mauch-Mani B, Turlings TCJ (2007) Priming by airborne signals boosts direct and indirect resistance in maize. Plant J 49:16–26
Tooker JF, Hanks LM (2006) Tritrophic interactions and reproductive fitness of the prairie perennial *Silphium laciniatum* Gillette (Asteraceae). Environ Entomol 35:537–545
Tumlinson JH, Paré PW, Turlings TCJ (1999) Plant production of volatile semiochemicals in response to insect-derived elicitors. In: Chadwick DJ, Goode JA (eds) Insect–plant interactions and induced plant defence. Wiley, Chichester, pp 95–109
Turlings TCJ, Ton J (2006) Exploiting scents of distress: the prospect of manipulating herbivore-induced plant odours to enhance the control of agricultural pests. Curr Opin Plant Biol 9:421–427
Turlings TCJ, Tumlinson JH (1992) Systemic release of chemical signals by herbivore-injured corn. Proc Natl Acad Sci USA 89:8399–8402
Turlings TCJ, Tumlinson JH, Lewis WJ (1990) Exploitation of herbivore-induced plant odors by host-seeking parasitic wasps. Science 250:1251–1253
Turlings TCJ, Loughrin JH, McCall PJ, Röse USR, Lewis WJ, Tumlinson JH (1995) How caterpillar-damaged plants protect themselves by attracting parasitic wasps. Proc Natl Acad Sci USA 92:4169–4174
Turlings TCJ, Lengwiler UB, Bernasconi ML, Wechsler D (1998) Timing of induced volatile emissions in maize seedlings. Planta 207:146–152
Vallat A, Gu HN, Dorn S (2005) How rainfall, relative humidity and temperature influence volatile emissions from apple trees in situ. Phytochemistry 66:1540–1550
Van den Boom CEM, Beek TA van, Posthumus MA, Groot A de, Dicke M (2004) Qualitative and quantitative variation among volatile profiles induced by *Tetranychus urticae* feeding on plants from various families. J Chem Ecol 30:69–89
Van Loon JJA, Boer JG de, Dicke M (2000) Parasitoid–plant mutualism: parasitoid attack of herbivore resistance increases plant reproducion. Entomol Exp Appl 97:219–227
Van Rijn PCJ, Tanigoshi LK (1999) The contribution of extrafloral nectar to survival and reproduction of the predatory mite *Iphiseius degenerans* on *Ricinus communis*. Exp Appl Acarol 23:281–296

Voelckel C, Baldwin IT (2004) Herbivore-induced plant vaccination. Part II. Array-studies reveal the transience of herbivore-specific transcriptional imprints and a distinct imprint from stress combinations. Plant J 38:650–663

Wäckers FL, Bezemer TM (2003) Root herbivory induces an above-ground indirect defence. Ecol Lett 6:9–12

Wäckers FL, Zuber D, Wunderlin R, Keller F (2001) The effect of herbivory on temporal and spatial dynamics of foliar nectar production in cotton and castor. Ann Bot 87:365–370

Walling LL (2000) The myriad plant responses to herbivores. J Plant Growth Regul 19:195–216

Walling LL (2001) Induced resistance: from the basic to the applied. Trends Plant Sci 6:445–447

Walter DE (1996) Living on leaves: Mites, tomenta, and leaf domatia. Annu Rev Entomol 41:101–114

Wasternack C, Parthier B (1997) Jasmonate-signalled plant gene expression. Trends Plant Sci 2:302–307

Webber BL, Abaloz BA, Woodrow IE (2007) Myrmecophilic food body production in the understorey tree, *Ryparosa kurrangii* (Achariaceae), a rare Australian rainforest taxon. New Phytol 173:250–263

Wei JN, Kang L (2006) Electrophysiological and behavioral responses of a parasitic wasp to plant volatiles induced by two leaf miner species. Chem Senses 31:467–477

Wickers S (1997) Study of nectariferous secretion in a pioneer plant, *Inga thibaudiana*, in relation with ants. Acta Bot Gall 144:315–326

Yan ZG, Wang CZ (2006) Identification of *Mythmna separata*-induced maize volatile synomones that attract the parasitoid *Campoletis chlorideae*. J Appl Entomol 130:213–219

Yu DW, Davidson DW (1997) Experimental studies of species-specificity in *Cecropia*-ant relationships. Ecol Monogr 67:273–294

Zhu JW, Park KC (2005) Methyl salicylate, a soybean aphid-induced plant volatile attractive to the predator *Coccinella septempunctata*. J Chem Ecol 31:1733–1746

Zimmerli L, Jakab G, Métraux J-P, Mauch-Mani B (2000) Potentiation of pathogen-specific defense mechanisms in *Arabidopsis* by beta-aminobutyric acid. Proc Natl Acad Sci USA 97:12920–12925

Zoghbi MDB, Andrade EHA, Carreira LMM, Maia JGS (2000) Volatile constituents of the flowers of *Wulffia baccata* (L. f.) Kuntze. and *Zinnia elegans* Jacq. (Asteraceae). J Essent Oil Res 12:415–417

Functional Differences in Soil Water Pools: a New Perspective on Plant Water Use in Water-Limited Ecosystems

Ronald J. Ryel (✉), **Carolyn Y. Ivans**, **Michael S. Peek**, **and A. Joshua Leffler**

Contents

Abstract Arid and semi-arid ecosystems cover roughly half of the earth's surface. Significant changes in vegetation cover combined with climate change have increased concern over the future of these lands, which have considerable economic importance. Much research has focused on plant–soil water relations in these systems, yet many mechanisms and significance of water use patterns are not well understood. Here we describe a new conceptual model that considers two pools of soil water accessed by plants: a growth pool that is located in shallow soil layers, and a maintenance pool that is often in deeper soil layers. While they may be spatially and

Ronald J. Ryel
Department of Wildland Resources and the Ecology Center, Utah State University, Logan, UT 84322-5230, USA
e-mail: ron.ryel@usu.edu

U. Lüttge et al. (eds.), *Progress in Botany 69.*

temporally separated when used, these pools are largely separated by function and have different dynamics and competitive interactions. The growth pool is characterized by rapid use and high competition among plants, is exploited during the period of high resource acquisition and growth, and couples directly to nutrient availability. The maintenance pool is characterized by low competition, persistence, and is largely decoupled from nutrient resources. We argue that, by assessing the functional importance of soil water pools, we are able to more effectively assess how changes in hydrology related to climate, vegetation or land-use alterations may affect plant communities and ecosystem function. We also explore the mechanisms plants use to conserve water and discuss the significance of feedforward and feedback control.

1 Introduction

Arid and semi-arid ecosystems cover approximately 50% of the land surface of the earth (Bailey 1996) and are characterized by low, but highly variable annual precipitation (Noy-Meir 1973). Although productivity in arid and semi-arid landscapes is often less than in more mesic ecosystems, net primary productivity is 30–35% of the annual total for all terrestrial ecosystems (Field et al. 1998). With primary productivity closely related to precipitation, alterations to vegetation structure and climate change may significantly affect soil water dynamics (Schlesinger et al. 1990; Jackson et al. 2000). Vegetation changes in arid and semi-arid ecosystems throughout the world has already been documented (Archer et al. 2001), and the hypothesized causes range from grazing practices, altered climate and fire regimes, non-native species, elevated atmospheric CO_2, and N deposition (Archer 1995; Van Auken 2000).

The influence of temporal and spatial patterns of water availability on the survival and reproduction of plants in arid and semi-arid communities has been the focus of research for decades. Semi-arid and arid lands are characterized by annual evapotranspiration demands that exceed annual precipitation, and plant productivity is often directly related to the total precipitation (Rosenzweig 1968; Chong et al. 1993). Yet, our understanding of plant–soil–water interactions is far from complete, especially in terms of how plants with different water-use strategies compete for water, affect soil water dynamics and ecosystem function, and how management and climate change may affect these dynamics. In addition, there is recognition that water availability and competition for water in semi-arid and arid ecosystems is not necessarily the same thing, and competition for water in these systems may not always be greater than in other systems (Casper and Jackson 1997).

Here we review current perspectives on plant–soil–water interactions in arid and semi-arid ecosystems and propose a new conceptual model for soil water partitioning that differentiates between water used primarily for growth when conditions are favorable and water used for maintenance during periods of drought. We feel better understanding of the dynamics of maintenance water is critical for more fully understanding the distribution of vegetation in arid and semi-arid ecosystems, and for helping to assess vegetation changes due to anthropogenic and climatic alterations of hydrology.

2 Models of Soil–Plant Interactions

2.1 Two-Layer Model

Walter (1971) proposed a two-layer model of plant-water use ("Walter's two-layer model") that focuses on the vertical distribution of water and predicts that the coexistence of herbaceous and woody plants in a community is a consequence of niche separation. He hypothesized that vertical stratification of soil water would minimize competition between grasses and shrubs; herbaceous plants relying primarily on shallow soil water and woody plants relying primarily on deeper soil water. Root distributions would reflect these water use patterns with grasses having shallow rooting distributions, while shrubs would be more deeply rooted. He proposed this coexistence of grasses and woody plants in semi-arid savannahs depended upon summer precipitation. A number of studies support this model (e.g., Walker and Noy-Meir 1982; Soriano and Sala 1983; Knoop and Walker 1985; Gordon et al. 1989; Liang et al. 1989; Sala et al. 1989; Nobel 1997; Schenk and Jackson 2002; Fargione and Tilman 2005), and the model has been used as the basis in modeling to compare vegetation responses and ecosystem function across ecosystems (Lauenroth et al. 1993; Breshears and Barnes 1999). However, despite its appeal and long-term consideration, this model does not have universal relevance. Several studies have shown that woody species are able to extract soil water from a variety of soil layers (e.g., Flanagan et al. 1992; Donovan and Ehleringer 1994; Evans and Ehleringer 1994; Le Roux et al. 1995; Breshears et al. 1997; Le Roux and Bariac 1998; Yoder et al. 1998), and that the model does not necessarily fit the observed patterns of water use (Belsky 1994; Briones et al. 1996; Weltzin and McPherson 1997; Dodd et al. 1998; Reynolds et al. 1999, 2000; Ryel et al. 2007). In addition, Ogle and Reynolds (2004) point out that, under this paradigm, woody plants should dominate when precipitation is higher (>250 mm $year^{-1}$) due to deeper recharge of soils, and grasses would dominate when precipitation is lower (<250 mm $year^{-1}$), the exact opposite they found from their reevaluation of data from Paruelo and Lauenroth (1995) for 49 grass and shrub communities in the western USA.

Schwinning and Ehleringer (2001) presented a theoretical assessment of water use trade-offs of different plant life-forms using two soil layers, a pulse-dominated shallow layer and a deeper, more continuous, less than saturated layer. They found that plants of different life forms may use water in both layers, but that the importance of the soil layer depended on the life-form and the timing of the available water resource in each layer.

2.2 Desertification

Another pervasive perspective is the model of desertification from Schlesinger et al. (1990). In contrast to Walter's two-layer model, Schlesinger's model focuses on the horizontal heterogeneity of soil water and nutrients. It postulates that disturbance-induced reductions in herbaceous vegetation coupled with intercanopy soil compaction lead to increases in runoff from intercanopy areas which further favor long-term increases in the proportion of woody plants. This focus on horizontal distributions of available soil water as opposed to vertical distributions leads to alternative predictions of plant community compositions at a given site (Belsky 1994; Reynolds et al. 1997; Scholes and Archer 1997; Breshears and Barnes 1999). The model predicts that desertification will be exacerbated by global warming with subsequent changes in global biogeochemical cycling (Schlesinger et al. 1990). This model, however, is limited to processes in altered environments where the distribution and effectiveness of precipitation for productivity is reduced or altered to favor woody vegetation.

2.3 Pulse–Reserve Model

In addition to conceptual models focused on the vertical and horizontal distribution of soil water in arid ecosystems, much attention has been given to plant responses to "pulses" of precipitation (e.g., Goldberg and Novoplanksy 1997; Schwinning and Ehleringer 2001; Ivans et al. 2003; Schwinning and Sala 2004; Chesson et al. 2004). The pulse–reserve hypothesis (Noy-Meir 1973) centers on within-year variability of precipitation by considering biologically significant rainfall events (pulses) that stimulate plant processes, primarily growth and reproduction. The hypothesis focuses on the fact that resource availability varies not only in space but that in arid regions especially, precipitation events and their effects may be viewed as discrete events in time (Noy-Meir 1973; Bilbrough and Caldwell 1997; Goldberg and Novoplansky 1997). An underlying assumption of the pulse-related model is that the sparse and variable nature of precipitation in arid and semi-arid regions can play a strong role in species composition (e.g., Chesson et al. 2004). Reynolds et al. (2004) point out, however, that this paradigm is too general to effectively predict plant responses to individual precipitation events. They propose the variability in response of plants to a pulse is too highly dependent upon interactions between precipitation, antecedent soil water content, and plant functional type and phenology to generalize among deserts and seasons.

2.4 Threshold Delay Model

Given the limitations in the two-layer and pulse–reserve models, Ogle and Reynolds (2004) proposed the "threshold delay model" to more effectively predict how plants would respond to the timing and duration of precipitation pulses. This model was designed to account for root plasticity, delayed responses to rainfall, precipitation thresholds for functional response, and phenology when considering plant physiological responses (e.g., photosynthesis, growth, transpiration) to precipitation pulses. The model is dynamic in that its current state is dependent upon the previous state and contains six parameters: (a) maximum rate of physiological process, (b) a variable describing the reduction in the rate over time, (c) maximum rate response to a precipitation pulse, (d, e) upper and lower thresholds for precipitation events that will elicit a rate response (above upper threshold: no additional response; below lower threshold: no response), and (f) delay length before a response to the pulse occurs. Although the model can be parameterized for computing actual rates, they propose the model as primarily a conceptual tool to evaluate how cooccurring plants may differ in physiological response to specific patterns of precipitation pulses. While relatively new as a tool, this model was successful in predicting general responses of 11 trees and shrubs to a single large precipitation event in southern Australia (Burgess 2006), and time courses of growth for five functional types of plant growing in the desert southwest USA (Ogle and Reynolds 2004).

2.5 *Models of Soil Water Dynamics*

Movement of water in soils has been long studied and effective models for this movement are well known and used (see, e.g., Kemp et al. 1997; Jury 2004; Kirkham 2005; Warrick 2006). In unsaturated soils, matric (soil water) potential gradients and gravity are primarily responsible for water movement. Once the larger pore spaces in unsaturated soil become drained via gravity and preferential flow, subsequent water movements occur almost entirely through capillary and film flow. These mechanisms of infiltration move water from wetter to dryer regions with movements occurring vertically as well as horizontally. The rate and direction of water movement depends on gradients in matric potential and soil characteristics. Water extraction for transpiration by plants is related to root distribution and density, root and soil hydraulic conductivities, and atmospheric and plant properties that affect transpiration potential.

More recently, it has been found that plants are able to transfer water between soil compartments via roots (for reviews, see Caldwell et al. 1998; Ryel 2004). This hydraulic redistribution (Burgess et al. 1998) of water occurs when roots span soil of different water potentials, and it has been documented in at least 50 species (Jackson et al. 2000). The upward movement of water from relatively moist to more dry soils (Richards and Caldwell 1987; Caldwell and Richards 1989; Dawson 1993; Yoder and Nowak 1999) is thought to increase nutrient availability to shallow roots (Richards and Caldwell 1987), maintain root–soil hydraulic contact (Caldwell et al. 1998), increase transpiration rates (Ryel et al. 2002), increase water availability to shallow rooted plants growing in the vicinity of deep rooted plants that are redistributing soil water (Dawson 1993), and may buffer myccorhizal fungi against severe soil drying (Querejeta et al. 2007). The rapid downward redistribution of rainwater to dry subsoils by roots (Burgess et al. 1998; Schulze et al. 1998; Smith et al. 1999; Scholz et al. 2002; Hultine et al. 2003; Ryel et al. 2003) is thought to enhance the growth of roots to reach deep groundwater (Caldwell et al.

1998; Schulze et al. 1998; Burgess et al. 1998; Hultine et al. 2003), to allow plants to conserve water during drought periods (Ryel et al. 2004), and it may greatly alter the dynamics of water penetration to depth in soils and water storage in ecosystems (Ryel et al. 2003). Hydraulic redistribution is most prevalent when soil water content is sufficiently low to greatly limit movement by infiltration (Caldwell et al. 1998; Ryel et al. 2002).

3 Partitioning Soil Water by Function

The above review highlights the breadth and perspectives for modeling plant responses to soil water distributions in arid and semi-arid environments. Many have attempted with varying degrees of success to generalize patterns of plant response to precipitation and soil water dynamics. While it is apparent that no single conceptual or simulation model approach addresses all questions concerning plant performance as related to soil water, there are critical elements still imploring further illumination.

Plants in arid and semi-arid environments are subject to periods of water limitations, when soil water availability is less than the potential demand for transpiration. During periods when all necessary resources are available for growth (water, nutrients, light, temperature) we would expect plants to have relatively high rates of transpiration and significant competition for resources. If a plant does not use the water at a sufficiently rapid rate, a competitor can use it and perhaps gain a biomass advantage over a low-water-using plant species (Cohen 1970). This is likely the case regardless of life-form or functional type. Thus, we would expect plants during periods favorable to growth to use water at rates permitting high physiological function, and not conserve water.

When drought becomes significant in length, plants of different functional type divide in their response to limited water resources. Annuals produce seeds and die. Some perennial species may senesce shortly after the onset of drought, while others may continue their physiological activity for somewhat longer periods of time before senescing. Other perennial species maintain some level of physiological activity throughout the drought. For species which continue physiological activity during some or all of a drought period, exhaustive use of water may quickly result in losses of this critical resource to levels below that to sustain the plant (Holbrook et al. 1995), even with greatly reduced physiological activity. Thus, we would expect these plants to have low water use (at levels to at least maintain sufficient physiological activity for survival), not high water use.

A common problem in assessing water availability and plant water use is the lack of consideration of soil water distribution, and in concert, the distribution of nutrients. Nutrients are largely concentrated in shallow soil layers (Jobbágy and Jackson 2001; Embaye et al. 2005; Swamy et al. 2006; Hooker et al. 2007), often in the upper 20 cm. Water is the primary vehicle for nutrient movement to plants, largely through diffusion processes in non-saturated soils as plant uptake creates depletion zones that cause a diffusion of nutrients from zones of higher concentration to zones of lower concentration at the root surface. As the soil dries, nutrient availability to the plant is reduced due to reduced rates of diffusion and mass flow (Nye and Tinker 1977), even if there are substantial concentrations of nutrients in the soil. Nutrient diffusion has been found to effectively cease when soils reach water potentials well above water potentials for plant function for many semi-arid species (Jackson and Caldwell 1996; Ivans et al. 2003). However, when nutrients are no longer (or only minimally) available due to drought, plant growth often ceases or is minimal, as cell elongation (Nonami et al. 1997) and protein synthesis (Lambers et al. 1998) are affected. At this point, plants can enter a maintenance mode, senesce or die. If plants have roots only in the shallow soils, then water resources are now also limiting for significant transpiration, and senescence or death are the only options.

However, for plants with roots in deeper soil layers, there may be significant water resources available (Schlesinger et al. 1987; Schwinning and Ehleringer 2001). Unlike water in shallow soils, this soil water is not in concert with significant nutrient resources, which are concentrated in shallow soil layers. Thus, during periods when plants can exploit only these deep water resources, their nutrient resources would be insufficient for significant growth. If the drought period is predictably short, a plant that maintains physiological activity can use the deep soil water pool at mostly any rate with minimal danger of running out of water. However, if the drought is likely long and the available resources not replenished, a plant not conserving water may run out of water before the end of the drought period and may risk serious physiological damage or death. To minimize such a possibility, we would expect these plants to employ water conservation measures to ensure that the water resource is sufficient for the plant to maintain physiological activity during the duration of the drought.

It has been know for some time that plants use water from different portions of the soil column during different time periods (Walter and Stadelmann 1974; Smith and Nobel 1986; Schwinning et al. 2002; Cox et al. 2005; Warren et al. 2005; Ryel et al. 2007). Some trees in riparian areas are noted for changing from water in near-surface soils to groundwater (Smith et al. 1991; Busch et al. 1992; Zhang et al. 1999; Horton et al. 2003; Cox et al. 2005), while in semi-arid systems shrubs and trees may initially use water in shallow soil layers in the spring and deeper soil water later in the season (Ehleringer et al. 1991; Leffler et al. 2002; Ryel et al. 2002; Zou et al. 2005). In both cases, there is a temporal, and in part spatial, bifurcation in water use: high rates during growth in the spring and early summer using near-surface water pools, and conservation measures during dry and hot summer using deeper pools of water. We refer to the former, shallow soil water pool as the "growth pool" and the other, often deeper pool as the "maintenance pool". In the remainder of this chapter, we present evidence for this behavior and discuss mechanisms that allow plants to effectively utilize these different pools of water. In addition, we discuss the characteristics of these two water pools and how changes in their dynamics could cause substantial changes in species composition.

4 Two-Pool Soil Water Model

4.1 Plant Functional Types and Root Distribution

Plants in semi-arid and arid ecosystems are of multiple life forms. These include annuals, herbaceous perennials, semi-shrubs, shrubs, trees, and stem succulents (Rundel and Nobel 1991; Smith et al. 1997). Annuals and stem succulents are characterized by shallow root systems (Schenk and Jackson 2002), the former able to only extract water from soils with a relatively high matric potential. Annuals persist during periods when shallow soils have sufficient water for physiological function, then senesce. Many non-succulent perennials extract water from deeper soils, but many of the herbaceous and drought deciduous shrub species eventually senesce in prolonged drought. Perennials, which do not senesce in periods of drought, are dependent upon water stores and often have abilities to extract water from soils of low matric potential, as low as −16 MPa in *Larrea tridentata* (Sperry 1995). These differences in phenology, physiology, and rooting distribution result in different water use patterns.

Despite the variety in life-history strategies, overall community rooting distributions in semi-arid or arid environments are similar, with many roots concentrated near the surface (Jackson et al. 1996) and with some species having roots to considerable depths (Canadell et al. 1996; Schenk and Jackson 2002). In water-limited ecosystems, significant positive relationships have been found between mean annual precipitation and rooting depth (Schenk and Jackson 2002), likely due to differences in the infiltration depth of precipitation (Dobrowolski et al. 1990; Reynolds et al. 2000). Regardless of the life forms, however, most plant species in these systems have roots to extract water from shallow soil layers (Noy-Meir 1973).

4.2 Water Availability – Two Pools

Water availability in arid and semi-arid environments has been described to occur in two soil zones (Schwinning and Ehleringer 2001). This includes a shallow zone which can have rapid and extreme fluctuations in water levels due to recharge by individual precipitation events and a deeper zone which is characterized by slower changes in water content and is recharged by a series of events (often over winter) that accumulate sufficient quantities of water to infiltrate deeper. The shallow zones may have individual water pulses following individual rainfall events or contain significant quantities of water following overwinter accumulation of multiple events. The effective depth of this zone depends on water availability, but is associated with most of the nutrient resources. The deeper zone of water accumulation may contain water at lower matric potentials, with sizable quantities spread out over greater depths and it can be relatively stable in water content (Andraski

1997; Scanlon et al. 1999). In many arid and semi-arid systems, water in this deeper soil zone comes from infiltration of surface water where deep infiltration and groundwater recharge do not occur (Seyfried et al. 2005). Groundwater may also be available to some arid and semi-arid species, especially phreotophytic species, which may have roots to depths of 10 m or more (Canadell et al. 1996) to reach groundwater resources.

4.3 Dynamics of Water Use

If the rate of soil water use is related to root density, then shallow soils are more rapidly dried than deeper soils due to differences in rooting distributions. In fact there are numerous examples of this soil water use pattern in arid and semi-arid environments (e.g., Noy-Meir 1973; Kemp et al. 1997; Booth et al. 2003; Warren et al. 2005; Zou et al. 2005). Ryel et al. (2007) also found this pattern for an annual grass, perennial tussock grass, evergreen shrub, and evergreen tree in a semi-arid steppe in west-central USA (Fig. 1). This study clearly showed the much more rapid use of soil water shallow layers (0–50 cm) than in deeper layers (>50 cm) with the onset of spring when temperatures became suitable for plant growth. The relatively rapid use of water occurred with all life forms. They also found that the primary growth period was related to the period of extraction of water from these shallow layers.

Once shallow soil water resources are depleted, however, the strategies of different life forms diverge. Annuals die, herbaceous species largely senesce, shrubs and trees often continue physiological activity at much reduced levels (Noy-Meir 1973; Ogle and Reynolds 2004). Ryel et al. (2007) found that the annual, *Bromus tectorum*, produced seeds and died when the upper soil resources available to this species were depleted, despite lower soil resources remaining nearly fully available. Three other perennial species greatly reduced physiological activity once the upper soil water was mostly depleted, but continued to use soil water at much reduced rates (Fig. 1).

4.4 Characteristics of the Two Water Pools

With its association with significant nutrient supplies, the "growth" pool in shallow soils would be characterized by high water use rates by individual plant species. In addition, there would likely be significant competition for water for transpiration by cooccurring individuals. Thus, we would expect this pool to decline rather rapidly after initiation of use with sufficient vapor-pressure deficit. The primary growth function of this pool would need to persist only as long as plants could effectively use this water in association with sufficient acquisition of nutrients. The growth pool would also need to persist for sufficient periods for decomposition and mineralization processes by the microbial community to replenish and recycle the supply of nutrients.

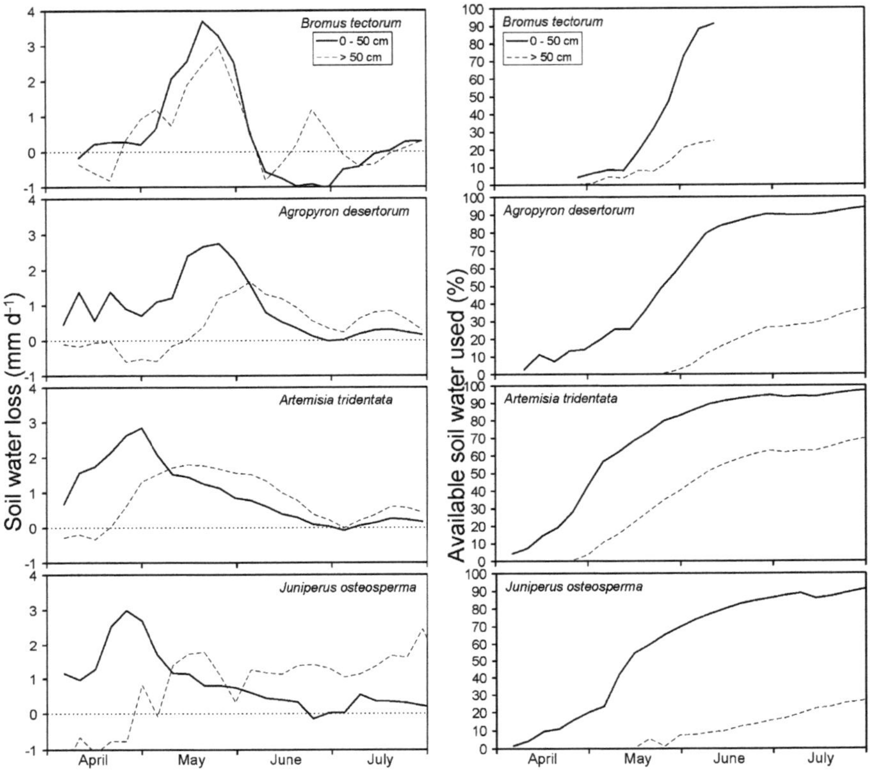

Fig. 1 Dynamics of soil shallow (0–50 cm depth) and deep (>50 cm depth) water pools for four species of different life forms (*Bromus tectorum*, annual grass; *Agropyron desertorum*, perennial grass; *Artemisia tridentata*, evergreen shrub; *Juniperus osteosperma*, evergreen tree) from near-monoculture stands in west-central Utah, USA. The *left panels* show evaoptranspiration (primarily transpiration) rates from the two depth pools during the spring and early summer growing season. *d* Day. The *right panels* show cumulative use of available soil water in each pool (modified from Ryel et al. 2007)

The "maintenance pool", in contrast, would need to persist throughout the drought periods faced by plants maintaining transpirational activity. To accomplish this, water in this portion of the soil could not go below matric potentials and supply rates insufficient to maintain transpirational and physiological activity for plant survival. For this to happen, water in these soil layers would need to be conserved. If water availability became insufficient, these perennial plants would suffer severe physiological damage or more likely, death. Thus, competition for this water would need to be minimal, with the improbable potential for full use of the available resource during periods of plant activity.

While association with the primary nutrient pool necessitates that the growth pool of water occurs in shallow soil layers, the maintenance pool need not always be relatively independent and in deeper soil layers. A deeper soil water pool with lower rates of water use by deep-rooted species was described by Noy-Meir (1973) and Schwinning and Ehleringer (2001) and may be the most prevalent site for storage of this maintenance pool. Reynolds et al. (2000), however, indicated that water for periods of high and low water use, vertical partitioning of soil layers does not really occur in the Chihuahuan Desert of the southwestern USA. To a degree, this can be seen in the semi-arid steppe discussed above (Fig. 1), where soil water from both shallow and deep layers is used during the extended summer drought period. The upper soil layers become sources for maintenance water once nutrient resource availability becomes greatly limited by reduced diffusion and mass flow due to low matric potentials. However, use of a deeper soil pool greatly increases the total quantity of water available for transpiration and also allows for separation of this resource from potential shallow-rooted competitors.

Relatively low competition is expected for the maintenance pool of water. We feel there are several reasons. First, if significant competition was occurring, the resource would frequently be depleted causing death of one or more species in the system. Thus, if this was the water pattern, except for stem succulents we would not see species that maintain transpirational and physiological activity during prolonged periods of drought. Yet, in most arid and semi-arid ecosystems, we find numbers of deep-rooted shrubs or trees that maintain physiological activity during drought periods (Weaver and Albertson 1943; Canadell et al. 1996). Second, deeper soil layers can have water stored at matric potentials below levels that are accessible by many annuals, herbaceous perennials or drought-deciduous species. As a result, species which can access water from soils of low matric potential have access to water that is inaccessible by other species. While this can occur in the deeper soil layers, it can also arise when the matric potential of shallow soils reaches a level where water is inaccessible by shallow-rooted species, which become senesced. Finally, the movement of maintenance water to roots can function as a slow-release mechanism, which limits the rate at which plants can actually use this water. When soils dry, hydraulic conductivity is greatly reduced and the movement via film flow becomes very slow. As a result, rapid use of soil water by plants becomes difficult.

For water exploited from the growth pool, plants use high root densities to more effectively and competitively access soil moisture during this period of high transpiration. Nutrient uptake is also enhanced. The use of this water can also be enhanced by high leaf area with high transpirational potential. However, to conserve the maintenance pool, plants utilize a variety of mechanisms to aid in storing and accessing this soil water to maintain physiological function during drought periods.

5 Mechanisms for Conserving Water

5.1 Root Distribution

Probably the most familiar mechanism for reducing soil water uptake by plants is fewer roots. As presented earlier, root distributions in arid and semi-arid ecosystems are skewed toward shallow soil depths and the soil zone of the growth pool.

The lower rooting zone, in contrast, is usually much less dense in distribution and can provide an effective mechanism for reducing the use of deeper water resources. In addition, roots in arid and semi-arid shrubs and trees often occur at depths below the maximum infiltration depth for precipitation (Noy-Meir 1973; Scanlon 1991; McCulley et al. 2004; Walvoord et al. 2004; Seyfried et al. 2005). Thus, they are effective in capturing all precipitation that infiltrates to depth within these soils (Hillel and Tadmor 1962; Seyfried et al. 2005), maximizing the size of the maintenance resource pool.

Countering this process, however, is the movement of water through hydraulic redistribution (Richards and Caldwell 1987) and from deeper soil layers to shallow soil layers when the upper soil layers are dryer. It has been hypothesized that plants move water via hydraulic redistribution, in part to maintain better root function of shallow roots (Caldwell et al. 1998; Ryel 2004) by reducing the decline in soil matric potential of the shallow soils (Meinzer et al. 2004). Water movement by lifting water via hydraulic redistribution can improve the effectiveness of the deeper roots for water uptake and transpiration (Caldwell and Richards 1989). Such movement, however, would be counter to water conservation.

To assess this potential dilemma, we assessed the effectiveness of a "top-skewed" root distribution to conserve water when hydraulic redistribution occurs. We employed the model of Ryel et al. (2002), which simulates soil water dynamics with plant water uptake and hydraulic redistribution. The model was expanded by Ryel et al. (2004) to estimate plant water potential as a function of the distribution soil moisture and roots. It was parameterized for *Artemisia tridentata* (big sagebrush), a shrub noted for significant hydraulic redistribution (Richards and Caldwell 1987; Williams et al. 1993; Ryel et al. 2002, 2003). Simulations were conducted for 100 days without precipitation for: (a) uniform soil water throughout the rooting zone, beginning at field capacity, and (b) for a measured early season soil moisture distribution from a site in western Utah, USA that was at or near field capacity in the upper 40 cm and then declined to levels where matric potential was approximately −2 MPa at the deepest rooting depth (1.6 m). Root distributions assessed were: as measured at the western Utah site for *A. tridentata*, uniform with the same quantity of roots, and uniform with four times the quantity of roots. The root distribution of the measured plants had 65% of their roots in the upper 50 cm of soil. The dynamics of upper (0–50 cm) and lower (>50 cm) water pools were simulated and considered to represent the growth and maintenance pools, respectively.

Simulated dynamics of the soil water pools were similar for both water distributions. As expected, without hydraulic redistribution, the measured root distribution was more effective than either uniform distribution in obtaining water from the growth pool and more effective in conserving the maintenance pool (Fig. 2a, c). With hydraulic redistribution, however, less water was conserved in the deeper soil pool (Fig. 2b, d) and this was most prevalent in the simulation beginning with uniform water content (Fig. 2b). In this situation, more water was stored in the shallow layers, which in part offset the lower water in the deep pool. In all simulations, the uniform root distribution had less water in the deeper pool and less water overall within the rooting zone at the end of the simulation. The simulation

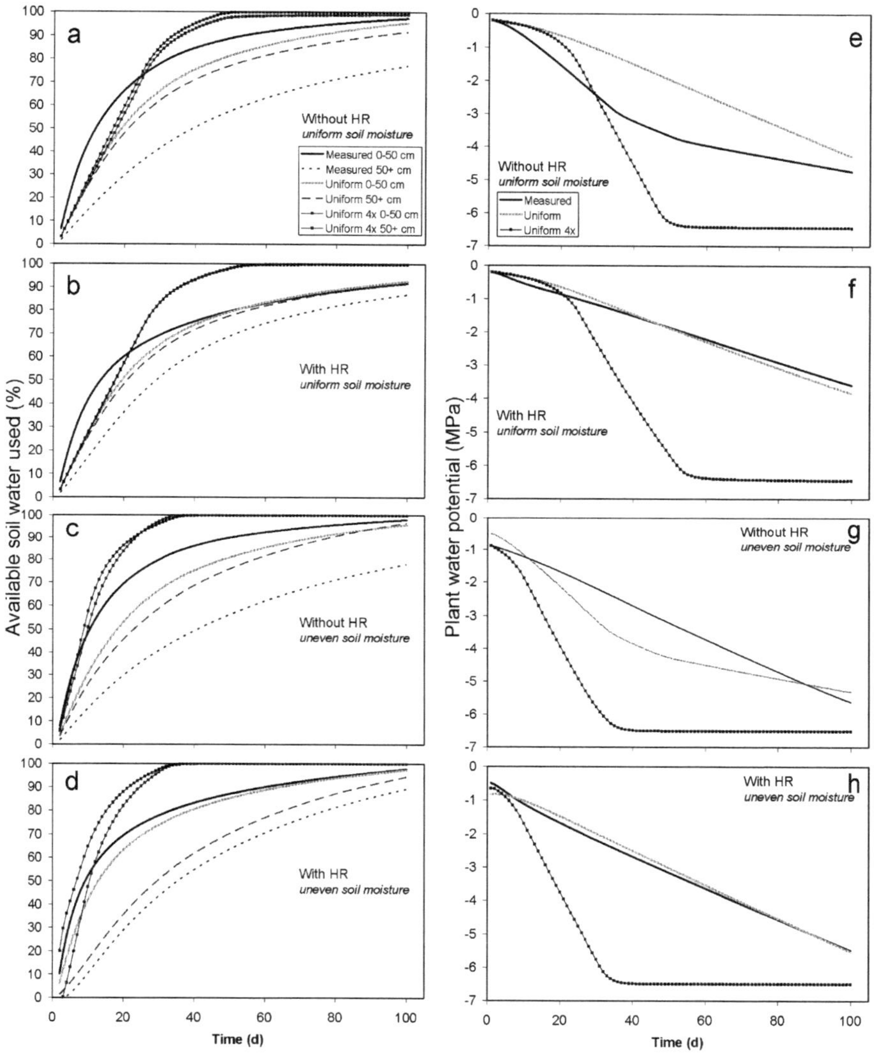

Fig. 2 Simulated soil water pools (*left panels*) and plant predawn water potential (*right panels*) for the evergreen shrub *Artemisia tridentata* with and without hydraulic redistribution operating and for uniform or top-skewed root distributions. Initial soil water distributions were assumed to be uniform at field capacity, and as measured and used in Ryel et al. (2004), where shallow soils (0–50 cm depth) had higher matric potentials than the deeper soils (>50 cm depth); matric potentials were near 0.0 at the surface declining to −2.0 MPa at 160 cm depth. Root distributions were as measured and used in Ryel et al. (2004), uniform with the same quantity of roots, and uniform with 4× the amount of roots. Simulated plant predawn water potentials were weighted (by root distribution) averages of soil matric potentials

with uniformly high root distributions (4×), quickly depleted both soil water pools, but was not more effective in depleting the upper, growth pool. While the presence of hydraulic redistribution reduced the amount of water in the deeper pool, plants with the measured root distribution had improved plant water potential over the situation without hydraulic redistribution (Fig. 2e–h) due to higher soil matric potentials in the upper soil zone. In fact, the predicted plant water potentials were similar to that of the uniform root distributions throughout the simulation period, despite the fact that plants with the measured root distribution were able to use the water in the upper soil layers more rapidly. The results also indicate, however, that high rates of hydraulic redistribution may result in more rapid depletion of maintenance pools and that other mechanisms may need to be employed by plants to help prolong this pool.

5.2 *Root Activity*

Another method for conserving the maintenance pool is to have differential root activity. As discussed above, there are numerous examples of plants using water from different soil layers during different time periods. In addition, Fernandez and Caldwell (1975) found delayed growth of deeper roots in three semi-arid shrubs in the western USA that they suggested may affect water extraction. Ryel et al. (2003, 2007) found delayed uptake of water below 60 cm depth in *A. tridentata* until upper soil layers reached matric potentials near those of the deeper soils. This behavior of roots allowed for minimal use of the maintenance pool and additionally minimizing the movement of water between soil layers via hydraulic redistribution. Since the matric potential of the deeper soils was less than the upper soil layers, limited activity of deeper roots reduced the downward movement of water during the period of maximum nutrient availability.

Potential differential root activity can be seen in the water use patterns for the three perennial species (tussock grass, shrub, tree) in Fig. 1. The lower maintenance water pool was not significantly accessed until at least half of the shallow growth pool was used, and maximum use of the lower pool did not occur until well after the peak of use of the shallow pool. The annual grass, however, used the lower pool at the same time as the upper pool, indicating that this species did not attempt to conserve water at depth. This behavior is expected for a species that senesces once nutrient resources are depleted and growth potential becomes minimal.

Differential root activity is also associated with phreatophytic species which access groundwater resources. These species can change usage from near surface soil water to groundwater stores after the upper soil pools are largely depleted (Smith et al. 1991; Zhang et al. 1999; Rood et al. 2003; Cox et al. 2005). The ability to turn “on and off” roots may allow these species to conserve groundwater stores by minimizing their use during exploitation of near surface water when nutrients are most available, and by reducing hydraulic redistribution that may more rapidly reduce the groundwater stores, by providing water to cooccurring shallow-rooted species.

The mechanism involved in turning roots "on and off" likely involves suberization of roots or seasonally shedding fine roots used in water uptake (or perhaps regulation of aquaporins, see below). Suberization and shedding of roots under soil dry soil conditions and increased conductivy in existing roots and production of new roots with rewetting of the soil has been found in the succulents *Agave deserti* and *Ferocactus acanthodes* (Nobel and Huang 1992). BassiriRad et al. (1999) produced evidence that production of new roots occurred in *Prosopis glandulosa* following a 3 cm rainfall event during drought. Identifying specific triggers and mechanisms involved requires additional research to elicit how plants are able to fine-tune root activity to help in the management of water resources.

5.3 Storage

Effective storage of water is another strategy to maintain physiological activity during drought. Storage can occur within the plant (succulents) and within the soil. Both strategies reduce the potential for other species to use this resource.

5.3.1 Within Plant Storage

Succulent plants store water in a variety of organs: leaves, stems or tuberous roots. Even small precipitation events can be captured by the shallow rooting systems of these plants. During periods of drought, stored water is transferred from water-storage cells to photosynthetic cells, to allow the plant the necessary water for low levels of photosynthesis. With these storage structures, succulent plants have little need to access deeper water, effectively collecting water in shallow soils (Hunt and Nobel 1987; Nobel 1988; Schwinning and Ehleringer 2001), even with competition for water in this shallow soil pool with other species (Nobel 1997).

5.3.2 Hydraulic Redistribution

Water hydraulically redistributed to depth has been shown to be an effective way for plants to store water in the soil (Ryel et al. 2004). By moving accumulated water from shallow soil layers during periods when physiological activity is reduced or vapor pressure is low (e.g., after summer or late fall rain events), plants may reduce the rate at which the redistributed water can be used for transpiration by spreading it out over a greater soil volume. Reduced transpiration rates occur because hydraulic conductivity of water in soil is not linearly related to soil water content. For example, the hydraulic conductivity of soil with half the soil moisture content is less than half the rate of the same soil with twice the water content. By redistributing water to depth, plants can prolong higher transpiration rates during drought periods (Fig. 3).

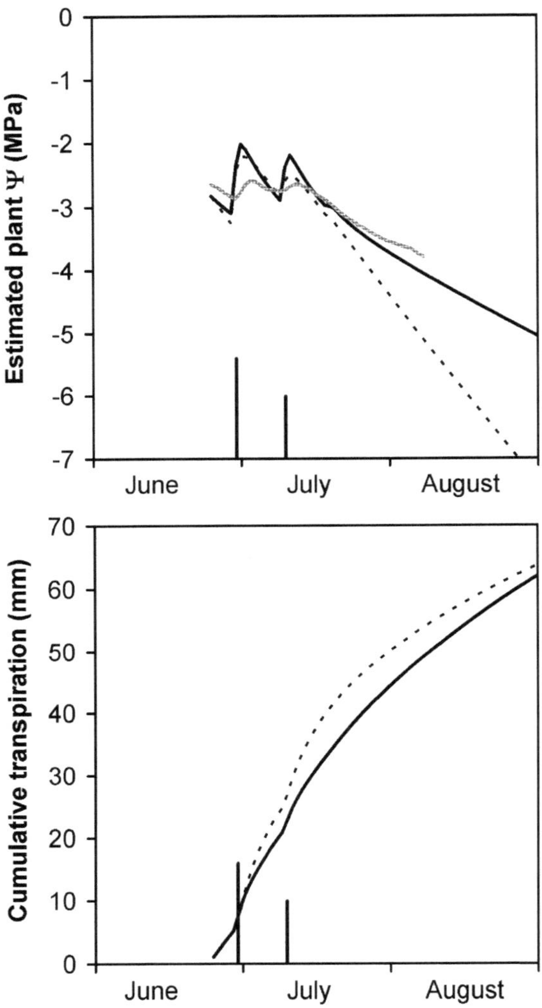

Fig. 3 Simulated and measured (*gray line*) predawn plant water potential (*upper panel*) and transpiration rates (*lower panel*) for a monotypic stand of the evergreen shrub, *Artemisia tridentata*, before and after two rain events during summer 2000 with (*black line*) and without (*dotted line*) hydraulic redistribution by roots. Rainfall events are shown as *vertical bars* (16.0 and 9.9 mm; modified from Ryel et al. 2004)

While nutrient acquisition deep in the soil column has been proposed for rooting depths that are significantly below infiltration depths for precipitation (McCulley et al. 2004), water storage may be a more effective use of these deep roots. Not only is water stored at matric potentials that slow plant water uptake, low matric potentials and deep soil depth make this water unavailable for many cooccurring species. This storage of water is most appropriate for species in systems which predominately have surface recharge of soil moisture, the situation in many arid and semi-arid ecosystems (Seyfried et al. 2005). In addition, the deep roots may increase the potential for deeper penetration of precipitation through preferential flow (Seyfried 1991; Devitt and Smith 2002), a process that would also help deep-rooted species to store water at depth.

5.3.3 Loss of Transpirational Surface

Plants may reduce transpirational demands by shedding leaf material. This strategy is common in grasses and drought-deciduous species where portions (to nearly all) of the leaf material are shed or senesced to reduce potential transpirational losses with the onset of drought in arid and semi-arid environments. Whether this is truly a mechanism to reduce transpiration (and thus store water) or is simply a way to reduce the embolism in specific xylem paths depends on whether the shedding of leaf material actually restricts water use by the plant. To accomplish this, the remaining foliage would have to have near-maximal rates of transpiration, but the plant total transpiration rate would need to be lower than with all foliage intact. In fact the opposite was found, where leaf drop in drought deciduous species was related to drought-induced embolism (Sobrado 1993, 1997; Brodribb and Holbrook 2003).

5.3.4 Chemical Signals and Stomatal Conductance

Considerable research focuses on the role of chemical signals in plants growing under drying soil conditions (Jackson 1997). Much of the focus is on the effects of abscisic acid (ABA), a phytohormone that accumulates in plant tissues when plant cell dehydrate. ABA production is stimulated by roots in contact with dry soil and ABA is transported in the xylem to leaves, where high concentrations are found to reduce stomatal aperture (Davies et al. 1994; Hartung et al. 1999). ABA is also known as a growth inhibitor and is implicated in induction of senescence (Davies 1995; Voesenek and Blom 1996; Jackson 1997).

By reducing stomatal conductance, ABA has the potential to reduce transpiration rates of leaves to levels below that regulated by hydraulic processes. Such a mechanism would be a "feedforward" process whereby the plant would sense drought and conserve water through stomatal control at levels greater than would be necessary to maintain plant hydraulic conductivity. However, the importance of ABA in controlling transpiration is the subject of much debate (Augé et al. 2000; Buckley and Mott 2001; Lovisolo et al. 2002). In arid and semi-arid ecosystems, low soil water content exerts considerable hydraulic limitations on water movement from the soil to roots such that the stomatal behavior is much more likely limited by transpiration rate (Mott and Parkhurst 1991). Davies et al. (1994) indicate that the concentration of xylem ABA cannot explain all of the variability in stomatal behavior, and that the sensitivity of stomata to ABA may also be affected by leaf water potential. The sensitivity may result in different responses to ABA, even within a single day, as leaf water potential changes during the day.

More recently, the importance to water uptake of aquaporins was illuminated. These water channel proteins regulate hydraulic conductance of cell membranes (Morillon and Chrispeels 2001) and mediate mass flow of water in the plant (Luu and Maurel 2005). There is evidence that ABA may regulate genes that encode aquaporins (Mariaux et al. 1998; Ramanjulu and Bartels 2002), suggesting that ABA may play an indirect role in plant hydraulic conductance through regulating production of aquaporins. In addition, evidence exists that the abundance of aquaporins can be regulated by environmental factors (Maurel et al. 2002) such as salt, chilling, and hypoxia, and that aquaporin gene expression may also be in part under circadian regulation (Luu and Maurel 2005), the latter perhaps limiting water movement at night. Assessing the ecological importance of these responses of aquaporins abundance to environmental factors is still in its infancy. While it is unclear how the abundance of aquaporins may allow plants to conserve water during drought period, this is an exciting new area of research.

6 Discussion

As can be seen, the model approaches that define water dynamics as related to vegetation use are varied. Missing among these approaches is a perspective of soil water pools that are defined by temporal and spatial differences in plant use patterns and competitive interactions. These assessments become important as we attempt to assess the effects of climate and vegetation change to soil water and plant community dynamics (Jackson et al. 2000). We have attempted to fill this void by developing a model which elucidates two pools, one related to periods of significant resource acquisition and growth and one related to maintenance. The growth water pool is most subject to rapid exploitation and competition among species, even different life forms. It is coupled to nutrient availability both in time and space where diffusion and mass flow rates are affected by soil water availability. The maintenance pool is primarily used by species which maintain transpiration and physiological activity during periods of prolonged drought. This pool would not be subject to significant competition as it must persist for these plants to survive, and competition would ensure mortality for the competing individuals or species. This pool is largely decoupled from nutrients, both in time and in space, as deeper soil water is often not associated with significant concentrations of nutrients, or occurs at matric potentials too low for significant nutrient diffusion. When only shallow-rooted species are present (annuals, or succulents), the deeper water pools are not used, and water can actually build up in the deeper soil column (Cline et al. 1977; Ryel et al. 2007). The growth and maintenance pools can be separated both in time and space (as in Fig. 1), but can also be separated by only time in the case of drying of shallow soils that are exploited by species that can extract water from soils with low matric potential. In the latter case, deeper soil pools are often used as shallow soils have limited capacity (Fig. 1).

By assessing the functional importance of soil water pools, we are able to more effectively assess how changes in hydrology due to climate change or land use may affect plant communities. Changes in precipitation patterns affect the inputs to one or both soil water pools. For example, if changes in the timing or magnitude precipitation patterns reduce the size of the deeper maintenance pool, these pools may become insufficient for the survival of deep-rooted species as competition for the now limited resource becomes a dominant mechanism.

We can also more effectively assess the importance of competition (or non-competition) between individuals of the same or different species. For example, changes in species or life-form composition may also affect water use patterns in these two pools. While a mixture of perennial species of different life forms (e.g., grasses, forbs, shrubs) may use water in the upper growth soil pool at similar rates to monocultures of these same species, the difference in use of the deeper maintenance pool (e.g., Fig. 1) may make survival more likely for the deep-rooted shrub species. This might occur as the density of shrubs in the mixed community would be less than in a monoculture, and more water would be available per individual shrub during drought periods when herbaceous species have senesced. Thus, a dense monoculture of deep-rooted shrubs that induce significant competition for the maintenance pool would result in high mortality of these shrubs. This may be one explanation for the unusual, but rapid mortality of near-monoculture stands of the evergreen shrub *Artemisia tridentata* in the Great Basin of west-centeral USA that recently occurred (US Department of the Interior, Bureau of Land Management, unpublished data). Historically, these stands were, in many cases, mixtures of shrubs, grasses, and forbs, but land management practices that greatly reduced the cover of grasses and forbs (e.g., heavy grazing of livestock or lack of fire) resulted in the dominance of these sites by dense stands of the evergreen shrub (Huber-Sannwald and Pyke 2005). Vegetation change may also alter the rate of water use of the shallow soil growth pool. Introduced annuals such as *Bromus tectorum* in the Great Basin, USA have higher water use rates than the native vegetation, which may affect the recruitment of native perennials by changing the dynamics of the shallow soil water pool (Harris 1967; Cline et al. 1977; Booth et al. 2003). Both of these examples indicate how vegetation composition change may affect the dynamics by making the resource pools more "volatile" (S. Archer, P.W. Barnes, R.J. Ryel, unpublished conceptual model). That is, increased rates of consumption of a resource pulse results in shorter duration of resource availability. Increased volatility of the resource can have profound effects on physiological activity and survival.

Potential mechanisms that allow plants to more effectively prolong water availability in the maintenance pool were presented. They involve long-term (root distribution, seasonal root activity), mid-term (within season root activity, leaf senescence) and short-term (hydraulic redistribution, chemical signals) methods that may conserve water. It is often difficult to assess whether plants actively or passively employ these methods to conserve water, or whether the methods simply correlate with periods of low water use. Overall root distribution patterns and seasonal differences in root activity are likely the result of long-term natural selection, although root distributions can be altered by environmental conditions (e.g., Nobel 1997; Wan et al. 2002). Top-skewed root distributions were shown above to passively result in slower rates of water use in deeper soil layers where root density was low, even with hydraulic redistribution increasing the depletion of these zones. Hydraulic constraint on conductivity resulting from water hydraulically redistributed by roots to a greater soil volume was also shown to be a passive mechanism for water conservation. In addition, seasonal differences in root activity can minimize use of deep-water pools until shallow soil water resources become limited (e.g., Fig. 1). All of these mechanisms may be considered to be a feedforward response, but at the timescale of evolution. Thus, the plant conserves water because these mechanisms were beneficial under selective pressures. Feedforward responses would be expected when plants are not competing for a resource, as would be the case with the maintenance pool. In contrast, within-season root activity, leaf senescence, and chemical signaling are primarily feedback mechanisms in arid and semi-arid plants, as they are simply responses to environmental conditions of reduced water availability (Schulze 1986). In competitive situations, short-term feedforward mechanisms would reduce the ability of the plant to obtain a limiting resource, benefiting competitors (Cohen 1970). However, some of these mechanisms could become feedforward responses when plants are exploiting maintenance water pools, as competition for these water resources would be minimal. Documenting the existence of short-term feedforward mechanisms under conditions of low competition for water in arid and semi-arid ecosystems is a subject for future study.

While we do not propose a conceptual model of soil water use by plants that explains all observed behavior of plants in arid and semi-arid ecosystems, we do provide a model that illuminates mechanisms that help explain soil water dynamics and vegetation response that cannot easily be explained by other models. Part of the reason we feel that consideration of functionally different water pools has not been pursued is that documenting plant water use of soils is difficult. Most studies that quantified differential water use monitored soil moisture sensors buried at depth. These sensors are often difficult to install and can disrupt the soil column and roots during installation. Nondestructive methods that employ stable isotopes of water and oxygen (Ogle et al. 2004; Ogle and Barber 2007) are a promising approach to reconstructing plant root area and water uptake profiles.

7 Concluding Remarks

More fully understanding below-ground processes, both physical and biological, are challenges for ecologists. Yet, such knowledge is essential for predicting and evaluating potential effects of anthropogenic intervention into and effects upon ecosystems. The conceptual model presented here is one more tool for providing a perspective on the observed behaviors of plant communities in arid and semi-arid ecosystems that will help us evaluate and understand ecosystem dynamics and function in a changing world.

Acknowledgements This work was funded by the National Science Foundation (DEB-9807097) and the Utah Agricultural Experiment Station.

References

Andraski BJ (1997) Soil-water movement under natural-site and waste-site conditions: a multiple-year field study in the Mojave Desert, Nevada. Water Resour Res 33: 1901–1916

Archer S (1995) Tree-grass dynamics in a *Prosopis*-thornscrub savanna parkland: reconstructing the past and predicting the future. Ecoscience 2:83–99

Archer S, Boutton TW, Hibbard KA (2001) Trees in grasslands: biogeochemical consequences of woody plant expansion. In: Schulze E-D, Heimann M, Harrison S, Holland E, Lloyd J, Prentice I, Schimel D (eds). Global biogeochemical cycles in the climate system. Academic, San Diego. pp 115–138

Augé RM, Green CD, Stodola JW, Saxton AM, Olinick JB, Evans RM (2000) Correlations of stomatal conductance with hydraulic and chemical factors in several deciduous tree species in a natural habitat. New Phytol 145:483–500

Bailey RG (1996) Ecosystem geography. Springer, Berlin Heidelberg New York

BassiriRad H, Tremmel DC, Virginia RA, Reynolds JF, Soyza AG de, Brunell MH (1999) Short-term patterns in water and nitrogen acquisition by two desert shrubs following a simulated summer rain. Plant Ecol 145:27–36

Belsky AJ (1994) Influences of trees on savanna productivity: tests of shade, nutrients and tree–grass competition. Ecology 75:922–932

Bilbrough CJ, Caldwell MM (1997) Exploitation of springtime ephemeral N pulses by six Great Basin plant species. Ecology 78:231–243

Booth MS, Caldwell MM, Stark JM (2003) Overlapping resource use in three Great Basin species: implications for community invasibility and vegetation dynamics. J Ecol 91:36–48

Breshears DD, Barnes FJ (1999) Interrelationships between plant functional types and soil moisture heterogeneity for semiarid landscapes within the grassland/forest continuum: a unified conceptual model. Landsc Ecol 14:465–478

Breshears DD, Meyers OB, Johnson SR, Meyer CW, Martens SN (1997) Differential water use of heterogeneous soil moisture by two semiarid woody species: *Pinus edulis* and *Juniperus monosperma*. J Ecol 85:289–299

Briones O, Montana C, Ezcurra E (1996) Competition between three Chihuahuan desert species: evidence from plant size-distance relations and root distribution. J Veg Sci 7:453–460

Brodribb TJ, Holbrook NM (2003) Changes in leaf hydraulic conductance during leaf shedding in seasonally dry tropical forest. New Phytol 158:295–303

Buckley TN, Mott KA (2001) Stomatal water relations and the control of hydraulic supply and demand. Prog Bot 63:309–325

Burgess SSO (2006) Measuring transpiration responses to summer precipitation in a Mediterranean climate: a simple screening tool for identifying plant water-use strategies. Physiol Plant 127:404–412

Burgess SSO, Adams MA, Turner NC, Ong CK (1998) The redistribution of soil water by tree root systems. Oecologia 115:306–311

Busch DE, Ingraham NL, Smith SD (1992) Water uptake in woody riparian phreatophytes of the southwestern United States: a stable isotope study. Ecol Appl 2:450–459

Caldwell MM, Richards JH (1989) Hydraulic lift: water efflux from upper roots improves effectiveness of water uptake by deep roots. Oecologia 79:1–5

Caldwell MM, Dawson TE, Richards JH (1998) Hydraulic lift: consequences of water efflux from the roots of plants. Oecologia 113:151–161

Canadell J, Jackson RB, Ehleringer JR, Mooney HA, Sala OE, Schulze E-D (1996) Maximum rooting depth of vegetation types at the global scale. Oecologia 108:583–595

Casper BB, Jackson RB (1997) Plant competition underground. Annu Rev Ecol Syst 28: 545–570

Chesson P, Gebauer RLE, Schwinning S, Huntly N, Wiegand K, Ernest MSK, Sher A, Novoplansky A, Weltzin JF (2004) Resource pulses, species interactions, and diversity maintenance in arid and semi-arid environments. Oecologia 141:236–253

Chong DLS, Mougin E, Gastellu-Etchegorry JP (1993) Relating the global vegetation index to net primary productivity and actual evapotranspiration over Africa. Int J Remote Sensing 14:1517–1546

Cline JF, Uresk DW, Rickard WH (1977) Comparison of soil water used by a sagebrush–bunchgrass and a cheatgrass community. J Range Manage 30:199–201

Cohen D (1970) The expected efficiency of water utilization in plants under different competition and selection regimes. Isr J Bot 19:50–54

Cox G, Fischer D, Hart SC, Whitham TG (2005) Nonresponse of native cottonwood trees to water additions during summer drought. West N Am Nat 65:175–185

Davies WJ (ed) (1995) Plant hormones and their role in plant growth and development. Kluwer, Dordrecht

Davies WJ, Tardieu F, Trejo CL (1994) How do chemical signals work in plants that grow in drying soil? Plant Physiol 104:309–314

Dawson TE (1993) Hydraulic lift and water use by plants: implications for water balance, performance and plant-plant interactions. Oecologia 95:565–574

Devitt DA, Smith SD (2002) Root channel macropores enhance downward movement of water in a Mojave Desert ecosystem. J Arid Environ 50:99–108

Dobrowolski JP, Caldwell MM, Richards JH (1990) Basin hydrology and plant root systems. In: Osmond CB, Pitelka LF, Hidy GM (eds) Plant biology of the basin and range. Springer, Berlin Heidelberg New York, pp 243–292

Dodd MB, Lauenroth WK, Welker JM (1998) Differential water resource use by herbaceous and woody plant life-forms in a shortgrass steppe community. Oecologia 117:504–512

Donovan LA, Ehleringer JR (1994) Waterstress and use of summer precipitation in a Great Basin shrub community. Funct Ecol 8:289–297

Ehleringer JR, Phillips SL, Schuster WSF, Sandquist DR (1991) Differential utilization of summer rains by desert plants. Oecologia 88:430–434

Embaye K, Weih M, Ledin S, Christersson L (2005) Biomass and nutrient distribution in a highland bamboo forest in southwest Ethiopia: implications for management. For Ecol Manage 204:159–169

Evans RD, Ehleringer JR (1994) Water and nitrogen dynamics in an arid woodland. Oecologia 99:233–242

Fargione J, Tilman D (2005) Niche differences in phenology and rooting depth promote coexistence with a dominant C_4 bunchgrass. Oecologia 143:598–606

Fernandez OA, Caldwell MM (1975) Phenology and dynamics of root growth of three cool semi-desert shrubs under field conditions. J Ecol 63:703–714

Field CB, Behrenfeld MJ, Randerson JT, Falkowski P (1998) Primary production of the biosphere: integrating terrestrial and oceanic components. Science 281:237–240

Flanagan LB, Ehleringer JR, Marshall JD (1992) Differential uptake of summer precipitation among co-occurring trees and shrubs in a pinyon-juniper woodland. Plant Cell Environ 15:831–836

Goldberg D, Novoplansky A (1997) On the relative importance of competition in unproductive environments. J Ecol 85:409–418

Gordon DR, Welker JM, Menke JW, Price KJ (1989) Competition for soil water between annual plants and blue oak (*Quercus douglasii*) seedlings. Oecologia 79:533–541

Harris G (1967) Some competitive relationships between *Agropyron spicatum* and *Bromus tectorum*. Ecol Monogr 37:89–111

Hartung W, Peuke A, Davies WJ (1999) Abscisic acid – a hormonal long-distance stress signal in plants under drought and salt stress. In: Pessarakali M (ed) Handbook of crop stress. Dekker, New York, pp 737–747

Hillel D, Tadmor NH (1962) Water regime and vegetation in the Central Negev highlands of Israel. Ecology 43:33–41

Holbrook NM, Whitbeck JL, Mooney HA (1995) Drought responses of neotropical dry forest trees. In: Bullock SH, Mooney HA, Medina E (eds) Seasonally dry tropical forests. Cambridge University Press, Cambridge

Hooker TD, Norton U, Ivans C, Leffler AJ, Peek M, Ryel RJ, Stark JM (2007) Distribution of ecosystem C and N within contrasting vegetation types in a semiarid rangeland in the Great Basin, USA. Ecol Appl (in review)

Horton JL, Hart SC, Kolb TE (2003) Physiological condition and water source use of Sonoran Desert riparian trees at the Bill Williams River, Arizona, USA. Isot Environ Health S 39:69–82

Hultine KR, Cable WL, Burgess SSO, Williams DG (2003) Hydraulic redistribution by deep roots of a Chihuahuan desert phreatophyte. Tree Physiol 23:353–360

Hunt ER Jr, Nobel PS (1987) Non-steady-state water flow for three desert perennials with different capacitances. Aust J Plant Physiol 14:363–375

Huber-Sannwald E, Pyke DA (2005) Establishing native grasses in a big sagebrush-dominated site: an intermediate restoration step. Restor Ecol 13:292–301

Ivans CY, Leffler AJ, Spaulding U, Stark JM, Ryel RJ, Caldwell MM (2003) Root responses and nitrogen acquisition by *Artemisia tridentata* and *Agropyron desertorum* following small summer rainfall events. Oecologia 134:317–324

Jackson M (1997) Hormones from roots as signals for the shoots of stressed plants. Trends Plant Sci 2:22–28

Jackson RB, Caldwell MM (1996) Integrating resource heterogeneity and plant plasticity: modelling nitrate and phosphate uptake in a patchy soil environment. J Ecol 84:891–903

Jackson RB, Canadell J, Ehleringer JR, Mooney HA, Sala OE, Schulze E-D (1996) A global analysis of root distributions for terrestrial biomes. Oecologia 108:389–411

Jackson RB, Sperry JS, Dawson TE (2000). Root water uptake and transport: using physiological processes in global predictions. Trends Plant Sci 5:482–488

Jobbágy EG, Jackson RB (2001) The distribution of soil nutrients with depth: global patterns and the imprint of plants. Biogeochemestry 53:51–77

Jury WA (2004) Soil physics. Wiley, Hoboken, 384 pp

Kemp PR, Reynolds JF, Pachepsky Y, Chen J-L (1997) A comparative modeling study of soil water dynamics in a desert ecosystem. Water Resour Res 33:73–90

Kirkham MB (2005) Principles of soil and plant water relations. Elsevier, Amsterdam, 500 pp

Knoop WT, Walker BH (1985) Interactions of woody and herbaceous vegetation in a southern African Savanna. J Ecol 73:235–253

Lambers H, Chapin III F, Pons TL (1998) Plant physiological ecology. Springer, Berlin Heidelberg New York, 540 pp

Lauenroth WK, Urban DL, Coffin DP, Parton W, Shugart HH, Kirchner TB, Smith TM (1993) Modeling vegetation structure–ecosystem process interactions across sites and ecosystems. Ecol Model 67:49–80

Le Roux X, Bariac T (1998) Seasonal variations in soil, grass and shrub water status in a West African humid savanna. Oecologia 113:456–466

Le Roux X, Bariac T, Mariotti A (1995) Spatial partitioning of the soil water resource between grass and shrub components in a West African humid savanna. Oecologia 104:147–155

Leffler AJ, Ryel RJ, Hipps L, Ivans S, Caldwell MM (2002) Carbon acquisition and water use in northern Utah *Juniperus osteosperma* (Utah juniper) population. Tree Physiol 22:1221–1230

Liang YM, Hazlett DL, Lauenroth WK, (1989) Biomass dynamics and water use efficiencies of five plant communities in the shortgrass steppe. Oecologia 80:148–153

Lovisolo C, Hartung W, Schubert A (2002) Whole-plant hydraulic conductance and root-to-shoot flow of abscisic acid are independently affected by water stress in grapevines. Funct Plant Biol 29:1349–1356

Luu DT, Maurel C (2005) Aquaporins in a challenging environment: molecular gears for adjusting plant water status. Plant Cell Environ 28:85–96

Mariaux JB, Bockel C, Salamini F, Bartels D (1998) Desiccation- and abscisic acid-responsive genes encoding major intrinsic proteins (MIPS) from the resurrection plant *Craterostigma plantagineum*. Plant Mol Biol 38:1089–1099
Maurel C, Javot H, Lauvergeat V, Gerbeau P, Tournaire C, Santoni V, Heyes J (2002) Molecular physiology of aquaporins in plants. Int Rev Cytol 215:105–148
McCulley RL, Jobbágy EG, Pockman WT, Jackson RB (2004) Nutrient uptake as a contributing explanation for deep rooting in arid and semi-arid ecosystems. Oecologia 4:620–628
Meinzer FC, Brooks JR, Bucci S, Goldstein G, Scholz FG, Warren JM (2004) Converging patterns of uptake and hydraulic redistribution of soil water in contrasting woody vegetation types. Tree Physiol 8:919–928
Morillon R, Chrispeels MJ (2001) The role of ABA and the transpiration stream in the regulation of the osmotic water permeability of leaf cells. Proc Natl Acad Sci USA 98:14138–14143
Mott KA, Parkhurst DF (1991) Stomatal responses to humidity in air and helox. Plant Cell Environ 14:509–515
Nobel PS (1988) Environmental biology of agaves and cacti. Cambridge University Press, New York
Nobel PS (1997) Root distribution and seasonal production in the northwestern Sonoran Desert for a C-3 subshrub, a C-4 bunchgrass, and a CAM leaf succulent. Am J Bot 84:949–955
Nobel PS, Huang BR (1992). Hydraulic and structural-changes for lateral roots of 2 desert succulents in response to soil drying and rewetting. Int J Plant Sci 153:S163–S170
Nonami H, Wu YJ, Boyer JS (1997) Decreased growth-induced water potential – primary cause of growth inhibition at low water potentials. Plant Physiol 2:501–509
Noy-Meir I (1973) Desert ecosystems: environment and producers. Annu Rev Ecol Syst 4:25–41
Nye PH, Tinker PB (1977) Solute movement in the soil-root system. University of California Press, Berkeley
Ogle K, Barber JJ (2007) Bayesian data–model integration in plant physiological and ecosystem ecology. In: Lüttge U, Beyschlag W (eds) Progress in botany, vol 69. Springer, Berlin Heidelberg New York (in press)
Ogle K, Reynolds JF (2004) Plant responses to precipitation in desert ecosystems: integrating functional types, pulses, thresholds and delays. Oecologia 141:282–294
Ogle K, Wolpert RL, Reynolds JF (2004) Reconstructing plant root area and water uptake profiles. Ecology 85:1967–1978
Paruelo JM, Lauenroth WL (1995) Regional patterns of normalized difference vegetation index in North American shrublands and grasslands. Ecology 76:1888–1898
Querejeta JI, Egerton-Warburton LM, Allen MF (2007) Hydraulic lift may buffer rhizosphere hyphae against the negative effects of severe soil drying in a California Oak savanna. Soil Biol Biochem (in press)
Ramanjulu S, Bartels D (2002) Drought- and desiccation-induced modulation of gene expression in plants. Plant Cell Environ 25:141–151
Reynolds JF, Virginia RA, Schlesinger WH (1997) Defining functional types for models of desertification. In: Smith TM, Shugart HH, Woodward FI (eds) Plant functional types: their relevance to ecosystem properties and global changes. Cambridge University Press, New York, pp 195–216
Reynolds JF, Virginia RA, Kemp PR, Soyza AG de, Tremmel DC (1999) Impacts of drought on desert shrubs: effects of seasonality and degree of resource island development. Ecol Monogr 69:69–106
Reynolds JF, Kemp PR, Tenhunen JD (2000) Effects of long-term rainfall variability on evapotranspiration and soil water distribution in the Chihuahuan Desert: a modeling analysis. Plant Ecol 150:145–159
Reynolds JF, Kemp PR, Ogle K, Fernandez RJ (2004) Modifying the 'pulse-reserve' paradigm for deserts of North America: precipitation pulses, soil water, and plant responses. Oecologia 141:194–210
Richards JH, Caldwell MM (1987) Hydraulic lift: substantial nocturnal water transport between soil layers by *Artemisia tridentata* roots. Oecologia 73:486–489

Rood SB, Braatne JH, Hughes FMR (2003) Ecophysiology of riparian cottonwoods: stream flow dependency, water relations and restoration. Tree Physiol 23:1113–1124
Rosenzweig ML (1968) Net primary productivity of terrestrial communities: prediction from climatological data. Am Nat 102:67–74
Rundel PW, Nobel PS (1991) Rundel structure and function in desert root systems. In: Atkinson D (ed) Plant root growth: an ecological perspective. (Br Ecol Soc Spec Publ 10) Blackwell, Oxford, pp 349–377
Ryel RJ (2004) Hydraulic redistribution. Prog Bot 65:413–435
Ryel RJ, Caldwell MM, Yoder CK, Or D, Leffler AJ (2002) Hydraulic redistribution in a stand of *Artemisia tridentata*: evaluation of benefits to transpiration assessed with a simulation model. Oecologia 130:173–184
Ryel RJ, Caldwell MM, Leffler AJ, Yoder CK (2003) Rapid soil moisture recharge to depth by roots in a stand of *Artemisia tridentata*. Ecology 84:757–764
Ryel RJ, Leffler AJ, Peek MS, Ivans CY, Caldwell MM (2004) Water conservation in *Artemisia tridentata* through redistribution of precipitation. Oecologia 141:335–345
Ryel RJ, Ivans CY, Peek MS, Leffler AJ (2007) Water use patterns of four Great Basin species. J Arid Environ (in review)
Sala OE, Golluscio RA, Lauenroth WK, Soriano A (1989) Resource partitioning between shrubs and grasses in the Patagonian steppe. Oecologia 81:501–505
Scanlon BR (1991) Evaluation of moisture flux from chloride data in desert soils. J Hydrol 128:137–156
Scanlon BR, Langford RP, Goldsmith RS (1999) Relationship between geomorphic settings and unsaturated flow in an arid setting. Water Resour Res 35:983–999
Schenk HJ, Jackson RB (2002) The global biogeography of roots. Ecol Monogr 72: 311–328
Schlesinger WH, Fonteyn PJ, Marion GM (1987) Soil-moisture content and plant transpiration in the Chihuahuan Desert of New Mexico. J Arid Environ 12:119–126
Schlesinger WH, Reynolds JF, Cunningham GL, Huenneke LF, Jarrell WM, Virginia RA, Whitford WG (1990) Biological feedbacks in global desertification. Science 247:1043–1048
Scholes RJ, Archer SR (1997) Tree-grass interactions in savannas. Annu Rev Ecol Syst 28:517–544
Scholz FS, Bucci SJ, Goldstein G, Meinzer FC, Franco AC (2002) Hydraulic redistribution of soil water by neotropical savannah trees. Tree Physiol 22:603–612
Schulze E-D (1986) Carbon-dioxide and water-vapor exchange in response to drought in the atmosphere and in the soil. Annu Rev Plant Phys 37:247–274
Schulze E-D, Caldwell MM, Canadell J, Mooney HA, Jackson RB, Parson D, Scholes R, Sala OE, Trimborn P (1998) Downward flux of water through roots (i.e. inverse hydraulic lift) in dry Kalahari sands. Oecologia 115:460–462
Schwinning S, Ehleringer JR (2001) Water use trade-offs and optimal adaptations to pulse-driven arid ecosystems. J Ecol 89:464–480
Schwinning S, Sala OE (2004) Hierarchy of responses to resource pulses in and and semi-arid ecosystems. Oecologia 141:211–220
Schwinning S, Davis K, Richardson L, Ehleringer JR (2002) Deuterium enriched irrigation indicates different forms of rain use in shrub/grass species of the Colorado Plateau. Oecologia 130: 345–355
Seyfried MS (1991) Infiltration patterns from simulated rainfall on a semiarid rangeland soil. Soil Sci Soc Am J 55:1726–1734
Seyfried MS, Schwinning S, Walvoord MA, Pockman WT, Newman BD, Jackson RB, Phillips EM (2005) Ecohydrological control of deep drainage in arid and semiarid regions. Ecology 86:277–287
Smith DM, Jackson NA, Roberts JM, Ong CK (1999) Reverse flow of sap in tree roots and downward siphoning of water by *Grevillea robusta*. Functional Ecol 13:256–264
Smith SD, Nobel PS (1986) Deserts. In: Baker NR, Long SP (eds) Photosynthesis in contrasting environments. Elsevier, Amsterdam, pp 13–62

Smith SD, Wellington AB, Nachlinger JL, Fox CA (1991) Functional response of riparian vegetation to streamflow diversion in the eastern Sierra Nevada. Ecol Appl 1:89–97

Smith SD, Monson RK, Anderson JE (1997) Physiological ecology of North American desert plants. Springer, Berlin Heidelberg New York

Sobrado MA (1993) Trade-off between water transport efficiency and leaf life span in a tropical dry forest. Oecologia 96:19–23

Sobrado MA (1997) Embolism vulnerability in drought-deciduous and evergreen species of a tropical dry forest. Acta Oecol 18:383–391

Soriano A, Sala OE (1983) Ecological strategies in a Patagonian arid steppe. Vegetatio 56:9–15

Sperry JS (1995) Limitations on stem water transport and their consequences. In: Gartner BL (ed) Plant stems. Physiology and functional morphology. Academic, San Diego, pp 105–124

Swamy SL, Mishra A, Puri S (2006) Comparison of growth, biomass and nutrient distribution in five promising clones of *Populus deltoides* under an agrisilviculture system. Bioresour Technol 97:57–68

Van Auken OW (2000) Shrub invasions of North American semiarid grasslands. Annu Rev Ecol Syst 31:197–215

Voesenek LACJ, Blom CWPM (1996) Plants and hormones: an ecophysiological view on timing and plasticity. J Ecol 84:111–119

Walker BH, Noy-Meir I (1982) Aspects of the stability and resilience of savanna ecosystems. In Huntley BJ, Walker BH (eds) Ecology of tropical savannas. Springer, Berlin, Heidelberg, New York, pp 556–590

Walter H (1971) Natural savannahs as a transition to the arid zone. In: Oliver and Boyd (ed) Ecology of tropical and subtropical vegetation. Oliver and Boyd, Edinburgh, pp 238–265

Walter H, Stadelmann E (1974) A new approach to the water relations of desert plants. In: Brown R (ed) Desert biology vol 2. Academic, New York, pp 213–310

Walvoord MA, Stonestrom DA, Andraski BJ, Striegl RG (2004) Constraining the inferred paleo-hydrologic evolution of a deep unsaturated zone in the Amargosa Desert. Vadose Zone J 3:502–512

Wan CG, Yilmaz I, Sosebee RE (2002). Seasonal soil-water availability influences snakeweed, root dynamics. J Arid Environ 51:255–264

Warren JM, Meinzer FC, Brooks JR, Domec JC (2005) Vertical stratification of soil water storage and release dynamics in Pacific Northwest coniferous forests. Agr For Meteorol 130:39–58

Warrick AW (2006) Soil water dynamics. Oxford University Press, Oxford, 416 pp

Weaver JE, Albertson FW (1943) Resurvey of grasses, forbs, and underground plant parts at the end of the great drought. Ecol Monogr 13:63–117

Weltzin JF, McPherson GR (1997) Spatial and temporal soil moisture resource partitioning by trees and grasses in a temperate savannah, Arizona, USA. Oecologia 112:156–164

Williams K, Caldwell MM, Richards JH (1993) The influence of shade and clouds on soil water potential: the buffered behavior of hydraulic lift. Plant Soil 157:83–95

Yoder CK, Nowak RS (1999) Hydraulic lift among native plant species in the Mojave Desert. Plant Soil 215:93–102

Yoder CK, Boutton TW, Thurow TL, Midwood AJ (1998) Differences in soil water use by annual broomweed and grasses. J Range Manage 51:200–206

Zhang H, Morison JIL, Simmonds LP (1999) Transpiration and water relations of poplar trees growing close to the water table. Tree Physiol 19:563–573

Zou CB, Barnes PW, Archer S, McMurtry CR (2005) Soil moisture redistribution as a mechanism of facilitation in Savanna tree–shrub clusters. Oecologia 145:32–40

Plant Herbivore Interactions at the Forest Edge

Rainer Wirth (✉), Sebastian T. Meyer, Inara R. Leal, and Marcelo Tabarelli

Contents

Abstract An ever-increasing proportion of the global forested landscape is in close proximity to edges and edge effects have been shown to represent key forces affecting both organisms and ecological processes. Despite increasing recognition of edge effects on species interactions, a systematic review devoted to plant-herbivore interactions along forest edges has not yet been performed. Here we synthesize published research attempting to detect patterns of herbivore densities and herbivory at forest edges, identify the underlying mechanisms generating these patterns, and explore their potential impacts for the forest edge as an ecosystem. Key conclusions are that herbivores, especially generalists, profoundly benefit from forest edges, often due to favourable microenvironmental conditions, an edge-induced increase in food quantity/quality, and (less well documented) disrupted top-down regulation of herbivores. Finally, we present evidence and causal explanations that edge-associated herbivores, via a range of direct and indirect impacts, may alter species interactions, delay successional processes at the edge, and amplify the often human-induced changes on forest biota.

Rainer Wirth
Plant Ecology & Systematics, University of Kaiserslautern, PO Box 3049, D-67663 Kaiserslautern, Germany
e-mail: wirth@rhrk.uni-kl.de

U. Lüttge et al. (eds.), *Progress in Botany 69*.

1 Introduction

An edge can be defined as a zone of transition between different types of habitats (Ries et al. 2004). Historically, most research has addressed changes in biotic variables, vegetational transition, and patterns of species diversity in natural edge zones (Fagan et al. 1999). One of the most important patterns documented by this approach is the tendency of increased species diversity in natural edges and ecotones (the "edge effect principle", sunsu Odum 1971). This is now a fundamental concept of ecology and it has been largely applied by wildlife biologists over the past half a century, particularly as a guideline in terms of habitat management in temperate ecosystems (Harris 1988).

During the past two decades, escalating levels of habitat loss and fragmentation, particularly in the tropical rain forests, have greatly revived the interest in the topic of edges (for a review, see Ries et al. 2004). Researchers are now concerned with the fact that the creation of forest edges by human activities (artificial or anthropogenic edges) can act as a huge catalytic source of ecosystem degradation and biodiversity loss in fragmented forests (Fagan et al. 1999; Tabarelli and Gascon 2005). The diverse physical and biotic changes resulting from the occurrence of abrupt and artificial margins are referred to as edge effects (Bierregaard et al. 2001), and they greatly result from differences in the structural complexity and biomass between the juxtaposed ecosystems, the "patch contrast", sensu Harper et al. (2005).

Habitat fragmentation, and the consequent creation of artificial forest edges, is one of the most pervasive and disturbing results of present-day human land use dynamics (Whitmore 1997; Aguilar et al. 2006). Most forest biota tend to be reduced to archipelagos of small fragments or edge-affected habitats (increased edge-to-interior ratios), which remain immersed within environmentally hostile matrices, such as pastures, croplands, and urban areas (Tabarelli et al. 2004). Although the whole effect produced by habitat fragmentation results from a myriad of processes (see Fahrig 2003), much of the ecological alterations faced by fragmented forests (e.g. disruption of species interactions, species loss, biomass collapse, exotic species invasion) can be assigned to edge-driven changes or edge effects (D'Angelo et al. 2004; Tabarelli and Gascon 2005). Consequently, patterns of biological organization from organism to ecosystem level may be completely altered or disrupted in those portions of the remnants under the influence of edge effects (see Foggo et al. 2001) – hereafter forest edge or edge habitat in contrast to forest interior or core area habitat (i.e. those portions showing no detectable edge influence).

Interactions between plants and herbivores are considered to be one of the key processes shaping life history traits and ecosystem function. By consuming plant tissues, herbivores are able to reduce plant survival and fitness (Crawley 1983; Marquis 1984) and limit plant population size (i.e. consumer regulation). In a comprehensive meta-analysis Bigger and Marvier (1998) concluded that herbivores have a strong negative impact on the biomass of the plants they feed on, and their impact might be as high as, or higher than, competition within the plant community. As a consequence, herbivores may greatly affect the organization of plant communities and their patterns of species richness (Hulme 1996). Differential herbivory among species is known to influence the pace and trajectory of plant succession, via suppression of early or late successional species (Brown 1990; Zamora et al. 1999). Finally, as ecosystems are structured by the flow of energy from plants through consumers, predators, and decomposers, herbivores can drastically affect ecosystem level properties such as primary productivity, resilience, species richness, diversity of food webs, and patterns of nutrient cycling (Crawley 1983; Huntly 1991).

Recent reviews (e.g. Yahner 1988; Saunders et al. 1991; Murcia 1995; Foggo et al. 2001; Laurance et al. 2002; Hobbs and Yates 2003; Harper et al. 2005) revealed an extensive suite of edge effects on forest biota with clear detrimental consequences for ecosystem functioning, particularly the long-term retention of biodiversity. Given the increasing recognition of the global importance of edges as a dominating habitat (Yahner 1988; Ries et al. 2004), it is remarkable that minor attention has been devoted to plant-herbivore interactions in fragmentation-related experiments (for reviews, see Debinski and Holt 2000; Hobbs and Yates 2003). To exemplify, one of the most influential reviews on edge effects by Murcia (1995) cited only one article to document herbivory as one of six categories of species interactions disrupted at edges. Since then, there has been an increasing sense of urgency and awareness that edge-induced disruptions of species interactions is a fundamental issue for a more complete and profound perspective on the impacts of habitat fragmentation and the ecology of fragmented landscapes (Fagan et al. 1999; Hunter 2002).

Here we synthesize key findings regarding edge-related changes in patterns of herbivore occurrence and herbivory at the edges of forest ecosystems (excluding seed predation and xylophagia). The present review was conducted to:

1. Detect whether herbivores or herbivory are negatively, positively or neutrally affected by edges;
2. Identify the mechanisms that are proposed to explain these patterns;
3. Explore how plant-herbivore interactions may impact forest edge ecosystems, particularly abiotic conditions, the plant community, and higher trophic levels.

2 What Edges Induce in Forest Remnants

Collectively, cascades of abiotic, direct biotic, and indirect biotic edge effects (sensu Murcia 1995) are expected to dramatically alter the nature of biological organization along forest edges and eventually drive the entire remnant to collapse (Gascon et al. 2000). With potential impacts on herbivore populations (see Section 3.2), three groups of edge-related effects must be highlighted: (a) physical and structural changes of the forest edge, (b) changes in resource availability and quality including the hyper proliferation of pioneer plants and environmentally induced changes of foliage quality, and (c) changes in animal responses to forest edges, including the assemblage of natural enemies.

Compared with the forest interior, forest edges are more illuminated, desiccated, and wind-exposed habitats (Laurance 2001). This promotes elimination of emergent trees, an increased number of treefalls, and a collapse of aboveground biomass (Laurance et al. 1998, 2000), thereby leading to lower canopy cover and a reduced vertical stratification close to forest edges (Chen et al. 1992; Harper et al. 2005). It is likely that such features turn forest edges much more susceptible to desiccation and environmental extremes through greater variability in microclimatic parameters (Kapos 1989; Saunders et al. 1991). Increased light availability fuels the proliferation of pioneer or successional plants (trees, lianas, shrubs), while populations of some shade-tolerant trees are expected to face strong declines on forest edges (Tabarelli et al. 2004; Martinez-Garza et al. 2005). Over time, pioneers can account for over 80% of species and individuals of tree assemblages in this habitat (Oliveira et al. 2004; Laurance et al. 2006). In addition to this biased ratio of pioneers to shade-tolerant plants, foliage nutritional value may be enhanced at forest edges (e.g. Fortin and Mauffette 2001).

Like plants, many groups of animals are reported to present increased or decreased abundance in forest edges and fragmented landscapes, including those that prey on herbivores. In both tropical and temperate forests, large-area-demanding species and forest habitat specialists (particularly vertebrates) are expected to be extirpated from severely fragmented landscapes, including forest edges (Terborgh 1992). Decreased densities in forest edges are documented, e.g. for some insectivorous birds (edge-avoiders), ants (Carvalho and Vasconcelos 1999), and parasitic insects (Laurance et al. 2002; Tscharntke et al. 2002). In contrast, canopy-foraging insectivorous birds (Strelke and Dickson 1980; Lovejoy et al. 1986), small folivorous/omnivorous mammals, and other generalist vertebrate predators (especially matrix users) appear to be favoured and reach increased abundances at forest edges or concentrate their foraging activities in this habitat (Laurance 1994; Malcolm 1997; Laurance et al. 2002).

Finally, the magnitude and intensity of edge effects (sensu Harper et al. 2005) are greatly modulated by edge evolution, orientation, and structure, as well as by the patch contrast (Chen et al. 1995; Murcia 1995; Didham and Lawton 1999). Most edge effects described in the literature are reported to penetrate less than 100 m and few as far as 400 m (Matlack 1993; Laurance et al. 2002; Harper et al. 2005). Such extension of impacts appears to be enough to turn forest edges into a simplified and impoverished habitat in comparison to the forest interior in terms of: (a) plant species richness (particularly trees), (b) taxonomic and ecological composition of flora and fauna (i.e. ecological groups based on life history traits), (c) microhabitat heterogeneity, and (d) trophic structure and complexity (loss of trophic levels and trophic interactions).

3 Plant-Herbivore Interactions at the Edge: Patterns and Processes

3.1 *Patterns*

Most revisions about edge effects emphasize a positive effect of edge creation on herbivores and their damage to vegetation (e.g. Laurance et al. 1997; Foggo et al. 2001; Ries et al. 2004), although few studies in fact present empirical data, and responses are reported to be positive, negative, or neutral. We found a total of 55 articles that investigated whether forest edges affect herbivores and/or herbivory – 28 of them dealing with edge effects on density, abundance, and richness of herbivores (Table 1), and 32 evaluate edge effects on herbivory patterns (Table 2). These articles cover several levels of biological organization and a variety of forest types, ranging from single species interactions (e.g. McGeoch and Gaston 2000) to studies on the community level (e.g. Barbosa et al. 2005), and from tropical (e.g. Urbas et al. 2007) to boreal forests (e.g. Roland 1993), with a predominance of investigations in temperate regions (e.g. Cadenasso and Pickett 2000).

In the case of herbivore responses (e.g. abundance, species richness), the majority of studies (23 articles, 82.1%) presented evidence that edge creation promotes positive effects, but we also registered studies showing negative (four articles, 14.3%) and neutral (five articles, 17.8%) responses (note that the total exceeds 100% because some studies presented several herbivores with differential edge responses; Table 1). There was a clear tendency for studies focusing on insect assemblages (e.g. Barbosa et al. 2005), addressing groups such as the Lepidoptera (Harris and Burns 2000), Coleoptera (Ozanne et al. 2000), and Homoptera (Martinez-Garza et al. 2005), but studies with vertebrates such as deer (Alverson et al. 1988), voles (Ostfeld et al. 1997), or marsupials (Laurance 1990) were also recorded.

Herbivore responses to edge creation were often species-specific (e.g. herbivorous beetle species; Harris and Burns 2000), but a general trend emerged – the increased abundance of generalist herbivores such as deer (Alverson et al. 1988), Homoptera (Ozanne et al. 2000), Orthoptera (Knight and Holt 2005), and leaf-cutting ants (Wirth et al. 2007). Among the more specialized groups, such as Lepidoptera, leaf-miners, and gall-forming insects, responses to edge creation were more variable, including positive, negative, and neutral patterns (e.g. respectively Martinez et al. 2005; Valladares et al. 2006; Julião et al. 2004).

A less clear picture emerged with respect to herbivory rates at forest edges. Of the articles in Table 2, 50% (16 out of 32) reported positive effects of edges (i.e. increased herbivory rates), 11 (34.4%) detected negative, and 10 (31.2%) neutral responses. Paralleling the above response of herbivores, the variation of herbivory rates at forest edges appears to be both species- and site-specific, and studies were biased toward insects and seedling damage (e.g. Meiners et al. 2000; Benitez-Malvido and Lemus-Albor 2005). Studying herbivore damage to seedlings in a deciduous forest in the USA, for example, Cadenasso and Pickett (2000)

Table 1 Edge effects upon abundance, density, richness, or performance of herbivores in forested ecosystems. References are displayed in chronological order. Herbivores are classified as specialists (S) or generalists (G). Three groups of mechanisms causing increased (↑), reduced (↓), or neutral (↔) responses are identified: environmental conditions (C), resources (R), and natural enemies (E). Reports for these mechanisms are distinguished as to whether they are supported by original data (italics), derived from secondary information (normal text), or only hypothetical (in brackets)

Herbivore	Type	Response	Mechanism	Ecosystem	Reference
Positive effects					
Herbivorous insect community	S/G	↑ Abundance	*C* (E)	Temperate forest	Mac Garvin et al. (1986)
White-tailed deer	G	↑ Density	R	Temperate forest	Alverson et al. (1988)
Gypsy moth	S	↑ Egg mass	C	Temperate forest	Bellinger et al. (1989)
Coppery brushtail and green ringtail possums	G	↑ Abundance	R	Tropical rain forest	Laurance (1990)
Herbivorous insect community	S/G	↑ Abundance		Amazonian rain forest	Fowler et al. (1993)
Sun-loving butterflies	S	↑ Species richness	C R	Amazonian rain forest	Brown and Hutchings (1997)
Hypothyris euclea (Ithomiinae)	S	↑ Abundance (1980–1986)	R	Amazonian rain forest	Brown and Hutchings (1997)
Birch tube maker	S	↑ Larval density at paper birch	*C* (R) (E)	Boreal forest	Cappuccino and Martin (1997)
Lepidopteran community	S	↑ Abundance	E	Atlantic rain forest	Bragança et al. (1998)
Arthropod herbivorous	S/G	↑ Abundance	C	Boreal forests	Jokimäki et al. (1998)
Holly leaf-miner	S/G	↑ Prevalence	C	Temperate forest	McGeoch and Gaston (2000)
Forest tent caterpillar	S	↑ Larval performance	*E*	Boreal forest	Rothman and Roland (1998)
Forest tent caterpillar	S	↑ Larval performance	*R*	Boreal forest	Fortin and Mauffette (2001)
Forest tent caterpillar	S	↑ Larval performance	*C R*	Boreal forest	Levesque et al. (2002)
Two species of herbivorous beetles	-	↑ Abundance		Podocarp forest	Harris and Burns (2000)
Whilst aphid species	G	↑ Density	(C)	Pine plantation	Ozanne et al. (2000)
Beaver	G	↑ Herbivory activity	(C) E	Riparian forest	Barnes and Mallik (2001)
Female mule deer	G	↓ Smaller home-range		Dry forest	Kie et al. (2002)
Herbivorous insect community	S/G	↑ Abundance and richness	(C) *R* (E)	Temperate forest	Major et al. (2003)
Herbivorous insect community	S/G	↑ Abundance and richness	R	Atlantic rain forest	Barbosa et al. (2005)

Orthopteran community	G	↑ Abundance of species		Sandhill forest	Knight and Holt (2005)
Gall-inducing aphids	S	↑ Species richness	*R*	Mediterranean woodland	Martinez et al. (2005)
Leaf-cutting ants	G	↑ Colony density	*R* E	Atlantic rain forest	Urbas et al. (2007)
Leaf-cutting ants	G	↑ Colony density	R	Atlantic rain forest	Wirth et al. (2007)
Negative effects					
Shade-loving butterflies	S	↓ Species richness	C	Amazonian rain forest	Brown and Hutchings (1997)
Hypothyris euclea (Ithomiinae)	S	↓ Abundance (>1990)	E	Amazonian rain forest	Brown and Hutchings (1997)
Four species of herbiv. Coleoptera	-	↓ Abundance		Podocarp forest	Harris and Burns (2000)
Herbivorous Coleoptera community	S/G	↓ Abundance and richness	(C)	Pine plantation	Ozanne et al. (2000)
Lepidopteran community	S	↓ Abundance	(C)	Pine plantation	Ozanne et al. (2000)
Neutral					
White pine weevil	S	↔ Performance		Norway spruce plantations	Lavallée et al. (1996)
Red crabs	G	↔ Density		Tropical rain forest	Green et al. (2004)
Gall-forming insects	S	↔ Species richness		Pantanal wetland	Julião et al. (2004)
Herbivorous insect community	S/G	↔ Abundance		Boreal forest	Deans et al. (2005)
Herbivorous insect community	S/G	↔ Abundance and richness		Bottomland hardwood forest	Ulyshen et al. (2005)

Table 2 Edge effects upon herbivory level or damage rates in forested ecosystems. References are displayed in chronological order. Three groups of mechanisms causing increased (↑), reduced (↓) or neutral (↔) responses are identified: environmental conditions (C), resources (R), and natural enemies (E). Reports for these mechanisms are distinguished as to whether they are supported by original data (italics), derived from secondary information (normal text), or only hypothetical (in brackets)

Herbivore	Response	Mechanism	Ecosystem	Reference
Positive effects				
Herbivorous animals	↑ Herbivory		Temperate forest	Sork (1983)
Herbivorous insect community	↑ Damage level on bracken	*C* (E)	Temperate forest	MacGarvin et al. (1986)
White-tailed deer	↑ Browsing of woody and herbaceous plants	R	Temperate forest	Alverson et al. (1988)
Grey squirrel	↑ Damage to beech and sycamore individuals		Temperate forest	Gill (1992)
Forest tent caterpillar	↑ Duration of outbreaks	C *E*	Boreal forest	Roland (1993)
White pine weevil	↑ Herbivory of Norway spruce		Norway spruce plantations	Lavallée et al. (1996)
Red-necked pademelon	↑ Herbivory and ↓ growth rate of seedlings	R E	Tropical rain forest	Wahungu et al. (1999)
Red-necked pademelon	↑ Herbivory and ↓ growth rate of seedlings	R E	Tropical rain forest	Wahungu et al. (2002)
Meadow vole	↑ Herbivory of seedlings		Mesic deciduous forest	Cadenasso and Pickett (2000)
Herbivorous insect community	↑ Herbivory of *Acer rubrum* seedlings		Temperate forest	Meiners et al. (2000)
Australian brushtail possum	↑ Leaf herbivory of mistletoe	(R)	*Nothofagus* forest	Bach and Kelly (2004)
Invertebrate species	↑ Herbivory of holly	C	Holly woodland	Arrieta and Suarez (2005)
Chewing animals	↑ Damage of *Angophora costata* leaf		Sandstone gully and ridgetop woodland	Christie and Hochuli (2005)
Orthopteran community	↑ Herbivory in eight plant species		Sandhill forest	Knight and Holt (2005)
Gall-inducing aphid	↑ Occupancy on *Pistacia palaestina* plants		Mediterranean woodland	Martinez et al. (2005)
Leaf-cutting ants	↑ Herbivory rate	*R* (E)	Atlantic rain forest	Urbas et al. (2007)
Negative effects				
Herbivorous insect community	↓ Herbivory of dogwood	*C*	Temperate forest	Dudt and Shure (1994)
Herbivorous insect community	↓ Herbivory of *Eucalyptus camaldulensis*	*C*	River red gum forest	Stone and Bacon (1994)

Herbivorous insect community	↓ Herbivory of *Eucalyptus camaldulensis*	*C*	River red gum forest	Stone and Bacon (1995)
White-tailed deer	↓ Herbivory of seedlings		Mesic deciduous forest	Cadenasso and Pickett (2000)
Herbivorous animals	↓ Herbivory on seedlings of *Micropholis venulosa*	C	Amazonian rain forest	Benitez-Malvido (2001)
Lepidopteran community	↓ Herbivory of saplings	C	Tropical rain forest	Arnold and Asquith (2002)
Herbivorous insects and mammals	↓ Damage and ↑ survival of *Quercus rubra* seedlings		Temperate forest	Meiners et al. (2002)
Red crabs	↓ Herbivory of *Clausena excavate* plants		Tropical rain forest	Green et al. (2004)
Herbivorous insects and mammals	↓ Herbivory of *Quercus germana* and *Q. xalapensis* seedlings		Tropical montane forest	Guzman-Guzman and Williams-Linera (2006)
Herbivorous mammals	↑ Survivorship of tussock grass	E	Chaparral shrubland	Lambrinos (2006)
Leaf-miners species	↓ Herbivory rate	C *E*	Chaco serrano	Valladares et al. (2006)
Neutral				
Meadow voles	↔ Seedling predation		Temperate forest	Ostfeld et al. (1997)
Invertebrate species	↔ Seedling herbivory		Mesic deciduous forest	Cadenasso and Pickett (2000)
Herbivorous insect community	↔ Herbivory of *Fraxinus americana* seedlings		Temperate forest	Meiners et al. (2000)
Herbivorous insect community	↔ Herbivory		Oak-pine forest	Patel and Rapport (2000)
Red crabs	↔ *Adenanthera pavonia* survival		Tropical rain forest	Green et al. (2004)
Herbivorous insect community	↔ Leaf herbivory of mistletoe		*Nothofagus* forest	Bach and Kelly (2004)
Herbivorous mammals	↔ Seedling herbivory		Tropical rain forest	Asquith and Mejia-Chang (2005)
Herbivorous insect community	↔ Seedling damage		Tropical rain forest	Benitez-Malvido and Lemus-Albor (2005)
Cattle and large indigenous mammals	↔ Woody seedling abundance and richness		Afromontane forest	Lawes et al. (2005)
Herbivorous insects and rodents	↔ Leaf-damage of five *Quercus* species		Tropical montane forest	Lopez-Barrera et al. (2006)

recorded higher herbivory by meadow vole (*Microtus pennsylvanicus*), lower herbivory by white-tailed deer (*Odocoileus virginianus*), and no differences in herbivory by invertebrates. At other study sites, however, herbivory damage by the white-tailed deer was observed to increase (Alverson et al. 1988).

Despite the occurrence of species-specific responses, two groups of dominant herbivores are well documented to benefit from forest fragmentation and increased proportion of edge habitats: deer in temperate forests of North America and leaf-cutting ants in the Neotropics. Deer, especially the edge-loving white-tailed deer, have shown increased population densities since the nineteenth century, greatly because of extensive timber harvesting, continued habitat fragmentation, creation of "wildlife openings", and reduced hunting (Alverson et al. 1988). Recently, we demonstrated that leaf-cutting ant populations (*Atta* spp) strongly benefit from edges in the Brazilian Atlantic forest. The colony density of *Atta cephalotes* was about 8.5 times higher in the first 50-m edge zone than in the >100-m forest interior (Wirth et al. 2007). This edge proliferation of leaf-cutting ants went along with an increased rate of annual herbivory per colony. Equally sized *A. cephalotes* colonies located at the forest edge removed about twice as much leaf area from their foraging grounds as interior colonies (14.3% vs 7.8% $colony^{-1}$ $year^{-1}$). This higher herbivory rate within the forest edge zone was a consequence of reduced foraging areas (0.9 ha vs 2.3 ha $colony^{-1}$ $year^{-1}$) and a moderately lower leaf area index in this habitat, while harvest rates were the same (Urbas et al. 2007).

3.2 *Processes*

Regarding causalities for the above patterns, we extracted three main complexes of factors: (a) changes resulting from edge-induced shifts in environmental conditions, (b) resource-based aspects, and (c) responses related to natural enemies (Tables 1, 2). One primary and most noteworthy finding of this review is the scarcity of works that deliberately and directly addressed functional issues or aimed to provide mechanistic explanations for the observed patterns. Only a few studies reporting changes in herbivory along the forest edge included an a priori design to assess the potential causal factors, and most of the explanations given in the literature were based on indirect evidence or deductive reasoning (Tables 1, 2). In the following we try to synthesize the essential mechanisms arising from this review, our own studies on leaf-cutting ants in the Brazilian Atlantic forest, and theoretical considerations.

3.2.1 Microenvironment-Related Processes

Some of the more obvious changes following edge creation are abiotic (see Section 2) and mechanisms derived from organisms' responses to abiotic factors are best understood (Foggo et al. 2001). While it is reasonable to assume that herbivores, like other groups of organisms, should be directly affected by the considerable microenvironmental alterations in the forest edge zone (Kapos 1989), there is little concrete evidence for the relevance of this factor. Only in two cases was the improved edge performance of lepidopteran herbivores directly attributed to the increased temperature of boreal forest edges (Cappuccino and Martin 1997; Levesque et al. 2002). Nevertheless, reported changes in herbivore densities and/or herbivore damage were reasonably related to the edge environment. Herbivores that respond negatively include shade-loving (sub)tropical forest interior species (Brown and Hutchings 1997; Arnold and Asquith 2002; Valladares et al. 2006), which may be especially sensitive to deleterious edge conditions (see Section 2).

Increasing herbivory as a result of positive abiotic edge effects is consistently observed in temperate or boreal forests (Tables 1, 2), where edges represent warmer and sunnier habitats than the stand interiors (Matlack 1993; Chen et al. 1995) and may therefore promote a wide range of insect herbivores (for contrasting patterns among species, see MacGarvin et al. 1986). This might be due to two aspects: reduced developmental time for larvae (Cappuccino and Martin 1997) or preferential oviposition on sunlit hosts (Moore et al. 1988). The resulting edge aggregation of adults and early life stages appears to be a more general consequence of edge microclimate and is proposed by several authors (Roland 1993; Jokimäki et al. 1998; McGeoch and Gaston 2000; Tables 1, 2). In addition, microclimate-related edge properties are responsible for higher-order effects on both herbivores and plants via the alteration of trophic interactions, as will be detailed below.

3.2.2 Trophic Processes

Whether the extent of plant-herbivore interactions is enhanced or reduced in a given environmental context depends on the functioning of bottom-up and top-down forces, which act in concert to influence the abundance and distribution of herbivores (Hunter et al. 1997). Bottom-up regulation is hypothesized to affect herbivore populations through food limitation (Mattson and Addy 1975). In contrast, top-down forces are proposed to regulate herbivores through predators and parasites (Hairston et al. 1960). The degree to which plant-herbivore interactions are structured by these two basic processes is a classic debate in ecology (Dyer and Letourneau 1999) which, in the past decade, received new impetus from the study of habitat fragmentation (Tscharntke and Brandl 2004). Research documenting that habitat loss and isolation may lead to the truncation of food chains, especially the release of herbivores from top-down control, is particularly influential (Kareiva 1987; Kruess and Tscharntke 1994; Terborgh et al. 2001; Kondoh 2003), but the role of edges remains largely obscure. In contrast, the trophic regulation of herbivores along forest edges has long been related to resource availability (e.g. Leopold 1933), despite a remarkable paucity of solid empirical data.

Changes in resource availability/quality

Enhanced availability and/or quality of resources are the most commonly invoked processes to causally link the edge habitat with increasing herbivory. The adjustment of herbivore distribution and feeding behaviour to food value and quantity is reflected in the resource concentration hypothesis (Kareiva 1983) and is based upon general assumptions of optimal foraging theories (see Pyke et al. 1977). Of the 55 papers taken into account, 14 advocated this bottom-up view, but direct evidence is rather limited (five articles, Tables 1, 2). Several resource-related factors accounting for elevated edge herbivory come into question.

One of the most prominent and well documented edge effect in forested ecosystems is a shift in plant community composition towards fast-growing, early-successional species (see also Section 2), which are generally known to be the preferred food of herbivores because of lower levels of quantitative defences and higher nutritive value than shade-tolerant species (Coley 1980). Following this pattern, diversity, abundance, or damage level of herbivores is greater in pioneer-dominated forest stands like, e.g. treefall gaps (Richards and Coley 2007). For forest edges, this or variations of this mechanism have been repeatedly proposed (Leopold 1933; Lovejoy et al. 1986; Murcia 1995; Brown and Hutchings 1997; Matlack and Litvaitis 1999), but only rarely addressed explicitly (Barbosa et al. 2005; Urbas et al. 2007; Tables 1, 2).

The drastic (>8.5-fold) increase in colony density and annual herbivory rate of leaf-cutting ants in the Brazilian Atlantic forest (see Section 3.1) gives further support to the relevance of compositional shifts in the plant community as a driving bottom-up mechanism. Increasing edge herbivory due to decreasing foraging areas (see above) can be explained by the greater abundance of highly palatable pioneer species in the edge zone of the studied area in terms of both species and stem number (Oliveira et al. 2004; Grillo 2005). Such an inverse relationship between resource density and foraging area is consistent with predictions of the optimal foraging theory (Ford 1983) and is well documented (e.g. Leal and Oliveira 2000).

Greater herbivore pressure at the forest edge than in the forest interior may result from a range of edge-mediated effects on the palatability of food plants other than those associable with the dominance of early-successional forest plants. There is some evidence that microenvironmental alterations near edges (see Section 2) affect plant performance to the extent that leaves exposed to the edge taste better. For example, Fortin and Maufette (2001) showed that forest tent caterpillars perform better on sugar maple trees at the edge, because their foliage contains higher concentrations of nitrogen and soluble sugars. Likewise, intraspecific and within-plant leaf quality may be improved by edge-induced drought stress (Laurance et al. 2001). Food plant preference of leaf-cutting ants, for example, was linked to drought stress via increased osmolyte concentrations within the leaves (Meyer et al. 2006). Given the fact that folivory was shown to intensify the drought stress of a plant by uncontrolled water loss from wounds (Aldea et al. 2005), it is plausible that edge conditions initiate a positive feedback loop, where stressed plants are attractive to leaf-cutting ants, therefore experiencing higher levels of herbivory damage, which in turn intensifies their drought stress, thus making them even more attractive (Meyer et al. 2006).

Even in the absence of drought stress, edge environments may render individual plants and plant parts more susceptible to herbivory. As a consequence of increased light availability at exposed forest edges (Davies-Colley et al. 2000), more leaves are sunlit, and thus the proportion of sun versus shade leaves is supposed to be higher than in the forest interior. The preferences for and/or greater survival on sun-leaves is a well known feature in various groups of insect herbivores (for a review, see White 1984) and is often related to their increased nutritive value (Nichols-Orians 1991). We are not aware of studies that attempted to directly correlate herbivory at forest edge versus interior with the availability/quality of sun-leaves (but see MacGarvin et al. 1986). However, a decrease in herbivory has never been associated with edge-induced changes in food availability/quality (Table 2), though there is some circumstantial evidence pointing to the possibility of light-induced plant defence (e.g. Shure and Wilson 1993).

Forest boundaries can have a positive effect on the palatability of resources by modulating fluxes of nutrients. Based on studies in various systems, Pickett et al. (2000) concluded that the enhancement of nitrogen in the forest edge zone (e.g. as fertilizer drift from adjacent agriculture) is a more general process. Since fertilized plants are often more palatable to herbivores than unfertilized plants (Bryant et al. 1987), this impact may well be responsible for increased edge herbivory (see also Major et al. 2003; Christie and Hochuli 2005).

The fact that edges represent high-resource environments for forest plants in terms of light and nutrient availability is likely to lead to a general increase of (leaf) productivity (Laurance et al. 2003) and may thus represent another component of edge-induced herbivore attraction (Major et al. 2003). This is documented for treefall gaps and forest canopies, where highly productive vegetation and increased leaf-turnover generate a persistent source of young, soft leaves, which are frequently preferred by herbivores (Basset 1992; Richards and Coley 2007).

Top-down-related processes

Given the popularity of top-down-related mechanisms in the study of fragmentation effects (Terborgh et al. 2001; Thies et al. 2003), our knowledge about performance, abundance, or predation/infection efficiency of natural enemies in forest edge zones is surprisingly limited. Of 55 articles, only three presented original data to demonstrate that edges influence the top-down control of herbivores, one in which herbivory was lower and two in which it was higher in edge habitats (Tables 1, 2). In a formal respect the latter studies fit the basic pattern uncovered in fragmented landscapes, where the elimination of natural enemies is followed by the subsequent release of lower trophic levels, such as herbivores (Kareiva 1987; Kruess and Tscharntke 2000; Terborgh et al. 2001). Causal explanations are largely related to area loss and isolation, although it is conceivable that island biogeographic studies are inevitably confounded by edge effects (Malcolm 1994). In fact, we suspect that edges may also add to consumer-driven changes in herbivory, as illustrated by the following observations.

The most convincing evidence for a negative edge-influence on top-down processes stems from studies with the tent caterpillar, an important North American herbivore pest, and its natural enemy, a nuclear polyhedrosis virus (Roland 1993; Roland and Kaupp 1995; Rothman and Roland 1998). Pathogen-induced mortality of the herbivore was considerably reduced at the forest edge because virus particles do not remain infectious under high light (particularly

UV-light) and low humidity. Although not properly examined, such adverse environmental edge conditions may also be responsible for reduced parasitism of moisture-loving phorid flies (a specialist parasitoid of leaf-cutting ants) on edge colonies of *A. cephalotes* in the Brazilian Atlantic forest. Almeida (2004) found that both the number of flies attracted to nests and their attack rates were significantly lower in edge habitats.

Edges are called important population sinks, especially for large carnivores (Laurance 2000), some of which may be important control agents of herbivore species. Other than in area-based extinction models, this kind of animal loss is especially due to the tendency of large predators to range outside the boundaries of nature reserves combined with the increased hunting pressure along patch interfaces with human-dominated landscapes (Woodroffe and Ginsberg 1998; Laurance 2000). This can be exemplified by an edge-associated decrease in armadillo predation on leaf-cutting ants along forest fragments in the Brazilian Atlantic forest, where armadillos are subjected to human hunting (Fernandes 2003). During a year-round monitoring of ten *A. cephalotes* colonies, Darrault (2005) found a marginally significant lower number of armadillo burrows in edge colonies (33.0 ± 6.91 colony^{-1}) compared with interior colonies (45.6 ± 7.56 colony^{-1}).

Generally, there is ample support for negative edge impacts on several groups of predators that have been shown to be vulnerable to edge creation (see also Section 2). However, we still lack studies to evaluate how these changes affect herbivorous edge communities and whether they may be compensated or surpassed by opposing edge effects of increased predator intensities.

In fact, despite a scarcity of published data (Tables 1, 2), there is substantial indirect evidence suggesting that edge herbivory may be decreased by natural enemies. To give some examples, greater predatory impact may arise from:

1. Cross-edge spillover of subsidized predators from the anthropogenic matrix to the natural forest habitat (Rand et al. 2006);
2. Predator avoidance behaviour of herbivores to escape increased mortality risks along the forest edge (Lambrinos 2006);
3. Increased predation/parasitism efficiency – e.g. because of improved visual host location (Valladares et al. 2006).

We recommend future studies to elucidate whether such phenomena are more broadly relevant for the population dynamic of edge herbivores.

4 Impact of Herbivory on the Forest Edge

To date no studies have specifically evaluated the impact that herbivores have on the forest edge at the ecosystem level; and studies on both the population or community level are still scarce (Sork 1983; Jules and Rathcke 1999; Meiners et al. 2000; Benitez-Malvido and Lemus-Albor 2005). Strong evidence on the effects of a high herbivory pressure on forested ecosystems was compiled in temperate forests on the effects of ungulates – mainly deer (for reviews, see Alverson et al. 1988; Rooney and Waller 2003) and more recently in a tropical forest on land bridge islands with high densities of howler monkeys and leaf-cutting ants (Rao et al. 2001; Terborgh et al. 2001, 2006).

This section aims to explore and propose possible effects based on the general impact of herbivory as demonstrated for different ecosystems and the uncovered patterns of herbivory at the forest edge as described above. It will discuss how the community of herbivores impacts ecosystems on three levels: the abiotic conditions, the plant community, and higher trophic levels.

4.1 Abiotic Conditions

4.1.1 Vegetation Structure and Light Regime

Removing leaves or leaf tissue of plants opens the canopy, allowing a higher proportion of light to penetrate the forest. Thus, herbivory can potentially cause similar effects as the creation of edges per se (see Section 2). In forests, an increased light availability following herbivory has rarely been documented. Wirth et al. (2003) found patches of high light transmittance on the forest floor and increased sunfleck frequency in a tree canopy within the foraging territory of leaf-cutting ants (*Atta colombica*) in a tropical lowland forest in Panama. An even stronger impact was demonstrated for colonies of *Atta cephalotes* whose foraging areas presented 18% less vegetation cover and 40% more diffuse light compared with control sites in the Brazilian Atlantic forest (Correa 2006).

Herbivores influence vegetation structure by more than the removal of foliage, i.e. herbivory per se, for example by trampling and crushing plants, burrowing, forming trails, and nesting (Huntly 1991). Elephants (*Loxodonta africana*) were shown to induce damages in African savannas by breaking and uprooting trees (Augustine and McNaughton 2004) and leaf-cutting ants by creating understory gaps of up to 150 m^2 (Meyer, unpublished data), predominantly close to the forest edge (Wirth et al. 2007). Preliminary data on the microclimatic effects caused by nest clearings of *A. cephalotes* indicated that about three times as much light reaches the nest compared to the surrounding forest floor accompanied by higher maximum air and soil temperatures, as well as by lower air humidity plus stronger daily fluctuations in all these parameters (Meyer, unpublished data).

Summarizing, increases in light availability induced by herbivores may promote heterogeneity and micro-site diversity in a relatively homogeneous environment, like the interior of a continuous forest (Wirth et al. 2003). In contrast, at the forest edge herbivores enhance the high light availability typical for the edge and ultimately homogenize the microclimate.

4.1.2 Nutrient Cycling

The second important resource for plants are nutrients, the availability of which can be altered by the activity of herbivores. Generally, herbivory increases the cycling rate of macronutrients, because they are often released more rapidly from the faeces of herbivores than from decomposing plants (Zamora et al. 1999). Recent studies in forest systems indicated that herbivores can cause heterogeneity in nutrient concentrations and might influence nutrient availability and plant productivity. Wirth et al. (2003) found leaf-cutting ants to concentrate nutrients derived of leaves gathered in large areas in their refuse dumps and Feeley (2005) showed that clumped defaecation of howler monkeys leads to a local concentration of nitrogen and phosphor. Higher biomass of fine roots within these resource patches (leaf cutting ant nests: Moutinho et al. 2003; howler monkey latrines: Feeley 2005) indicates nutrient exploitation by the surrounding vegetation.

The concentration of herbivores at forest edges (Table 1) could, in addition to accelerated nutrient cycling, lead to an accumulation of nutrients at edges. In combination with the observed nutrient drift from the surrounding matrix (fertilizer from agricultural systems: Pickett et al. 2000; pollution from urban areas: Christie and Hochuli 2005), forest edges should be generally less nutrient-limited than the forest interior. Possible consequences of such relaxed nutrient limitation were demonstrated on isolated islands under high herbivory pressure for plant productivity (Feeley and Terborgh 2005), with resulting cascading impacts on the whole ecosystem (Terborgh et al. 2006).

4.2 Plant Community

Few herbivores have the potential to alter vegetation communities by killing adult trees. A prominent exception of this occurs during insect outbreaks – a common phenomenon in fragmented forests (see Kondoh 2003, and references therein). Such defoliation calamities were also reported from forest edges (Roland 1993). Which plant species experience the greatest impact depends on the feeding preferences of herbivores and on differences in the plant's ability to tolerate herbivory damage, as discussed below.

4.2.1 Differential Feeding Preference

All herbivores – even highly polyphagous ones – show preferences for certain plant taxa (Bernays and Chapman 1994). Since damage by herbivores is thus not homogeneously distributed across the plant community, herbivores can alter community composition (e.g. Vasconcelos and Cherrett 1997; Terborgh et al. 2001). Especially the preference of dominant herbivores has been demonstrated to strongly impact forests. For example, damage to shrub canopies by elephants cause large reductions in cover of *Acacia mel-*

lifera and *Grewia tenax*, but lesser reductions in cover of *A. etbaica* (Augustine and McNaughton 2004); and deer browsing in temperate forests suppresses tree regeneration of *Abies alba* and *Acer pseudoplatanus* (Ammer 1996).

Generally, pioneer plant species dominating forest edges are believed to be more attractive food sources for herbivores (Coley 1980; see also Section 3.2). Therefore, pioneer and early successional species at the forest edge should be more heavily attacked than later successional species, but the impact of herbivory also depends on the plants' ability to tolerate damage.

4.2.2 Differential Tolerance of Herbivory

Slow-growing plants are believed to be less able to compensate for biomass losses than fast-growing plants and are likely to be more susceptible to herbivory if attacked, especially in habitats with high resource availability (Coley et al. 1985). A community of pioneer plant species at the forest edge should therefore better compensate for losses to herbivory than less abundant later successional species. For example, the two pioneer species investigated by Khan and Tripathi (1991) showed higher growth rates and, consequently, a higher increase in biomass despite higher herbivory rates compared with two shade tolerant species.

4.2.3 Regeneration and Succession at the Forest Edge

While the feeding preference of herbivores seems to discriminate pioneer plant species, their capacity to tolerate damage should favour them. What will be the resulting overall pattern for the species composition at the edge? Forest edges face a dramatic shift in plant species composition subsequent to edge creation. They typically display an impoverished set of species compared with the forest interior and are largely dominated by pioneers (see Section 2). This first change in species composition renders edges more prone to higher-order changes imposed by foliage removal (Harper et al. 2005). Under very high herbivory pressures forest regeneration might be suppressed, resulting in a plant community comprised mostly of resistant, non-palatable species. This was documented, for example, on small land-bridge islands, where leaf-cutting ant (*Atta* spp) populations reached 10–100 times their mainland densities (Rao et al. 2001; but see White 2007). While we have demonstrated that the majority of studies on herbivory at forest edges detect an increase in abundance of and damage by herbivores, documentation of destructive foraging by overabundant herbivores is relatively rare (but see Gilham 1955; Alverson et al. 1988; Roland 1993; Rao et al. 2001). Consequently, it is reasonable to expect some disruption, not a thorough inhibition, of the regeneration process at forest edges.

Succession across the forest edge takes place as: (a) structural development of the edge and (b) forest recovering by recolonization processes in the adjacent matrix. Such recolonization of the matrix occurs when the disturbance, which created the edge, has ceased and can be influenced by herbivory (Meiners et al.

2000). Succession and forest recovering requires edge effect amelioration and the consequent successful establishment of more shade-tolerant or later successional tree species (Gascon et al. 2000). Increased herbivory may depress the recruitment of shade-tolerant trees, while favouring the establishment of pioneer plants in two ways: directly by changing microclimate (particularly light availability) and indirectly by reducing competition with shade-tolerant plants. More pioneers support more herbivores, which in turn promotes the persistence of pioneers: this kind of "positive feedback loop" was proposed to explain replacement of heather by grasses in British uplands exposed to sheep grazing (Palmer et al. 2005). Similarly, elephants prefer to feed on pioneer species, so that open areas offer them more feeding opportunities and the damage they cause, in turn, maintains or increases openness (Sheil and Salim 2004). In the Shimba Hills in Kenya, Höft and Höft (1995) concluded that elephant activity arrestes forest succession as their damage promotes regeneration of the most browsed early successional and fast-growing species.

In our study system (see Section 3.1), *Atta cephalotes* concentrates foraging on pioneer species at the forest edge (Falcão 2004; see also Section 3.1). The resulting high herbivory rate is believed to increase light availability and, therefore, enhance regeneration of pioneer plants. In addition, ant nest-related canopy gaps and the resulting changes in microclimate were shown to negatively impact germination and seedling establishment of a typical, shade tolerant forest tree. With increasing distance from nests, *Chrysophyllum viride* showed higher proportions of germination (nest centre = 3.3 ± 1.6%, nest edge = 4.6 ± 2.3%, 5 m = 5.3 ± 1.9%, 25 m = 6.8 ± 1.6%; Correa 2006). In conclusion, leaf-cutting ants profit from the creation of forest edges that they colonize in high densities and help to maintain edge microclimate and pioneer dominance by high herbivory rates and the creation of nest clearings.

4.3 Higher Trophic Levels

Herbivores represent the interface between primary production and higher trophic levels within food webs. An increased abundance of herbivores at the forest edge (see Section 3.1) can attract predators and parasitoids (Matlack and Litvaitis 1999), in a similar way as high densities of palatable pioneer plants offer more resources for herbivores. This is in accordance with current theory predicting that, when primary productivity is high, herbivore populations increase to levels that can support substantial predator populations (Fretwell 1977; Oksanen et al. 1981). Communities of insectivorous birds, for example, concentrate their foraging on forest edges (Strelke and Dickson 1980; Lewke 1982; Lovejoy et al. 1986). Predator concentrations at the forest edge may in turn propagate through the food web affecting other members of the faunal community. Such cascading effects are now widely recognized (Terborgh et al. 2001). Ries and Fagan (2003), for example, reported an edge-induced increase of the (predatory) mantid *Stagmomantis limbata* leading to higher bird predation rates. Thus, altered plant-herbivore interactions imply not only direct changes in population and community level processes but also indirect effects on edge ecosystem functions.

5 Conclusions and Outlook

This review provides ample evidence that herbivores profoundly benefit from forest edges. This positive edge effect seems more pronounced for herbivore densities rather than herbivore damage, possibly because high plant productivity at the edge masks the true extent of foliage removal. Moreover, generalist herbivores appear to be particularly favoured by edges, while decreased edge abundance of herbivores frequently involves specialists. Increased herbivory pressure at forest edges likely results from a combination of favourable microenvironmental conditions for herbivores and increased food quantity and quality (relaxation of bottom-up control), caused inter alia by a shift in plant composition at forest edges. The impact of natural enemies on edge herbivory is less well documented, but there are good examples for an edge-mediated disruption of top-down regulation of herbivores. Although there is strong indirect support for reduced edge herbivory via enhanced top-down control in forest edges, we still lack more conclusive evidence. These general findings emerged despite the relatively low number of available reports (55) which exhibit significant biases toward insects, few plant species, seedling damage and survivorship, and population level impacts. Additional limits of current literature include a noticeable lack of long-term studies and those addressing key herbivores (great foliage consumers), herbivory rates on community level, as well as magnitude and intensity of herbivory-related edge effects.

In addition to amplifying edge-induced microclimatic changes and influencing trophic cascades by the attraction of predators, high densities of herbivores may potentially affect forest edges by speeding up an already accelerated turnover rate of matter and energy. Being directly favoured by the abundance of pioneers and indirectly favouring pioneer plants, herbivores may ultimately delay succession or regeneration at forest edges, thereby amplifying biota modifications triggered by habitat fragmentation. These proposed impacts of increased herbivory on ecosystem structure and function following edge creation are an example of how human activity not only directly disturbs habitats and drives species loss, but also alters species interactions in a way that amplifies disturbance, reduces resilience, and increases the risk of secondary species loss.

Based upon the above findings, future research should: (a) validate and assess the generality of both the patterns and underlying mechanisms proposed here regarding herbivore occurrence and herbivory at forest edges and (b) address more ecosystem-level effects driven by high herbivory loads on forest edges (e.g. changes in primary productivity, nutrient cycling, succession, ecosystem resilience), as such effects are in theory more pervasive in ecological and geographic terms.

Acknowledgements We are grateful to W. Beyschlag for his invitation to prepare this review. Some fundamental ideas emerged from research projects in Brazil, for which financial support was provided in Germany by the Deutsche Forschungsgemeinschaft (grant No. WI 1959/1-2) and the AFW Schimper Foundation and in Brazil by CAPES (No. 007/01) and CNPq (No. 540322/01-6). We thank Walkiria R.Almeida, Manuel Vieira Araújo Jr, Veralucia Barbosa, Olivier Darrault, Christoph Dohm, Poliana F. Falcão, Simone Jürgens, Clarissa M. Knoechelmann, Meike Passlack, Úrsula A. Silveira, and Pille Urbas for their great help in data collection.

References

Aguilar R, Ashworth L, Galetto L, Aizen MA (2006) Plant reproductive susceptibility to habitat fragmentation: review and synthesis through a meta-analysis. Ecol Lett 9:968–980

Aldea M, Hamilton JG, Resti JP, Zangerl AR, Berenbaum MR, DeLucia EH (2005) Indirect effects of insect herbivory on leaf gas exchange in soybean. Plant Cell Environ 28:402–411

Almeida WR (2004) Fragmentação florestal: Redução no controle da formiga cortadeira *Atta laevigata* por moscas parasitóides. MSc thesis, Universidade Federal de Pernambuco, Recife

Alverson WS, Waller DM, Solheim SL (1988) Forests too deer – edge effects in northern Wisconsin. Conserv Biol 2:348–358

Ammer C (1996) Impact of ungulates on structure and dynamics of natural regeneration of mixed mountain forests in the Bavarian Alps. For Ecol Manage 88:43–53

Arnold AE, Asquith NM (2002) Herbivory in a fragmented tropical forest: patterns from islands at Lago Gatun, Panama. Biodivers Conserv 11:1663–1680

Arrieta S, Suarez F (2005) Spatial patterns of seedling emergence and survival. For Ecol Manage 205:267–282

Asquith NM, Mejia-Chang M (2005) Mammals, edge effects, and the loss of tropical forest diversity. Ecology 86:379–390

Augustine DJ, McNaughton SJ (2004) Regulation of shrub dynamics by native browsing ungulates on East African rangeland. J Appl Ecol 41:45–58

Bach CE, Kelly D (2004) Effects of forest edges on herbivory in a New Zealand mistletoe, *Alepis flavida*. N Z J Ecol 28:195–205

Barbosa VS, Leal IR, Iannuzzi L, Almeida-Cortez J (2005) Distribution pattern of herbivorous insects in a remnant of Brazilian Atlantic forest. Neotrop Entomol 34:701–711

Barnes DM, Mallik AU (2001) Effects of beaver, *Castor canadensis*, herbivory on streamside vegetation in a northern Ontario watershed. Can Field Nat 115:9–21

Basset Y (1992) Influence of leaf traits on the spatial-distribution of arboreal arthropods within an overstory rain-forest tree. Ecol Entomol 17:8–16

Bellinger RG, Ravlin FW, Mcmanus ML (1989) Forest edge effects and their influence on gypsy-moth (Lepidoptera, Lymantriidae) egg mass-distribution. Environ Entomol 18:840–843

Benitez-Malvido J (2001) Regeneration in tropical rainforest fragments. In: Bierregaard RO Jr, Gascon C, Lovejoy TE, Mesquita RCG (eds) Lessons from Amazonia: the ecology and conservation of a fragmented forest. Yale University Press, New Haven, pp 136–146

Benitez-Malvido J, Lemus-Albor A (2005) The seedling community of tropical rain forest edges and its interaction with herbivores and pathogens. Biotropica 37:301–313

Bernays EA, Chapman RF (1994) Host–plant selection by phytophagous insects. Chapman & Hall, New York

Bierregaard RO Jr, Laurance WF, Gascon C, Benitez-Malvido J, Fearnside PM, Fonesca CR, Ganade G, Malcolm JR, Martins MB, Mori S, Oliveira M, Rankin-de Mérona J, Scariot A, Spironello W, Willianson B (2001) Principles of forest fragmentation and conservation in the Amazon. In: Bierregaard RO Jr, Gascon C, Lovejoy TE, Mesquita RCG (eds) Lessons from Amazonia: the ecology and conservation of a fragmented forest. Yale University Press, New Haven, pp 371–385

Bigger DS, Marvier MA (1998) How different would a world without herbivory be? A search for generality in ecology. Integr Biol 1:60–67

Bragança MAL, Zanuncio JC, Picanco M, Laranjeiro AJ (1998) Effects of environmental heterogeneity on Lepidoptera and Hymenoptera populations in *Eucalyptus* plantations in Brazil. For Ecol Manage 103:287–292

Brown KS Jr, Hutchings RW (1997) Disturbance, fragmentation, and the dynamics of diversity in Amazonian forest butterflies. In: Laurance WF, Bierregaard RO Jr (eds) Tropical forest remnants: ecology, management, and conservation of fragmented communities. University of Chicago Press, Chicago, pp 91–110

Brown VK (1990) Insect herbivory and its effect on plant sucession. In: Burden JJ, Leather SR (eds) Pests, pathogens and plant communities. Blackwell Scientific, Oxford, pp 275–288

Bryant JB, Clausen TP, Reichardt MC, McCarthy MC, Werner RA (1987) Effect of nitrogen fertilization up on the secundary chemistry and nutritional value of quaking aspen (*Populus tremuloides*, Michx.) leaves for the large aspen tortix (*Christoneura conflicatana*, Walker). Oecologia 73:513–517

Cadenasso ML, Pickett STA (2000) Linking forest edge structure to edge function: mediation of herbivore damage. J Ecol 88:31–44

Cappuccino N, Martin MA (1997) The birch tube-maker *Acrobasis betulella* in a fragmented habitat: the importance of patch isolation and edges. Oecologia 110:69–76

Carvalho KS, Vasconcelos HL (1999) Forest fragmentation in central Amazonia and its effects on litter-dwelling ants. Biol Conserv 91:151–157

Chen JQ, Franklin JF, Spies TA (1992) Vegetation responses to edge environments in old-growth Douglas-fir forests. Ecol Appl 2:387–396

Chen JQ, Franklin JF, Spies TA (1995) Growing-season microclimatic gradients from clear-cut edges into old-growth Douglas-fir forests. Ecol Appl 5:74–86

Christie FJ, Hochuli DF (2005) Elevated levels of herbivory in urban landscapes: are declines in tree health more than an edge effect? Ecol Soc 10:10

Coley PD (1980) Effects of leaf age and plant life-history patterns on herbivory. Nature 284:545–546

Coley PD, Bryant JP, Chapin FS (1985) Resource availability and plant antiherbivore defense. Science 230:895–899

Correa MM (2006) Formigas cortadeiras (Hymenoptera, Formicidae) como agentes modificadores da disponibilidade de luz e da estrutura da comunidade vegetal em Floresta Atlântica nordestina. PhD thesis, Universidade Federal de Pernambuco, Recife

Crawley MJ (1983) Herbivory: the dynamics of animal plant interactions. Blackwell Science, Oxford

D'Angelo SA, Andrade ACS, Laurance SG, Laurance WF, Mesquita RCG (2004) Inferred causes of tree mortality in fragmented and intact Amazonian forests. J Trop Ecol 20:243–246

Darrault OPG (2005) Efeito da fragmentação no controle top-down das formigas cortadeiras. MSc thesis, Universidade Federal de Pernamuco, Recife

Davies-Colley RJ, Payne GW, Elswijk M van (2000) Microclimate gradients across a forest edge. N Z J Ecol 24:111–121

Deans AM, Malcolm JR, Smith SM, Bellocq MI (2005) Edge effects and the responses of aerial insect assemblages to structural-retention harvesting in Canadian boreal peatland forests. For Ecol Manage 204:249–266

Debinski DM, Holt RD (2000) A survey and overview of habitat fragmentation experiments. Conserv Biol 14:342–355

Didham RK, Lawton JH (1999) Edge structure determines the magnitude of changes in microclimate and vegetation structure in tropical forest fragments. Biotropica 31:17–30

Dudt JF, Shure DJ (1994) The influence of light and nutrients on foliar phenolics and insect herbivory. Ecology 75:86–98

Dyer LA, Letourneau DK (1999) Relative strengths of top-down and bottom-up forces in a tropical forest community. Oecologia 119:265–274

Fagan WE, Cantrell RS, Cosner C (1999) How habitat edges change species interactions. Am Nat 153:165–182
Fahrig L (2003) Effects of habitat fragmentation on biodiversity. Annu Rev Ecol Syst 34:487–515
Falcão PF (2004) Efeito da fragmentação florestal na diversidade de plantas cortadas pela formiga cortadeira *Atta laevigata*. MSc thesis, Universidade Federal de Pernambuco, Recife
Feeley K (2005) The role of clumped defecation in the spatial distribution of soil nutrients and the availability of nutrients for plant uptake. J Trop Ecol 21:99–102
Feeley KJ, Terborgh J (2005) The effects of herbivore density on soil nutrients and tree growth in tropical forest fragments. Ecology 86:116–124
Fernandes ACA (2003) Censo de mamíferos em fragmentos de Floresta Atlântica no Nordeste do Brasil. MSc thesis, Universidade Federal de Pernambuco, Recife
Foggo A, Ozanne CMP, Speight MR, Hambler C (2001) Edge effects and tropical forest canopy invertebrates. Plant Ecol 153:347–359
Ford RG (1983) Home range in a patchy environment: optimal foraging predictions. Am Zool 23:315–326
Fortin M, Mauffette Y (2001) Forest edge effects on the biological performance of the forest tent caterpillar (Lepidoptera: Lasiocampidae) in sugar maple stands. Ecoscience 8:164–172
Fowler HG, Silva CA, Venticinque E (1993) Size, taxonomic and biomass distributions of flying insects in central amazonia – forest edge vs understory. Rev Biol Trop 41:755–760
Fretwell SD (1977) Regulation of plant communities by food-chains exploiting them. Perspect Biol Med 20:169–185
Gascon C, Williamson GB, Fonseca GAB da (2000) Ecology – receding forest edges and vanishing reserves. Science 288:1356–1358
Gilham ME (1955) Ecology of the Pembrokeshire Islands III: The effect of grazing on the vegetation. J Ecol 43:172–206
Gill RMA (1992) A review of damage by mammals in north temperate forests: 2. Small mammals. Forestry 65:281–308
Green PT, Lake PS, O'Dowd DJ (2004) Resistance of island rainforest to invasion by alien plants: influence of microhabitat and herbivory on seedling performance. Biol Invasions 6:1–9
Grillo AS (2005) As implicações da fragmentação e da perda de habitats sobre a assembléia de árvores na floresta Atlântica ao norte do São Francisco. PhD thesis, Universidade Ferderal de Pernambuco, Recife
Guzman-Guzman J, Williams-Linera G (2006) Edge effect on acorn removal and oak seedling survival in Mexican lower montane forest fragments. New For 31:487–495
Hairston NG, Smith FE, Slobodkin LB (1960) Community strucutre, population control, and competition. Am Nat 94:421–425
Harper KA, Macdonald SE, Burton PJ, Chen JQ, Brosofske KD, Saunders SC, Euskirchen ES, Roberts D, Jaiteh MS, Esseen PA (2005) Edge influence on forest structure and composition in fragmented landscapes. Conserv Biol 19:768–782
Harris LD (1988) Edge effects and conservation of biotic diversity. Conserv Biol 2:330–332
Harris RJ, Burns BR (2000) Beetle assemblages of kahikatea forest fragments in a pasture-dominated landscape. N Z J Ecol 24:57–67
Hobbs RJ, Yates CJ (2003) Impacts of ecosystem fragmentation on plant populations: generalising the idiosyncratic. Aust J Bot 51:471–488
Höft R, Höft M (1995) The differential effects of elephants on rain forest communities in the Shimba Hills, Kenya. Biol Conserv 73:67–79
Hulme PE (1996) Herbivory, plant regeneration, and species coexistence. J Ecol 84:609–615
Hunter MD (2002) Landscape structure, habitat fragmentation, and the ecology of insects. Agric For Entomol 4:159–166
Hunter MD, Varley GC, Gradwell GR (1997) Estimating the relative roles of top-down and bottom-up forces on insect herbivore populations: a classic study revisited. Proc Natl Acad Sci USA 94:9176–9181

Huntly N (1991) Herbivores and the dynamics of communities and ecosystems. Annu Rev Ecol Syst 22:477–503

Jokimäki J, Huhta E, Itämies J, Rahko P (1998) Distribution of arthropods in relation to forest patch size, edge, and stand characteristics. Can J For Res 28:1068–1072

Jules ES, Rathcke BJ (1999) Mechanisms of reduced *Trillium* recruitment along edges of old-growth forest fragments. Conserv Biol 13:784–793

Julião GR, Amaral MEC, Fernandes GW, Oliveira EG (2004) Edge effect and species-area relationships in the gall-forming insect fauna of natural forest patches in the Brazilian Pantanal. Biodivers Conserv 13:2055–2066

Kapos V (1989) Effects of isolation on the water status of forest patches in the Brazilian Amazon. J Trop Ecol 5:173–185

Kareiva P (1983) Influence of vegetation texture on herbivore populations: ressource concentration and herbivore movement. In: Denno RF, McClure MS (eds) Variable plants and herbivores in natural and managed systems. Academic, New York, pp 259–289

Kareiva P (1987) Habitat fragmentation and the stability of predator prey interactions. Nature 326:388–390

Khan ML, Tripathi RS (1991) Seedling survival and growth of early and late successional tree species as affected by insect herbivory and pathogen attack in subtropical humid forest stands of north-east India. Acta Oecol Int J Ecol 12:569–579

Kie JG, Bowyer RT, Nicholson MC, Boroski BB, Loft ER (2002) Landscape heterogeneity at differing scales: effects on spatial distribution of mule deer. Ecology 83:530–544

Knight TM, Holt RD (2005) Fire generates spatial gradients in herbivory: an example from a Florida sandhill ecosystem. Ecology 86:587–593

Kondoh M (2003) Habitat fragmentation resulting in overgrazing by herbivores. J Theor Biol 225:453–460

Kruess A, Tscharntke T (1994) Habitat fragmentation, species loss, and biological control. Science 264:1581–1584

Kruess A, Tscharntke T (2000) Species richness and parasitism in a fragmented landscape: experiments and field studies with insects on *Vicia sepium*. Oecologia 122:129–137

Lambrinos JG (2006) Spatially variable propagule pressure and herbivory influence invasion of chaparral shrubland by an exotic grass. Oecologia 147:327–334

Laurance WF (1990) Comparative responses of 5 arboreal marsupials to tropical forest fragmentation. J Mammal 71:641–653

Laurance WF (1994) Rain-forest fragmentation and the structure of small mammal communities in tropical Queensland. Biol Conserv 69:23–32

Laurance WF (2000) Do edge effects occur over large spatial scales? Trends Ecol Evol 15:134–135

Laurance WF (2001) Fragmentation and plant communities: synthesis and implications for landscape management. In: Bierregaard RO Jr, Gascon C, Lovejoy TE, Mesquita RCG (eds) Lessons from Amazonia: the ecology and conservation of a fragmented forest. Yale University Press, New Haven, pp 158–168

Laurance WF, Laurance SG, Ferreira LV, Rankin-de Merona JM, Gascon C, Lovejoy TE (1997) Biomass collapse in Amazonian forest fragments. Science 278:1117–1118

Laurance WF, Ferreira LV, Rankin-de Merona JM, Laurance SG (1998) Rain forest fragmentation and the dynamics of Amazonian tree communities. Ecology 79:2032–2040

Laurance WF, Delamonica P, Laurance SG, Vasconcelos HL, Lovejoy TE (2000) Conservation – rainforest fragmentation kills big trees. Nature 404:836–836

Laurance WF, Williamson GB, Delamonica P, Oliveira A, Lovejoy TE, Gascon C, Pohl L (2001) Effects of a strong drought on Amazonian forest fragments and edges. J Trop Ecol 17:771–785

Laurance WF, Lovejoy TE, Vasconcelos HL, Bruna EM, Didham RK, Stouffer PC, Gascon C, Bierregaard RO Jr, Laurance SG, Sampaio E (2002) Ecosystem decay of Amazonian forest fragments: a 22-year investigation. Conserv Biol 16:605–618

Laurance WF, Rankin-de Merona JM, Andrade A, Laurance SG, D'Angelo S, Lovejoy TE, Vasconcelos HL (2003) Rain-forest fragmentation and the phenology of Amazonian tree communities. J Trop Ecol 19:343–347

Laurance WF, Nascimento HEM, Laurance SG, Andrade AC, Fearnside PM, Ribeiro JEL, Capretz RL (2006) Rain forest fragmentation and the proliferation of successional trees. Ecology 87:469–482

Lavallée R, Archambault L, Morissette J (1996) Influence of drainage and edge vegetation on levels of attack and biological performance of the white pine weevil. For Ecol Manage 82:133–144

Lawes MJ, Lamb BCC, Boudreau S (2005) Area- but no edge-effect on woody seedling abundance and species richness in old Afromontane forest fragments. J Veg Sci 16:363–372

Leal IR, Oliveira PS (2000) Foraging ecology of attine ants in a Neotropical savanna: seasonal use of fungal substrate in the cerrado vegetation of Brazil. Insect Soc 47:376–382

Leopold A (1933) Game management. Scribners, New York

Levesque KR, Fortin M, Mauffette Y (2002) Temperature and food quality effects on growth, consumption and post-ingestive utilization efficiencies of the forest tent caterpillar *Malacosoma disstria* (Lepidoptera : Lasiocampidae). Bull Entomol Res 92:127–136

Lewke RE (1982) A comparison of foraging behavior among permanent, summer, and winter resident bird groups. Condor 84:84–90

Lopez-Barrera F, Manson RH, Gonzalez-Espinosa M, Newton AC (2006) Effects of the type of montane forest edge on oak seedling establishment along forest-edge-exterior gradients. For Ecol Manage 225:234–244

Lovejoy TE, Bierregaard RO Jr, Rylands AB, Malcom JR, Quintela CE, Harper LH, Brown KS Jr (1986) Edge and other effects of isolation on Amazon forest fragments. In: Soule ME (ed) Conservation biology: the science of scarcity and diversity. Sinauer, Sunderland, pp 257–285

MacGarvin M, Lawton JH, Heads PA (1986) The herbivorous insect communities of open and woodland bracken – observations, experiments and habitat manipulations. Oikos 47:135–148

Major RE, Christie FJ, Gowing G, Cassis G, Reid CAM (2003) The effect of habitat configuration on arboreal insects in fragmented woodlands of south-eastern Australia. Biol Conserv 113:35–48

Malcolm JR (1994) Edge effects in central Amazonian forest fragments. Ecology 75:2438–2445

Malcolm J (1997) Biomass and diversity of small mammals in Amazonian fragments. In: Laurance WF, Bierregaard RO Jr (eds) Tropical forests remnants: ecology, management, and conservation of fragmented communities. University of Chicago Press, New York, pp 207–221

Marquis RJ (1984) Leaf herbivores decreased fitness of a tropical plant. Science 226:537–539

Martinez JJI, Mokady O, Wool D (2005) Patch size and patch quality of gall-inducing aphids in a mosaic landscape in Israel. Landsc Ecol 20:1013–1024

Martinez-Garza C, Pena V, Ricker M, Campos A, Howe HF (2005) Restoring tropical biodiversity: leaf traits predict growth and survival of late-successional trees in early-successional environments. For Ecol Manage 217:365–379

Matlack GR (1993) Microenvironment variation within and among forest edge sites in the eastern United States. Biol Conserv 66:185–194

Matlack GR, Litvaitis JA (1999) Forest edges. In: Hunter JML (ed) Maintaining diversity in forest ecosystems. Cambridge University Press, Cambridge, pp 210–233

Mattson WJ, Addy ND (1975) Phytophagous insects as regulators of forest primary production. Science 190:515–522

McGeoch MA, Gaston KJ (2000) Edge effects on the prevalence and mortality factors of *Phytomyza ilicis* (Diptera, Agromyzidae) in a suburban woodland. Ecol Lett 3:23–29

Meiners SJ, Handel SN, Pickett STA (2000) Tree seedling establishment under insect herbivory: edge effects and interannual variation. Plant Ecol 151:161–170

Meiners SJ, Pickett STA, Handel SN (2002) Probability of tree seedling establishment changes across a forest-old field edge gradient. Am J Bot 89:466–471

Meyer ST, Roces F, Wirth R (2006) Selecting the drought stressed: effects of plant stress on intraspecific and within-plant herbivory patterns of the leaf-cutting ant *Atta colombica*. Funct Ecol 20:973–981

Moore LV, Myers JH, Eng R (1988) Western tent caterpillars prefer the sunny side of the tree, but why? Oikos 51:321–326

Moutinho P, Nepstad DC, Davidson EA (2003) Influence of leaf-cutting ant nests on secondary forest growth and soil properties in Amazonia. Ecology 84:1265–1276

Murcia C (1995) Edge effects in fragmented forests: application for conservation. Trends Ecol Evol 10:58–62

Nichols-Orians CM (1991) The effects of light on foliar chemistry, growth and susceptibility of seedlings of a canopy tree to an attine ant. Oecologia 86:552–560

Odum EP (1971) Fundamentals of ecology, 3rd edn. Saunders, Philadelphia

Oksanen L, Fretwell SD, Arruda J, Niemela P (1981) Exploitation ecosystems in gradients of primary productivity. Am Nat 118:240–261

Oliveira MA, Grillo AS, Tabarelli M (2004) Forest edge in the Brazilian Atlantic forest: drastic changes in tree species assemblages. Oryx 38:389–394

Ostfeld RS, Manson RH, Canham CD (1997) Effects of rodents on survival of tree seeds and seedlings invading old fields. Ecology 78:1531–1542

Ozanne CMP, Speight MR, Hambler C, Evans HF (2000) Isolated trees and forest patches: Patterns in canopy arthropod abundance and diversity in *Pinus sylvestris* (Scots Pine). For Ecol Manage 137:53–63

Palmer SCF, Gordon IJ, Hester AJ, Pakeman RJ (2005) Introducing spatial grazing impacts into the prediction of moorland vegetation dynamics. Landsc Ecol 19:817–827

Patel A, Rapport DJ (2000) Assessing the impacts of deer browsing, prescribed burns, visitor use, and trails on an oak-pine forest: Pinery Provincial Park, Ontario, Canada. Nat Areas J 20:250–260

Pickett STA, Cadenasso ML, Jones CG (2000) Generation of heterogeneity by organism: creation, maintenance and transformation. In: Hutchings MJ, John EA, Stewart AJA (eds) The ecological consequences of environmental heterogeneity. Blackwell Science, Oxford, pp 33–52

Pyke GH, Pulliam HR, Charnov EL (1977) Optimal foraging – selective review of theory and tests. Q Rev Biol 52:137–154

Rand TA, Tylianakis JM, Tscharntke T (2006) Spillover edge effects: the dispersal of agriculturally subsidized insect natural enemies into adjacent natural habitats. Ecol Lett 9:603–614

Rao M, Terborgh J, Nunez P (2001) Increased herbivory in forest isolates: implications for plant community structure and composition. Conserv Biol 15:624–633

Richards LA, Coley PD (2007) Seasonal and habitat differences affect the impact of food and predation on herbivores: a comparison between gaps and understory of a tropical forest. Oikos 116:31–40

Ries L, Fagan WF (2003) Habitat edges as a potential ecological trap for an insect predator. Ecol Entomol 28:567–572

Ries L, Fletcher RJ, Battin J, Sisk TD (2004) Ecological responses to habitat edges: mechanisms, models, and variability explained. Annu Rev Ecol Evol System 35:491–522

Roland J (1993) Large-scale forest fragmentation increases the duration of tent caterpillar outbreak. Oecologia 93:25–30

Roland J, Kaupp WJ (1995) Reduced transmission of forest tent caterpillar (Lepidoptera, Lasiocampidae) nuclear polyhedrosis-virus at the forest edge. Environ Entomol 24:1175–1178

Rooney TP, Waller DM (2003) Direct and indirect effects of white-tailed deer in forest ecosystems. For Ecol Manage 181:165–176

Rothman LD, Roland J (1998) Forest fragmentation and colony performance of forest tent caterpillar. Ecography 21:383–391

Saunders DA, Hobbs RJ, Margules CR (1991) Biological consequences of ecosystem fragmentation: a review. Conserv Biol 5:18–32

Sheil D, Salim A (2004) Forest tree persistence, elephants, and stem scars. Biotropica 36:505–521

Shure DJ, Wilson LA (1993) Patch-size effects on plant phenolics in successional openings of the southern Appalachians. Ecology 74:55–67

Sork VL (1983) Distribution of pignut hickory (*Carya glabra*) along a forest to edge transect, and factors affecting seedling recruitment. Bull Torrey Bot Club 110:494–506

Stone C, Bacon PE (1994) Relationships among moisture stress, insect herbivory, foliar cineole content and the growth of river red gum *Eucalyptus camaldulensis*. J Appl Ecol 31:604–612
Stone C, Bacon PE (1995) Leaf dynamics and insect herbivory in a *Eucalyptus camaldulensis* forest under moisture stress. Aust J Ecol 20:473–481
Strelke WK, Dickson JG (1980) Effect of forest clear-cut edge on breeding birds in east Texas. J Wildl Manage 44:559–567
Tabarelli M, Gascon C (2005) Lessons from fragmentation research: improving management and policy guidelines for biodiversity conservation. Conserv Biol 19:734–739
Tabarelli M, Da Silva MJC, Gascon C (2004) Forest fragmentation, synergisms and the impoverishment of neotropical forests. Biodivers Conserv 13:1419–1425
Terborgh J (1992) Maintenance of diversity in tropical forests. Biotropica 24:283–292
Terborgh J, Lopez L, Nunez P, Rao M, Shahabuddin G, Orihuela G, Riveros M, Ascanio R, Adler GH, Lambert TD, Balbas L (2001) Ecological meltdown in predator-free forest fragments. Science 294:1923–1926
Terborgh J, Feeley K, Silman M, Nunez P, Balukjian B (2006) Vegetation dynamics of predator-free land-bridge islands. J Ecol 94:253–263
Thies C, Steffan-Dewenter I, Tscharntke T (2003) Effects of landscape context on herbivory and parasitism at different spatial scales. Oikos 101:18–25
Tscharntke T, Brandl R (2004) Plant–insect interactions in fragmented landscapes. Annu Rev Entomol 49:405–430
Tscharntke T, Steffan-Dewenter I, Kruess A, Thies C (2002) Characteristics of insect populations on habitat fragments: a mini review. Ecol Res 17:229–239
Ulyshen MD, Hanula JL, Horn S, Kilgo JC, Moorman CE (2005) Herbivorous insect response to group selection cutting in a southeastern bottomland hardwood forest. Environ Entomol 34:395–402
Urbas P, Araújo MV Jr, Leal IR, Wirth R (2007) Cutting more from cut forests – edge effects on foraging and herbivory of leaf-cutting ants. Biotropica 39:489–495
Valladares G, Salvo A, Cagnolo L (2006) Habitat fragmentation effects on trophic processes of insect–plant food webs. Conserv Biol 20:212–217
Vasconcelos HL, Cherrett JM (1997) Leaf-cutting ants and early forest regeneration in central amazonia – effects of herbivory on tree seedling establishment. J Trop Ecol 13:357–370
Wahungu GM, Catterall CP, Olsen MF (1999) Selective herbivory by red-necked pademelon *Thylogale thetis* at rainforest margins: factors affecting predation rates. Aust J Ecol 24:577–586
Wahungu GM, Catterall CP, Olsen MF (2002) Seedling predation and growth at a rainforest–pasture ecotone, and the value of shoots as seedling analogues. For Ecol Manag 162:251–260
White TCR (1984) The abundance of invertebrate herbivores in relation to the availability of nitrogen in stressed food plants. Oecologia 63:90–105
White TCR (2007) Flodded forests: death by drowning, not herbivory. J Veg Sci 18:147–148
Whitmore TC (1997) Tropical forest disturbance, disappearance, and species loss. In: Laurance WF, Bierregaard RO Jr (eds) Tropical forest remnants: Ecology, management, and conservation of fragmented communities. University of Chicago Press, Chicago, pp 3–12
Wirth R, Herz H, Ryel R, Beyschlag W, Hölldobler B (2003) Herbivory of leaf-cutting ants – a case study on *Atta colombica* in the tropical rainforest of Panama. Springer, Berlin Heidelberg New York
Wirth R, Meyer ST, Almeida WR, Araújo MV Jr, Barbosa VS, Leal IR (2007) Increasing densities of leaf-cutting ants (*Atta* spp.) with proximity to the edge in a Brazilian Atlantic forest. J Trop Ecol 23:501–505. DOI 10.1017/S0266467407004221
Woodroffe R, Ginsberg JR (1998) Edge effects and the extinction of populations inside protected areas. Science 280:2126–2128
Yahner RH (1988) Changes in wildlife communities near edges. Conserv Biol 2:333–339
Zamora R, Hodar JA, Gomez JM (1999) Plant–herbivore interaction: beyond a binary vision. In: Pugnaire FI, Valladares F (eds) Handbook of functional plant ecology. Dekker, New York, pp 677–718

Getting Plant–Soil Feedbacks out of the Greenhouse: Experimental and Conceptual Approaches

Andrew Kulmatiski (✉) and Paul Kardol

Contents

Abstract Plant-soil feedbacks (PSFs) contain two steps: (a) different plant species must create different soils and (b) these different soils must alter subsequent plant growth. Two-phase experiments are used to demonstrate PSF. In Phase I, soils are cultivated by known plants. In Phase II, plants are used as phytometers to measure soil differences created in Phase I. Ideally, a third phase, Phase III, is performed in which soil analyses and inoculation experiments are used to identify mechanisms of PSFs. A review of the literature reveals 25 studies that performed phases I and II. Of these 25 studies, 22 used grassland species, 21 were performed in a greenhouse, 19 were published between 2004 and 2006, and 12 used non-native species. Across studies, 271 feedback values were reported. Negative feedbacks represented 147 of the 271 values and produced, on average, a 37% reduction in plant biomass. Positive feedbacks represented 41 tests and produced, on average,

Andrew Kulmatiski
Department of Wildland Resources and the Ecology Center Utah State University
Logan, UT 84322, USA
e-mail: andrewkulmatiski@hotmail.com

a 43% increase in biomass. The remaining 83 tests showed no significant PSFs. This review reveals that: (a) PSFs can be important to plant growth and community development relative to competing explanations of plant growth, (b) PSFs may be most important in invaded plant communities and during successional species replacements, and (c) greenhouse experiments using grassland species dominate the PSF literature. We suggest that future research should measure PSFs under field conditions in a wide variety of ecosystems. We discuss conceptual and experimental advances that are needed to extend PSF research to field conditions. As an example, we present a conceptual model, the contingent niche-creation model, that explores the interaction between disturbance and PSF. Research opportunities in this field are abundant and likely to provide key insights into plant growth and community assembly.

1 Introduction

For thousands of years farmers have known that plants grow better on some soils than others (Ehrenfeld et al. 2005). It is clear that some soil factors increase growth (i.e., nutrients, symbionts) while other factors decrease growth (i.e., fine-root grazers, pathogens). In the past century, it was recognized that plants can change the biology, chemistry, and physical nature of soils through mechanisms such as root exudation, litter deposition, and susceptibility to herbivores. Plant–soil feedbacks (PSFs) can be inferred from these two components, but determining that a PSF exists requires that both plant effects on soils *and* soil effects on plants are demonstrated.

1.1 The PSF Model and Approach

1.1.1 Individual PSFs

Ideally, PSFs are measured by growing plants in the field to cultivate soils using known plant species: this is called Phase I. The effects of soil cultivation are then tested in Phase II by contrasting plant growth in 'self-cultivated' (self) soils to growth in soils that were not self-cultivated (Bever 1994). Finally, in Phase III, the soil conditions (e.g., symbionts, pathogens, nutrient status) found to be associated with significant changes in plant growth are identified and recreated to determine the mechanism of PSF (Bever 1994; Klironomos 2002; Kardol et al. 2007). This field-based approach, however, requires large amounts of time to execute, lacks control over factors external to the experiment, and increases the difficulty of examining component effects responsible for any observed PSFs. Because of these limitations, no feedback study to date has been performed in this way.

To avoid the time required to cultivate field soils (Phase I), researchers often use a natural experiment approach, where field soils with known plant growth histories are used as the cultivated soil. These soils can be used in field plots, pots in the field, or pots in greenhouses. A strength of this approach is that it uses soils conditioned by known plants for long periods of time under natural field conditions. A weakness of this approach is that soils are assumed to only reflect previous plant growth and not site conditions, such as parent material. To control for differences that are not from previous plant growth, field soils can be collected and used as inocula as opposed to whole soil. This approach isolates plant–microbial feedbacks from plant–physical and plant–chemical feedbacks. The inocula approach often uses disturbed, fertilized soils in microcosms under greenhouse conditions because sterilized soils are required. Plant growth on these soils is often rapid, but may not reflect plant growth under field conditions (Bonanomi and Mazzoleni 2005).

An alternative to the natural experiment approach is to have known plants grow on soils, Phase I, as a part of the experiment. This approach allows better control over Phase I, but in practice homogenized, disturbed soils are often used. We refer to this approach as the cultivation experiment approach.

Once soils with known plant growth histories are obtained, the effects of these plant growth histories are determined using plants as phytometers (Phase II). Several approaches have been used to assess plant growth in Phase II. Plant growth on self-cultivated ('self') soil can be compared to growth on soil cultivated by other species ('other'; Bever 1994). We refer to this as a self-other, individual PSF; this approach has also been referred to as a 'home vs away' comparison (Klironomos 2002). It is important that a large number of 'other' species is used to reduce bias in the calculation of feedback and to make results more general. The self-other technique is good for identifying how a plant will grow on soils cultivated by other species, but if few 'other' species are used results are limited.

Plant growth on self-cultivated soil can also be compared to growth on control soils (soils that have not grown plants during the experiment) or soils that have been sterilized. Importantly, soil sterilization increases nutrient availability, which should be controlled (Troelstra et al. 2001). We refer to PSFs calculated in these ways as self-control and self-sterilized individual PSFs, respectively. These techniques are good for identifying the effect of a plant on itself, but reveal little about how that plant will grow on soils cultivated by other species.

Plant growth at the end of Phase II is typically measured as above-ground biomass, though other measures can be used, such as below-ground biomass, productivity, seed production, or survivorship. While above-ground biomass is most commonly used, other measures may be more sensitive or appropriate to specific research questions (Bever 1994; De Deyn et al. 2004a; Bezemer et al. 2006b). More specifically, above-ground measures are likely to be more sensitive in mesic environments where light is limiting; belowground measures are likely to be more sensitive in semi-arid and arid environments where water and soil resources are limiting.

Individual PSFs can be reported as absolute or standardized values (e.g., [self-sterilized] or [{self-sterilized}/self], respectively). Both values reflect an integrated measure of the plant–soil interaction and can be either positive or negative.

Unfortunately, standardized values are constrained by $-\infty$ and +1, making it difficult to compare a −10 value to a +0.8 value, for example. To avoid this problem values can be standardized by dividing by the largest growth value (e.g., either self or sterilized). This calculation constrains all feedback values to between −1 and +1 and presents a proportional change in biomass. Using this calculation, positive and negative PSF values are comparable within and between studies.

1.1.2 Net-Pairwise PSFs

Using assumptions common to population modeling, Bever et al. (1997) incorporated individual PSFs into spatially explicit continuous growth models that predict the population dynamics of more than one plant species (Fig. 1; Bever et al. 1997; Bever 1999, 2003). These models utilize net-pairwise PSFs (net PSFs) to predict plant growth as a function of plant–soil–plant interactions. Net PSFs are sometimes also called indirect feedbacks because the effect of one plant on another via soil is measured. When determining net PSFs, a self-other individual PSF value is determined for each species. Using this approach, Bever et al. (1997) demonstrated that the interaction coefficient I (i.e., $I = [\alpha_A - \beta_A - \alpha_B + \beta_B]$; see Fig. 1) of their growth model can predict both stable coexistence (a negative interaction coefficient) and competitive exclusion (a positive interaction coefficient; Fig. 1).

The model developed by Bever et al. (1997) demonstrated that net PSFs can be more important than the direction or magnitude of individual PSFs (Mills and Bever 1998; Bever 2003; Reynolds et al. 2003). For example, a plant A that has a positive PSF (e.g., decreases fine-root grazing) may be expected to improve its own growth. However, if competing plant B suffers more from fine-root grazing than plant A, then plant A (by reducing fine-root grazing) may benefit the growth of plant B more than it benefits itself (Eppinga et al. 2006). Under these conditions plant B could outcompete plant A. But there could be stable coexistence, if plant B benefits plant A more than it benefits itself.

Essentially, the model demonstrates that, all else being equal, interspecific effects must be greater than intraspecific effects for two species to coexist. An

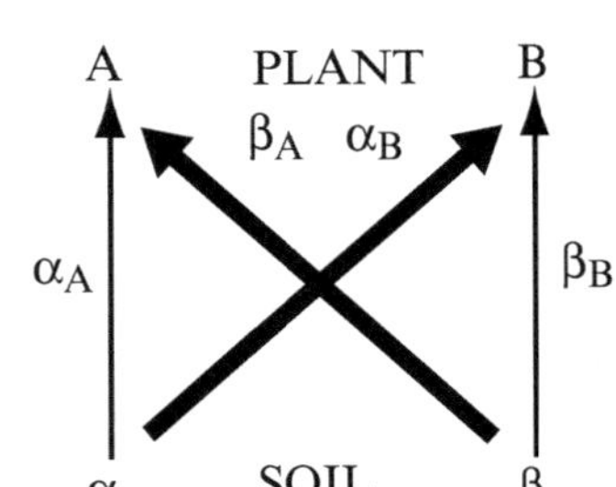

Fig. 1 An example of a net PSF where plant A cultivates a soil type α and plant B cultivates a soil type β (Bever et al. 1997). If the soil type cultivated by one plant species is persistent and has measurable effects on subsequent plant growth, then this effect can be measured as α_A, α_B, β_A, or β_B. A plant that grows better on soils cultivated by heterospecifics (α_B or β_A) than on soils cultivated by conspecifics (α_A or β_B) is said to have a negative individual feedback. This effect is quantified using differences in plant growth. Coexistence of competitively equivalent species requires that interspecific effects (α_B or β_A) are greater than intraspecific effects (α_A or β_B)

important caveat to this rule of thumb is that, if both interspecific and intraspecific effects favor one species over another, there will not be stable coexistence, even if the interaction coefficient is negative. For example, Bonanomi and Mazzoleni (2005) found that a plant (A) grew to 8 g on self soil and 16 g on the soil of another plant (B). They also found that plant B grew to 25 g on the soil of A and 21 g on its own soil. The sum of these growth values (i.e., I) is −12 g. Despite the negative interaction coefficient, in this case a stable equilibrium does not exist because plant B always grows better than plant A. Note that net PSFs are not standardized and reflect the measure of growth used.

Net-pairwise PSFs highlight the importance of species identity to plant and soil community development (Porazinska et al. 2003; De Deyn et al. 2004b), yet the model used to predict net-pairwise PSFs assumes that: (a) the species of interest are competitively equivalent, (b) changes in the soil community produce linear responses in plant growth, and (c) there is no density dependence within plant populations (Bever et al. 1997). While convenient for modeling, it is likely that these assumptions are violated. Competitive inequality and density-dependent growth among plants have been shown to impose theoretical constraints on the role of PSFs (Fig. 2; Bever 2003). However, perhaps more importantly, the same model demonstrated that negative net PSFs can allow coexistence between plant species

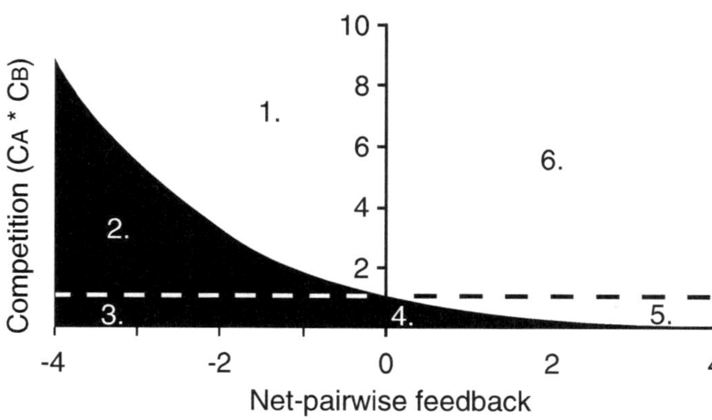

Fig. 2 Competition and feedback conditions under which two plant species (A, B) are predicted to coexist (*shaded area*). The *y*-axis represents the product of the competition coefficients for species A and B. The *x*-axis represents the net feedback value (i.e., $I = [\alpha_A - \beta_A - \alpha_B + \beta_B]$). According to the PSF model, species coexist when the net feedback value is negative (i.e., left of the *y*-axis). According to a Lotka–Volterra competition model, the two plant species coexist when the product of the competition coefficients is less than one (i.e., under the *hatched line*). Strong interspecific competition prevents coexistence in area *1*. Net negative PSFs allow species that would otherwise competitively exclude each other to coexist in area *2*. Strong intraspecific competition allows species with a net positive PSF to coexist in area *3*. Strong intraspecific competition and strong net negative PSFs allow species coexistence in area *4*. Strong positive net PSFs prevent coexistence despite strong intraspecific competition in area *5*. Interspecific competition and positive net PSF prevent coexistence under the conditions in area *6*. Figure adapted from Bever (2003)

that would otherwise competitively exclude one another. The effects of non-linear plant–soil interactions remain unresolved.

One additional factor that can change predictions of the net PSF model is spatial scale. When examining any particular location, positive net PSFs are predicted to result in competitive exclusion. When looking at larger spatial scales, however, positive net PSFs, have been shown, theoretically, to produce persistent single-species patches that improve species richness (Molofsky et al. 2001). This has not been demonstrated experimentally. It may be important, therefore, to consider the spatial scale over which PSFs are considered.

2 Literature Review

2.1 Experimental Approaches

To collect information on PSFs, we performed a search on *Web of Science* and contacted researchers who perform PSF research to obtain datasets that were 'in press'. Six PSF values from one unpublished dataset were also included (Kulmatiski, unpublished data). We accepted studies containing data of plant growth on 'self-sterilized' or 'self-other' soils. Studies that explicitly tested the effects of pathogens or symbionts were not included. Studies that included several PSF values for the same species (e.g., PSFs developed on different soils or with different plants) were included. This search produced 25 studies (Table 1). These studies reported 271 individual feedback values, 134 of which were from only three studies (Klironomos 2002; Van der Stoel et al. 2002; Kardol et al. 2006). Feedback values reported throughout this chapter are standardized individual PSF values, not net PSF values.

2.1.1 The Natural-Experiment and Cultivation-Experiment Approaches

The natural-experiment approach was used in 15 studies, the cultivation-experiment approach was used in eight studies, and both approaches were used in two studies. These two approaches appeared to produce different results. The absolute PSF values averaged 0.34 and 0.21 for the natural-experiment and the cultivation-experiment approaches, respectively. These values were significantly different in a *t*-test, but this should be interpreted with caution because the number of studies was small and other aspects of the experimental designs, as described above, were not controlled. Two studies used both the natural-experiment and the cultivation-experiment approach (Callaway et al. 2004a; Van der Putten et al. 2007). Callaway et al. (2004a) found that feedbacks determined using field soils as inocula, i.e., the natural-experiment approach) produced individual PSF values of −0.5 and −0.2 for the short-lived perennial forb *Centaurea maculosa* in native and non-native soils, respectively. Feedbacks determined using experimentally-cultivated soils produced

Table 1 Studies reporting 'self-other' or 'self-sterilized' plant growth. *EC* Experimental cultivation, *NE* natural experiment

Reference	Ecosystem	Approach (EC/NE)	Approach (calculation)	Species (origin)	Setting	Soil	Number of species	Phase I	Phase II
Agrawal et al. (2005)	Grassland	EC	Self-other	Native/non-native	Greenhouse	Whole soil	20	4 months	4 months
Bever (1994)	Grassland	EC	Self-other	Native	Greenhouse	Inoculum	3	15 months	3 months
Bezemer et al. (2006a)	Grassland	EC	Self-other	Native	Greenhouse	Inoculum	1	6 years	2.5 months
Bezemer et al. (2006b)	Grassland	NE	Self-other	Native	Greenhouse	Whole soil	11	Field soils	3.8 months
Bodelier et al. (2006)	Wetland	NE	Self-sterilized	Native	Greenhouse	Whole soil	1	Field soils	1.5 months
Bonanomi et al. (2005b)	Grassland	NE	Self-other	Native	Greenhouse	Whole soil	1	Field soils	2 months
Bonanomi and Mazzoleni (2005)	Grassland	EC	Self-other	Native	Greenhouse	Whole soil	3	3.5 months	2 months
Callaway et al. (2004a)	Grassland	EC	Self-other	Non-native	Greenhouse	Inoculum	1	2 years	3.8 months
Callaway et al. (2004b)	Grassland	EC/NE	Self-other and sterilized	Non-native	Greenhouse	Inoculum	1	Field soils, 4 months	5 months
De Deyn et al. (2004a)	Grassland	NE	Self-sterilized	Native	Greenhouse	Whole soil	Comm.	Field soils	6 months
Gustafson and Casper (2004)	Grassland	NE	Self-other	Native	Pots in field/green-house	Inoculum	3	Field soils	3 months
Kardol et al. (2006)	Grassland	EC	Self-other	Native	Greenhouse	Inoculum	12	5 months	5 months
Kardol et al. (2007)	Grassland	EC	Self-other	Native/non-native	Greenhouse	Inoculum	6	2 months	2 months

(continued)

Table 1 (continued)

Reference	Ecosystem	Approach (EC/NE)	Approach (calculation)	Species (origin)	Setting	Soil	Number of species	Phase I	Phase II
Klironomos (2002)	Grassland	EC	Self-other	Native/non-native	Greenhouse	Inoculum	71	5 months	2.5 months
Knevel et al. (2004)	Dunegrass	NE	Self-sterilized	Native/non-native	Greenhouse	Inoculum	1	Field soils	1.5 months
Kulmatiski et al. (2006)	Grassland	NE	Self-other	Native/non-native	Field	Whole soil	Comm.	Field soils	3 years
Kulmatiski (unpublished data)	Grassland	NE	Self-other	Native/non-native	Field	Whole soil	6	Field soils	3 years
Macel et al. (2007)	Grassland	NE	Self-other	Native	Pots in field	Whole soil	2	Field soils	2 years
Morris et al. (2006)	Grassland	NE	Self-other	Non-native	Greenhouse	Whole soil	1	Field soils	3 months
Peltzer (2001)	Grassland	NE	Self-other	Native	Pots in field	Whole soil	2	Field soils	1 year
Reinhart et al. (2003)	Forest	NE	Self-other and sterilized	Native/non-native	Greenhouse	Inoculum	1	Field soils	2 months
Reinhart and Callaway (2004)	Forest	NE	Self-other and sterilized	Native/non-native	Greenhouse	Inoculum	2	Field soils	1 year
Van der Putten (1997)	Dunegrass	NE	Self-other	Native	Greenhouse	Whole soil	3	Field soils	4 months
Van der Putten et al. (2007)	Grassland	EC/NE	Self-other and sterilized	Native/non-native	Greenhouse	Inoculum	2	Field and 1 month	2 months
Van der Stoel et al. (2002)	Dunegrass	NE	Self-sterilized	Native	Greenhouse	Whole soil	1	Field soils	2 months

individual PSF values of 0.0 and 0.4 for *C. maculosa* in native and non-native soils. In this case, experimental cultivation resulted in an accumulation of symbionts and/or a loss of pathogens relative to field soil inoculations. In a PSF experiment using the natural-experiment approach, Van der Putten et al. (2007) found that two native grasses demonstrated feedback values of 0.0 and a non-native grass demonstrated a feedback value of 0.4. When feedback values were determined for the same species using experimentally cultivated soils, the native grasses demonstrated negative feedbacks of –0.9 and –0.6 and the non-native grass demonstrated a feedback value of 0.0. In both studies, plants on non-native soils developed more positive PSFs than plants on native soils, but the values determined from the natural-experiment and cultivated-experiment approaches within each study were very different. It is unknown if the different results produced by the two techniques reflect external factors in field soils, an inability for the cultivation approach to replicate field conditions, or both.

2.1.2 'Self-Sterilized' and 'Self-Other' Approaches

Phase II plant growth can be assessed from self-sterilized or self-other comparisons. Four studies used the self-sterilized approach only. The remaining 21 studies used the self-other or both approaches. The self-sterilized approach appeared to produce larger effect sizes. The average of the absolute value of effects sizes was 0.46 and 0.21 for the self-sterilized and self-other approaches, respectively. The smaller effect size produced by the self-other technique suggested that there is some redundancy among soil communities in regards to their effects on plants. The 'self-sterilized' technique is more likely to produce statistically significant results, but the relevance to plant growth under field conditions is less obvious.

2.1.3 General Patterns in Experimental Approaches

Of the 25 studies reviewed, 14 (56%) isolated plant–microbial feedbacks by using soil inoculum as opposed to whole soils. Of the 25 studies reviewed, 21 were performed in a greenhouse, two were performed in pots in field soils, and two were performed in undisturbed field soils. The greenhouse approach can control for factors external to the research question, but typically involves a short growing phase (Phase I), disturbed soils, and fails to assess the many potential interactions that occur among plants and soils under field conditions. The use of the natural experiment approach and/or greenhouses allowed 17 of the studies to be conducted in less than a year. The remaining eight studies required between one and six years. All but three PSF studies were performed with grassland species. This is likely to reflect the time and space required to grow plants for two or more generations. Notable exceptions include work on *Prunus* and *Acer* tree species (Packer and Clay 2000; Reinhart et al. 2005a, b).

2.1.4 Plant–Soil Interactions that May not Create PSFs

Of the 271 PSF values reported, 83 (31%) were not significant, although a publishing bias could be expected to inflate the number of significant results reported. Of the 25 studies reviewed, 18 reported non-significant PSF values. However, 16 of the 25 studies were for multiple species and should be more insulated from a publishing bias; 81% of these studies reported non-significant PSF values. Of the nine single species studies, 33% presented non-significant PSF values. Only one single species study and one multiple species study reported only non-significant PSF values (Callaway et al. 2004; Macel et al. 2007).

2.2 Conceptual Approaches

2.2.1 Succession

Some of the earliest ecological research on PSFs formed around the observation of plant succession in coastal foredune communities (Van der Putten et al. 1988, 1993). PSF values of foredune species were particularly negative, which suggest that PSF plays a role in successional replacements. Since then, advances in this field have been slow, and only six of the 25 studies reviewed explicitly examined the role of PSFs in succession (Table 1). The direction of PSF (negative or positive) was suggested to vary across successional gradients (Reynolds et al. 2003). One of the best examples of this was published in 2006 (Kardol et al. 2006). This study demonstrated that, in secondary grassland succession, negative PSFs explained the replacement of early-successional species by late-successional species, and positive PSFs explained the replacement of mid-successional species by late-successional species. Fast growth and poor defense of early secondary succession species (ruderals, pioneers) potentially make them particularly vulnerable to natural enemies and therefore to negative PSF. In contrast, late-successional species may depend more on associations with beneficial soil organisms in their rhizosphere, such as mycorrhizal fungi and growth-promoting bacteria (Van Bruggen and Semenov 2000; De la Pena et al. 2006). Interestingly, the positive feedback values of late-successional species depended on soil origin and were highest in late-successional soil, which could be related to the status of mycorrhizal fungi in the soil.

Results obtained to date are promising, but our understanding of multiple plant species effects on soils and soils effects on multiple plant species limits our understanding of PSFs and succession. Moreover, effects of PSFs on plant community succession may be much more complex than previously thought. A recent study by Kardol et al. (2007) shows two interacting mechanisms by which early-successional PSFs can affect plant community succession: (a) through plant species accumulation of soil pathogens, and (b) through the biotic legacy effects that early-successional species provide. These legacy effects can lead to historical contingent effects in later successional plant community assembly.

2.2.2 Species Coexistence

Soon after work began on the role of PSFs in succession, the PSF conceptual model was applied to plant species coexistence (Bever 1994). As described above, negative net PSFs are expected to result in species coexistence. Of 271 feedback values obtained from the literature, 147 were negative and represented, on average, a 37% reduction in biomass. Positive feedbacks represented 41 tests and explained, on average, a 43% increase in biomass. The remaining 83 tests showed no significant difference among treatments, or a PSF value of zero. These findings suggest that individual feedbacks are commonly negative and are, therefore, likely to encourage species coexistence (Bever et al. 1997). In support of this, Klironomos (2002) observed that species that demonstrated negative PSFs were less abundant on the landscape. Theoretically, positive PSFs can also allow species coexistence by encouraging the growth of subordinate species (Bever 2003; Kardol et al. 2006). Two studies examined the effects of feedbacks and competition, but formal tests of Bever's (2003) model were not performed as a part of these studies (Bonanomi and Mazzoleni 2005;, Kardol et al. 2007; Fig. 2).

2.2.3 Biogeography

Given that PSFs reflect differences among plant species, it may be expected that differences in coevolution would create different PSFs for natives and non-natives (Callaway and Aschehoug 2000). Three mechanisms are suggested to create positive PSFs for non-natives: enemy release, enhanced mutualism, and indirect feedback. More specifically, soils in adoptive habitats are expected to be relatively enemy-free (enemy release hypothesis, ERH) or symbiont-rich (enhanced mutualism hypothesis, EMH) because root herbivores and pathogens have not coevolved to specialize on non-native species while vesicular arbuscular mycorrhiza (VAM) are common symbionts and are generalists (Reinhart and Callaway 2006). Alternatively, it is suggested that non-native species can accumulate pathogens but be resistant to these pathogens relative to natives, resulting in an indirect positive PSF (Eppinga et al. 2006). When non-natives perpetuate or accentuate low pathogen abundance (ERH), high symbiont abundance EMH), or high abundance of pathogens to which they are more resistant than native species (indirect positive PSF), a positive PSF results. For example, *Prunus serotina* in Europe demonstrated a positive PSF that resulted from a release of fungal pathogens (Reinhart et al. 2003). *Centaurea maculosa* in North America demonstrated a positive PSF that resulted from C and P transfers from native grasses to itself via VAM (Carey et al. 2004). The dune grass *Ammophila arenaria* accumulated pathogens, but these pathogens were proposed to decrease the growth of plant competitors more than it decreased its own growth (Eppinga et al. 2006).

Plant–soil feedbacks for species in their non-native ranges were examined in 12 of the 25 studies. Of the 271 PSF values found in the literature, 192 feedback values were determined for species in their native ranges and the average feedback value for these species was –0.19. The remaining 78 were determined for non-native species and the average feedback value for these species was –0.02. For species considered invasive, the average PSF value was 0.11. Explicit examination of a species in its home and adoptive ranges was examined in only four studies. In each of these studies, plants were found to have more positive PSFs in their adoptive ranges than in their home range (Reinhart et al. 2003; Knevel et al. 2004; Reinhart and Callaway 2004; Van der Putten et al. 2007). Studies on the role of PSF for non-natives can only draw inference on the effect of PSFs for non-native species that became successful, because data are not available for the non-native species that were introduced and failed to establish. Overall, these data suggest that relatively positive individual PSFs may help explain the growth of successful non-natives.

3 Directions for Future PSF Research

Experimental data suggest that PSFs may play an important role in plant growth and community development, but the great majority of this data was derived from greenhouse experiments using monocultures of grassland plant species. We suggest that the role of PSFs under field conditions will remain uncertain until the following topics are addressed:

1. The ability of plant communities to develop and respond to changes in soil communities (community-level PSFs; De Deyn et al. 2004a);
2. The effect of time on the development of PSFs (Bever 2003);
3. The spatial extent over which PSFs function (Eppstein et al. 2006; Levine et al. 2006);
4. the conditions on which PSF are dependent, such as above-ground herbivory, plant diversity, and productivity (Van der Putten et al. 2001);
5. The effects of soil history on PSFs (i.e., historical legacies such as microbial community size, tillage, or nitrogen deposition; Kulmatiski et al. 2006; Kardol et al. 2007).

We discuss these five topics in order in the next five sections.

3.1 Community-Level PSFs

Research has emphasized individual PSFs. A greater effort must be made to understand net-pairwise and multiple species PSFs. In the following section, we explore an underappreciated aspect of net-pairwise PSFs that should help guide research on these interactions. More specifically we demonstrate that, according to the PSF model, coexisting plants must improve each other's growth. This provides a testable prediction of the PSF model and also provides an alternative explanation for observed diversity/productivity relationships (Spehn et al. 2005).

As stated above, negative PSF species coexist if they demonstrate negative net PSFs (Fig. 1). Because negative individual PSFs are thought to be caused by soil pathogens and root grazers, soils cultivated by negative PSF species are likely to demonstrate large pathogen and root grazer populations. For negative PSF species to coexist, interacting plant species must grow better on interspecific soils than intraspecific soils. In other words, negative PSF species must grow better together than they do alone to coexist.

It is more difficult to understand how positive individual PSF species coexist. Three conditions may allow coexistence of positive PSF species. First, positive PSF species could form persistent patches of species monocultures and these patches could maintain diversity (Molofsky et al. 2001). Few data are available to test this hypothesis, but this condition does not appear to be likely, at least in some systems, because persistent patches of species monocultures are not common (Kulmatiski 2006). Second, competitive inequality could counteract the effects of a positive individual PSF (Fig. 2). This condition, competitive inequality among plants, is likely and research on this topic should be pursued, but we will focus on a third condition because it highlights an important, though unexplored, consequence for plant community development.

The third condition is that each species could benefit an interacting species more than it benefits itself (Bever et al. 1997; Fig. 3). Positive individual PSFs are

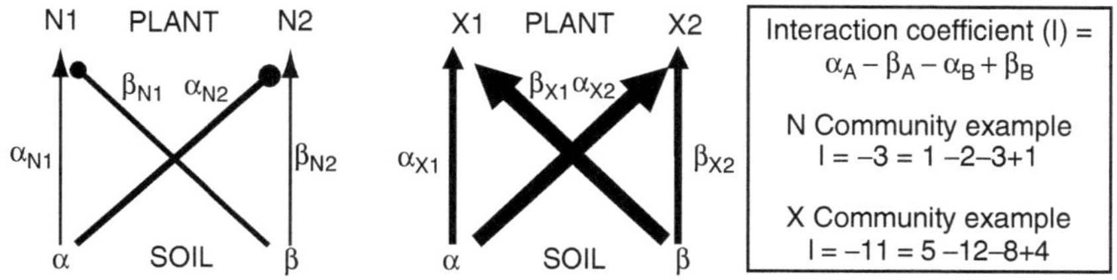

Fig. 3 Conceptual PSF model of two alternative state plant communities. Both communities must demonstrated negative net PSFs (i.e., a negative interaction coefficient), but the N plant community is demonstrate comprised of species with negative individual PSFs and the X plant community is comprised of species with positive individual PSFs. In both communities, interspecific effects via soils must benefit the growth of competitors in order to allow species coexistence. However, growth rates are small in the N community because of high pathogen loading and growth rates are large in the X community because each species must increase its competitor's growth even more than it increases its own growth. N species are unlikely to invade the X community because a plant with a negative PSF is unlikely to outcompete a plant with a positive individual PSF (*X1*) that grows even faster on soils cultivated by another species already present in that community (*X2*). X species could be expected to invade the N community unless the X species are highly susceptible to pathogen loading

thought to be caused by symbiotic associations, pathogen suppression, or nutrient accumulation. Negative net PSFs involving positive individual PSF species, therefore, are likely to create symbiont-rich, pathogen-poor soils with relatively high nutrient availability, and as a consequence, persistent plant populations with fast growth rates (Bever 2003; Suding et al. 2004; Vogelsang et al. 2006). Field data support these predictions. Many non-natives, which tend to create positive PSFs, create soils with relatively high nutrient availability (Ehrenfeld 2003) and dense, persistent plant populations (Stylinski and Allen 1999; Kulmatiski 2006). Furthermore, studies demonstrate greater plant growth on soils cultivated by late successional species, which tend to create positive PSFs, than on soil cultivated by early successional species (Bezemer et al. 2006b; Kardol et al. 2006).

The net PSF model therefore predicts that, for both negative and positive feedback species, a community of coexisting plants cultivates soils that allow faster growth rates of individual species than soils cultivated from any individual species in that community. In support of this, the average PSF value for 123 tests where plants were grown on soils cultivated by plant monocultures was −0.15, whereas the average PSF value for 75 tests where plants were grown on soils cultivated by plant communities was 0.08. In a more controlled test, Kardol et al. (2006) found that the average feedback value for individual species grown on soils cultivated by plant communities was 0.02, whereas the average feedback value when grown on soils cultivated by plant monocultures was −0.35 (Kardol et al. 2007). A second prediction of the net PSF model is that any community of coexisting species will grow faster than the average growth of each member of the community grown in monoculture. In support of this, there is extensive evidence demonstrating that productivity increases with plant diversity, but this literature has not considered the role of PSFs (Spehn et al. 2005).

3.2 Time

Current theoretical models of PSFs assume that changes in microbial community composition can be summarized by plant growth on a soil type or, similarly, by a multidimensional community value (e.g., a value on a principal coordinate axis). This value is assumed to change linearly between endpoints described as the soil communities associated with plant monocultures (Bever 2003). However, examples from agricultural settings suggest that non-linear plant–soil interactions are common (Larkin et al. 1993; Hamel et al. 2005). Some soils may be very responsive to changes caused by some plant species while other soils may be very resistant to change by the same species (Fig. 4; Murakami et al. 2000). Similarly, some plant species may be very responsive to small changes in soil community composition while other species may be resistant to changes in the soil community. The conceptual and theoretical models available to explain these non-linear interactions between plant communities and soils remain underdeveloped and future research should explore changes in PSFs over time (Bonanomi et al. 2005a).

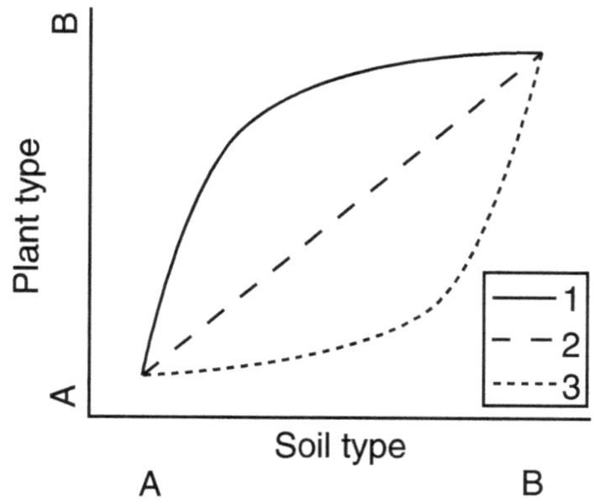

Fig. 4 Potential soil responses to plant growth and plant responses to soil communities that could change the timing of the outcome of positive plant–soil feedback. Plant–soil feedback models assume a linear relationship between plant and soil interactions (*line 2*). Sensitive soil responses to plant growth (*line 3*) or plant responses to soils (*line 1*) could hasten the development of plant–soil feedback. Threshold soil responses to plant growth (*line 1*) or plant responses to soils (*line 3*) could slow the development of plant–soil feedback. Plants and soils may also have no responses to changes in the other (data not shown)

3.3 Spatial Extent

It has been suggested that positive PSFs may not be important in species spread (Levine et al. 2006). An assumption of this model is that root growth, and therefore, the ability of a plant to create PSFs, is rarely greater than crown diameter (Levine et al. 2006). In support of this, it is easy to imagine that a trade-off exists between the ability to develop dense root systems over a small spatial area and diffuse root systems over a large area. However, we know little about the spatial extent of functional rooting patterns, but at least in some cases, they can be much larger than the crown diameter (Peek et al. 2005). Where plant roots do exert an influence over areas larger than their crown, PSFs are more likely to explain rapid spread.

Alternatively, dispersal may be important in determining spatial distribution of positive and negative PSF species. For example, it was found that short dispersal distances encourage the development of persistent patches of plant monocultures for species with positive PSFs (Molofsky et al. 2001). There is essentially no research that examined the spatial extent over which PSFs occur, yet this information may prove critical to the development of spatially-explicit models of plant growth and community development.

3.4 Interactions

Several attempts have been made to understand the interactions between above- and below-ground processes (Van der Putten et al. 2001). Soil microbial communities can be passengers to changes in above-ground processes: fungal endophytes and ungulate grazing were found to increase rhizodeposition and subsequent microbial activity (Matthews and Clay 2001; Frank et al. 2003; Van Hecke et al. 2005). Microbial communities can also be drivers in above-ground processes: plant–mycorhizal associations were found to nearly double pollination and seed set (Wolfe et al. 2005), and root-feeding nematodes were found to compete with aphids for plant resources (Bezemer et al. 2005). These are a few of the many examples that suggest that above- and below-ground processes interact but these effects on the development of PSFs have not been tested.

There is a clear need to understand the magnitude of PSFs relative to competing explanations of plant growth. Of the 23 PSF studies reviewed, just five compared multiple plant growth factors. One suggested that PSFs were more important to plant growth than competition, propagule pressure, shading, fungicide application, or soil disturbance (Kulmatiski et al. 2006). Three found that PSFs strongly interacted with other factors, for example climate (Bezemer et al. 2006b), soil nutrient availability (De Deyn et al. 2004a), and competition (Kardol et al. 2007). One found that PSFs were less important than competition (Peltzer 2001). It is likely that the full importance of PSFs will not be understood until we gain an understanding of how PSFs interact with other plant growth factors.

3.5 Soil History

Soil history, which can include, for example, grazing, tillage, fertilization, or pesticide application, is suggested to be important in the creation of PSFs (Kulmatiski et al. 2006, Kardol et al. 2007). Soil history is thought to be important because it can alter the soil microbial community, as well as soil chemistry and physical conditions. However, most studies demonstrating PSF were conducted using potted, highly disturbed, and fertilized soils, which do not allow measurement of historical effects on PSFs.

An intended consequence of the use of disturbed soils is that soils cultivated during the experiment strongly reflect the conditions created by the target plant species. An unintended consequence of this approach is that the microbial populations used in PSF experiments are likely to be very small relative to the microbial populations found in field soils. The use of small microbial populations may overestimate the effects of plant growth because the microbial predators and competitors of soil organisms that grow on the target plant may be removed (Van Bruggen and Semenov 2000; De la Pena et al. 2006). Alternatively, the use of small microbial populations may underestimate the effects of previous plant growth because small microbial communities may require a long lag phase of microbial growth. The use of small microbial populations in inoculation experiments may also underestimate the effects of previous plant growth when mycorrhizae or microbial communities responsible for nutrient cycling rates are important to plant growth, because laboratory experiments often rely on fertilization (De Deyn et al. 2004a; Hawkes et al. 2005).

Human disturbances create variations in soils on the landscape that are not reflected in soils in the greenhouse. These disturbances range from atmospheric N deposition to grazing to tillage. It has long been understood that these disturbances create conditions that favor some plant species (i.e., ruderals) over other plant species. A less recognized, but potentially critical, effect of disturbance is change to soil biology, chemistry, and structure. Fertilization as a result of N deposition or grazing is likely to reduce microbial diversity (Benizri and Amiaud 2005; Suguenza et al. 2006). Soil loosening is also known to reduce microbial diversity and activity (Calderon et al. 2000; Steenwerth et al. 2002; Ravit et al. 2003; Bossio et al. 2005; DeGrood et al. 2005; Potthoff et al. 2006). Research on the effects of disturbance on the development of PSF is likely critical in predicting future impacts of human-induced disturbances.

3.6 Disturbance-Contingent Niche Creation

A review of the literature revealed a dominant interest in the role of PSFs in succession and the growth of non-native species. This review also uncovered a need to improve our understanding of the role of PSFs under field conditions. Here we describe a conceptual model that explains how early-successional, non-native plants could establish dense, persistent populations in disturbed sites. This model is intended to provide testable predictions regarding the role of PSFs under field conditions.

Two conditions that prove most consistent in predicting non-native invasion are disturbance and early-successional plant traits (Reichard and Hamilton 1997; Sheley and Petroff 1998; Hansen and Clevenger 2005). Together these conditions provide a logical explanation for the establishment of non-natives: early-successional species are wide dispersers and fast growers and should be expected to grow well in disturbed, high resource conditions. More difficult to explain is the dominance and persistence of non-natives in many invaded sites (Stylinski and Allen 1999; Kulmatiski 2006).

We combine four conditions from disturbed, invaded communities to develop the disturbance-contingent niche creation model, a conceptual model that explains the success of early-successional, non-native plants:

1. Non-natives benefit from a release from specialist predators and pathogens (Mitchell and Power 2003; Reinhart and Callaway 2006).
2. Non-natives benefit from symbioses with generalists (Reinhart and Callaway 2006).
3. Soil disturbance suppresses soil biological activity (Bockus and Shroyer 1998; Steinkellner and Langer 2004; Belnap et al. 2005).
4. Non-natives are frequently early-successional species (Rejmanek and Richardson 1996; Reichard and Hamilton 1997; Fig. 5).

In native plant communities, non-natives are assumed to benefit from a release from specialist pathogens (Fig. 5A). This effect is expected to be greater than the

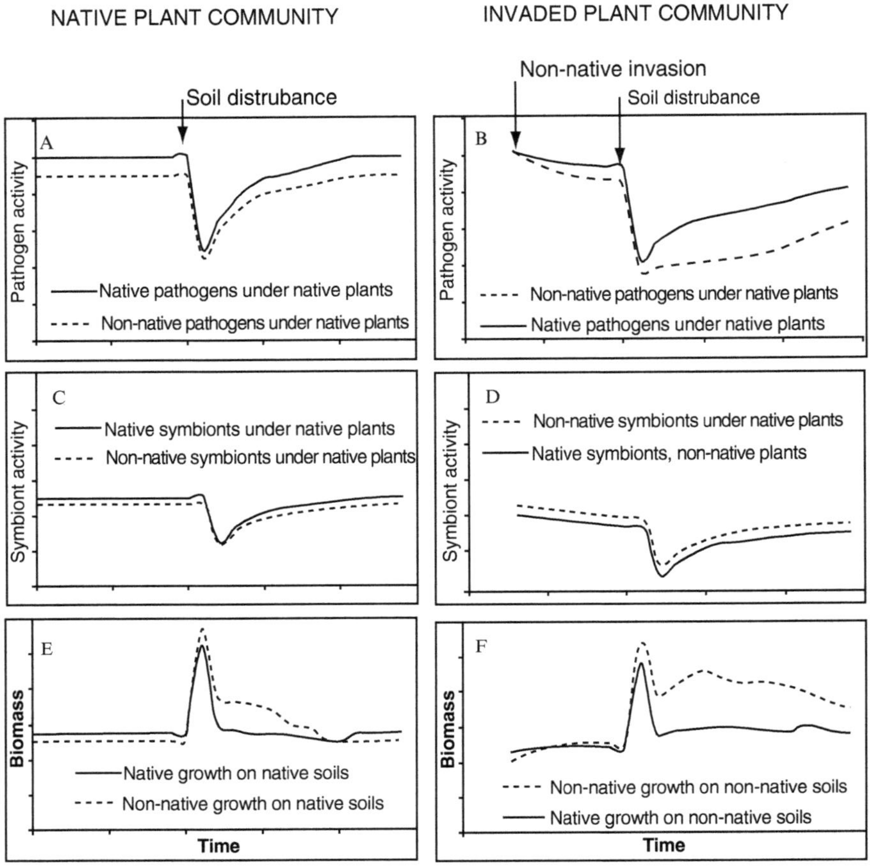

Fig. 5 Disturbance-contingent niche creation: proposed soil pathogen, soil symbiont, and soil fertility responses to non-native plant invasion and soil disturbance. **A**, **C**, **E** Soil responses to disturbance in native-dominated plant communities. **B**, **D**, **F** Soil responses to non-native plant invasion and disturbance in plant communities invaded by non-native plants. Decreases in native plant growth as a result of non-native plant invasion are expected to decrease the activity of specialized soil pathogens (**B**), but have less of an effect on generalist symbionts (**C**, **D**). Soil disturbances have been found to temporarily reduce pathogen and symbiont activity. Where native plants re-establish in disturbed soils, it is expected that original pathogen and symbiont populations will return (**A**, **C**). Where non-native plants establish in disturbed soils specialized pathogen populations are expected to disappear due to a loss of resources (i.e., native plants; **B**). In contrast, rapid growth of early-successional non-native plants is expected to result in a rapid recovery of soil symbiont populations (**D**). All else being equal, the loss of pathogens and increase in symbionts under non-native plants following disturbance is expected to produce fertile soils (**E**, **F**). Fertile soils are expected to provide a competitive advantage to plants with high relative growth rates (i.e., early-successional species)

effect of a loss from some specialized mutualists (Fig. 5C) because plant-mutualist interactions (i.e., mycorrhizae) are assumed to be less specialized than plant-pathogen interactions. We suggest that the effect of enemy release on non-native plant growth is not large enough to allow many early-successional species to outcompete established late-successional species (Kulmatiski 2006; Fig. 5E). Plant growth in Fig. 5E, F is calculated as the sum of negative effects on growth from pathogens and positive effects on growth from symbionts. Late-successional native species, however, are given a 10% growth advantage in all but immediate post-disturbance conditions.

Soil disturbances have been found to have large effects on soil communities relative to plant effects on soil communities (Calderon et al. 2000; Steenwerth et al. 2002; McKinley et al. 2005; Bossio et al. 2006; Ravit et al. 2006; Van der Wal et al. 2006). We infer from this that soil disturbance decreases pathogen and symbiont activity. Early-successional, non-native species are likely to realize rapid growth and a competitive advantage shortly following disturbance as a result of decreased pathogen activity and increased nutrient availability, but these species become less successful as pathogen abundance and competition from late-successional species increase (Fig. 5E; Kardol et al. 2006; 2007).

In many ecosystems, however, natives do not maintain dominance following soil disturbance (Sheley and Petroff 1998). In Figs. 5B, D, F we suggest a mechanism that may explain why. In communities invaded by non-native plants, fewer native plants are present to support populations of specialized pathogens and symbionts (Fig. 5B, D). As a consequence, pathogen and symbiont activity is lower in non-native than native plant communities. Furthermore, pathogen and symbiont activity are expected to decline over time as non-native abundance increases (Fig. 5B, D; Belnap et al. 2005). Declines in pathogen activity may allow fast-growing, early-successional species to outcompete slower-growing, late-successional species, as indicated in the growth curves prior to disturbance in Fig. 5F.

Disturbance is expected to reduce pathogen and symbiont activity even more than non-native plant growth. Where non-native plants are abundant there are few plants available with which soil organisms can associate, and microbial populations are expected to remain small (Belnap et al. 2005; Kulmatiski et al. 2006). As a consequence, post-disturbance soils have a low pathogen loading and high nutrient concentrations. These conditions benefit fast-growing species. The growth of non-native, fast-growing species perpetuate these fast-growth conditions by maintaining soils with a low pathogen loading. A low pathogen loading also benefits native species, but not as much as non-native species because pathogens found on the landscape are specialized to attack natives. Note that the positive feedback between low pathogen abundance and non-native plant growth results in greater biomass in non-native relative to native communities (Fig. 5E, F). Dispersal of pathogens that attack natives is expected to increase native pathogen activity over time (Fig. 5B). Non-natives may also serve as facultative hosts for pathogens that are more harmful to natives than non-natives (Fig. 5B; Eppinga et al. 2006).

Four simple conditions were used to describe the development of a positive PSF that is contingent upon soil disturbance. In this model, early-successional species growth is encouraged by a disturbance-induced decrease in soil pathogen activity; and soil pathogen activity is further decreased by non-native plant growth. Taken together, these conditions predict that disturbance and subsequent non-native plant growth results in the development of mutualist-rich, pathogen-poor soils that benefit the growth of early-successional species (Fig. 5F). Furthermore, the disturbance-contingent niche creation model suggests that native plant growth in disturbed soils may increase pathogen loading resulting in a competitive advantage for native plants relative to non-native plants (Fig. 5E). Model predictions are consistent with observations of persistent, early-successional, non-native plant communities and suggest that pathogen loading is necessary to induce a shift from non-native to native-dominated plant communities (Kulmatiski 2006).

4 Broadening the Scope of PSF Research

There are many opportunities to extend PSF research to interactions other than plant–soil microbial interactions in grasslands. In a recent review on PSFs, Ehrenfeld et al. (2005) presents plant–soil interactions, including physical and chemical interactions, across a wide variety of ecosystems and discusses the potential for these interactions to produce PSFs. Ehrenfeld's review clearly illustrates the potential for PSFs to be produced by a variety of interactions and to exist in a variety of ecosystems. However, the lack of formal PSF tests for these examples illustrates how much research remains to be conducted in this area.

5 Conclusions

In plant–soil systems, negative individual feedbacks are common. All else being equal, negative individual feedbacks are suggested to decrease plant biomass and to encourage species coexistence and succession, unless a negative PSF species harms another species more than it harms itself. Positive feedbacks are suggested to increase plant biomass and encourage competitive exclusion. However, positive individual PSFs can encourage species coexistence by counteracting the effects of competitive inequality, creating persistent patches of species monocultures, or improving interspecific growth more than intraspecific growth. Many PSF studies produce promising results regarding the explanation of plant growth and plant community development, yet much more can be learned by extending PSF research from greenhouse to field conditions. We provide an example of a conceptual model that integrates PSF with other site conditions (i.e., disturbance, non-native plant growth). This type of model will be needed to guide PSF research under field conditions. In particular, we suggest that the effects of plant communities, time, spatial extent, interactions with above-ground processes, and soil history in PSF be explored.

References

Agrawal AA, Kotanen PM, Mitchell CE, Power AG, Godsoe W, Klironomos J (2005) Enemy release? An experiment with congeneric plant pairs and diverse above- and belowground enemies. Ecology 86:2979–2989

Belnap J, Phillips SL, Sherrod SK, Moldenke A (2005) Soil biota can change after exotic plant invasion: does this affect ecosystem processes? Ecology 86:3007–3017

Benizri E, Amiaud B (2005) Relationship between plants and soil microbial communities in fertilized grasslands. Soil Biol Biochem 37:2055–2064

Bever JD (1994) Feedback between plants and their soil communities in an old field community. Ecology 75:1965–1977

Bever JD (1999) Dynamics within mutualism and the maintenance of diversity: inference from a model of interguild frequency dependence. Ecol Lett 2:52–62

Bever JD (2003) Soil community feedback and the coexistence of competitors: conceptual frameworks and empirical tests. New Phytol 157:465–473

Bever JD, Westover KM, Antonovics J (1997) Incorporating the soil community into plant population dynamics: the utility of the feedback approach. J Ecol 85:561–573

Bezemer TM, De Deyn GB, Bossinga TM, Dam NM van, Harvey JA, Putten WH van der (2005) Soil community composition drives aboveground plant-herbivore-parasitoid interactions. Ecol Lett 8:652–661

Bezemer TM, Harvey JA, Kowalchuk GA, Korpershoek H, Putten WH van der (2006a) Interplay between *Senecio jacobaea* and plant, soil, and aboveground insect community composition. Ecology 87:2002–2013

Bezemer TM, Lawson CS, Hedlund K, Edwards AR, Brook AJ, Igual JM, Mortimer SR, van der Putten WH (2006b) Plant species and functional group effects on abiotic and microbial soil properties and plant–soil feedback responses in two grasslands. J Ecol 94:893–904

Bockus WW, Shroyer JP (1998) The impact of reduced tillage on soilborne plant pathogens. Annu Rev Phytopathol 36:485–500

Bodelier PLE, Stomp M, Santamaria L, Klaassen M, Laanbroek HJ (2006) Animal–plant–microbe interactions: direct and indirect effects of swan foraging behaviour modulate methane cycling in temperate shallow wetlands. Oecologia 149:233–244

Bonanomi G, Mazzoleni S (2005) Soil history affects plant growth and competitive ability. Community Ecol 6:23–28

Bonanomi G, Giannino F, Mazzoleni S (2005a) Negative plant–soil feedback and species coexistence. Oikos 111:311–321

Bonanomi G, Rietkerk M, Dekker SC, Mazzoleni S (2005b) Negative plant–soil feedback and positive species interaction in a herbaceous plant community. Plant Ecol 181:269–278

Bossio DA, Girvan MS, Verchot L, Bullimore J, Borelli T, Albrecht A, Scow KM, Ball AS, Pretty JN, Osborn AM (2005) Soil microbial community response to land use change in an agricultural landscape of Western Kenya. Microb Ecol 49:50–62

Bossio DA, Fleck JA, Scow KM, Fujii R (2006) Alteration of soil microbial communities and water quality in restored wetlands. Soil Biol Biochem 38:1223–1233

Bruggen AHC van, Semenov AM (2000) In search of biological indicators for soil health and disease suppression. Appl Soil Ecol 15:13–24

Calderon FJ, Jackson LE, Scow KM, Rolston DE (2000) Microbial responses to simulated tillage in cultivated and uncultivated soils. Soil Biol Biochem 32:1547–1559

Callaway RM, Aschehoug ET (2000) Invasive plants versus their new and old neighbors: a mechanism for exotic invasion. Science 290:521–523

Callaway RM, Thelen GC, Barth S, Ramsey PW, Gannon JE (2004a) Soil fungi alter interactions between the invader *Centaurea maculosa* and North American natives. Ecology 85:1062–1071

Callaway RM, Thelen GC, Rodriguez A, Holben WE (2004b) Soil biota and exotic plant invasion. Nature 427:731–733

Carey EV, Marler MJ, Callaway RM (2004) Mycorrhizae transfer carbon from a native grass to an invasive weed: evidence from stable isotopes and physiology. Plant Ecol 172:133–141

De Deyn GB, Raaijmakers CE, Putten WH van der (2004a) Plant community development is affected by nutrients and soil biota. J Ecol 92:824–834

De Deyn GB, Raaijmakers CE, Ruijven J van, Berendse F, Putten WH van der (2004b) Plant species identity and diversity effects on different trophic levels of nematodes in the soil food web. Oikos 106:576–586

De la Pena E, Echeverria SR, Putten WH van der, Freitas H, Moens M (2006) Mechanism of control of root-feeding nematodes by mycorrhizal fungi in the dune grass *Ammophila arenaria*. New Phytol 169:829–840

DeGrood SH, Claassen VP, Scow KM (2005) Microbial community composition on native and drastically disturbed serpentine soils. Soil Biol Biochem 37:1427–1435

Ehrenfeld JG (2003) Effects of exotic plant invasions on soil nutrient cycling processes. Ecosystems 6:503–523

Ehrenfeld JG, Ravit B, Elgersma K (2005) Feedback in the plant–soil system. Annu Rev Environ Resour 30:75–115

Eppinga MB, Rietkerk M, Dekker SC, De Ruiter PC, Putten WH van der (2006) Accumulation of local pathogens: a new hypothesis to explain exotic plant invasions. Oikos 114:168–176

Eppstein MJ, Bever JD, Molofsky J (2006) Spatio-temporal community dynamics induced by frequency dependent interactions. Ecol Model 197:133–147

Frank DA, Gehring CA, Machut L, Phillips M (2003) Soil community composition and the regulation of grazed temperate grassland. Oecologia 137:603–609

Gustafson DJ, Casper BB (2004) Nutrient addition affects AM fungal performance and expression of plant/fungal feedback in three serpentine grasses. Plant Soil 259:9–17

Hamel C, Vujanovic V, Jeannotte R, Nakano-Hylander A, St-Arnaud M (2005) Negative feedback on a perennial crop: *Fusarium* crown and root rot of *Asparagus* is related to changes in soil microbial community structure. Plant Soil 268:75–87

Hansen MJ, Clevenger AP (2005) The influence of disturbance and habitat on the presence of non-native plant species along transport corridors. Biol Conserv 125:249–259

Hawkes CV, Wren IF, Herman DJ, Firestone MK (2005) Plant invasion alters nitrogen cycling by modifying the soil nitrifying community. Ecol Lett 8:976–985

Hecke MM van, Treonis AM, Kaufman JR (2005) How does the fungal endophyte *Neotyphodium coenophialum* affect tall fescue (*Festuca arundinacea*) rhizodeposition and soil microorganisms? Plant Soil 275:101–109

Kardol P, Bezemer TM, Putten WH van der (2006) Temporal variation in plant–soil feedback controls succession. Ecology Lett 9:1080–1088

Kardol P, Cornips NJ, Kempen ML van, Bakx-Shotman JM, Putten WH van der (2007) Microbe-mediated plant–soil feedback causes historical contingency effects in plant community assembly. Ecological Monographs 77:147–162

Klironomos JN (2002) Feedback with soil biota contributes to plant rarity and invasiveness in communities. Nature 417:67–70

Knevel IC, Lans T, Menting FBJ, Hertling UM, Putten WH van der (2004) Release from native root herbivores and biotic resistance by soil pathogens in a new habitat both affect the alien *Ammophila arenaria* in South Africa. Oecologia 141:502–510

Kulmatiski A (2006) Exotic plants establish persistent communities. Plant Ecol 187:261–275

Kulmatiski A., Beard KH, Stark JM (2006) Soil history as a primary control on plant invasion in abandoned agricultural fields. J Appl Ecol 43:868–876

Larkin RP, Hopkins DL, Maritin FN (1993) Effect of successive watermelon plantings on *Fusarium oxysporum* and other micro-organisms in soils suppressive and conducive to *Fusarium* wilt of watermelon. Phytopathology 83:1097–1105

Levine JM, Pachepsky E, Kendall BE, Yelenik SG, Lambers JHR (2006) Plant–soil feedbacks and invasive spread. Ecology Lett 9:1005–1014

Macel M, Lawson CS, Mortimer SR, Smilauerova M, Bischoff A, Cremieux L, Dolezal L, Edwards AR, Lanta V, Bezemer TM, Putten WH van der, Igual JM, Rodriguez-Barrueco C, Muller-Scharer H, Steinger T (2007) Climate vs. soil factors in adaptation of two common plant species. Ecology (in press)

Matthews JW, Clay K (2001) Influence of fungal endophyte infection on plant–soil feedback and community interactions. Ecology 82:500–509

McKinley VL, Peacock AD, White DC (2005) Microbial community PLFA and PHB responses to ecosystem restoration in tallgrass prairie soils. Soil Biol Biochem 37:1946–1958

Mills KE, Bever JD (1998) Maintenance of diversity within plant communities: soil pathogens as agents of negative feedback. Ecology 79:1595–1601

Mitchell CE, Power AG (2003) Release of invasive plants from fungal and viral pathogens. Nature 421:625–627

Molofsky J, Bever JD, Antonovics J (2001) Coexistence under positive frequency dependence. Proc R Soc Lond B Biol Sci 268:273–277

Morris C, Call CA, Monaco TA, Grossl PR, Dewey SA (2006) Evaluation of elemental allelopathy in *Acroptilon repens* (L.) DC. (Russian Knapweed). Plant Soil 289:279–288

Murakami H, Tsushima S, Shishido Y (2000) Soil suppressiveness to clubroot disease of Chinese cabbage caused by *Plasmodiophora brassicae*. Soil Biol Biochem 32:1637–1642

Packer A, Clay K (2000) Soil pathogens and spatial patterns of seedling mortality in a temperate tree. Nature 404:278–281

Peek MS, Leffler AJ, Ivans CY, Ryel RJ, Caldwell MM (2005) Fine root distribution and persistence under field conditions of three co-occurring Great Basin species of different life form. New Phytol 165:171–180

Peltzer DA (2001) Plant responses to competition and soil origin across a prairie-forest boundary. J Ecol 89:176–185

Porazinska DL, Bardgett RD, Blaauw MB, Hunt HW, Parsons AN, Seastedt TR, Wall DH (2003) Relationships at the aboveground–belowground interface: Plants, soil biota, and soil processes. Ecol Monogr 73:377–395

Potthoff M, Steenwerth KL, Jackson LE, Drenovsky RE, Scow KM, Joergensen RG (2006) Soil microbial community composition as affected by restoration practices in California grassland. Soil Biol Biochem 38:1851–1860

Ravit B, Ehrenfeld JG, Haggblom MM (2003) A comparison of sediment microbial communities associated with *Phragmites australis* and *Spartina alterniflora* in two brackish wetlands of New Jersey. Estuaries 26:465–474

Ravit B, Ehrenfeld JG, Haggblom MM (2006) Effects of vegetation on root-associated microbial communities: A comparison of disturbed versus undisturbed estuarine sediments. Soil Biol Biochem 38:2359–2371

Reichard SH, Hamilton CW (1997) Predicting invasions of woody plants introduced into North America. Conserv Biol 11:193–203

Reinhart KO, Callaway RM (2004) Soil biota facilitate exotic *Acer* invasions in Europe and North America. Ecol Appl 14:1737–1745

Reinhart KO, Callaway RM (2006) Soil biota and invasive plants. New Phytol 170:445–457

Reinhart KO, Packer A, Putten WH van der, Clay K (2003) Plant–soil biota interactions and spatial distribution of black cherry in its native and invasive ranges. Ecol Lett 6:1046–1050

Reinhart KO, Greene E, Callaway RM (2005a) Effects of *Acer* platanoides invasion on understory plant communities and tree regeneration in the northern Rocky Mountains. Ecography 28:573–582

Reinhart KO, Royo AA, Van der Putten WH, Clay K (2005b) Soil feedback and pathogen activity in *Prunus serotina* throughout its native range. J Ecol 93:890–898
Rejmanek M, Richardson DM (1996) What attributes make some plant species more invasive? Ecology 77:1655–1661
Reynolds HL, Packer A, Bever JD, Clay K (2003) Grassroots ecology: plant–microbe–soil interactions as drivers of plant community structure and dynamics. Ecology 84:2281–2291
Sheley RL, Petroff JK (1998) Biology and management of noxious rangeland weeds. Oregon State University Press, Corvallis
Spehn EM, Hector A, Joshi J, Scherer-Lorenzen M, Schmid B, Bazeley-White E, Beierkuhnlein C, Caldeira MC, Diemer M, Dimitrakopoulos PG, Finn JA, Freitas H, Giller PS, Good J, Harris R, Hogberg P, Huss-Danell K, Jumpponen A, Koricheva J, Leadley PW, Loreau M, Minns A, Mulder CPH, O'Donovan G, Otway SJ, Palmborg C, Pereira JS, Pfisterer AB, Prinz A, Read DJ, Schulze ED, Siamantziouras ASD, Terry AC, Troumbis AY, Woodward FI, Yachi S, Lawton JH (2005) Ecosystem effects of biodiversity manipulations in European grasslands. Ecol Monogr 75:37–63
Steenwerth KL, Jackson LE, Calderon FJ, Stromberg MR, Scow KM (2002) Soil microbial community composition and land use history in cultivated and grassland ecosystems of coastal California. Soil Biol Biochem 34:1599–1611
Steinkellner S, Langer L (2004) Impact of tillage on the incidence of *Fusarium* spp in soil. Plant Soil 267:13–22
Stoel CD van der, Putten WH van der, Duyts H (2002) Development of a negative plant–soil feedback in the expansion zone of the clonal grass *Ammophila arenaria* following root formation and nematode colonization. J Ecol 90:978–988
Stylinski CD, Allen EB (1999) Lack of native species recovery following severe exotic disturbance in southern Californian shrublands. J Appl Ecol 36:544–554
Suding KN, Gross KL, Houseman GR (2004) Alternative states and positive feedbacks in restoration ecology. Trends Ecol Evol 19:46–53
Suguenza C, Corkidi L, Allen EB (2006) Feedbacks of soil inoculum of mycorrhizal fungi altered by N deposition on the growth of a native shrub and an invasive annual grass. Plant Soil 286:153–165
Troelstra SR, Wagenaar R, Smant W, Peters BAM (2001) Interpretation of bioassays in the study of interactions between soil organisms and plants: involvement of nutrient factors. New Phytol 150:697–706
van der Putten WH (1997) Plant–soil feedback as a selective force. Trends Ecol Evol 12:169–170
van der Putten WH, Dijk C van, Troelstra SR (1988) Biotic soil factors effecting growth and development of *Ammophila arenaria*. Oecologia 76:313–320
van der Putten WH, Dijk C van, Peters BAM (1993) Plant-specific soil-borne diseases contribute to succession in foredune vegetation. Nature 362:53–56
van der Putten WH, Vet LEM, Harvey JA, Wackers FL (2001) Linking above- and belowground multitrophic interactions of plants, herbivores, pathogens, and their antagonists. Trends Ecol Evol 16:547–554
van der Putten WH Kowalchuk GA, Brinkman EP, Doodman GTA, Kaaij RM van der, Kamp AFD, Menting FBJ, Veenendaal EM (2007) Soil feedback of exotic savannah grass relates to the absence of pathogens and selective mycorhizal fungi. Ecology (in press)
Vogelsang KM., Reynolds HL, Bever JD (2006) Mycorrhizal fungal identity and richness determine the diversity and productivity of a tallgrass prairie system. New Phytol 172:554–562
Wal A van der, Veen JA van, Smant W, Boschker HTS, Bloem J, Kardol P, Putten WH van der, Boer W de (2006) Fungal biomass development in a chronosequence of land abandonment. Soil Biol Biochem 38:51–60
Wolfe BE, Husband BC, Klironomos JN (2005) Effects of a belowground mutualism on an aboveground mutualism. Ecology Lett 8:218–223

Index

Q

R

S